Bryn

4351

ORGANIC CHEMISTRY:

A SHORT COURSE

Fourth Edition

HOUGHTON MIFFLIN COMPANY

BOSTON
Atlanta
Dallas
Geneva, Ill.
Hopewell, N.J.
Palo Alto

ORGANIC CHEMISTRY:

A SHORT COURSE

Fourth Edition

HAROLD HART AND

ROBERT D. SCHUETZ

Michigan State University

Printed in the United States of America

Library of Congress Catalog Card Number: 72-165036

ISBN: 0-395-04574-6

ISBN: 0-395-13471-4 International Student Edition

PREFACE

It is nearly twenty years since the first edition of our text appeared. During that time student attitudes toward education have changed drastically; students are more sophisticated, less willing to accept what someone else thinks is good for them. They challenge the need for required courses, and want assurance that what they study is worthwhile and relevant to their goals. We have tried to meet this challenge in the present edition of this text.

This book is designed to serve students who must acquire a sound knowledge of organic chemistry but who do not plan to make chemistry their career. It is written for a one-semester or one-quarter course, usually taken by students in agriculture, nursing, home economics, veterinary and human medicine, dentistry and pharmacy. Many of these students proceed to a biochemistry course and have a strong interest in the chemistry of biological systems. For this reason, those aspects of organic chemistry pertinent to biochemistry have been stressed.

We have tried to avoid superficiality. Whenever possible, explanations for chemical properties and reactivity are presented. The mechanistic approach to organic chemistry is not only intellectually satisfying but helps to minimize memory work and provides some framework for facing new, previously unencountered chemistry.

We believe that students who take this course are as intellectually capable as any chemistry major but have other interests. In this book we have tried to engage their intellect with material that is, whenever possible, relevant to those other interests.

Teachers who used earlier editions of this text will find this a thorough revision. The entire text has been rewritten. Color has been introduced to highlight those portions of structures which undergo change during a reaction. Sections and subsections of chapters are numbered for convenient cross reference and for making reading assignments. Almost all of the questions at the end of each chapter are new, and in general they follow the subject matter sequence within the chapter. A study guide, with answers to all the questions, is available. The material on unsaturated hydrocarbons is now divided between two chapters, and a new chapter has been added on spectroscopy and structure. A few major changes include a thorough modernization of the chapter on stereochemistry, additional emphasis on heterocyclic compounds, a reorganization of the protein structure discussion, addition of the Merrifield peptide synthesis, a brief presentation of "primordial" organic chemistry, emphasis on biosynthesis in the chapter on natural products, and an expanded treatment of nucleic acids. Other changes pervade the book; the full extent of the revision probably will not be appreciated until one has taught from the text. We hope the book meets with the widespread approval experienced by previous editions.

We wish to thank here our many friends, both teachers and students, who have sent us helpful suggestions. Similar comments on this edition will be appreciated.

Harold Hart
Robert D. Schuetz

INTRODUCTION

Before beginning a systematic study of organic chemistry, one might like to know something in a general way about the scope of the subject and why it is important. The term organic suggests a relationship to living organisms, and indeed, as originally conceived, organic chemistry was a study of those substances which occur in living cells. The term was first used by the Swedish chemist J. J. Berzelius in 1808, and the first textbook on the subject was written less than 150 years ago, so organic chemistry is rather young as a separate discipline of science.

It soon became apparent, as various pure chemical substances were isolated from natural (i.e., animal or plant) sources during the late eighteenth and early nineteenth centuries, that the substances possessed certain common properties which distinguished them from compounds that originate in mineral sources. Most significant of these characteristics was the ubiquity of certain elements, notably carbon, hydrogen, oxygen, and nitrogen. Later, organic chemistry was to become defined in these terms—that is, the chemistry of carbon compounds.

For many years it was thought that some *vital force* was essential to the synthesis of compounds present in living things. This idea discouraged chemists from trying to synthesize such substances in the laboratory. In 1828 the German chemist Friedrich Wöhler accidentally synthesized urea, a well-known constituent of urine, from inorganic starting materials (sec 10.9). This and other experiments gradually discredited the vital force theory during the ensuing twenty years and opened the way for modern synthetic organic chemistry. At present the number of organic compounds (many of them unknown in nature) synthesized in research laboratories far exceeds the number isolated from natural sources, although the study of natural products is still an important part of organic chemistry. Indeed, recent years have seen a major rebirth of interest in natural substances, with the particular goal of determining precisely how they are synthesized by the organism (biosynthesis).

What do organic chemists do? In the main, they do three things— *determine structures, synthesize compounds,* and *study reaction mechanisms.* Although an organic compound usually contains only a few elements, it may contain several atoms of each kind, and they may be "connected" to one another, or arranged within the molecule, in a variety of ways. A complete "structure" determination would include an accurate three-dimensional description of the arrangement of atoms in the molecular framework. Since these arrangements determine the physical and chemical properties of the substance, knowledge of a compound's structure is a prime prerequisite to understanding its chemistry. To determine a structure, the chemist uses simple chemical tests together with spectroscopic methods (infrared, ultraviolet etc.); in the case of complex molecules he may degrade them through chemical reactions to simpler substances which are easier to study, and then deduce logically how these fragments may have been combined in the original compound. Current techniques for structure determination are quite sophisticated. Although 157 years elapsed between the isolation and the structure determination of cholesterol (sec 18.2a), the structures of molecules with equal or greater complexity may be determined today in a matter of weeks or months.

The synthetic chemist may have as his goal the synthesis of a natural product to verify its structure or perhaps to make it available more widely and at lower cost. Vitamins and amino acids (for use as food supplements), ethyl alcohol, camphor, and indigo are but a few examples of common compounds first isolated from nature that are now produced synthetically for commercial use. By synthesis the chemist may also create new substances which improve on the properties of those found in nature; examples from organic chemistry include the many synthetic fabrics (nylon, orlon), organic dyes, drugs (aspirin, barbiturates, Novocaine), detergents, insecticides, oral contraceptives, plastics (polyethylene, polystyrene), synthetic rubber, and innumerable others. Finally, the synthetic chemist may create molecules with new or unusual arrangements of atoms not found in nature to test chemical theories of bonding and structure. In any case, synthesis requires the assembling of simple molecules into more complex ones in a controlled fashion—each new bond between atoms must be introduced in a rational manner.

To degrade or synthesize molecules, one must have a thorough understanding of organic reaction mechanisms. This area of organic chemistry has blossomed during the last thirty years and is in large measure responsible for the facility with which new compounds of predetermined structure can now be synthesized. By a "reaction mechanism" one means a precise description of how reactants are transformed into products. One would like to know the manner in which two reactants approach one another and interact, the energy required to break old bonds and make new ones, the effect of changes in solvent or structure of reactants on the reaction rate, and so on. When these factors are well understood, one can determine experimental conditions which will achieve desired synthetic goals.

Organic chemistry impinges on our daily lives perhaps more than any other branch of the science. It is not only a professional discipline, or a tool for the physician, dentist, pharmacist, nurse, or agriculturalist. Some knowledge of organic chemistry, just as much as a knowledge of literature, is a sign of culture in our present technological age. The terms plastic, polymer, THC, LSD, polyethylene, octane number, polyunsaturated fat, heroin, nylon, nicotine, biodegradable detergent, DDT, cyclamates, ether, etc., are a part of our culture. To really understand what they mean and to discuss whether they should be banned, controlled by legislation, used, or avoided, it helps to know something of their structure, chemical properties, and effects on man and his environment.

TO THE STUDENT

Organic chemistry in many ways resembles a foreign language. A new and extensive vocabulary must be developed, together with rules (a grammar) for using it. It is important to understand each new term and concept as it is presented, since a large structure will be built from small beginnings. A thorough grasp of the foundation of organic chemistry is essential if the edifice to be constructed upon it is to have meaning. The subject matter is nonmathematical yet rigorously logical; an understanding of that logic will minimize the need for memorization and will provide intellectual satisfaction as the subject is studied.

CONTENTS

CHAPTER ONE

GENERAL PRINCIPLES

There are more than a million organic chemicals, each of which contains the tetravalent element carbon, usually combined with hydrogen, oxygen, nitrogen, or certain other elements. The unique property of the carbon atom which enables it to form so many different substances is its ability to combine with other carbon atoms through covalent bonds. Carbon atoms may join to form long chains, branched chains, or rings. The infinite possible arrangements account for the variety and large number of organic compounds. Many of these constitute the common material articles of everyday living, such as foodstuffs, clothing, fuels, and plastics.

Organic chemistry is the chemistry of substances which contain the element carbon. Although the term "organic" was used originally to designate compounds of plant or animal origin, contemporary usage includes many synthetic substances as well.

1.1 THE UNIQUE ROLE OF THE CARBON ATOM

Upon reading the above definition of organic chemistry, one's immediate reaction is to ask the question—why single out the element carbon? *What are the unique properties of this element, that the study of its compounds should constitute a separate branch of chemistry?*

The position of carbon in the periodic table throws some light on the reason for its particular importance (see Table 1-1). In the fourth group of the second period, with an atomic number of six, carbon has a total of six orbital electrons. Two of these comprise the first complete shell of electrons, leaving four electrons in the valence shell. Since we are concerned with valence electrons when considering the formation of compounds from elements, it is sometimes useful to represent carbon as

The letter **C** stands for the nucleus and the first shell of electrons, and the dots represent the valence electrons. Since the valence shell of carbon requires a total of eight electrons to acquire the stable, rare-gas configuration of neon, we expect carbon to have a valence of four.

Because of its central position in the periodic table, carbon is neither strongly **electronegative** (electron-attracting) nor strongly **electropositive** (electron-repelling). In fact, carbon forms bonds with other elements mainly by sharing electron pairs rather than by

Table 1-1
Periodic Classification of Elements of the First Three Periods

PERIOD				GROUP				
	I	II	III	IV	V	VI	VII	O
1	H							He
2	Li	Be	B	C	N	O	F	Ne
3	Na	Mg	Al	Si	P	S	Cl	A

complete gain or complete loss of electrons. The most striking property of carbon atoms, however, is that they, more than atoms of any other element, can share electrons with atoms of their own kind to form long chains.

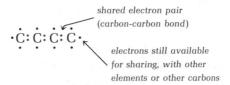

shared electron pair
(carbon-carbon bond)

·C:C:C:C·.

electrons still available
for sharing, with other
elements or other carbons

It is for this reason that carbon can form such a large number of compounds.

1.2 TYPES OF BONDS

The manner in which atoms join together to form compounds has a great deal to do with the chemical and physical properties of those compounds. In general, there are two important types of chemical bonds. **Ionic bonds** are usually formed between elements of widely different electronegativity and involve a *transfer of valence electrons* from one atom to another. Sodium chloride is a typical example of a compound formed by such a process.

$$\text{Na} \cdot \overset{\frown}{+} \cdot \overset{\cdot\cdot}{\underset{\cdot\cdot}{\text{Cl}}} : \longrightarrow \text{Na}^+ + : \overset{\cdot\cdot}{\underset{\cdot\cdot}{\text{Cl}}} :^- \tag{1-1}$$

In the crystal, the sodium and chloride ions are held together by the electrostatic attractive force between the positive and negative ions (see Figure 1-1).

Figure 1-1
Sodium chloride (Na^+Cl^-), an ionic crystal. The colored spheres represent sodium ions, Na^+, and the gray spheres chloride ions, Cl^-.

Of more interest to the organic chemist is the **covalent bond,** formed by a *mutual sharing of electron pairs* between atoms. A simple example is the formation of a hydrogen molecule from two hydrogen atoms.

$$\text{H}\cdot + \text{H}\cdot \longrightarrow \text{H}:\text{H} \qquad (1\text{-}2)$$

Each hydrogen atom has one valence electron in a spherical **atomic orbital** (called a 1s orbital). When the atoms combine, the two electrons occupy a **molecular orbital** which surrounds the two nuclei (Figure 1-2). The molecular orbital can be considered to have been formed by the *overlap* of the two atomic orbitals. In general, the greater the overlap the stronger the bond. The bonds shown in Figure 1-2 are called **sigma (σ) bonds;** they are symmetric about the axis which joins the two nuclei.

Figure 1-2
The formation of covalent bonds in hydrogen and hydrogen chloride.

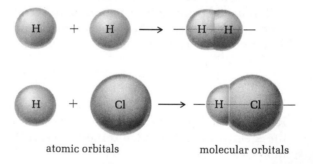

atomic orbitals   molecular orbitals

Heat is liberated when two hydrogen atoms combine to form a molecule; conversely, heat (i.e., energy) would have to be supplied to a hydrogen molecule to break it into two atoms. One concludes that the molecule constitutes a more stable arrangement of the nuclei and electrons than the separate atoms. The main reason for this is that in the molecule each electron is attracted by two positive nuclei rather than one. However, there is also a counterbalancing repulsive force between the two like-charged nuclei and also between the two like-charged electrons. A balance between these opposing forces is attained; the nuclei do not fuse nor do they fly apart, but they remain bound to one another at an equilibrium distance, called the **bond length.** For the hydrogen molecule, this is 0.74 Å (1 Å, or angstrom unit $= 10^{-8}$ centimeters).

Whereas the ionic bond in sodium chloride is formed between atoms of widely different electronegativity, the covalent bond of the hydrogen molecule is formed between atoms of identical electronegativity. Between these extremes, there are many compounds in which the bonds involve *unequal sharing* of electron pairs between atoms of intermediate difference in electronegativity. These are called **polar bonds.** In the hydrogen chloride molecule, the hydrogen and chlorine atoms are held together by a covalent bond. However, because chlorine is more electronegative than hydrogen, the electron pair is not shared equally. It is drawn toward the chlorine, which has a nucleus with a larger positive charge. The chlorine is slightly negative with respect to the hydrogen. This is sometimes indicated by an arrow, the head of which is negative and the tail marked with a plus sign. Alternatively, a partial charge ($\delta+$ or $\delta-$, to be read "delta plus" or "delta minus") may be shown.

$$\overset{\longrightarrow}{H : \overset{..}{\underset{..}{Cl}} :} \qquad or \qquad \overset{\delta+ \quad \delta-}{H : \overset{..}{\underset{..}{Cl}} :} \qquad \begin{array}{l} \textit{bonding electron pair} \\ \textit{shared unequally} \end{array}$$

To determine which end of a polar covalent bond is more negative and which is more positive, one can generally rely on the periodic table (Table 1-1). As we proceed from left to right within a given period (across the table), the elements become increasingly electronegative, due to their increasing atomic number or nuclear charge. As we proceed from the top to the bottom of the table within a given group (down the column), the elements become less electronegative, because the valence electrons are shielded from the nucleus by increasing numbers of inner-shell electrons. From these generalizations, one might predict that the atom on the right in each of the following bonds would be negative with respect to the atom on the left:

$$\begin{array}{ll} \overset{\longrightarrow}{C-Cl} & \overset{\longrightarrow}{C-Br} \\ H-O & C-N \\ C-O & S-O \end{array}$$

The carbon-hydrogen bond, so prevalent in organic compounds, requires special mention. Carbon and hydrogen have nearly the same electronegativity (note that the atoms of each of these elements need twice as many electrons as they have in order to complete their valence shells); therefore, the C—H bond is an almost purely covalent bond.

In some molecules, one atom of a pair of covalently bound atoms supplies *both* of the shared electrons. Such a bond is called a **coordinate covalent bond.** Examples of such bonding are found in nitric acid and in the oxonium ion illustrated below. The atom which

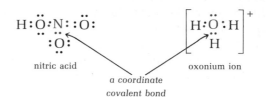

nitric acid   oxonium ion

*a coordinate
covalent bond*

donates the electron pair is shown in color. The polarity of the
resulting bond is determined by the same principles just outlined
for all other covalent bonds. In both examples, the oxygen atom,
being more electronegative than either nitrogen or hydrogen, forms
the negative end of the polar covalent bond.

The chemical behavior of organic compounds can often be predicted if the direction in which the bonds are polarized is known.
For example, since oxygen is much more electronegative than carbon, one might predict that compounds which have a hydrogen atom
attached to oxygen will be more acidic than those with only carbon-
hydrogen bonds. The O—H bond can be easily broken to furnish

$$-\ddot{\underset{\cdot\cdot}{O}}\!:\!H \;\rightleftharpoons\; -\ddot{\underset{\cdot\cdot}{O}}\!:^- + H^+ \tag{1-3}$$

$$\overset{\diagdown}{\underset{\diagup}{C}}\!:\!H \;\rightleftharpoons\; \overset{\diagdown}{\underset{\diagup}{C}}\!:^- + H^+ \tag{1-4}$$

protons (acidic behavior), the electron pair remaining with the electron-attracting oxygen atom. But C—H bonds are usually not acidic
unless a very strong base is used to remove the proton, or unless
the molecule contains special structural features which stabilize the
negative carbon ion **(carbanion).**

1.3 STRUCTURAL FORMULAS

Although organic compounds generally contain only a few elements,
they may contain many atoms of these elements per molecule. For
this reason, molecular formulas are of little use in organic chemistry.
For example, there are thirty-five known organic chemicals with the
molecular formula C_9H_{20}. Each of these compounds has its own
characteristic boiling and freezing points, as well as other properties
which distinguish it from all the others; yet the molecular formulas
are identical. The only possible explanation for these differences
must be that, although the kind and number of atoms are identical,
the *arrangement of the atoms within the molecule* must be different.
We can illustrate these differences in arrangement by using **structural
formulas.**

In writing chemical formulas it is frequently convenient to allow a dash to stand for a pair of electrons shared between two atoms (a chemical bond). Thus the water molecule may be represented as

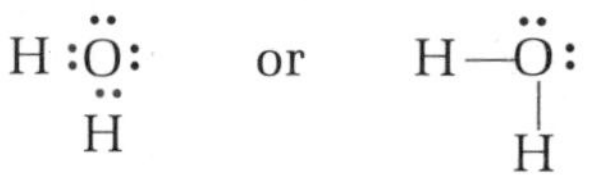

$$H : \ddot{O} : \qquad \text{or} \qquad H—\ddot{O} :$$
$$\quad H \qquad\qquad\qquad\quad H$$

The valence of an element is indicated by the number of bonds attached to it. In the formula for water, each hydrogen atom has one bond; the oxygen atom, two.

As stated previously, carbon, more than any other element, has the ability to unite or make bonds with itself. These bonds may involve the formation of a *continuous* chain of carbon atoms

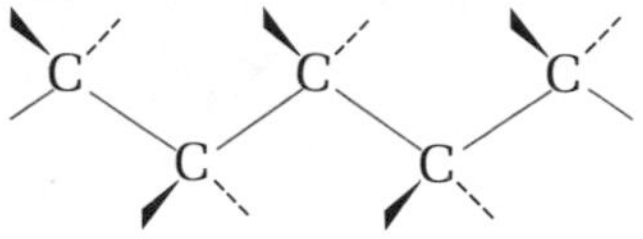

or a *branched* chain

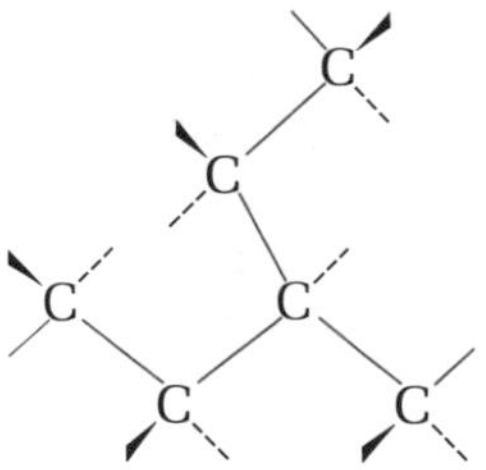

or a *ring*

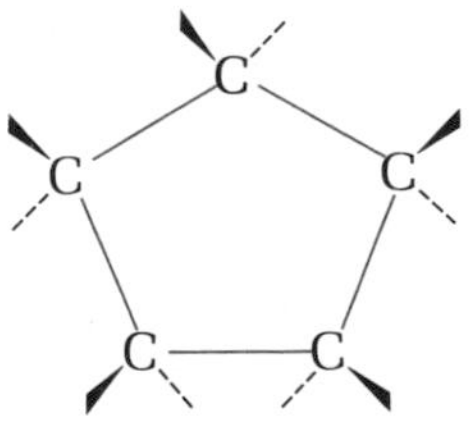

Note that in each skeletal formula, every carbon atom has four bonds. These structural formulas give much more information about the kinds of bonds present and the molecular shape than do molecular formulas.

We are now in a position to account for the existence of more than one compound with the same molecular formula.

1.4 ISOMERISM

The existence of several different compounds that have the same molecular formula is called **isomerism;** compounds having the same molecular formula but differing in structural formulas are called **structural** or **constitutional isomers.**

Consider the molecular formula C_2H_6O. If one arranges these atoms according to the ordinary rules of valence which indicate that carbon can be bonded to four other atoms, oxygen to two others, and hydrogen to one other, one soon finds that there are two, and only two, ways in which these atoms can be arranged.

A   B

Corresponding to these two possible arrangements, two and only two substances have been found with the formula C_2H_6O. One is ethyl alcohol (grain alcohol), a liquid at ordinary room temperature; the other is dimethyl ether, a gas. Because of the different arrangement of the atoms within the molecules, these compounds exhibit different physical and chemical properties. The *structural* formulas **A** and **B** tell us that these substances are different, whereas the *molecular* formula C_2H_6O does not. The three-dimensional structures of these molecules are indicated in Figure 1-3. How, without being

---

Figure 1-3
Three-dimensional models of ethyl alcohol (left) and dimethyl ether (right). Compare these models with formulas **A** *and* **B.**

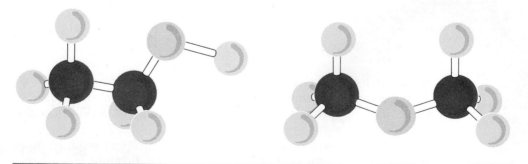

---

able to examine the individual molecules, one is able to decide which of the above formulas is that of ethyl alcohol and which is dimethyl ether is part of the beautiful logic of organic chemistry. In this case, the matter is rather simply determined. Ethyl alcohol reacts with metallic sodium to liberate one-sixth of its hydrogen as a gas; dimethyl ether is inert to sodium. The only reasonable interpretation of these experimental facts is that ethyl alcohol is represented by formula **A,** where one hydrogen is acidic and different (bound to oxygen) from the other five (bound to carbon). Sometimes the distinction is not so obvious, but diligence and ingenuity go not unrewarded in determining structures.

Another example of isomerism is illustrated in Figure 1-4.

Figure 1-4
Three-dimensional models of five carbon atoms arranged in three possible ways to give three isomeric structures.

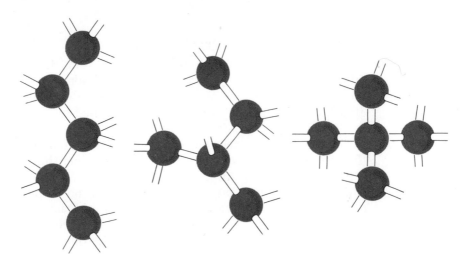

1.5 MULTIPLE BONDS BETWEEN ATOMS

Atoms may share more than two electrons between them. If they share *two pairs* of electrons, the bond is said to be a *double bond;* if *three pairs,* a *triple bond.* For example, the bond between carbon and oxygen in carbon dioxide is a double bond.

$$:\overset{..}{O}::C::\overset{..}{O}: \quad \text{or} \quad :\overset{..}{O}{=}C{=}\overset{..}{O}:$$

Of the sixteen valence electrons available (4 from carbon and 6 from each oxygen) eight are present as unshared pairs on oxygen and eight form the two double bonds. Each nucleus has eight valence electrons around it, and if one counts bonds as a measure of valence, carbon still has a valence of four, although only two atoms are bound to it (and oxygen has its usual valence of two).

Carbon-carbon multiple bonds are common. Examples of skeletal formulas which show four bonds (with eight electrons) surrounding each carbon atom are:

$$\text{>C=C<} \qquad \text{and} \qquad \text{—C≡C—}$$

When atoms are held together by more than one electron pair, the nuclei are pulled closer together than when only a single electron pair binds them. The distance between nuclei (or the bond length) is shorter. The lengths of the C—C, C=C and C≡C bonds are 1.54, 1.34, and 1.20 Å, respectively.

1.6 THE GEOMETRY OF CARBON BONDS

We have seen that carbon may be bound to four, three, or only two other atoms, depending upon whether the bonds are single or multiple. The geometric arrangement depends on the number of atoms attached to carbon and is therefore determined by the orbitals used for bonding.

The four valence electrons of an unbound carbon atom, although they are in a single valence shell, do not in fact have exactly the same energies. An electron may be in an *s* orbital which is spherically symmetrical around the nucleus, or it may be in one of three *p* orbitals which are mutually perpendicular and shaped as shown in Figure 1-5. An electron in the *s* orbital has a somewhat lower energy (i.e., is closer to the nucleus, and therefore more stable) than an electron in the *p* orbital.

Toward the latter part of the nineteenth century, chemists deduced that, when a carbon atom had four other atoms attached to it, these atoms had to be located at the corners of a tetrahedron with the carbon atom at its center. The experiments and reasoning which led to that deduction will be discussed in Chapters 2 and 15. Much later, theoreticians showed that this tetrahedral concept was consistent with and a logical deduction from theories of chemical bonding. When four atoms are attached to carbon (by four pairs of electrons), and one electron of each pair comes from carbon, it can be shown that the best bonding (i.e., the most stable molecule) will result if all four electrons are in identically shaped orbitals which are hybrids (combinations) of one *s* and three *p* orbitals (called *sp*^{3} hybrid orbitals). These orbitals are directed from the carbon nucleus toward the corners of a regular tetrahedron (Figures 1-6 and 1-7).

Figure 1-5
The shapes of s and p orbitals used by the valence electrons of carbon. The nucleus is at the center of the spherical s orbital and at the origin of the three coordinate axes of the p orbitals.

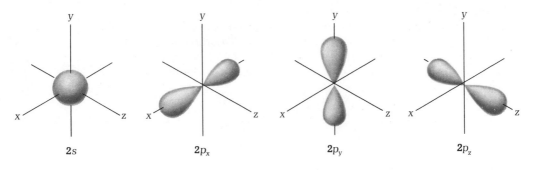

Figure 1-6
An sp³ orbital. A bond can be formed in the direction of the arrow.

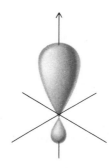

Figure 1-7
Four sp³ orbitals are directed toward the corners of a regular tetrahedron. (The small back lobes of each orbital have been omitted to simplify this figure.)

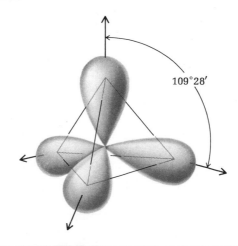

Chemical bonds are then formed by the overlap of suitably shaped orbitals (Figures 1-8, 1-9, and 1-10). The angles between these bonds are 109.5°, the angle obtained when lines are drawn from the center of a regular tetrahedron to two of its corners. The numerical value of this angle has been confirmed experimentally by studying the

---

Figure 1-8
A C—H bond formed by the overlap of an sp³ carbon orbital with a hydrogen s orbital (one electron from each atom). The direction of the bond is on a line joining the two nuclei. This type of bond, symmetrical about the axis joining the atoms, is called **a sigma (σ) bond.**

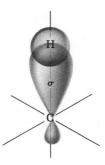

Figure 1-9
A C—C bond (also a σ bond) formed by the overlap between two carbon sp³ orbitals. It has the same symmetry as the bond in Figure 1-8.

Figure 1-10
A molecule of methane, CH_4, formed by the overlap of the four sp^3 carbon orbitals with the s orbitals of four hydrogen atoms. The resulting molecule has the geometry of a tetrahedron.

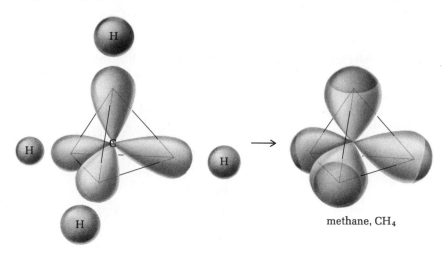

methane, CH_4

diffraction of X-rays by crystals, or the diffraction of an electron beam by gaseous molecules. However, the angle may be a little greater or less than 109.5°, depending on the particular atoms attached to carbon.

The nature of the bonding and its geometric consequences when only three or two atoms are bound to carbon will be taken up later (Chapters 3 and 4).

1.7   REPRESENTATION OF ORGANIC FORMULAS

The true representation of molecules requires three dimensions, but their geometry can sometimes be indicated in a two-dimensional figure. For example, a chain of five carbon atoms might be represented by

where the colored bonds are in the plane of the paper, dash-line bonds extend behind the plane of the paper, and black wedge-shaped bonds extend in front of the plane of the paper toward the reader. This representation will occasionally be useful, but, for most purposes, to conserve space such a formula can be written as

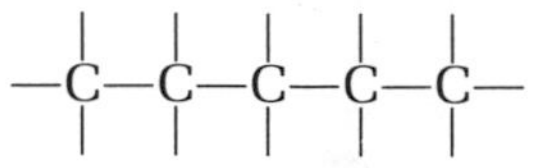

$$-\overset{|}{\underset{|}{C}}-\overset{|}{\underset{|}{C}}-\overset{|}{\underset{|}{C}}-\overset{|}{\underset{|}{C}}-\overset{|}{\underset{|}{C}}-$$

One must bear in mind, however, that this is an oversimplification. If the molecule is C_5H_{12}, it may be indicated as

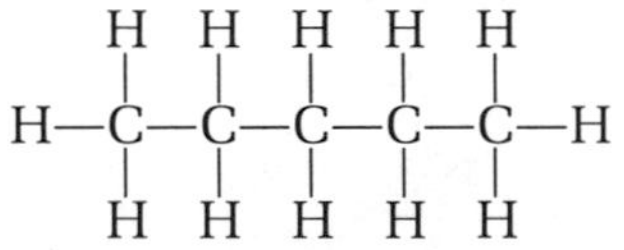

$$\begin{array}{ccccc} H & H & H & H & H \\ | & | & | & | & | \\ H-C-&C-&C-&C-&C-H \\ | & | & | & | & | \\ H & H & H & H & H \end{array}$$

or abbreviated still further to

$$CH_3{-}CH_2{-}CH_2{-}CH_2{-}CH_3$$

or even to

$$CH_3CH_2CH_2CH_2CH_3$$

or

$$CH_3(CH_2)_3CH_3$$

A molecule which is an isomer of this compound may be represented in any of the following ways:

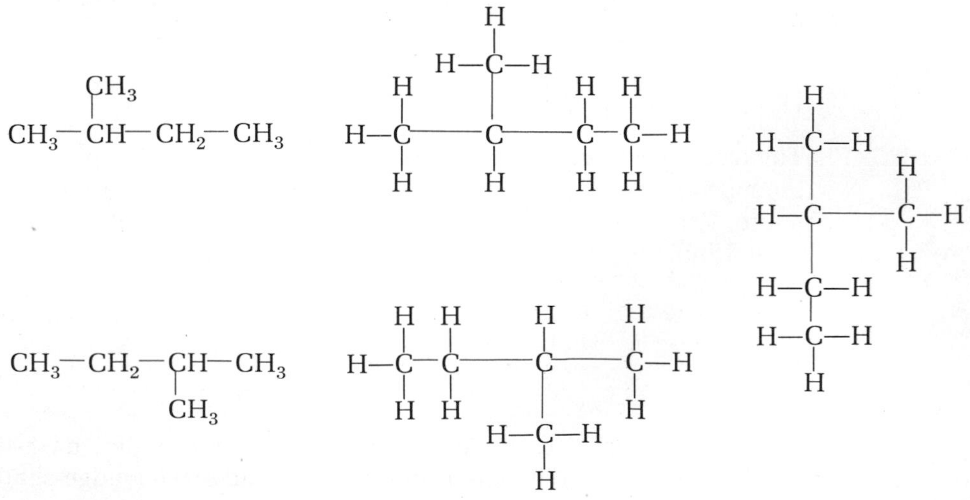

The reader will recognize that all of the above structures are
equivalent and represent only one chemical compound, but any one
of these structures is different from that of its isomer in which the
five carbons form a continuous chain (compare the colored bonds).

$$CH_3-CH_2-CH_2-CH_2-CH_3$$

No matter how the atoms are displayed in a formula, the same
compound is represented as long as the atoms are attached in the
same order or sequence.

1.8   POSSIBLE MECHANISMS FOR COVALENT BOND CLEAVAGE

Chemical reactions generally involve the breaking and making of
bonds. Let us consider the three possible ways in which a covalent
bond A:B may be broken.

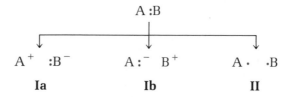

When the bond breaks according to **Ia** or **Ib**, the products are *charged*
particles, or **ions**, whereas a cleavage according to **II** yields *neutral*
fragments called **radicals**, each with an odd number of electrons.
Any one of these types of cleavage may occur in organic reactions.
If A is a carbon with three other groups attached to it, the three
possible fragments are a **carbonium ion**, a **carbanion**, or a **free
radical**.

These species are quite reactive, and are usually produced as
intermediates during a reaction, rather than as end products. Free
radicals and carbonium ions seek one or two electrons, respectively,
to complete the shell of eight electrons around carbon. They are
electrophiles, and react with substances that can supply electrons.
Carbanions, on the other hand, are **nucleophiles;** they seek out
reactants, sometimes positively charged, with which they can share
the unshared electron pair. Examples of stable carbonium ions,

carbanions, and free radicals are known. However, the groups attached to the central carbon atom must have special structures which allow the charge, or odd unpaired electron, to be stabilized, or they must be produced in a solvent or reaction medium with which they do not react.

   If the A part of an A:B bond is a carbon fragment, bond cleavage is most likely to produce a carbonium ion (A⁺) if B has a strong attraction for electrons, that is, if B⁻ is a stable anion, such as chloride ion. On the other hand, if B were a metal, the A:B bond would more likely break to give a carbanion (A⁻) and the positive metallic ion (say, Li⁺). If A and B are similar in electronegativity, the chance that the electron pair would be split to give two radicals A· and B· would be increased. We will see examples in our study of organic reactions of all three types of bond breaking. Ionic reactions will be favored by solvents which are polar (water, alcohols) whereas free radical reactions are relatively less susceptible to solvent polarity.

1.9 THE ORGANIZATION OF ORGANIC CHEMISTRY

A branch of science which concerns itself with a million or more compounds might be very difficult to study were it not possible to classify these compounds into a relatively small number of groups, based upon the **reactive parts** or **functional groups** within the molecules.

   Let us recall an illustration from inorganic chemistry. Chloride ion will react with silver ion to form, in aqueous solution, a white, practically insoluble substance, silver chloride. This reaction is characteristic of the chloride ion and enables us to know a little bit of the chemistry of any substance which contains chloride ions. Thus, although we may try the experiment in the laboratory using sodium chloride, we might predict that potassium chloride, barium chloride, magnesium chloride, cupric chloride, etc., would also give a white precipitate of silver chloride when treated in aqueous solution with silver nitrate.

   In organic chemistry, too, the presence of a particular functional group in an organic molecule imparts to it certain properties. If an organic molecule contains a hydroxyl group (—OH), it will exhibit characteristic properties of that group, whether the substance be

Each of the preceding formulas represents one member of a whole class of organic compounds called **alcohols.** Our study of organic chemistry begins with compounds of carbon and hydrogen only, then proceeds to substances with one functional group containing oxygen, nitrogen, or other elements. Next, we shall consider compounds which contain more than one functional group and study the ways they interact when present in the same molecule. This fundamental study will enable us to examine some of the multitude of complex and fascinating organic substances, such as dyestuffs, drugs, biologically important compounds, and plastics.

New Concepts, Facts, and Terms

1. Carbon is tetravalent. It can combine with other carbon atoms to form chains or rings of carbon atoms.
2. Electronegativity—the affinity of an atom for electrons. In general, the electronegativity of elements increases as one proceeds from left to right and from bottom to top in the periodic table.
3. Types of bonds
   a. ionic—transfer of electrons between atoms that differ widely in electronegativity.
   b. covalent—mutual sharing of electrons between atoms of similar electronegativity.
   c. polar—unequal sharing of electrons between atoms of intermediate difference in electronegativity.
   d. coordinate covalent—a covalent bond in which both electrons are supplied by one of the atoms.
4. Structural formulas—necessary to show the arrangement of atoms within molecules.
5. Isomers—compounds with the same molecular formula but different structural formulas.
6. Multiple bonds—may be formed by sharing two or three electron pairs between atoms.
7. The geometry of carbon bonds—is determined by the number of atoms attached. When a carbon atom is bound to four other atoms by single bonds, sp^3 hybrid orbitals are used. These extend from the carbon nucleus to the corners of a regular tetrahedron. Bonds are formed by the overlap of appropriately shaped orbitals, with two electrons per bond.
8. Organic reactions—usually involve covalent bonds. Carbonium ions, carbanions, or radicals may be reactive intermediates during a reaction. These have, respectively, positive, negative, or neutral carbon atoms, with only three groups attached.
9. Functional groups—the reactive parts of a molecule; can be used to classify the myriad of organic compounds for systematic study.

Exercises and Problems

1. When a solution of salt (sodium chloride) in water is treated with a silver nitrate solution, a white precipitate forms immediately; when carbon tetrachloride is shaken with aqueous silver nitrate, no such precipitate is produced. Explain these facts in terms of the types of bonds in the two chlorides.

2. Using position in the periodic table (Table 1-1) as a criterion of electronegativity, classify the following substances as ionic or covalent:
   a. NaF
   b. F_2
   c. $MgCl_2$
   d. P_2S_5
   e. S_2Cl_2 ✓
   f. LiCl ✓

3. Draw structural formulas for each of the following covalent molecules. Which bonds are polar? Indicate the direction of polarity by an arrow ($\longmapsto$) over each of the bonds, the head of the arrow representing the negative end:
   a. Cl_2
   b. CH_3F
   c. CO_2
   d. HBr
   e. CH_3OH
   f. CH_4

4. Using dots to represent valence electrons and remembering that each element has the number of valence electrons corresponding to the group (Table 1-1) in which it appears, write electronic formulas for each of the following substances:
   a. CH_3Cl
   b. C_3H_8
   c. C_2H_5F
   d. HCN
   e. CO_3^{-2}
   f. NO_3^{-1}
   g. CH_3NH_2
   h. Cl_2
   i. C_2H_5OH
   j. CH_3^+

5. Define and give an example of each of the following:
   a. orbital
   b. nonpolar covalent bond
   c. polar covalent bond
   d. unshared electron pair
   e. multiple bond
   f. sigma bond

6. The six valence electrons of oxygen are in the following orbitals: 2 in the 2s, 2 in the $2p_x$ and one each in the $2p_y$ and $2p_z$. Hydrogen atoms have one electron in the 1s orbital. Draw an orbital representation (like Figure 1-10) for the combination of one oxygen atom with two hydrogen atoms to form a water molecule. What H—O—H angle do you predict from this model? The observed angle is 105°—account for the difference between the predicted and observed angles.

7. Draw structures for each of the following, using atomic orbitals, appropriately hybridized when necessary. Show the expected geometry for each molecule
   a. HF
   b. H_2O_2
   c. NH_3
   d. NH_4^+
   e. BF_3

8. In the methyl carbonium ion, CH_3^+, carbon is bound to only 3 other atoms. Using the idea that there should be maximum separation of orbitals (i.e., minimum repulsion between the electrons in such orbitals), what geometry would you predict for this species?

9. Silicon is just below carbon in the periodic table (Table 1-1). Predict the geometry of silane, SiH_4.

10. Draw a plausible structural formula (similar to those in sec 1.7) for each of the following molecular formulas:
    a. C_4H_{10} e. $C_2H_3Br_3$
    b. C_2H_6O f. C_3H_6
    c. C_3H_8O g. C_3H_4
    d. C_2H_7N (nitrogen usually is tricovalent in organic compounds)

11. Draw all possible structural formulas for each of the following molecular formulas, using only the conventional valences ($C = 4, H = 1, O = 2$, $N = 3, Cl = 1$):
    a. C_6H_{14} c. C_4H_9Cl
    b. $C_4H_{10}O$ d. C_3H_9N

12. Which of the following molecular formulas are plausible and which are not? For those which are, draw all possible isomeric structures.
    a. C_3H_8 c. $C_2H_4Br_2$ e. C_4H_{11}
    b. C_4H_9 d. C_2H_5Cl f. CH_4O

13. Write a plausible molecular and structural formula for a compound which contains:
    a. C, H, and N   c. 3 C's, one O, and sufficient H's
    b. C, H, and S   d. C, H, O, and Cl

14. The combustion of 0.300 gram of an organic compound containing only C, H, and O gave 0.660 gram of CO_2 and 0.360 gram of H_2O. What is its empirical formula? Its molecular weight was determined to be about 60. What is the molecular formula? Suggest two possible structural formulas for the compound. (Atomic Weights $C = 12, H = 1, O = 16$.)

15. A white, crystalline solid isolated from human urine had a molecular weight of 60 and contained C, H, O, and N. Combustion of a 0.250 gram sample gave 0.178 gram of CO_2 and 0.146 gram of H_2O. Another 0.250 gram sample, when boiled with excess alkali, liberated all the nitrogen as ammonia. The ammonia generated was sufficient to neutralize 41.8 milliliters of 0.200 N hydrochloric acid. Calculate the correct molecular formula of the substance. (Atomic Weight $N = 14$.)

16. Group the following compounds according to those which might be expected to exhibit similar chemical behavior.
    a. CH_3OH d. C_5H_{12} g. $CH_3CH_2OCH_2CH_3$
    b. CH_3OCH_3 e. C_4H_9OH h. CH_2Cl_2
    c. $C_3H_5(OH)_3$ f. CH_2Br_2 i. C_3H_7OH
    Draw the structural formula for each.

CHAPTER TWO

SATURATED HYDROCARBONS

Petroleum and natural gas are resources which supply a large percentage of our energy—they are rich in organic chemicals. The main components are saturated hydrocarbons which, because they are relatively unreactive, were originally known as paraffins (meaning that they had "little affinity" for other chemicals). Through extensive research, however, many processes have been discovered for converting them into industrially important organic compounds. They are also the fuels used in automobiles, airplanes, diesel railroad engines and trucks, and in many industrial operations. Saturated hydrocarbons are also of interest as the parent compounds from which the structures of all other organic molecules can be derived.

Our study of organic chemistry begins with the simplest possible organic compounds from the point of view of structure, the **saturated hydrocarbons.** As the name implies, hydrocarbons contain only the elements carbon and hydrogen. The term "saturated" means that only single covalent bonds are present within the molecule. Since hydrocarbons contain only carbon and hydrogen, most of the more complex organic compounds may be thought of as having been derived from saturated hydrocarbons by replacement of hydrogen atoms with other atoms or groups of atoms.

In addition to being the roots of the family tree of organic chemistry, the saturated hydrocarbons are of considerable importance in their own right. *Natural gas,* which is used extensively for fuel purposes both in domestic gas ranges and in industry, consists primarily of methane, CH_4, the simplest saturated hydrocarbon. *Petroleum,* one of our natural resources which has contributed tremendously to our high standard of living, contains high percentages of hydrocarbons. It furnishes gasoline, lubricating oils, vaseline and paraffin wax, tar, and pitch, all of which are rich in saturated hydrocarbons. Petroleum also supplies raw materials for the manufacture of many useful products—plastics, synthetic fibers and rubber, alcohol, antifreeze, and a large number of petrochemicals.

2.1   STRUCTURES OF SOME SATURATED HYDROCARBONS

The simplest saturated hydrocarbon is methane (see Figure 2-1).

Figure 2-1
Three-dimensional models of a methane molecule. At left, the structural model has distorted bond lengths but shows the consecutive attachment of the atoms and the correct bond angles. At right, the scale model more closely represents the actual molecular shape which results from overlap of the appropriate atomic orbitals.

methane

By increasing the number of carbon atoms in a chain, one can develop additional compounds in this series (see Figures 2-2, 2-3, and 2-4).

Figure 2-2
Models of ethane.

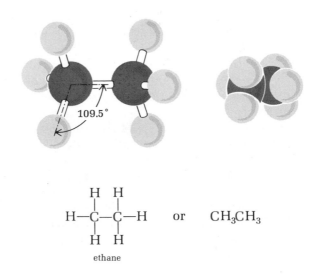

$$H-\underset{\underset{H}{|}}{\overset{\overset{H}{|}}{C}}-\underset{\underset{H}{|}}{\overset{\overset{H}{|}}{C}}-H \quad \text{or} \quad CH_3CH_3$$

ethane

Figure 2-3
Models of propane.

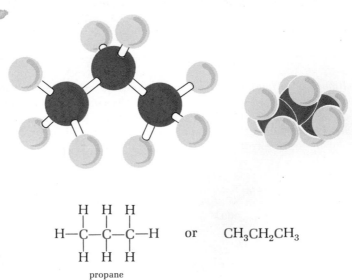

$$H-\underset{\underset{H}{|}}{\overset{\overset{H}{|}}{C}}-\underset{\underset{H}{|}}{\overset{\overset{H}{|}}{C}}-\underset{\underset{H}{|}}{\overset{\overset{H}{|}}{C}}-H \quad \text{or} \quad CH_3CH_2CH_3$$

propane

Figure 2-4
Models of butane.

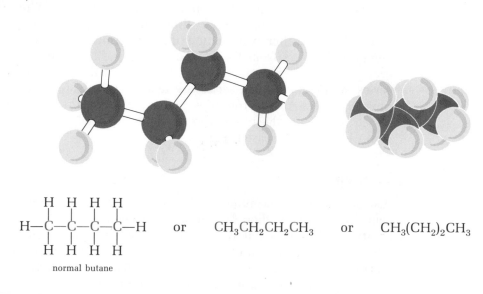

$$H{-}\underset{\displaystyle\overset{\displaystyle H}{|}}{\overset{\displaystyle \overset{H}{|}}{C}}{-}\underset{\displaystyle\overset{H}{|}}{\overset{\displaystyle\overset{H}{|}}{C}}{-}\underset{\displaystyle\overset{H}{|}}{\overset{\displaystyle\overset{H}{|}}{C}}{-}\underset{\displaystyle\overset{H}{|}}{\overset{\displaystyle\overset{H}{|}}{C}}{-}H \qquad or \qquad CH_3CH_2CH_2CH_3 \qquad or \qquad CH_3(CH_2)_2CH_3$$

normal butane

The names of the first four members of this series are common names. As the length of the chain increases, the root of the name is derived from the number of carbon atoms, and the **-ane** ending is retained.

pentane

heptane

hexane

octane

Compounds like these in which there is a single continuous chain of carbon atoms are sometimes called **normal,** or, in this case, normal paraffin hydrocarbons.

2.2 HOMOLOGOUS SERIES

Saturated hydrocarbons conform to a general formula C_nH_{2n+2}, in which n is the number of carbon atoms in the molecule. The reader should test this formula and note that it applies to all the compounds listed above. Further examination of these formulas shows that each compound differs from the next one by a $-CH_2-$, or **methylene** group. A series of compounds so related is called a **homologous series.** Members of such series are usually closely related in both physical and chemical properties. See, for example, the smooth curve which represents the boiling points of the normal paraffins as the chain length is increased (Figure 2-7).

2.3 NOMENCLATURE

Because of the large number of possible saturated hydrocarbons, some systematic method for naming these compounds became necessary. An international system of nomenclature has been devised which is recognized and used by organic chemists throughout the world. The system has been recommended by the International Union of Pure and Applied Chemistry, and is sometimes designated by the initials I.U.P.A.C. (or IUPAC, colloquially pronounced yew-pak).

The rules of the IUPAC system are few and simple to use:

1. The general name for paraffin or saturated hydrocarbons is **alkane.**

2. For branched chain hydrocarbons, determine the longest *continuous* chain of carbon atoms in the molecule. Use the name of the alkane corresponding to this number of carbon atoms as the basis for the name of the compound. Thus, the longest continuous chain of carbon atoms (colored) in the compound shown below has six carbon atoms; the last part of the name will therefore be hexane.

$$\overset{\displaystyle (CH_3)}{\underset{\underset{2}{|}}{CH_3}-\underset{2}{CH}-\underset{3}{CH_2}-\underset{4}{CH_2}-\underset{5}{CH_2}-\underset{6}{CH_3}}$$

3. Number the carbon atoms of the continuous chain, beginning at the end nearest the branching. This has been done in the preceding formula.

4. The substituents which are attached to the chain are given names and numbers. The name of the substituent is taken from the alkane with the same number of carbon atoms by changing the -*ane* ending to -*yl*. The substituent in the preceding formula, shown in

the circle, has only one carbon atom. It is derived from the hydro-carbon methane and is called the methyl group.

$$\underset{\text{methane}}{H-\overset{\displaystyle H}{\underset{\displaystyle H}{C}}-H} \qquad \underset{\text{methyl group}}{H-\overset{\displaystyle H}{\underset{\displaystyle H}{C}}-} \qquad \text{or} \qquad CH_3-$$

The number used to locate the substituent is the number of the carbon atom to which it is attached. Each substituent receives a name *and* a number. The correct name for the formula given above is 2-methylhexane.

5. The longest continuous chain is numbered in such a way that the substituents will have the lowest possible numbers. Thus, if the chain in the above formula were numbered from right to left, the name would be 5-methylhexane. This is incorrect according to the IUPAC rules because the name 2-methylhexane uses a lower number.

The following examples further illustrate the method in practice:

$$\underset{1}{CH_3}-\overset{\displaystyle \overset{2}{|}\,CH_3}{\underset{\displaystyle \underset{}{|}\,CH_3}{C}}-\overset{3}{CH_2}-\overset{4}{CH_3}$$

2,2-dimethylbutane

Here, two numbers are used, one for each substituent. The numbers are separated by a comma, and a hyphen is placed between the numbers and the name. There is no space between the names of the substituents and the name of the longest chain; they are written as a single word. The prefixes di-, tri-, tetra- are used to show the presence of 2, 3, or 4 identical substituents.

Sometimes, the longest chain may not be written on a single line, as in

$$\begin{array}{c}
\overset{2}{CH}-\overset{3}{CH_2}-\overset{4}{CH}-\overset{5}{CH_2}
\end{array}$$

2,4,6-trimethylheptane $\equiv$ 2,4,6-trimethylheptane

2.4 THE SIMPLE ALKYL GROUPS

Under Rule 4 of the IUPAC system a method was set up for naming substituents or fragments of saturated hydrocarbons. These fragments are called **alkyl groups** and are derived from the parent compound by removing one of the hydrogen atoms. The first two groups are

$$
\underset{\text{methyl group}}{H-\overset{\displaystyle H}{\underset{\displaystyle H}{C}}-} \quad \text{or} \quad CH_3-
\qquad\qquad
\underset{\text{ethyl group}}{H-\overset{\displaystyle H}{\underset{\displaystyle H}{C}}-\overset{\displaystyle H}{\underset{\displaystyle H}{C}}-} \quad \text{or} \quad CH_3CH_2- \quad \text{or} \quad C_2H_5-
$$

There are *two* possible groups which can be derived from propane, depending upon which type of hydrogen is removed.

$$
\underset{\text{propane}}{H-\overset{\displaystyle H}{\underset{\displaystyle H}{C}}-\overset{\displaystyle H}{\underset{\displaystyle H}{C}}-\overset{\displaystyle H}{\underset{\displaystyle H}{C}}-H}
$$

If a terminal primary* hydrogen (black) is removed, the group is

$$
\underset{\text{normal or } n\text{-propyl group}}{H-\overset{\displaystyle H}{\underset{\displaystyle H}{C}}-\overset{\displaystyle H}{\underset{\displaystyle H}{C}}-\overset{\displaystyle H}{\underset{\displaystyle H}{C}}-} \quad \text{or} \quad CH_3CH_2CH_2-
$$

whereas if one of the central secondary hydrogens (colored) is removed, the group is

$$
\underset{\text{isopropyl or } i\text{-propyl group}}{H-\overset{\displaystyle H}{\underset{\displaystyle H}{C}}-\overset{\displaystyle H}{C}-\overset{\displaystyle H}{\underset{\displaystyle H}{C}}-H} \quad \text{or} \quad CH_3CHCH_3 \quad \text{or} \quad (CH_3)_2CH-
$$

*The terms primary, secondary, and tertiary are used to distinguish hydrogens attached to carbon atoms which in turn are bound to one, two, or three other carbon atoms, respectively.

The names of the alkyl groups are used in both the common and IUPAC names of organic compounds. This is illustrated in the following formulas:

$$CH_3CH_2Br \qquad CH_3CHCH_3 \qquad \overset{1}{CH_3}\overset{2|}{CH}\overset{3}{CH}\overset{4|}{CH}\overset{5}{CH_2}\overset{6}{CH_3}$$

$$\begin{array}{ccc} & CH_3 & CH_3 \\ & | & | \\ & Cl & CH_2 \\ & & | \\ & & CH_3 \end{array}$$

ethyl bromide	isopropyl chloride	2,4-dimethyl-3-ethylhexane
(bromoethane)	(2-chloropropane)	

Sometimes it is desirable to have a symbol for an alkyl group, without specifying which one. For this purpose the symbol **R** (for alkyl) is used. Thus the formula

$$R—Cl$$

represents any member of the homologous series of alkyl chlorides (methyl chloride, ethyl chloride, etc.).

2.5 CONFORMATIONS AT SINGLE BONDS

The distance between two carbon atoms joined by a single electron pair is relatively independent of the remaining structure of the particular molecule and is usually about 1.54 Å. Since the bond is symmetrical about the axis joining the two nuclei, rotation about a single bond is possible.

At first it would seem as if an infinite number of ethane molecules might exist, depending upon the angle of rotation of one carbon atom with respect to the other. It is indeed true that an infinite number of **conformations** is possible for ethane. But at room temperature the ethane molecule has sufficient thermal energy so that rotation about the single bond from one conformation to another is easily possible. Therefore only one ethane (and no isomer of it) is known. However, there is a preference for staggered conformations over eclipsed ones, because this minimizes any repulsion due to the bonds to the hydrogen atoms on adjacent carbons (Figure 2-5).

With larger molecules, the number of possible conformations increases. Figure 2-6 shows sawhorse formulas for two eclipsed and two staggered conformations around the central carbon-carbon bond in *n*-butane.

Figure 2-5

Two of the possible conformations of ethane, eclipsed and staggered. Interconversion is easily possible by 60° rotation about the C—C bond, as shown by the curved arrows. The upper formulas show space-filling models. The lower left structure in each case is a "sawhorse" drawing. The lower right structure in each case is a Newman projection formula—an end-on view down the C—C axis.

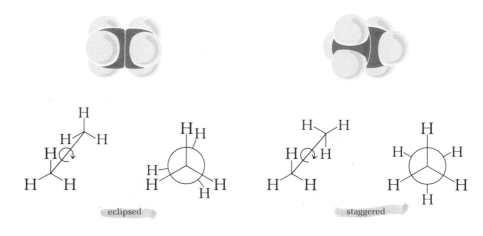

Figure 2-6

Four conformations about the C$_2$—C$_3$ bond of butane.

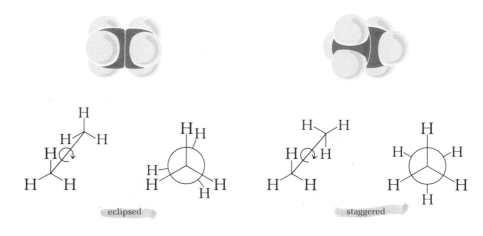

2.6 PHYSICAL PROPERTIES OF ALKANES

The alkanes are insoluble in water, and those which are liquid are considerably less dense than water and float on it. The attractive forces between alkane molecules are small, because the molecules contain no polar groups, multiple bonds, or unshared electron pairs. As a result, alkanes tend to have lower boiling points for a given molecular weight than any other hydrocarbon-derived organic compounds.

A summary of the boiling points of normal saturated hydrocarbons is given in Figure 2-7. The boiling point is directly related to the molecular weight or chain length. At ordinary temperatures and pressures, the first four members of the series are gases; those with more than twenty carbon atoms are wax-like solids. Those with intermediate numbers of carbons are colorless liquids.

Isomeric hydrocarbons (those having the same molecular formula but different structural formulas) have different physical properties. Usually the more branched the chain, the lower the boiling point. This is illustrated at the right in Figure 2-7 for the various isomers of pentane, C_5H_{12}.

Figure 2-7
The boiling points of the normal alkanes rise smoothly as the length of the carbon chain increases. With equal numbers of carbon atoms, chain branching causes a decrease in boiling point (see table).

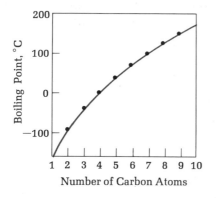

Name	Formula	Boiling Point
pentane	$CH_3CH_2CH_2CH_2CH_3$	36°
2-methylbutane (isopentane)	$CH_3CHCH_2CH_3$ $\mid$ CH_3	28°
2,2-dimethylpropane (neopentane)	CH_3 $\mid$ $CH_3—C—CH_3$ $\mid$ CH_3	10°

2.7   REACTIONS OF ALKANES

Saturated hydrocarbons are relatively inert; they have little affinity for most chemical reagents. They are almost insoluble in water and do not react with aqueous solutions of acids or alkalies. They are not readily oxidized by most chemical oxidizing agents, but they do react with halogens, with concentrated nitric acid, and with oxygen when ignited.

The members of a homologous series have similar chemical properties. Thus the reactions of ethane, C_2H_6, and pentane, C_5H_{12}, are similar. If we study the chemistry of one of these compounds, we can predict the chemistry of the other with some confidence.

2.7a HALOGENATION

As saturated hydrocarbons contain only covalent σ-bonds, there are no unshared electron pairs, multiple bonds, or vacant orbitals accessible to an attacking reagent. Reaction can only be initiated by paths which break C—H or C—C single bonds.

A mixture of an alkane and chlorine or bromine can be kept indefinitely at low temperatures in the dark; no reaction occurs. In sunlight or at high temperatures, however, hydrogen atoms in the alkane are replaced by halogen atoms. The reaction is called **halogenation** (chlorination or bromination) and is illustrated for methane in equation 2-1.

$$\text{Cl—Cl} + \text{H—}\underset{\underset{\text{H}}{|}}{\overset{\overset{\text{H}}{|}}{\text{C}}}\text{—H} \xrightarrow[\text{or heat}]{\text{sunlight}} \text{H—}\underset{\underset{\text{H}}{|}}{\overset{\overset{\text{H}}{|}}{\text{C}}}\text{—Cl} + \text{H—Cl} \qquad (2\text{-}1)$$

chlorine   methane   methyl chloride

The weakest bond in the reactants is the Cl—Cl bond. Reaction is initiated when chlorine molecules absorb energy and split into the very reactive chlorine atoms (eq 2-2). Although some chlorine

$$:\overset{..}{\underset{..}{\text{Cl}}}:\overset{..}{\underset{..}{\text{Cl}}}: \xrightarrow[\text{sunlight}]{\text{heat or}} 2 :\overset{..}{\underset{..}{\text{Cl}}}\cdot \qquad (2\text{-}2)$$

chlorine molecule   chlorine atoms

atoms may recombine, others may attack a hydrocarbon molecule and remove a hydrogen atom (eq 2-3). The resulting alkyl radical

$$\text{H—}\underset{\underset{\text{H}}{|}}{\overset{\overset{\text{H}}{|}}{\text{C}}}:\text{H} + :\overset{..}{\underset{..}{\text{Cl}}}\cdot \longrightarrow \text{H—}\underset{\underset{\text{H}}{|}}{\overset{\overset{\text{H}}{|}}{\text{C}}}\cdot + \text{H}:\overset{..}{\underset{..}{\text{Cl}}}: \qquad (2\text{-}3)$$

methane   chlorine atom   methyl radical   hydrogen chloride

may, if it collides with a chlorine molecule, abstract a chlorine atom to produce an alkyl halide and another chlorine atom (eq 2-4). The chlorine atom formed in equation 2-4 may react with another methane molecule as in equation 2-3 to repeat the process. This is called a *free radical chain reaction*. Radical combinations are rare, at least in the early stages of the reaction, and when they occur (for

$$\text{H—}\underset{\underset{\text{H}}{|}}{\overset{\overset{\text{H}}{|}}{\text{C}}}\cdot + :\overset{..}{\underset{..}{\text{Cl}}}:\overset{..}{\underset{..}{\text{Cl}}}: \longrightarrow \text{H—}\underset{\underset{\text{H}}{|}}{\overset{\overset{\text{H}}{|}}{\text{C}}}:\overset{..}{\underset{..}{\text{Cl}}}: + :\overset{..}{\underset{..}{\text{Cl}}}\cdot \qquad (2\text{-}4)$$

methyl radical   chlorine molecule   methyl chloride   chlorine atom example, if two methyl radicals combine to form ethane) they terminate the chain reaction.

In the overall reaction, a C—H and a Cl—Cl bond are broken (requiring 102 and 58 kcal/mole respectively) and a C—Cl and H—Cl bond are produced (furnishing 81 and 103 kcal/mole respectively). The process is thus favorable (exothermic) by 24 kcal/mole and the free radical chain reaction proceeds spontaneously, once it is initiated. Only one-half of the chlorine atoms with which one starts end in the organic product, methyl chloride. The other half form a compound (hydrogen chloride) with the atom (hydrogen) which was displaced or substituted.

Side reactions and by-products are often associated with substitution reactions. For example, the methyl chloride may react with chlorine, giving rise to the following products:

$$\text{Cl—Cl} + \text{H—}\underset{\underset{\displaystyle H}{|}}{\overset{\overset{\displaystyle H}{|}}{\text{C}}}\text{—Cl} \longrightarrow \text{Cl—}\underset{\underset{\displaystyle H}{|}}{\overset{\overset{\displaystyle H}{|}}{\text{C}}}\text{—Cl} + \text{HCl} \qquad (2\text{-}5)$$

dichloromethane
(methylene chloride)

$$\text{Cl—Cl} + \text{Cl—}\underset{\underset{\displaystyle H}{|}}{\overset{\overset{\displaystyle H}{|}}{\text{C}}}\text{—Cl} \longrightarrow \text{Cl—}\underset{\underset{\displaystyle Cl}{|}}{\overset{\overset{\displaystyle H}{|}}{\text{C}}}\text{—Cl} + \text{HCl} \qquad (2\text{-}6)$$

trichloromethane
(chloroform)

$$\text{Cl—Cl} + \text{Cl—}\underset{\underset{\displaystyle Cl}{|}}{\overset{\overset{\displaystyle H}{|}}{\text{C}}}\text{—Cl} \longrightarrow \text{Cl—}\underset{\underset{\displaystyle Cl}{|}}{\overset{\overset{\displaystyle Cl}{|}}{\text{C}}}\text{—Cl} + \text{HCl} \qquad (2\text{-}7)$$

tetrachloromethane
(carbon tetrachloride)

When the hydrocarbon has more than one carbon atom, the process may become even more complex. Thus the first substitution product of ethane is ethyl chloride.

$$\text{Cl—Cl} + \text{H—}\underset{\underset{\displaystyle H\ H}{|\ |}}{\overset{\overset{\displaystyle H\ H}{|\ |}}{\text{C—C}}}\text{—H} \xrightarrow[\text{sunlight}]{\text{heat or}} \text{H—}\underset{\underset{\displaystyle H\ H}{|\ |}}{\overset{\overset{\displaystyle H\ H}{|\ |}}{\text{C—C}}}\text{—Cl} + \text{HCl} \qquad (2\text{-}8)$$

ethane

chloroethane
(ethyl chloride)

When a second hydrogen is replaced, two alternatives are possible, and a mixture of both products is obtained.

$$Cl{-}Cl + \underset{\substack{\text{H} \quad \text{H} \\ \big| \quad \big| \\ \text{H} \quad \text{H}}}{H{-}C{-}C{-}Cl} \xrightarrow[\text{sunlight}]{\text{heat or}} \left\{ \begin{array}{c} \underset{\substack{\text{H} \quad \text{H}}}{\overset{\substack{\text{H} \quad \text{Cl}}}{H{-}C{-}C{-}Cl}} \\ \text{1,1-dichloroethane} \\[2mm] \underset{\substack{\text{H} \quad \text{H}}}{\overset{\substack{\text{H} \quad \text{H}}}{Cl{-}C{-}C{-}Cl}} \\ \text{1,2-dichloroethane} \end{array} \right\} + HCl \qquad (2\text{-}9)$$

When propane is halogenated, a mixture of products can be obtained, even at the first step:

$$Cl{-}Cl + CH_3CH_2CH_3 \xrightarrow[\text{or heat}]{\text{sunlight}} \left\{ \begin{array}{c} CH_3CH_2CH_2Cl \\ \text{1-chloropropane} \\ \text{(n-propyl chloride)} \\[1mm] + \\[1mm] \underset{\quad \overset{|}{Cl}}{CH_3CHCH_3} \\ \text{2-chloropropane} \\ \text{(isopropyl chloride)} \end{array} \right\} + HCl \qquad (2\text{-}10)$$

Because of the multiplicity of products, halogenation of saturated hydrocarbons may not always be a useful laboratory reaction. But cyclic hydrocarbons, because of their symmetry, can be converted to monosubstituted derivatives in good yield (eq 2-11).

$$\begin{array}{c} \underset{\substack{CH_2 \\ / \quad \backslash \\ CH_2{-}CH_2}}{\overset{CH_2}{CH_2}} + Cl_2 \xrightarrow[\text{sunlight}]{\text{heat or}} \underset{\substack{CH_2 \\ / \quad \backslash \\ CH_2{-}CH_2}}{\overset{\overset{\displaystyle Cl}{|}}{\underset{CH_2}{CH}}} + HCl \qquad (2\text{-}11) \\ \text{cyclopentane} \qquad\qquad \text{cyclopentyl chloride} \end{array}$$

A petroleum fraction consisting mainly of pentane and 2-methylbutane is chlorinated commercially to give a mixture of chlorinated pentanes which are not separated but used directly (for example, in the manufacture of solvents for automobile lacquers).

2.7b NITRATION

Another commercially important substitution reaction involves the reaction of saturated hydrocarbons with nitric acid vapor:

$$\underset{\substack{H \\ | \\ H-\overset{\displaystyle |}{\underset{\displaystyle |}{C}}-H \\ | \\ H}}{} + \underset{(HNO_3)}{HO-NO_2} \xrightarrow{400°} \underset{\substack{H \\ | \\ H-\overset{\displaystyle |}{\underset{\displaystyle |}{C}}-NO_2 \\ | \\ H}}{} + HOH \qquad (2\text{-}12)$$

$$\underset{\text{nitric acid}}{} \qquad \underset{\text{nitromethane}}{}$$

Nitration is somewhat different from halogenation in that only mononitration is observed and, with higher alkanes, breaking of carbon-carbon as well as carbon-hydrogen bonds occurs.

$$\underset{\text{propane}}{CH_3CH_2CH_3} \xrightarrow{HONO_2}$$

$$CH_3CH_2CH_2NO_2 + CH_3\underset{\substack{| \\ NO_2}}{CHCH_3} + CH_3CH_2NO_2 + CH_3NO_2 \qquad (2\text{-}13)$$

$$\underset{\text{1-nitropropane}}{} \qquad \underset{\text{2-nitropropane}}{} \qquad \underset{\text{nitroethane}}{} \qquad \underset{\text{nitromethane}}{}$$

The products are excellent solvents, ingredients of special racing and model engine fuels, as well as intermediates for the synthesis of drugs, insecticides, and explosives.

2.7c OXIDATION

Perhaps the most important use of hydrocarbons is as fuels. Hydrocarbons burn in an excess of oxygen to form carbon dioxide and water. The reaction is accompanied by the evolution of large quantities of heat.

$$CH_4 + 2\,O_2 \longrightarrow CO_2 + 2\,H_2O + 212.8 \text{ kcal/mole} \qquad (2\text{-}14)$$

$$2\,C_4H_{10} + 13\,O_2 \longrightarrow 8\,CO_2 + 10\,H_2O + 688.0 \text{ kcal/mole} \qquad (2\text{-}15)$$

Reactions of this type constitute the basis for the important use of hydrocarbons as sources of heat (natural gas) and power (gasoline in the internal combustion engine). The mechanism of combustion is complex and still under study. There is little doubt, however, that it involves a free radical chain reaction. Like halogenation, an initiation step is required—usually ignition by a spark or flame. Once initiated, the reaction proceeds spontaneously and exothermically.

Gasoline consists of hydrocarbons which boil from about 30° to 200° C* and contain from five to twelve carbon atoms per molecule. The equation for the combustion of these substances shows that the reaction involves the formation of large volumes of gases. It is this sudden expansion of gases that drives the pistons in an automobile engine.

Pyrofax, or "bottled gas," is mainly propane which, through pressure, has been condensed to a liquid so that it may be compactly stored and shipped in metal cylinders or tanks.

In the absence of sufficient oxygen for complete reaction, partial combustion may occur. The products may be carbon monoxide or even carbon.

$$2 \, CH_4 + 3 \, O_2 \longrightarrow 2 \, CO + 4 \, H_2O \qquad (2\text{-}16)$$

$$CH_4 + O_2 \longrightarrow C + 2 \, H_2O \qquad (2\text{-}17)$$

The effects of incomplete combustion are well known to every motorist in the form of carbon deposits in the head and on the pistons of the motor, and the toxic carbon monoxide in the exhaust fumes. Incomplete combustion of natural gas is sometimes carried out purposely to manufacture carbon blacks, particularly lampblack, a pigment for ink, and channel black, used in automobile tires.

2.7d   CRACKING

Hydrocarbons are generally very stable compounds which can be distilled without decomposition. But if they are heated to temperatures well above their boiling points *in the absence of air,* the molecules can be caused to disintegrate or "crack" into smaller fragments. Carbon-carbon, as well as carbon-hydrogen bonds are broken. Higher hydrocarbons yield a mixture of saturated and unsaturated hydrocarbons of lower molecular weight. This reaction is carried out extensively in the petroleum industry to convert hydrocarbons with about fifteen to eighteen carbon atoms into gasoline hydrocarbons with numbers of carbon atoms in the range of five to twelve atoms. The "cracking" process has made available a considerably larger quantity of motor fuel than is directly obtainable from crude petroleum. Catalysts have been developed which allow the process to be carried out at lower temperatures than was formerly possible (Figure 2-8).

Other reactions important to petroleum chemistry will be discussed in the next chapter.

---

*All temperatures in this book are centigrade unless otherwise stated.

Figure 2-8
The "cat cracker"—a catalytic cracking unit is shown here lighted for night operation in a petroleum refinery. In this unit the carbon chains of high-boiling hydrocarbons are broken into lower molecular weight compounds suitable for gasoline. (Courtesy of the Atlantic Refining Company.)

2.8 PREPARATION OF ALKANES

The most common source of alkanes is petroleum, from which they may be obtained by fractional distillation. However, a particular hydrocarbon sometimes has to be tailor-made or synthesized for research studies. To do this, several laboratory methods have been developed. Two of these will be described here.

2.8a THE WURTZ REACTION

In 1855 the French chemist Charles Adolphe Wurtz discovered that ethane was obtained when methyl iodide was treated with metallic sodium.

$$CH_3-I + 2\ Na + I-CH_3 \longrightarrow CH_3-CH_3 + 2\ Na^+I^- \quad (2\text{-}18)$$

It is likely that this reaction proceeds by way of an organosodium compound, formed as in equation 2-19. This then reacts with a second molecule of methyl iodide to form the final product (eq 2-20).

$$CH_3 \!:\! \ddot{\underset{\cdot\cdot}{I}} \!: \ + 2\ Na\cdot \longrightarrow CH_3 \!:\! ^-Na^+ + Na^+ \!:\! \ddot{\underset{\cdot\cdot}{I}} \!:\! ^- \quad (2\text{-}19)$$

methylsodium

$$CH_3:^-Na^+ + CH_3:\overset{..}{\underset{..}{I}}: \longrightarrow CH_3:CH_3 + Na^+:\overset{..}{\underset{..}{I}}:^- \qquad (2\text{-}20)$$

The reaction may be extended to more complex alkyl halides. For example, ethyl bromide gives a good yield of *n*-butane.

$$2\ CH_3CH_2{-}Br + 2\ Na \longrightarrow CH_3CH_2{-}CH_2CH_3 + 2\ Na^+Br^- \qquad (2\text{-}21)$$

Note that the hydrocarbon which is formed is *symmetrical* and has *twice* as many carbon atoms as the original alkyl halide. The carbon atoms that join are those which originally held the halogen atoms. The reaction can be summarized by the general equation

$$R{-}X + 2\ Na \longrightarrow R{-}R + 2\ Na^+X^- \qquad (2\text{-}22)$$

where R is an alkyl group and X is chlorine, bromine, or iodine.

One variation of the Wurtz reaction uses a 1,3-dihalide and zinc, to produce cyclopropane (a cycloalkane).

$$
\underset{\substack{\text{Br} \qquad \text{Br}}}{\overset{\substack{CH_2}}{\underset{CH_2 \quad CH_2}{\diagup\ \diagdown}}}
+ Zn \longrightarrow
\overset{CH_2}{\underset{CH_2{-}CH_2}{\diagup\ \diagdown}}
+ Zn^{+2}Br_2^{-1} \qquad (2\text{-}23)
$$

1,3-dibromopropane   cyclopropane

2.8b REDUCTION OF AN ALKYL HALIDE

Halogen compounds can be converted to hydrocarbons with the same number of carbon atoms by a variety of reducing agents which replace the halogen by hydrogen. The general equation is

$$R{-}X \xrightarrow{\text{[H]}^*} R{-}H \qquad (2\text{-}24)$$

Zinc and acetic acid make a good reducing combination for this purpose. An intermediate in the reaction may be an organozinc halide, which is then hydrolyzed by the acid (eq 2-25).

$$
\underset{\text{Br}}{CH_3CHCH_3} + Zn \longrightarrow \left[\underset{Zn^+Br^-}{CH_3CHCH_3}\right] \xrightarrow{H^+} \underset{H}{CH_3CHCH_3}
$$

$$+ Zn^{+2} + Br^{-1} \qquad (2\text{-}25)$$

*[H] signifies a reducing agent.

If the metal, such as zinc or magnesium, is allowed to react with the alkyl halide in a dry (anhydrous) solvent such as ether, a solution of the organometallic reagent is obtained. This can then be hydrolyzed in a separate step. If D_2O is used in place of H_2O, this is a good way of preparing isotopically labeled hydrocarbons. An example is shown in equation 2-26.

$$CH_3CH_2CH_2Br + Mg \xrightarrow[\text{ether}]{\text{anhydrous}} CH_3CH_2CH_2MgBr$$

n-propyl bromide   *n*-propylmagnesium

                            bromide   (2-26)

$$CH_3CH_2CH_2D + Mg^{+2}(OD)^{-1}Br^{-1} \xleftarrow{\text{D—O—D}}$$

labeled propane
(propane—1—d_1)

The organomagnesium halide is called a **Grignard reagent** after its discoverer, Victor Grignard. This type of reagent is useful in many organic syntheses, as we shall see throughout the text.

2.9   STRUCTURE PROOF

The Wurtz reaction affords an excellent example of how the organic chemist can use deductive reasoning to determine the structure of an organic compound. It is possible to obtain from natural petroleum two hydrocarbons with the formula C_4H_{10}. One boils at 0°; the other, at −10°. From valence considerations, we arrive at two possible formulas:

$$CH_3CH_2CH_2CH_3 \qquad\qquad CH_3CHCH_3$$

$$\qquad\qquad\qquad\qquad\qquad\qquad\qquad\qquad CH_3$$

                n-butane   isobutane

                                 (2-methylpropane)

How can the organic chemist determine which compound is which? The reasoning involved goes somewhat as follows: Ethane can form only one monobromination product, since all six hydrogens of ethane are equivalent (substitute bromine for chlorine in equation 2-8). Ethyl bromide, when it reacts with sodium in the Wurtz reaction (see equation 2-21), must give *n*-butane. The product obtained from the Wurtz reaction with ethyl bromide boils at 0°. Therefore, the isomer of C_4H_{10} which boils at 0° must be *n*-butane, and the one that boils at −10° must be isobutane. This is an example of proof of structure by synthesis.

2.10 CYCLOALKANES

Alkanes which contain one or more carbon rings are called **cyclo-alkanes.** The smallest possible ring contains three carbon atoms—cyclopropane. The most common ring sizes, for compounds which occur in nature, are 5- and 6-membered rings. Larger rings, with as many as 30 or more carbon atoms, are also known.

$$
\begin{array}{c}
\text{CH}_2 \\
\diagup \diagdown \\
\text{CH}_2\text{—CH}_2
\end{array}
\qquad
\begin{array}{c}
\text{CH}_2\text{—CH}_2 \\
| \qquad | \\
\text{CH}_2\text{—CH}_2
\end{array}
\qquad
\begin{array}{c}
\text{CH}_2 \\
\diagup \quad \diagdown \\
\text{CH}_2 \qquad \text{CH}_2 \\
\diagdown \qquad \diagup \\
\text{CH}_2\text{—CH}_2
\end{array}
\qquad
\begin{array}{c}
\text{CH}_2 \\
\diagup \quad \diagdown \\
\text{CH}_2 \qquad \text{CH}_2 \\
| \qquad | \\
\text{CH}_2 \qquad \text{CH}_2 \\
\diagdown \quad \diagup \\
\text{CH}_2
\end{array}
$$

cyclopropane   cyclobutane   cyclopentane   cyclohexane

$$
\begin{array}{c}
\overset{3}{\text{CH}_2} \\
\diagup \diagdown \\
\overset{1}{\text{CH}}\text{—}\overset{2}{\text{CH}} \\
| \qquad | \\
\text{CH}_3 \quad \text{CH}_3
\end{array}
\qquad
\begin{array}{c}
\text{CH}_3 \quad \text{CH}_3 \\
\diagdown \diagup \\
\overset{1}{\text{C}} \\
\diagup \quad \diagdown \\
\overset{5}{\text{CH}_2} \qquad \overset{2}{\text{CH}_2} \\
| \qquad\qquad | \\
\overset{4}{\text{CH}_2}\text{—}\overset{3}{\text{CH}}\text{—CH}_3
\end{array}
$$

1,2-dimethylcyclopropane   1,1,3-trimethylcyclopentane

Although the general formula for these compounds is C_nH_{2n}, the structures are analogous to the open-chain saturated hydrocarbons, all bonds being single. Except for the first two members of the series (C_3 and C_4), the reactions of cycloalkanes are very similar to those of their open-chain analogs.

Cyclopropane is a planar molecule. The C—C—C angles are appreciably smaller than the normal tetrahedral angle, and for this reason the ring is under some strain. In chemical reactions, there is a tendency to relieve this strain by opening the ring. Thus, although cyclopropane can be chlorinated in the presence of ultraviolet light to give cyclopropyl chloride, it reacts with many reagents to give ring-opened products. Some examples are:

$$
\begin{array}{c}
\text{CH}_2 \\
\diagup \diagdown \\
\text{CH}_2\text{—CH}_2
\end{array}
+ \text{HBr} \longrightarrow \text{H—CH}_2\text{—CH}_2\text{—CH}_2\text{—Br}
$$

n-propyl bromide

$$
\xrightarrow[\text{(H}_2\text{SO}_4)]{\text{H—OSO}_3\text{H}} \text{H—CH}_2\text{—CH}_2\text{—CH}_2\text{—OSO}_3\text{H} \qquad (2\text{-}27)
$$

n-propyl hydrogen sulfate

Despite the strain, the cyclopropane ring occurs in many natural products, including chrysanthemums, caraway seeds, and the proteins of pear juices. The cyclobutane ring, also strained, nevertheless occurs in natural products, perhaps the most common being pine oil (Chapter 18).

Five- and six-membered rings are relatively free of strain; cyclopentane is nearly planar, with just a slight puckering to accommodate the difference between the tetrahedral angle and the internal angle of 108° in a flat pentagon. In cyclohexane, the difference (120° vs. 109° 28′) can be accommodated by a somewhat larger puckering of the ring. Two possible arrangements, called the **chair** and the **boat** forms, because of their shapes, are shown in Figures 2-9 and 2-10. The chair form is much preferred, because in it all the hydrogens are in staggered conformations with respect to their nearest neighbors, very much like the staggered conformation of ethane. In the boat form, some hydrogens are eclipsed; also, two of the "bowsprit"

Figure 2-9

The chair conformation of cyclohexane. The upper formulas show stick-and-ball and space-filling models. In the lower left, one sees that six of the hydrogens (black) lie roughly in the carbon ring plane, and the other six hydrogens (colored) lie along an axis perpendicular to that plane with three above and three below the plane. These are called **equatorial** *and* **axial** *hydrogens, respectively. In the lower right, a Newman projection formula, looking down the C_2—C_3 and C_5—C_6 bond axes, shows the staggered arrangement of hydrogens on adjacent carbons.*

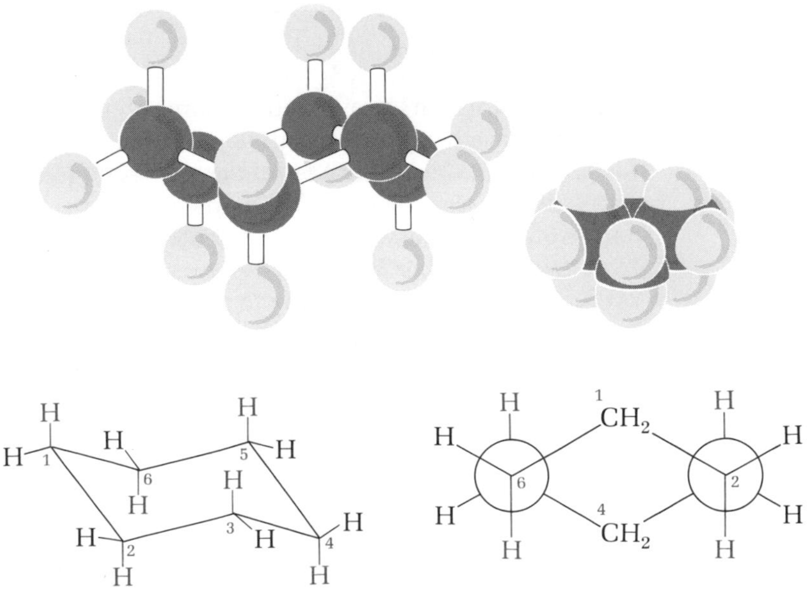

Figure 2-10

The boat form of cyclohexane. The upper structures depict stick-and-ball and space-filling models. In the lower left, the bowsprit hydrogens which come very close to one another are shown in color. In the lower right, a slightly distorted Newman projection formula shows the eclipsing of hydrogens (in color) on C_2-C_3 and C_5-C_6.

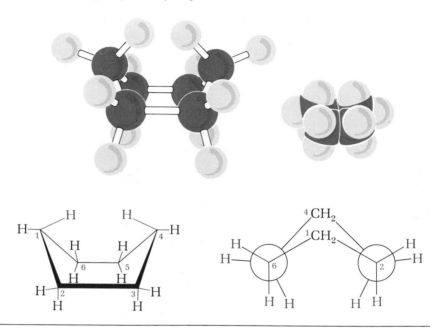

hydrogens (on carbons 1 and 4) come very close to one another.

At ordinary temperatures, the cyclohexane ring exists in the chair form. Since axial groups on alternate carbons come very close to one another (see especially the space-filling model in Figure 2-9), substituted cyclohexanes prefer structures in which the substituent is equatorial rather than axial. Thus the equilibrium for methyl-cyclohexane (eq 2-28) lies far to the right.

(2-28)

Since many natural products contain the cyclohexane ring, an understanding of these conformational effects has been of paramount importance in elucidating their chemistry.

2.11 UNUSUAL RING STRUCTURES

In recent years, much has been learned about the nature of chemical bonding by studying molecules with unusual ring structures. Several classes of compounds with more than one ring can be delineated. Compounds with one carbon atom common to two rings are called **spiro** (rhymes with gyro) compounds. An example is spiropentane, in which the two three-membered rings are mutually perpendicular and carbon-1 (colored) is common to both rings.

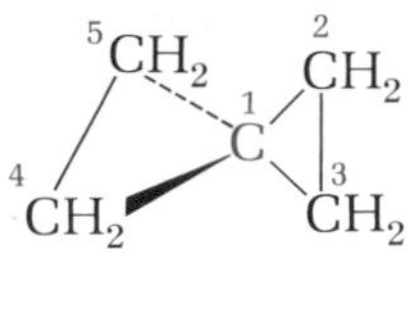

spiropentane 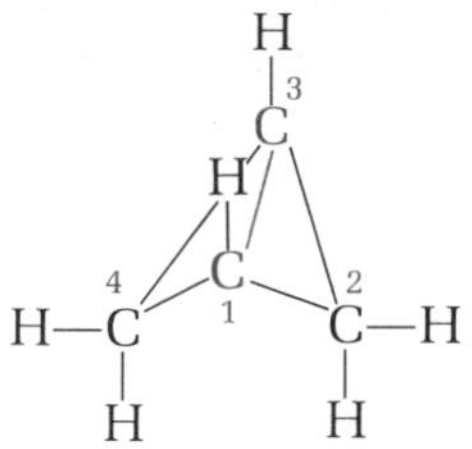   bicyclobutane 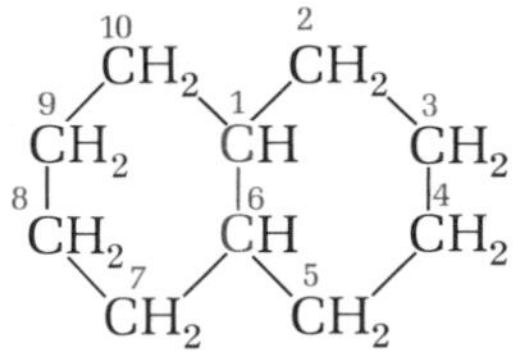   decalin

If two adjacent carbon atoms are shared by two rings, the rings are said to be **fused.** Examples include *bicyclobutane* and *decalin;* the carbon atoms which fuse the rings are shown in color.

Several molecules with **cage-like** structures have been prepared. *Cubane* is a strained molecule, with eight carbons located at the

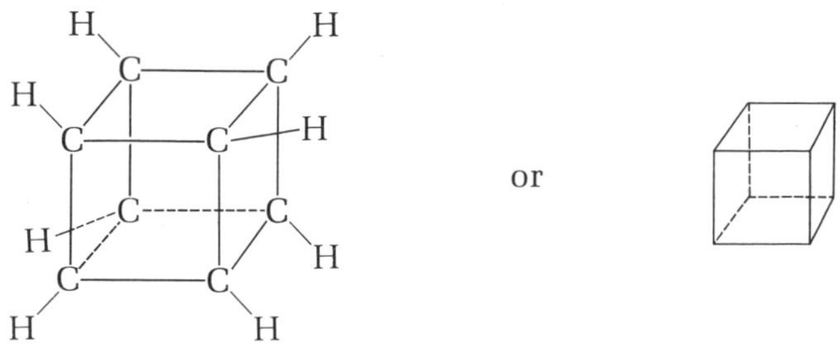

or

cubane

corners of a cube—it contains fused four-membered rings. *Adamantane,* first isolated from petroleum but now readily synthesized in the laboratory, has four fused cyclohexane rings, each in the chair conformation.

CH
CH₂
CH₂   CH₂
CH
CH₂   CH₂
CH   CH
CH₂

or

H

H---   ---H

H

adamantane

(For clarity, only the CH hydrogens are shown in the three-dimensional formula of adamantane; the other ring positions are CH₂ groups.) Because of their highly symmetric structures, these compounds pack very well into crystals and have unusually high melting points for their molecular weights (cubane, mp 131°; adamantane, mp 268°). If the formula for adamantane is extended in all possible directions, one obtains a lattice work of carbon atoms with a diamond structure.

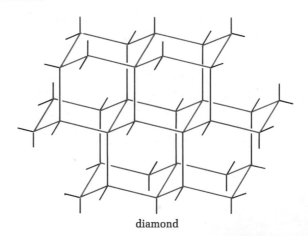

diamond

Intrigued with molecular shapes, organic chemists have also synthesized in recent years compounds in which two parts of a molecule are permanently linked by their geometry, but not by any chemical bonds. Two examples are shown:

catenane   rotaxane

In the first, two large rings are interlocked but not chemically bound. In the second, large groups at the ends of a chain prevent the chain from slipping through the loop. Clearly much ingenuity has gone into devising syntheses of such compounds; as yet, no practical uses for them are known.

New Concepts, Facts, and Terms

1. Saturated hydrocarbons, paraffins, alkanes:
   a. They contain only carbon and hydrogen, and only single covalent bonds.
   b. The simplest members are methane, ethane, propane, butane.
   c. Their general formula is C_nH_{2n+2}.

2. Homologous series—compounds of the same type which differ only by —CH_2— groups

3. IUPAC—the International Union of Pure and Applied Chemistry, which devised a system for naming organic compounds. The rules for nomenclature of saturated hydrocarbons are in sec 2.3.

4. Alkyl groups—derived from saturated hydrocarbons, by removing one of the hydrogen atoms. The -*ane* ending is changed to -*yl*. Methyl from methane, ethyl from ethane, normal propyl and isopropyl from propane. Useful in naming organic compounds.

5. Physical properties of alkanes—the distance between carbon atoms joined by single bonds is 1.54 Å. Rotation of one atom with respect to another leads to many conformations, but since rotation from one conformation to another is facile, no new readily separable isomers are produced. Staggered conformations are preferred over eclipsed conformations.

6. Reactions of alkanes:
   a. substitution reactions—displacement of one atom or group by another—halogenation, nitration. Free radical chain reactions.
   b. oxidation
   c. cracking

7. Preparation of alkanes
   a. Wurtz reaction (alkyl halide + sodium)
   b. reduction of halogen compounds; the Grignard reagent.

8. Cycloalkanes—saturated hydrocarbons in which the carbon atoms form a ring. The small rings are strained, but five- and six-membered rings are much less so. Cyclohexane prefers the chair conformation to the boat, and substituents prefer equatorial to axial positions. Spiro, fused, cage and other unusual ring structures are possible.

Exercises and Problems

1. Write structural formulas for the following compounds:
   a. 3-methylpentane
   b. 2,3-dimethylbutane
   c. 3,3-dimethyl-4-ethylhexane
   d. 2-chloro-3-methylpentane
   e. 2,2,3-trimethylbutane
   f. 2-bromopropane
   g. 1,1-dichlorocyclopropane
   h. 1,3-dimethylcyclohexane
   i. 1,1,3,3-tetrachloropropane
   j. 2,5-dimethyloctane

2. Write expanded formulas for and name the following compounds using the IUPAC system:
   a. $CH_3(CH_2)_3CH_3$
   b. $CH_3CH(CH_3)CH_2CH_3$
   c. $CH_3CH_2C(CH_3)_2CH_2CH_3$
   d. $CH_3(CH_2)_2C(CH_3)_3$
   e. $CH_3CH_2CHBrCH_3$
   f. $CH_3CCl_2CBr_3$
   g. $CF_3(CF_2)_2CF_3$
   h. CH_2ClCH_2Br
   i. $CH_2BrCH(CH_3)CH(CH_3)_2$
   j. $(CH_2)_5$

3. Give a common and a IUPAC name for the following compounds:
   a. CH_3I
   b. CH_3CH_2Cl
   c. CH_2Cl_2
   d. $CHBr_3$
   e. $CH_3CH_2CH_2Cl$
   f. $(CH_3)_2CHBr$
   g. $CHCl_3$
   h. $C_6H_{13}Br$
   i. CBr_4
   j. $CH_2{-}CH{-}Cl$ / $CH_2{-}CH_2$

4. In each of the following lists of compounds, which structural formulas represent identical substances?

A	B
a. $CH_3CH(CH_3)CH_2CH_2CH_3$	a. $(CH_3)_2CClCH_2CH_3$
b. $CH_3(CH_2)_2CH(CH_3)_2$	b. $(CH_3CH_2)_2CHCl$
c. $CH_3CH_2CH(CH_3)CH_2CH_3$	c. $CH_3CH(CH_2Cl)CH_2CH_3$
d. $(CH_3CH_2)_3CH$	d. $(CH_3)_2CHCH_2CH_2Cl$
e. $CH_3CH(CH_2CH_3)CH_2CH_3$	e. $CH_2ClCH(CH_3)CH_2CH_3$
f. $(CH_3)_2CHCH_2CH_2CH_3$	f. $CH_3CH_2CHClCH_2CH_3$
g. $CH_3CH_2CH(CH_2CH_3)_2$	g. $CH_3CH_2CH(CH_2Cl)CH_3$
h. $(CH_3CH_2)_2CHCH_3$	h. $(CH_3)_2CHCHClCH_3$

5. Write a structure for each of the compounds listed, explain why the given name is objectionable and give a correct name in each case.
   a. 4-methylpentane
   b. 2-ethylbutane
   c. 2,3-dichloropropane
   d. 1,4-dimethylcyclobutane
   e. 2,2-diethylbutane
   f. 4,5-dimethylhexane
   g. 1,1,3-trimethylpropane
   h. 3-bromo-2-methylpropane
   i. 1,3,3-trimethylcyclopentane
   j. 1,2,2-trichloroethane

6. Write the structural formulas for all the isomers (numbers indicated in parentheses) for each of the following compounds, and name each isomer by the IUPAC system.
   a. C_4H_{10} (2)
   b. C_4H_9Br (4)
   c. C_6H_{14} (5)
   d. $C_3H_6Br_2$ (4)
   e. $C_4H_8Cl_2$ (9)
   f. C_3H_6BrCl (5)

7. Name the alkane which has the indicated molecular weight and which can have only the number of substitution products shown. Name the substitution products.

   a. 44; 2 monobromo
   b. 58; 2 monochloro
   c. 42; 1 monochloro
   d. 72; 1 monochloro
   e. 72; 4 monochloro
   f. 70; 2 monochloro

8. In the industrial chlorination of pentanes, which of the following procedures would be expected to give the best yield of monochlorinated products? Why?

   a. C_5H_{12} + excess Cl_2
   b. Cl_2 + excess C_5H_{12}
   c. addition of the Cl_2 to the C_5H_{12}
   d. addition of the C_5H_{12} to the Cl_2

9. Number the four conformations of butane shown in Figure 2-6 in sequence from left to right. Then place these in order of decreasing stability. Explain the reasons for the sequence you have chosen.

10. Do you expect the energy difference between staggered and eclipsed conformations of propane to be less than, greater than, or equal to the same difference in ethane? Explain.

11. Arrange the following hydrocarbons in order of increasing boiling points without referring to tables:

    a. 2-methylhexane
    b. n-heptane
    c. 3,3-dimethylpentane
    d. n-hexane
    e. 2-methylpentane

12. When methane is chlorinated, small amounts of ethane and chlorinated ethanes may be detected amongst the products. Explain this observation and its relevance to the proposed free-radical chain mechanism for chlorination.

13. Using structural formulas, write equations for the following reactions and name each organic product. Indicate necessary catalysts and conditions.

    a. the complete combustion of pentane
    b. the complete combustion of cyclopentane
    c. the vapor phase nitration of propane
    d. the monochlorination of methylcyclopentane

14. The energy required to break each of the following bonds into the corresponding atoms is indicated below in kilocalories per mole.

C—H 87.3	H—Cl 102.7	Cl—Cl 57.8	O=O 119.1
C—Cl 66.5	H—Br 87.3	Br—Br 46.1	
C—Br 54.0	H—I 71.4	I—I 36.2	
C—I 45.5	H—O 110.6	C=O 192.0	

Would you expect the reaction of chlorine with methane to give methyl chloride and hydrogen chloride to release or absorb energy? How much? Can you explain why chlorination and bromination of methane are possible, but iodination is not?

15. Using the bond energies given in Problem 14, determine whether energy would be released or absorbed in the conversion of chloromethane to dichloromethane; dichloromethane to chloroform; chloroform to carbon tetrachloride. Could the reaction of excess chlorine with methane be explosive? Explain.

16. An alternative mechanism for the chlorination of methane to that given in equations 2-3 and 2-4 involves the chain:

$$Cl\cdot + CH_4 \longrightarrow CH_3Cl + H\cdot; \; H\cdot + Cl_2 \longrightarrow HCl + Cl\cdot$$

Calculate the energy released or absorbed for each step. Why is this sequence less likely to occur than the accepted one?

17. Would a rocket fueled with liquid CH_4 and liquid O_2, burning the methane to CO_2 and H_2O, be more or less efficient than a rocket fueled with liquid CH_4 and liquid Cl_2, converting the methane to CCl_4 and HCl? Explain your answer. (Hint: use the bond energies given in Problem 14 to calculate the amount of energy released in each reaction).

18. What hydrocarbons would be expected from the reaction of the following alkyl halides with sodium (Wurtz reaction).
    a. *n*-propyl iodide   c. cyclopentyl bromide
    b. 1-bromopentane   d. a mixture of methyl and ethyl iodides

19. *n*-Hexane can be prepared from three different combinations of alkyl bromides, employing the Wurtz reaction. Write an equation for each and explain which of the three is preferred.

20. An alkyl bromide reacts with magnesium in ether to yield an alkyl-magnesium bromide. Subsequent hydrolysis of this solution produced isobutane. Treatment of the alkyl bromide with sodium (Wurtz reaction) gave a hydrocarbon identified as 2,5-dimethylhexane. Deduce the structure of the alkyl bromide, and write an equation for each reaction.

21. Using a formula similar to that in the lower left part of Figure 2-9, draw the most stable conformation of each of the following:
    a. methylcyclohexane
    b. 1,4-dichlorocyclohexane
    c. 1,3-dichlorocyclohexane

22. In the chlorination of propane, four isomeric products with the formula $C_3H_6Cl_2$ were isolated, designated A–D. Each was separated, and further chlorinated to give one or more trichloropropanes, $C_3H_5Cl_3$. A and B gave three trichloro compounds, C gave one, and D gave two. Deduce the structures of C and D.

    One of the products from A was identical with the product from C. Deduce structures for A and B.

CHAPTER THREE

UNSATURATED HYDROCARBONS: ALKENES

Carbon atoms may share more than one pair of electrons between them. Molecules formed in this manner are said to be unsaturated. The cracking of petroleum produces large quantities of unsaturated hydrocarbons which are highly reactive substances. They combine with many reagents: halogens, acids, saturated hydrocarbons, and even with themselves. Antiknock fuels, synthetic rubber, plastics of various types, solvents, and permanent antifreeze are among the many useful products into which they may readily be transformed.

Unsaturated hydrocarbons are distinguished from alkanes by the presence of multiple bonds between carbon atoms. The **alkenes,** or **olefins,*** contain a carbon-carbon *double* bond, whereas the **alkynes,** or **acetylenes,** have a carbon-carbon *triple* bond. Multiple bonds impart greater reactivity to these compounds than is common for alkanes.

In this chapter we discuss alkenes. **Dienes** (compounds with two double bonds), **polyenes** (compounds with many double bonds), and alkynes will be described in Chapter 4.

3.1 NOMENCLATURE OF ALKENES

Alkenes have the general formula C_nH_{2n}. The simplest members of the series usually are best known by their common names, which have roots derived from the corresponding saturated hydrocarbons and the ending *-ylene*.

$$CH_2{=}CH_2 \qquad CH_3{-}CH{=}CH_2 \qquad CH_3{-}\underset{\underset{CH_3}{|}}{C}{=}CH_2$$

  ethylene   propylene   isobutylene

Rules of the IUPAC system are similar to those for the alkanes, with a few additions to include the double bond in the name. The general class name is alkene. The ending *-ene* is added to the root derived from the corresponding alkane. The carbon atoms in the *longest chain containing the double bond* are numbered in a manner such that the *carbon atoms of the double bond have the lowest possible numbers.* The position of the double bond is indicated by the number of the *lower* numbered carbon atom of the double bond. This number, when necessary, is placed before the name of the compound, as in the following examples:

$$\overset{1}{C}H_2{=}\overset{2}{C}H{-}\overset{3}{C}H_2{-}\overset{4}{C}H_3 \qquad \overset{1}{C}H_3{-}\overset{2}{C}H{=}\overset{3}{C}H{-}\overset{4}{C}H_3$$

       1-butene   2-butene

$$\overset{1}{C}H_3{-}\overset{2}{C}H{=}\overset{3}{C}H{-}\underset{\underset{CH_3}{|}}{\overset{4}{C}H}{-}\overset{5}{C}H_3$$

       4-methyl-2-pentene   3-chlorocyclopentene

*The word *olefin* means "oil-forming," a name originally given to ethylene because it formed an oil when treated with chlorine. The term "olefin" now applies to all hydrocarbons with carbon-carbon double bonds.

3.2   THE DOUBLE BOND – CIS-TRANS ISOMERISM

As can be seen from the formulas above, a carbon atom which is part of a double bond is attached to only three other atoms (a **trigonal** carbon atom). Since carbon has one 2s and three 2p orbitals available for bonding, the three equivalent orbitals necessary for trigonal bonding can be formed by blending the 2s orbital with two 2p orbitals to make three sp^2 hybrid orbitals (one-third s and two-thirds p). These orbitals lie in a plane and are directed to the corners of an equilateral triangle. The angle between them is 120°. One electron is placed in each hybrid orbital, and the fourth electron occupies the remaining 2p orbital, whose axis is perpendicular to the plane formed by the hybrid orbitals (Figure 3-1).

Figure 3-1
*A trigonal carbon showing three sp^2 orbitals in a plane with a 120°
angle between them. The remaining p orbital is perpendicular to the hybrid sp^2 orbitals. There is a small back lobe to each sp^2 orbital which
has been omitted for ease in representation.*

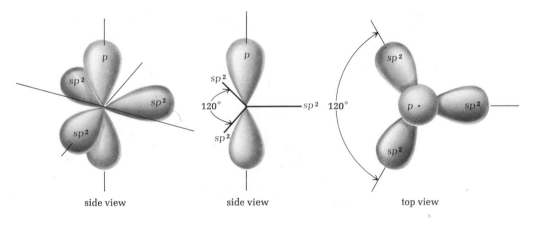

side view   side view   top view

The joining of two trigonal carbon atoms to form a double bond is pictured schematically in Figure 3-2. Overlap of two sp^2 orbitals results in one bond, a sigma (σ) bond, analogous to an ordinary single bond. If the two carbons are properly oriented so as to place the two p orbitals *parallel* with respect to one another, the p orbitals may also overlap to form a second bond, called a **pi (π) bond,** in which the electron cloud lies above and below the plane formed by the σ bonds.

Figure 3-2
Schematic formation of a carbon-carbon double bond. Two sp² carbons form a sigma (σ) bond (overlap of two sp² orbitals) and a pi (π) bond (overlap of properly aligned p orbitals).

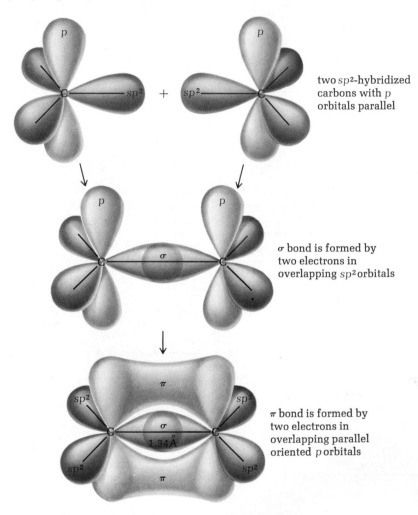

two sp²-hybridized carbons with p orbitals parallel

σ bond is formed by two electrons in overlapping sp²orbitals

π bond is formed by two electrons in overlapping parallel oriented p orbitals

There are several important consequences of this model for a double bond. The carbon-carbon double bond is 0.2 Å shorter than a single bond, the distance between nuclei being 1.34 Å. This is because two pairs of electrons draw the two nuclei closer together than only one pair.

The relatively free rotation possible for singly bound carbons is not possible in compounds with a double bond. Rotation of one carbon with respect to another would require the π bond to be broken (Figure 3-3). This does not occur ordinarily, although it can

Figure 3-3
Rotation of one sp^2 carbon 90° with respect to another orients the p orbitals so that no overlap (and therefore no π bond) is possible.

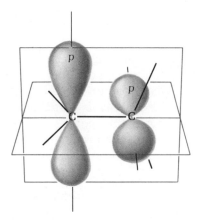

Figure 3-4
Three models of ethylene, each showing that the four atoms attached to a carbon-carbon double bond lie in a single plane.

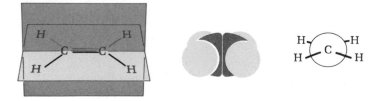

be brought about by supplying the molecule with enough energy to break the pi bond (about 62 kcal/mole), by heating the compound to a high temperature, or by exposing it to ultraviolet radiation.
As a result of this geometric restriction on a double bond, the carbon atoms involved and the four atoms attached to them must all lie in a single plane (Figure 3-4). If the two atoms attached to each carbon are different, two isomers are possible; in one of these,

Figure 3-5

Two isomers of 1,2-dichloroethene are possible because rotation around the double bond is restricted. This is an example of geometric isomerism.

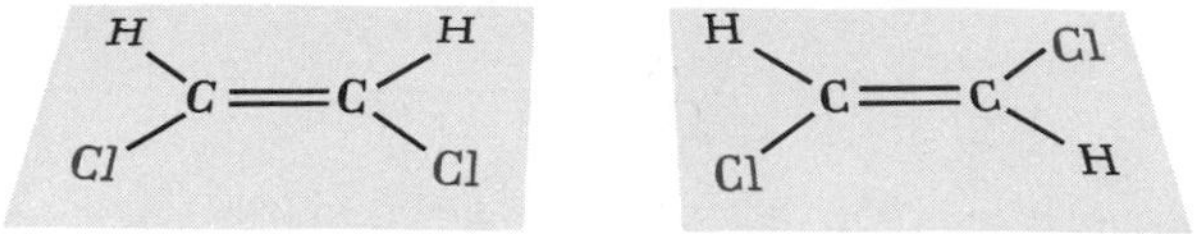

| *cis* -1,2-dichloroethene | *trans* -1,2-dichloroethene |

similar groups are on the same "side" of the double bond, and in the other, they are on opposite "sides." An example is seen in Figure 3-5. This phenomenon is called **geometric** or **cis-trans isomerism** (*cis* = on the same side; *trans* = across). Such isomers always have different physical properties and may also show different chemical behavior, since such properties depend not only on the number and kinds of atoms and bonds present, but also on their arrangement in space.

It should be pointed out that, if two of the groups attached to one of the doubly-bound carbon atoms are identical, as in I, only *one* isomer is possible. It is seen that I is identical with II by rotating it 180° out of the plane of the paper. *The two groups attached to each carbon must be different for cis-trans isomerism to be possible.*

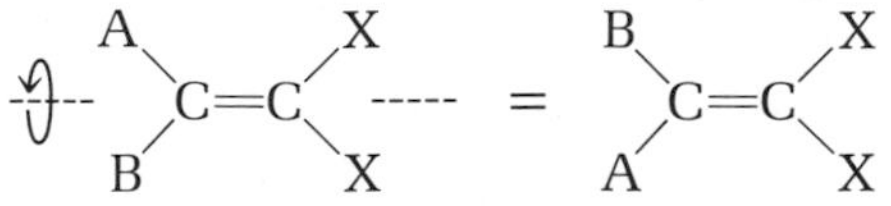

Since free rotation is impossible for carbon atoms which are part of ordinary sized rings, geometric isomerism is also possible in such molecules. This is true not only for cyclopropanes (III and IV) where the ring carbons necessarily form a plane, but also in larger, more flexible rings (V and VI).

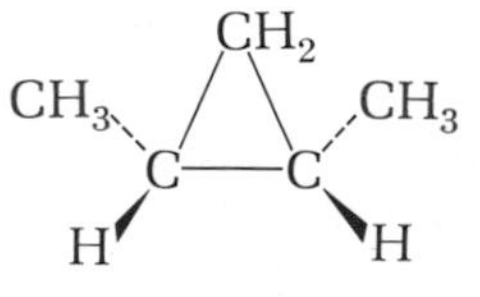

| cis-1,2-dimethylcyclopropane | trans-1,2-dimethylcyclopropane |
| III | IV |

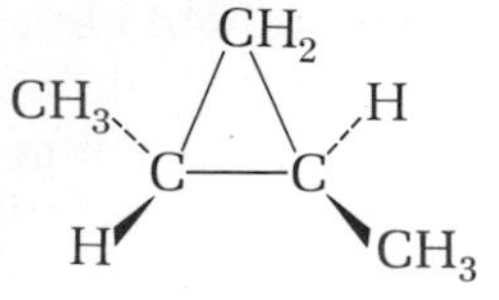

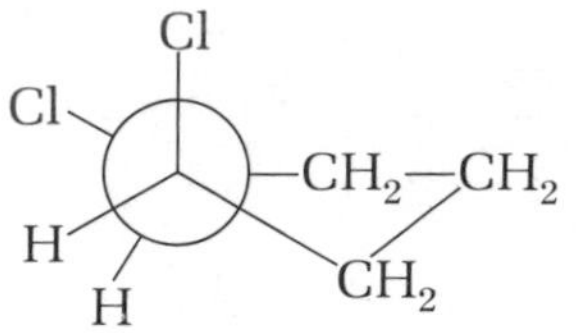

cis-1,2-dichlorocyclopentane

V

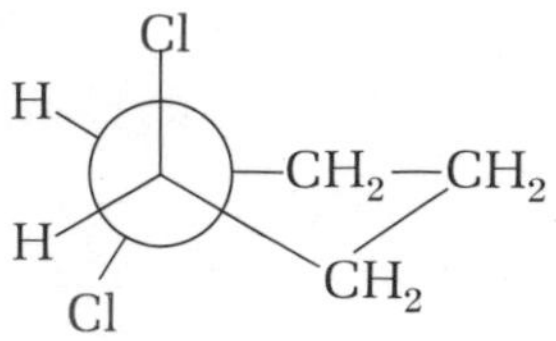

trans-1,2-dichlorocyclopentane

VI

3.3 PHYSICAL PROPERTIES OF ALKENES

The alkenes are similar to the alkanes in physical properties. Those with four carbon atoms or less are colorless gases, whereas the pentenes and higher homologs are liquids. Some properties of a few alkenes are given in Table 3-1. Note the different physical properties of the two geometric isomers of 2-butene.

3.4 REACTIONS OF ALKENES

The most important distinction between saturated and unsaturated hydrocarbons is the general type of reaction which each undergoes. Saturated hydrocarbons react by substitution, whereas *the characteristic reaction of alkenes is addition to the double bond.* Contrast,

Table 3-1
Physical Constants of Some Alkenes

IUPAC NAME	STRUCTURE	MP, °C	BP, °C	DENSITY (LIQUID)
Ethene	$CH_2{=}CH_2$	− 169.4	− 102.4	0.610
Propene	$CH_3CH{=}CH_2$	− 185	− 47.7	0.610
1-Butene	$CH_2{=}CHCH_2CH_3$	− 130	− 6.5	0.626
cis-2-Butene	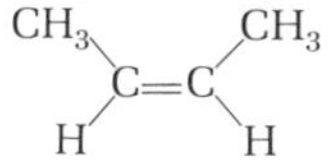	− 139	3.73	0.621
trans-2-Butene	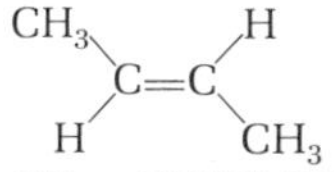	− 106	0.96	0.604
1-Pentene	$CH_2{=}CHCH_2CH_2CH_3$	− 138	30.1	0.643
1-Hexene	$CH_2{=}CHCH_2CH_2CH_2CH_3$	− 138	63.5	0.675
Cyclohexene	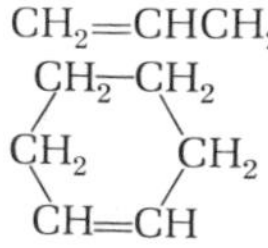	− 103.7	83.1	0.810

for example, the reactions of ethane and ethylene with bromine.

$$CH_3—CH_3 + Br_2 \xrightarrow[\text{(light)}]{h\nu} CH_3CH_2Br + HBr \quad \text{substitution} \qquad (3\text{-}1)$$

bromoethane

$$CH_2{=}CH_2 + Br_2 \longrightarrow \underset{\underset{Br}{|}}{CH_2}{-}\underset{\underset{Br}{|}}{CH_2} \quad \text{addition} \qquad (3\text{-}2)$$

1,2-dibromoethane

The addition reaction proceeds at a very rapid rate, even at low temperatures, whereas the substitution reaction frequently requires a catalyst and elevated temperatures.

3.4a   MECHANISM OF ADDITION TO DOUBLE BONDS

The π electrons of a double bond, being further from the carbon nuclei than the σ electrons, are less firmly bound to them. A double bond is, in effect, electron rich and can act as a supplier of electrons to a reagent which may seek them (an **electrophilic reagent**). Electrophilic reagents are electron deficient species. They may be positive ions *capable of forming a covalent bond with carbon* (for example H⁺, but not Na⁺) or they may be radicals seeking but one electron to complete their electronic structure. In the following discussion, it will be assumed for convenience that the reaction is ionic (i.e., that the electrophile is seeking a *pair* of electrons), but a similar sequence can be written for free radicals.

Reaction between a carbon-carbon double bond begins by attack of the electrophile on the π electrons in a direction perpendicular to the plane formed by the two carbons and the four atoms attached to them.

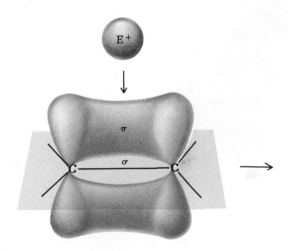

Ultimately the electrophile approaches sufficiently close to use the π electrons to form a covalent bond with one or the other of the two carbon atoms.

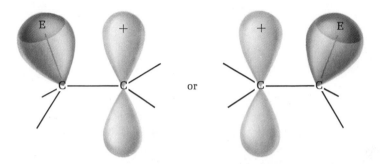

or

This appropriation of the π electrons to form the C—E bond leaves the other carbon atom with only six electrons and an empty p orbital. It is therefore positively charged (a carbonium ion) and combines with a **nucleophile,** usually a negative ion which can supply it with an electron pair. The nucleophile could come in from the top or bottom side of the plane; in most cases, it reacts from the side opposite the entering electrophile. The over-all reaction is the addi-

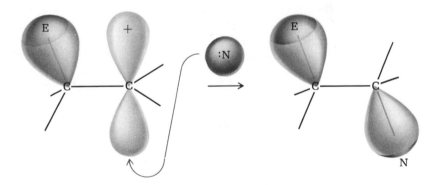

tion of two groups to the double bond, the electrophile to one carbon atom, and the nucleophile to the other.

$$\text{C=C} + E\{-N \longrightarrow \overset{E}{\underset{}{\text{C}}-\overset{+}{\text{C}} \quad :N^- \longrightarrow \overset{E}{\underset{}{\text{C}}-\text{C}}\underset{N}{} \tag{3-3}$$

To illustrate this general mechanism with a specific example, consider the addition of bromine to ethylene. In the first step, ethylene supplies an electron pair to a bromine molecule, causing a bromide ion to depart. This leaves a positive charge on one of the original doubly-bound carbons.

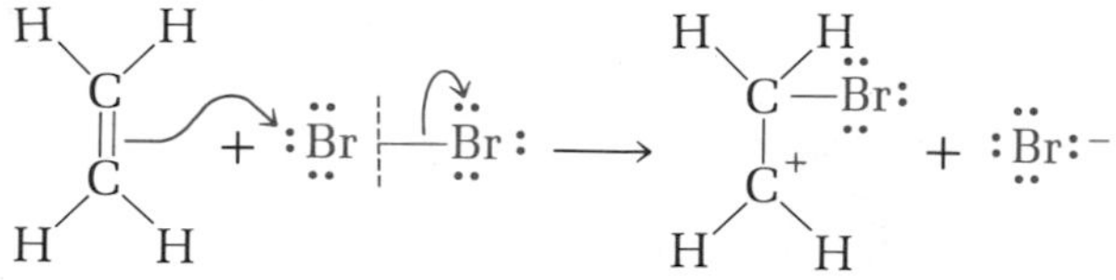

$$\text{(3-4a)}$$

carbonium ion

The carbonium ion produced in equation 3-4a is deficient in electrons and therefore is a strong electrophile. It can complete its octet of electrons by reacting either with a bromide ion (3-4b) or a bromine molecule (3-4c).

$$BrCH_2CH_2^+ + Br^- \longrightarrow BrCH_2CH_2Br \qquad \text{(3-4b)}$$

$$BrCH_2CH_2^+ + :\!\ddot{B}r\!-\!\ddot{B}r: \longrightarrow$$

$$BrCH_2CH_2\!-\!Br + \ddot{B}r:^+ \qquad \text{(3-4c)}$$

bromonium
ion

In the latter event, the Br$^+$ (bromonium ion) is an electrophile which can react with the alkene to form the carbonium ion again, thus continuing the reaction chain (3-4d).

$$CH_2\!=\!CH_2 + \ddot{B}r:^+ \longrightarrow {}^+CH_2CH_2\!-\!Br \qquad \text{(3-4d)}$$

The over-all equation (sum (3-4a) and (3-4b), or (3-4d) and (3-4c)) is simply

$$CH_2\!=\!CH_2 + Br\!-\!Br \longrightarrow \underset{\underset{Br}{|}}{CH_2}\!-\!\underset{\underset{Br}{|}}{CH_2} \qquad \text{(3-5)}$$

One can readily prove experimentally that in aqueous solution the addition is stepwise. If the bromination of ethylene is carried out in a solution which contains chloride ions (Cl$^-$), a mixture of 1,2-dibromoethane and 1-bromo-2-chloroethane is obtained. This can be explained by the formation of the intermediate carbonium ion as in equation 3-4a above, followed by reaction of this positive ion with either of the negative ions, Br$^-$ or Cl$^-$, present in the solution.

$$Br{-}Br + CH_2{=}CH_2 \xrightarrow{-Br^-} \left[Br{-}CH_2{-}\overset{+}{C}H_2 \right] \overset{Br^-}{\underset{Cl^-}{\rightleftharpoons}} \begin{array}{l} Br\,CH_2CH_2Br \\ \text{and} \\ Br\,CH_2CH_2Cl \end{array} \Bigg\} \times Cl^- \quad (3\text{-}6)$$

carbonium ion

The alternative explanation, that the bromo-chloro compound is formed by reaction of the dibromide with chloride ion, is ruled out by a separate experiment which shows that the dibromide does not react with chloride ion under the experimental conditions.

Some product containing a hydroxyl group and a bromine atom is also formed. Once again, this comes about by reaction of the intermediate carbonium ion with a nucleophile, this time the water molecule.

$$\left[\begin{array}{c} Br \\ \diagdown \\ CH_2{-}\overset{+}{C}H_2 \end{array} \right] + :\overset{\cdot\cdot}{O}{:}H \longrightarrow \begin{array}{c} Br \\ \diagdown \\ CH_2{-}CH_2 \\ \mid \\ :\overset{+}{O}{-}H \\ \mid \\ H \end{array} \longrightarrow \begin{array}{c} Br \\ \diagdown \\ CH_2{-}CH_2 \\ \mid \\ :\overset{\cdot\cdot}{O}H \end{array} + H^+ \quad (3\text{-}7)$$

An alternative mechanism for bromine addition involving a free radical chain reaction may describe the reaction path, especially in the absence of a polar solvent like water. The ideas are analogous to those used for the ionic mechanism outlined above, except that bromine *atoms* take only one of the two π electrons from the alkene.

$$CH_2{\cdots}CH_2 + :\overset{\cdot\cdot}{B}r\overset{\cdot\cdot}{B}r: \longrightarrow \cdot CH_2{-}CH_2{:}\overset{\cdot\cdot}{B}r + :\overset{\cdot\cdot}{B}r\cdot \quad (3\text{-}8a)$$

$$BrCH_2CH_2\cdot + :\overset{\cdot\cdot}{B}r\cdot \longrightarrow BrCH_2CH_2{-}\overset{\cdot\cdot}{B}r: \quad (3\text{-}8b)$$

$$BrCH_2CH_2\cdot + :\overset{\cdot\cdot}{B}r\overset{\cdot\cdot}{B}r: \longrightarrow BrCH_2CH_2{-}\overset{\cdot\cdot}{B}r: + :\overset{\cdot\cdot}{B}r\cdot \quad (3\text{-}8c)$$

$$CH_2{\cdots}CH_2 + :\overset{\cdot\cdot}{B}r\cdot \longrightarrow \cdot CH_2CH_2{-}\overset{\cdot\cdot}{B}r: \quad (3\text{-}8d)$$

The addition of bromine to alkenes is frequently used as a simple **qualitative test for unsaturation.** Bromine is dark red-brown, whereas the alkene and the alkene dibromide are both colorless. Thus a solution of bromine in some inert solvent such as carbon tetrachloride is rapidly decolorized when added to an alkene.

1,2-Dibromoethane (commonly known as ethylene dibromide) is manufactured commercially from ethylene and bromine; it is added to gasolines which contain tetraethyl lead in order to prevent the accumulation of lead in the engine. The lead is converted to volatile lead bromide and eliminated through the exhaust. This, unfortunately, contributes to air pollution, and efforts to design engines which can operate efficiently without lead have become important.

3.4b OTHER ADDITION REACTIONS

Many substances other than bromine are capable of adding to the double bond of alkenes. In general, the reactions proceed by mechanisms analogous to those described in detail for bromine.

Of the other halogens, *chlorine* adds readily,

$$CH_2{=}CH_2 + Cl_2 \longrightarrow \underset{\underset{Cl}{|}}{CH_2}{-}\underset{\underset{Cl}{|}}{CH_2} \qquad (3\text{-}9)$$

but *iodine* does not, and the reaction with *fluorine* is in most cases too vigorous to be of use.

Hydrogen, in the presence of an appropriate catalyst (finely divided platinum, palladium, or nickel), transforms alkenes to the corresponding alkanes.

$$CH_2{=}CH_2 + H_2 \xrightarrow{\text{Pt,Pd or Ni}} \underset{\underset{H}{|}}{CH_2}{-}\underset{\underset{H}{|}}{CH_2} \qquad (3\text{-}10)$$

ethene   ethane

Addition of both atoms of hydrogen, which are adsorbed on the catalyst surface, occurs from the same side of the double bond (*cis*-addition). With active catalysts, the reaction occurs readily at

room temperature and low hydrogen pressures (1 atmosphere). In other cases, higher temperatures or pressures may be required.

Hydrogenation is an important industrial process which is used in the manufacture of high octane automobile and aviation fuels, in the conversion of natural vegetable oils to solid cooking fats, and in the industrial preparation of synthetic detergents.

Most acids (H—A) can add to the carbon-carbon double bond; the proton adds to one carbon, and the group A which is attached adds to the other carbon. The proton (H⁺) is a good electrophile which readily attacks the π electrons of a double bond.

Hydrogen halides add with increasing ease as we proceed down the periodic table (HI > HBr > HCl).

$$CH_2{=}CH_2 + H{\overset{+}{\underset{\cdot}{\subset}}}Cl^- \longrightarrow \underset{\underset{H}{|}}{CH_2}{-}\underset{\underset{Cl}{|}}{CH_2} \qquad (3\text{-}11)$$

ethyl chloride

Ethyl chloride is used as a local anesthetic, causing anesthesia by freezing, as a result of its rapid evaporation from the surface of the skin.

Sulfuric acid adds, to produce an alkyl hydrogen sulfate. As will be seen later, this is the first step in the industrial manufacture of ordinary alcohol.

$$CH_2{=}CH_2 + H{\overset{+}{\underset{\cdot}{\subset}}}\bar{O}SO_2OH \longrightarrow \underset{\underset{H}{|}}{CH_2}{-}\underset{\underset{OSO_2OH}{|}}{CH_2} \qquad (3\text{-}12)$$

sulfuric acid   ethyl hydrogen sulfate

The *hypohalous acids* add as HO—Cl, HO—Br, or HO—I; there is no tendency to add as H—OX. Reaction is initiated by the positive halogen.

$$CH_2{=}CH_2 + \overset{-}{HO}{\overset{\cdot}{\underset{\cdot}{-}}}\overset{+}{Cl} \longrightarrow \underset{\underset{OH}{|}}{CH_2}{-}\underset{\underset{Cl}{|}}{CH_2} \qquad (3\text{-}13)$$

hypochlorous   ethylene
acid   chlorohydrin

3.4c   ADDITION TO UNSYMMETRICAL DOUBLE BONDS

When an alkene is symmetrical about the double bond as ethylene is, the product formed in an addition reaction is the same no matter which way the addend becomes attached to the alkene.

$$CH_2{=}CH_2 + HCl \longrightarrow CH_3{-}CH_2Cl$$
$$CH_2{=}CH_2 + HCl \longrightarrow ClCH_2{-}CH_3$$
identical

If, however, *both the alkene and the addend are unsymmetrical*, two alternatives are possible:

$$CH_2{=}CH{-}CH_3 + HCl \overset{(a)}{\underset{(b)}{\diagdown}} \begin{array}{l} CH_2{-}CH_2{-}CH_3 \\ \qquad Cl \quad \text{n-propyl chloride} \\[4pt] CH_3{-}CH{-}CH_3 \\ \qquad Cl \end{array} \qquad (3\text{-}14)$$

propylene   main
(b)   reaction   isopropyl chloride

Experimentally, one finds that isopropyl chloride is the major product of the addition of hydrogen chloride to propylene.

After studying a large number of such addition reactions, the Russian chemist Vladimir Markownikoff formulated the following rule which generalizes the experimental observations. *When an unsymmetrical reagent adds to an unsymmetrical double bond, the positive part of the addend becomes attached to the double-bonded carbon atom which bears the greater number of hydrogen atoms.* The following examples illustrate the use of this rule.

$$CH_2{=}CH{-}CH_3 + \overset{+}{H}\overset{-}{O}SO_2OH \longrightarrow CH_3{-}\underset{\underset{\displaystyle OSO_2OH}{|}}{CH}{-}CH_3 \tag{3-15}$$

propylene

isopropyl hydrogen sulfate

$$CH_3{-}\overset{\overset{\displaystyle CH_3}{|}}{C}{=}CH_2 + \overset{+}{H}\overset{-}{Br} \longrightarrow CH_3{-}\overset{\overset{\displaystyle CH_3}{|}}{\underset{\underset{\displaystyle Br}{|}}{C}}{-}CH_3 \tag{3-16}$$

2-methylpropene

2-bromo-2-methylpropane

$$CH_2{=}CH{-}CH_2{-}CH_3 + \overset{-}{H}\overset{+}{O}Cl \longrightarrow \underset{\underset{\displaystyle Cl}{|}}{CH_2}{-}\underset{\underset{\displaystyle OH}{|}}{CH}{-}CH_2CH_3 \tag{3-17}$$

1-butene

1-chloro-2-butanol

Markownikoff's rule can be rationalized in terms of modern organic chemical theory. Consider the addition of an acid to propylene. Reaction is initiated by addition of the **proton,** yielding two possible carbonium ions. At this point then, the direction of addition is determined, for the position taken by the nucleophile in the final step must be at the carbon atom bearing the positive charge. Since experimentally one obtains a product corresponding to the isopropyl rather than the n-propyl structure, one must conclude that the carbonium ion with the isopropyl structure is most readily formed and more stable.

$$CH_3CH{=}CH_2 + H^+ \underset{b}{\overset{a}{\rightleftharpoons}} \begin{matrix} \overset{a}{\nearrow} CH_3\overset{+}{C}HCH_3 & \text{isopropyl cation} \\[4pt] \underset{b}{\searrow} CH_3CH_2{-}\overset{+}{C}H_2 & \text{n-propyl cation} \end{matrix} \tag{3-18}$$

From the study of many reactions which proceed via carbonium ions it has become clear that the more highly substituted carbonium ions are the more stable. Carbonium ions may be classified as tertiary, secondary, or primary, depending on whether three, two, or only one alkyl group is attached to the carbonium carbon atom. Their stability decreases in the order shown.

$$
\begin{array}{ccc}
\underset{\substack{|\\ \text{R}}}{\overset{\text{R}}{\text{R}-\text{C}^+}} & > & \underset{\substack{|\\ \text{R}}}{\text{R}-\text{CH}^+} & > & \text{R}-\text{CH}_2{}^+
\end{array}
$$

$$
\begin{array}{ccc}
\text{tertiary(3°)} & \text{secondary(2°)} & \text{primary(1°)} \\
\text{(most stable)} & & \text{(least stable)}
\end{array}
$$

Markownikoff's rule may now be restated; *addition of an unsymmetrical reagent to an unsymmetrical double bond proceeds in such a direction as to involve the more stable intermediate carbonium ion.*

Markownikoff's rule is a specific example of a general principle true of all chemical reactions—a reaction will proceed by the best possible path available to it. That is, reactants will pass through the most stable possible transition state or intermediate on their way toward products. Free radical intermediates show the same relative order of stability as carbonium ions (i.e., 3° > 2° > 1°), whereas for carbanions this order is reversed.

3.4d POLYMERIZATION

When *many* small molecules unite, a giant molecule called a **polymer** is produced. These polymers are the basis of the plastics, synthetic textiles, and other industries. Among the important commercial polymers are those of the simple olefins. Ethylene, for example, when heated under pressure in the presence of certain catalysts, may unite with itself to form chains eight hundred or more carbon atoms in length. The product is known as *polyethylene*. The tough but flexible plastic produced in this way has been useful as an electrical insulator as well as in the fabrication of nonbreakable plastic cups, refrigerator dishes, waste bags, and innumerable other articles (Figure 3-6). Annual production of polyethylene in the United States exceeds two billion pounds.

High-pressure catalytic polymerization of ethylene, typical of many polymerizations, is a free radical chain reaction. Reaction is initiated by a catalyst which is a thermally unstable organic compound that readily yields free radicals on heating (organic peroxides are the most common catalyst type).

$$
\text{RO}-\text{OR} \xrightarrow{\text{heat}} 2\text{RO}\cdot \tag{3-19}
$$

Figure 3-6
The photo at the left shows the polymer polyisobutene, obtained from
polymerizing isobutene at temperatures as low as 135 degrees below
zero. One of the main applications of polyisobutene is in adhesives such
as on band-aids. Another common plastic is polyethylene. A familiar
use of it is the squeeze bottle. (Left, courtesy of Esso Research and
Engineering Company; right, courtesy of Monsanto.

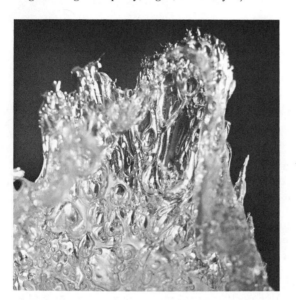

Polymerization is initiated when a catalyst radical attacks the ethylene double bond, taking one of its electrons, to produce a carbon
radical.

$$RO\cdot + CH_2\text{::}CH_2 \longrightarrow RO\text{---}CH_2CH_2\cdot \quad \text{initiation} \quad (3\text{-}20)$$

This leaves an odd electron on one of the carbon atoms which can
then attack another ethylene molecule in a similar fashion to propa-
gate the polymerization.

$$ROCH_2CH_2\cdot + CH_2\text{::}CH_2 \longrightarrow ROCH_2CH_2\text{---}CH_2CH_2\cdot \text{propagation}$$

$$(3\text{-}21)$$

This process continues until the chain is terminated in some way, for example by combination of two radicals.

$$R'\cdot + R''\cdot \longrightarrow R'-R'' \quad \text{termination} \tag{3-22}$$

Free radicals can also abstract hydrogen atoms from any C—H bond; this destroys one radical, but creates another one. This is called chain transfer.

$$R\cdot + R'-\overset{\overset{\displaystyle H}{|}}{\underset{\underset{\displaystyle H}{|}}{C}}-R'' \longrightarrow R-H + R'-\overset{\overset{\displaystyle \cdot}{}}{\underset{\underset{\displaystyle H}{|}}{C}}-R'' \tag{3-23}$$

A growing polymer chain may take a hydrogen atom from its own back, so to speak, causing chain-branching.

$$\tag{3-24}$$

The over-all result is a giant molecule with long and short carbon chains.

Recently, a new class of catalysts was discovered which allows ethylene to be polymerized at low pressures and temperatures; the product is nearly linear (that is, essentially no 'back-biting' occurs), and corresponds to the formula

$$+CH_2-CH_2\!\!\!+_n$$

where n is a large integer. Low-pressure polyolefins produced this new way have higher melting points and greater strength than the older type of polymer. A mixture of trialkylaluminums, R_3Al, and titanium tetrachloride, $TiCl_4$, is one of the catalyst combinations which produces low-pressure polyethylene.

3.4e OXIDATION

The π electrons of a double bond may be readily attacked by oxidizing agents. This property distinguishes alkenes from alkanes. Alkenes are oxidized by potassium permanganate. In the process, the permanganate, which is purple in color, is reduced to the brown manganese dioxide, MnO_2. This color change indicates that oxidation has occurred. Most saturated hydrocarbons are not oxidized under these

$$3 \; \diagdown C{=}C \diagup + \; 2 \, K^+MnO_4^- + 4 \, H_2O \longrightarrow \; 3 \; \diagdown \underset{OH}{\overset{}{C}}{-}\underset{OH}{\overset{}{C}} \diagup \; +$$

purple

$$2 \, MnO_2{\downarrow} + 2 \, KOH \quad (3\text{-}25)$$

brown

mild conditions, and thus show no change in the purple permanganate color. The test is sometimes called the **Baeyer test** after its discoverer, Adolph von Baeyer.

Ozone, which can be prepared by passing oxygen through an electric discharge, reacts rapidly and quantitatively with alkenes. The initially formed products are called *ozonides*. These are usually not isolated, for they are often explosive, but they may be decomposed by hydrolysis or reduction (eq 3-26).

$$\diagdown C{=}C \diagup + \; :\overset{+}{\underset{}{O}}: \quad :\overset{\cdot\cdot}{\underset{O}{O}}: \; \longrightarrow \; \left[\begin{array}{c} \diagdown C{-}C \diagup \\ \underset{O}{O} \quad O \end{array} \right]$$

O_3
(the net equation)

(an unusual rearrangement)   (3-26)

$$\diagdown C{=}O + O{=}C \diagup \; \underset{Zn + H^+}{\overset{H_2O}{\longleftarrow}} \; \diagdown C \overset{O}{\underset{O{-}O}{\diagup\diagdown}} C \diagup$$

ozonide

The over-all result of ozonolysis is the breaking of the alkene at the double bond into two fragments, each of which has an oxygen atom doubly-bound to the original olefinic carbons.

Ozonization can be used to locate the position of a double bond. Consider, for example, the isomeric butenes

$$CH_2{=}CH{-}CH_2{-}CH_3 \qquad CH_3{-}CH{=}CH{-}CH_3$$
1-butene   2-butene

Both compounds have the same percentage composition; both de-colorize bromine and give a positive Baeyer test. How, then, to distinguish them? Ozonolysis provides the answer, for one of these olefins will give two products ($CH_2=O$ and $O=CHCH_2CH_3$) whereas the other gives only one ($CH_3CH=O$).

3.4f SUBSTITUTION

Although alkenes commonly react by addition, they do have C—H bonds and, under special conditions, substitution may occur. This reaction is the basis of some important commercial processes. Ethyl-ene can be chlorinated at high temperatures to vinyl chloride, a starting material for preparing some important polymers.

$$CH_2=CH_2 + Cl_2 \xrightarrow{400°} CH_2=CHCl + HCl \qquad (3\text{-}27)$$
1-chloroethene
(vinyl chloride)

(The **vinyl group** is $CH_2=CH-$.) The polymer made from this chlo-ride is a plastic which can be molded or extruded; it can also be used as a lacquer for coating metals.

Substitution of hydrogens on a carbon adjacent to a double bond is also possible.

$$CH_3CH=CH_2 + Cl_2 \xrightarrow{heat} \underset{\underset{\displaystyle Cl}{|}}{C}H_2CH=CH_2 + HCl \qquad (3\text{-}27a)$$
allyl chloride

The $CH_2=CH-CH_2-$ group is known as the **allyl group** and has some special structural features (sec 4.8).

3.5 PREPARATION OF ALKENES

A double bond is usually introduced in a saturated molecule by removing two elements or groups from adjacent carbon atoms. One group (X) leaves with its bonding electron pair, the other (Y), without. The pair of electrons left behind then forms an additional bond between the carbon atoms involved.

$$R-\underset{\underset{\displaystyle X}{|}}{C}H-\underset{\underset{\displaystyle Y}{|}}{C}H-R' \longrightarrow R-CH=CH-R' + X-Y \qquad (3\text{-}28)$$

The compound X—Y is usually a small inorganic molecule such as water (H—OH) or a hydrogen halide (H—Br or H—Cl). The entire process is called an **elimination reaction,** because the molecule X—Y is eliminated; elimination is the reverse of the addition reaction.

3.5a   DEHYDRATION OF ALCOHOLS

When molecule X—Y in equation 3-28 is water, the process is called **dehydration,** and the starting material is an **alcohol.** Alcohols can be dehydrated by being heated with strong acids (sulfuric, phosphoric), as shown in equation 3-29.

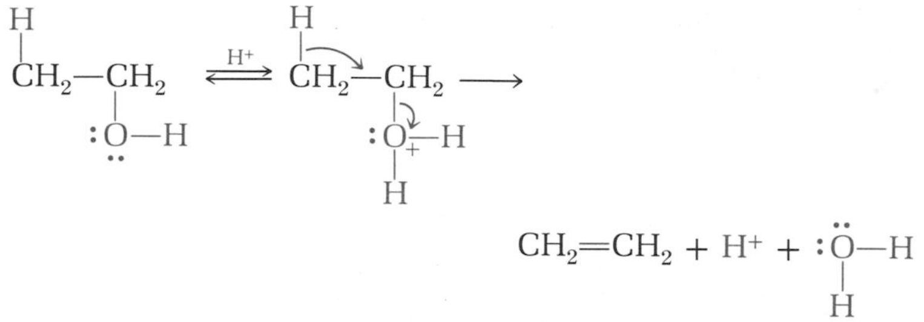

$$CH_2=CH_2 + H^+ + :\overset{..}{O}-H \quad (3\text{-}29)$$

The alcohol is reversibly protonated by the acid catalyst. In this case, X is a proton from a carbon atom adjacent to the hydroxyl-bearing carbon, and the leaving group Y is a water molecule.

For some alcohols, elimination can proceed in more than one direction giving a mixture of alkenes. Usually, the alkene with the most substituted double bond predominates (eq 3-30).

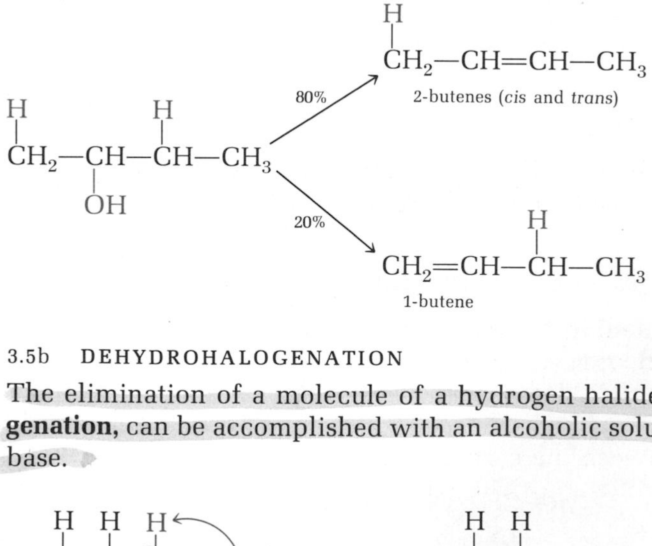

$$(3\text{-}30)$$

2-butenes (cis and trans)

1-butene

3.5b   DEHYDROHALOGENATION

The elimination of a molecule of a hydrogen halide, **dehydrohalogenation,** can be accomplished with an alcoholic solution of a strong base.

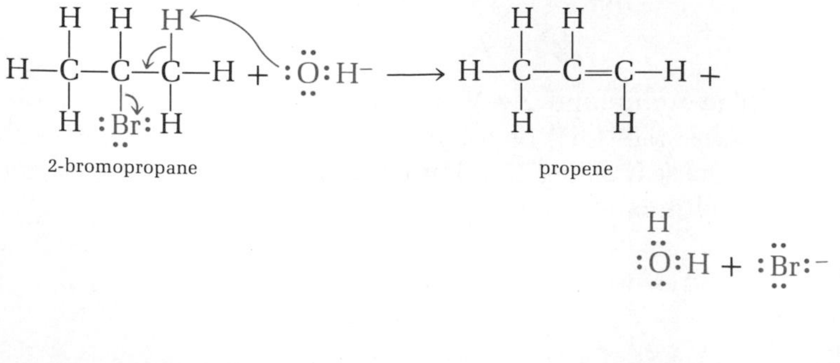

2-bromopropane

propene

$$:\overset{..}{O}:H + :\overset{..}{\underset{..}{Br}}:^- \quad (3\text{-}31)$$

As with the dehydration of alcohols, when dehydrohalogenation can occur in more than one direction a mixture of alkenes will be formed.

3.5c OLEFINS FROM CRACKING

Cracking of petroleum produces large quantities of mixtures of olefins. These result from fundamental changes in the structure of saturated hydrocarbons, brought about by high temperatures and catalysts (sec 2.7d). Some of the possible reactions which may occur are indicated for normal butane as follows:

$$CH_3 \overset{(a)}{\diagup} CH_2 \overset{(b)}{\diagup} CH_2 - CH_3 \underset{(b)}{\overset{(a)}{\diagup}} \begin{matrix} CH_4 + CH_2 = CH - CH_3 & \left(80\%\right) \\ \\ CH_3 - CH_3 + CH_2 = CH_2 & \left(20\%\right) \end{matrix} \qquad (3\text{-}32)$$

When the cracking stock consists of hydrocarbons in the C_{10} to C_{16} range, a very wide variety of products is obtained.

Cracking is the major source of ethylene and propylene, raw materials for the industrial synthesis of alcohols, permanent antifreeze, polyethylene, ethylene dibromide, and numerous other products. Cracking also furnishes butenes, which may be converted to high-octane fuels and synthetic rubber.

3.6 PETROLEUM REFINING

Petroleum, as it comes from the oil well, is a crude and complex mixture of hydrocarbons, with small amounts of oxygen-, nitrogen- and sulfur-containing compounds. It is not useful as such, but can be refined to give useful products with much narrower specifications (boiling range, combustibility). The first step in this refining process is distillation, a physical process. Chemical processes like cracking, isomerization, alkylation, and aromatization have been developed, mainly to increase the yield of motor and aviation gasoline from crude oil.

3.6a ANTIKNOCK FUELS

In order to compare gasolines, it was necessary to set up an arbitrary knock-rating scale. Isoöctane (2,2,4-trimethylpentane),

$$\begin{matrix} & CH_3 & & CH_3 \\ & | & & | \\ CH_3 - & \!\!\!C\!\!\! & -CH_2 - & \!\!\!CH\!\!\! & -CH_3 \\ & | & & \\ & CH_3 & & \end{matrix}$$

a pure hydrocarbon with excellent knock properties, was assigned an **octane number** of 100, and normal heptane,

$$CH_3CH_2CH_2CH_2CH_2CH_2CH_3$$

a very poor gasoline fuel, was assigned the octane number 0. The octane number of a gasoline, then, is the percentage of isoöctane in a mixture of isoöctane and n-heptane that has equivalent knock properties to the gasoline, when both are tested in a standard engine. Just as with temperature scales, once the units have been assigned, the scale can be extended above and below the original limits.

A fuel can be improved in two ways. Small amounts of "additives" (such as tetraethyl lead, in the case of gasolines) may be introduced, or the fuel itself may be improved by increasing the percentage of molecules with superior burning properties. Both techniques are employed. Diesel and jet fuels can use large percentages of straight-chain hydrocarbons, whereas high-octane fuels are improved by chain branching. Isomerization, alkylation, and aromatization are important in making high-octane gasoline.

3.6b ISOMERIZATION

Straight-chain saturated hydrocarbons can be isomerized to branched-chain isomers by certain catalysts. For example, aluminum chloride at 100° isomerizes n-pentane to isopentane.

$$CH_3CH_2CH_2CH_2CH_3 \xrightarrow{\text{AlCl}_3} CH_3\underset{\underset{CH_3}{|}}{C}HCH_2CH_3 \qquad (3\text{-}33)$$

The mechanism, which is complex, probably involves carbonium ions. The reaction is important in the manufacture of gasolines, but is not generally practical as a laboratory synthetic method.

3.6c ALKYLATION OF OLEFINS

In the presence of appropriate catalysts (aluminum chloride, sulfuric acid, or hydrofluoric acid), certain hydrocarbons add to the double bond of an alkene. The process is called **alkylation,** since an alkyl group is added to the alkene. The product is a highly branched hydrocarbon (eq 3-34).

$$CH_3-\underset{\underset{CH_3}{|}}{\overset{\overset{CH_3}{|}}{C}}=CH_2 + H-\underset{\underset{CH_3}{|}}{\overset{\overset{CH_3}{|}}{C}}-CH_3 \xrightarrow{H_2SO_4} CH_3-\underset{\underset{CH_3}{|}}{\overset{\overset{H}{|}}{C}}-CH_2-\underset{\underset{CH_3}{|}}{\overset{\overset{CH_3}{|}}{C}}-CH_3 \quad (3\text{-}34)$$

2,2,4-trimethylpentane

The reaction, which proceeds by a carbonium ion mechanism, is initiated by addition of a proton from the catalyst to the alkene.

$$CH_3-\underset{\underset{CH_3}{|}}{C}\!\!=\!\!CH_2 + H^+ \longrightarrow CH_3-\underset{\underset{CH_3}{|}}{\overset{+}{C}}-CH_3 \qquad (3\text{-}34a)$$

t-butyl cation

The resulting carbonium ion adds, according to Markownikoff's rule, to another alkene molecule forming a new carbonium ion.

$$CH_3-\underset{\underset{CH_3}{|}}{\overset{\overset{CH_3}{|}}{C}}{}^+ + CH_2\!\!=\!\!\underset{\underset{CH_3}{|}}{C}-CH_3 \longrightarrow CH_3-\underset{\underset{CH_3}{|}}{\overset{\overset{CH_3}{|}}{\underset{3}{C}}}-CH_2-\underset{\underset{CH_3}{|}}{\overset{+}{\underset{1}{C}}}-CH_3 \qquad (3\text{-}34b)$$

The chain is completed by a very fast reaction, the abstraction of hydrogen with its electron pair (hydride ion, H⁻) from the saturated hydrocarbon.

$$CH_3-\underset{\underset{CH_3}{|}}{\overset{\overset{CH_3}{|}}{C}}-CH_2-\underset{\underset{CH_3}{|}}{\overset{+}{C}}-CH_3 + H\!:\!\underset{\underset{CH_3}{|}}{\overset{\overset{CH_3}{|}}{C}}-CH_3 \longrightarrow$$

$$CH_3-\underset{\underset{CH_3}{|}}{\overset{\overset{CH_3}{|}}{C}}-CH_2-\underset{\underset{CH_3}{|}}{\overset{\overset{H}{\cdot\cdot}}{C}}-CH_3 + {}^+\underset{\underset{CH_3}{|}}{\overset{\overset{CH_3}{|}}{C}}-CH_3 \quad (3\text{-}34c)$$

The *t*-butyl cation produced in this step adds to the alkene (as in eq 3-34b) to continue the reaction chain.

3.6d IONIC DIMERIZATION AND POLYMERIZATION OF ALKENES

When isobutylene is treated with cold sulfuric acid, two molecules of the alkene combine by addition to form octenes.

$$CH_3-\underset{\underset{CH_3}{|}}{\overset{\overset{CH_2}{\|}}{C}} + CH_2\!\!=\!\!\underset{\underset{CH_3}{|}}{C}-CH_3 \xrightarrow[10^\circ]{H_2SO_4}$$

isobutylene

$$CH_3-\underset{\underset{CH_3}{|}}{\overset{\overset{CH_3}{|}}{C}}-CH_2-\underset{\underset{CH_3}{|}}{C}\!\!=\!\!CH_2 + CH_3-\underset{\underset{CH_3}{|}}{\overset{\overset{CH_3}{|}}{C}}-CH\!\!=\!\!\underset{\underset{CH_3}{|}}{C}-CH_3 \quad (3\text{-}35)$$

2,4,4-trimethyl-1-pentene   2,4,4-trimethyl-2-pentene

These highly branched alkenes may be used directly as fuels or may be hydrogenated to give 2,2,4-trimethylpentane, the same product which was obtained by the alkylation of isobutylene with isobutane (eq 3-34).

Dimerization therefore provides an alternative route to high octane fuels. The reaction proceeds by a carbonium ion mechanism. The first two steps are the same as those in equations 3-34a and 3-34b. Since there is no saturated hydrocarbon to furnish a hydride ion, the carbonium ion produced in equation 3-34b stabilizes itself by losing a proton (H^+) from either carbon 1 or 3 to yield the products shown in equation 3-35.

By appropriate control of reaction conditions, it is possible to make higher polymers of alkenes this way. For example, propylene can be converted to a tetramer, a mixture of several isomeric dodecenes (12-carbon atoms). Propylene tetramer

$$4 \ CH_3CH{=}CH_2 \ \xrightarrow{\ H^+\ } \ CH_3CHCH_2CHCH_2CHCH{=}CHCH_3 \qquad (3\text{-}36)$$
$$\underset{CH_3 \qquad CH_3 \qquad CH_3}{}$$

one isomer of
propylene tetramer

is used in large quantities in the manufacture of synthetic detergents (Chapter 11).

3.6e   AROMATIZATION

Alkanes, when heated over special catalysts, can be cyclized and dehydrogenated to produce a class of substances called **aromatic hydrocarbons** (Chapter 5) which have good octane ratings and are important industrial synthetic raw materials as well.

$$CH_3CH_2CH_2CH_2CH_2CH_2CH_3 \ \xrightarrow[\text{heat}]{\substack{\text{platinum} \\ \text{catalyst}}} \ \text{toluene} \ + \ 4 \ H_2 \qquad (3\text{-}37)$$

heptane

toluene

This type of reaction, called **hydroforming,** is an important part of petroleum refining operations. From the deep-seated structural changes which occur, it is clear that the mechanism for this reaction must be quite complex.

3.6f GASOLINE BLENDING

The above processes provide the raw materials for gasoline which, however, must be blended properly for efficient use. Sufficient aromatics and branched-chain compounds are included to provide a high octane number, and tetraethyl lead may be added to improve knock rating. Ethylene dibromide or dichloride are included to remove the lead oxide after combustion. Because of concern with atmospheric pollution, oil companies now market improved unleaded gasolines, and engines have been modified to use them without knocking. Additives which reduce fouling of spark plugs, lubricate engine parts, and prevent carburetor icing in winter, and a dye to identify the gasoline brand may be included. In winter, "low-boilers" (butanes) help ignition, but in summer they are omitted for they evaporate. Gasoline today is a complex blended mix of hydrocarbons and additives designed for the modern engine.

New Concepts, Facts, and Terms

1. Alkenes, or olefins, are unsaturated hydrocarbons which contain a carbon-carbon double bond.
2. Nomenclature of alkenes, sec 3.1
3. Geometry of the double bond, sigma (σ) and pi (π) orbitals, sp^2 hybrid orbitals. *Cis-trans* or *geometric* isomerism, and restricted rotation around double bonds.
4. Reagents which add to unsaturated hydrocarbons
   a. chlorine, bromine
   b. hydrogen
   c. mineral acids (H—X, H—OSO$_2$OH)
   d. hypohalous acids (HO—X)
6. Markownikoff's rule gives the correct orientation for the addition of unsymmetrical reagents to unsymmetrical alkenes. The most stable carbonium ion intermediate is formed.
7. Polymerization; initiation, propagation, and termination steps; polyethylene
8. Oxidation; Baeyer test, ozonolysis
9. Substitution; vinyl and allyl groups
10. Preparation of alkenes; elimination reaction
    a. dehydration of alcohols
    b. dehydrohalogenation of alkyl halides
    c. cracking
11. Petroleum refining
    a. fuels; octane number
    b. isomerization
    c. alkylation of olefins; carbonium ions; hydride ion transfer.
    d. dimerization and ionic polymerization of alkenes; propylene tetramer
    e. aromatization of alkanes
    f. gasoline blending

Exercises and Problems

1. Write structural formulas and IUPAC names of all possible alkenes with the indicated molecular formulas:
   a. C_4H_8 c. C_6H_{12}
   b. C_5H_{10} d. C_5H_8

2. Name the following compounds by the IUPAC system:

   a. $CH_3CH_2CH{=}CHCH_3$
   b. $(CH_3)_2C{=}CHCH_3$
   c. $CHBr{=}CHCH(CH_3)_2$
   d.

   $$\begin{array}{c} CH_2{-}CH \\ \diagup \quad\quad \| \\ CH_2 \quad\quad CH \\ \diagdown\;\diagup \\ CH_2 \end{array}$$

   e.
   $$\begin{array}{c} CH_3\diagdown \quad\quad\quad \diagup CH_2CH_2CH_3 \\ C{=}C \\ \diagup \quad\quad\quad \diagdown \\ H \quad\quad\quad\quad H \end{array}$$

   f.
   $$\begin{array}{c} CH_3\diagdown \quad\quad\quad \diagup H \\ C{=}C \\ \diagup \quad\quad\quad \diagdown \\ H \quad\quad\quad CH_2CH_2CH_3 \end{array}$$

   g.
   $$\begin{array}{c} CH_2 \\ \diagup \quad\quad \diagdown \\ CH \quad\quad CH{-}CH_3 \\ \| \quad\quad\quad\quad \\ CH \quad\quad CH_2 \\ \diagdown \quad\quad \diagup \\ CH_2 \end{array}$$

3. Write structural formulas for the following compounds:
   a. 3-octene
   b. 4-methyl-2-pentene
   c. cyclobutene
   d. 1,3-dibromo-2-butene
   e. 3-methylcyclopentene
   f. *trans*-2-octene
   g. 2,3-dimethyl-2-butene
   h. vinyl bromide

4. Explain why the following names are incorrect and give a correct name in each case.
   a. 3-butene
   b. 3-pentene
   c. 2-ethyl-1-propene
   d. 2-chloro-2-ethyl-3-butene
   e. 2-methylcyclopentene
   f. 2,2-dimethyl-4-pentene
   g. 1-ethyl-2-cyclobutene
   h. 3,5-dimethyl-3-hexene

5. Write the structural formula and name of the product when each of the following reacts with one mole of bromine:
   a. 2-butene
   b. 3-methyl-3-hexene
   c. vinyl chloride
   d. 1-methylcyclopentene

6. Write equations for the reaction of 1-butene with the following reagents:
   a. chlorine
   b. hydrogen chloride
   c. hypobromous acid
   d. sulfuric acid
   e. ozone
   f. potassium permanganate
   g. bromine, in aqueous solution containing potassium chloride

7. Illustrate with a structural formula or equation the meaning of the following terms:
   a. alkene
   b. addition reaction
   c. substitution reaction
   d. Markownikoff's rule
   e. electrophile
   f. ozonide
   g. geometric isomerism
   h. polymerization

8. Use reactions studied in this and the previous chapter to show how each of the following could be prepared from propene:
   a. propane
   b. 2-bromopropane
   c. 1,2-dichloropropane
   d. 1-chloro-2-propanol
   e. 2,3-dimethylbutane
   f. 2-propyl hydrogen sulfate

9. Would the bromination of propene in aqueous solution containing sodium chloride be expected to yield (in addition to 1,2-dibromopropane) 1-bromo-2-chloropropane or 2-bromo-1-chloropropane? Explain.

10. What products might be expected from the addition of bromine to a solution of cyclohexene in methanol, $CH_3\ddot{O}H$? Explain.

11. Each of the following conversions requires the combination, in the proper sequence, of several reactions described in this chapter. Write equations to show how each could be executed.
    a. *n*-propyl bromide to propene to 1,2-dibromopropane
    b. isopropyl alcohol to propene to 2-iodopropane
    c. 1-bromobutane to 2-chlorobutane
    d. ethyl bromide to 2-chloroethanol
    e. 2-butanol to butane
    f. bromocyclopentane to 1,2-dibromocyclopentane

12. How many grams of bromine are required to saturate the following? (Atomic weights: $C = 12$, $H = 1$, $Br = 80$.)
    a. 14 grams of 2-butene
    b. 41 grams of cyclohexene

13. A 10 gram mixture of butane and 2-butene reacts with just 8.0 grams of bromine (dissolved in carbon tetrachloride). Calculate the percentage of butane in the original mixture.

14. Which of the following compounds can exist as geometric (*cis-trans*) isomers? If such isomerism is possible, draw the structures in a way which clearly illustrates the geometry.
    a. 1-pentene
    b. 2-pentene
    c. 2-methyl-2-butene
    d. 1,2-dibromoethene
    e. 1-chloropropene
    f. 3-chloropropene
    g. 1,2-dibromocyclopentene
    h. 1,2-dibromocyclopentane
    i. 1,1-dimethylcyclopropane
    j. 3-bromo-4-methyl-3-hexene

15. Explain why there is only one cyclohexene but there are two cyclo-decenes (geometric isomers).

16. Which alkene would react with what reagent, to form each of the following compounds:
   a. $CH_3CH_2CHBrCH_2Br$
   b. $CH_3CH(OH)CH(Cl)CH_3$
   c. $(CH_3)_2CHCH_3$
   d. $CH_3CHBrCH_2CH_2CH_3$
   e. $(CH_3)_2CHOSO_2OH$

17. Give the structural formulas of the alcohols which, on dehydration, will give *only*:
   a. isobutylene
   b. 3-heptene
   c. cyclopentene
   d. 4-methylcyclohexene

18. Give the formulas of the alkenes which on ozonolysis give:
   a. only $CH_3CH_2CH{=}O$
   b. $(CH_3)_2C{=}O$ and $CH_3CH{=}O$
   c. $CH_2{=}O$ and $(CH_3)_2CHCH{=}O$
   d. $O{=}CHCH_2CH_2CH{=}O$

19. Write out all the steps in the ionic mechanism for addition of HBr to 1-methylcyclopentene (to give 1-bromo-1-methylcyclopentane). Explain why the reaction proceeds in the observed direction.

20. Write equations for each step in the mechanism of the conversion of propylene to its tetramer (eq 3-36). What other isomers are also likely to be present in the product?

CHAPTER FOUR

UNSATURATED HYDROCARBONS: ALKYNES, DIENES, AND POLYENES

Acetylene, the simplest alkyne, is an unsaturated hydrocarbon with a carbon-carbon triple bond. Although some of its reactions are analogous to those of alkenes, others are unique. It is an important raw material for such diverse products as orlon, dry cleaners, and synthetic rubber. In this chapter we will also describe compounds with two or more double bonds, and the special properties which ensue when these functions are sufficiently close in the same molecule to interact.

4.1 ALKYNES

The gaseous hydrocarbon acetylene has a molecular formula C_2H_2 and may be readily transformed into ethylene or ethane by addition of hydrogen. This suggests that its structural formula is

$$H\!:\!C\!:\!:\!:\!C\!:\!H \qquad \text{or} \qquad H\!-\!C\!\equiv\!C\!-\!H$$

and that the conversion to ethylene or ethane can be expressed as

$$H-C\equiv C-H \xrightarrow{\ H-H\ } \begin{array}{c} H \\ \diagdown \\ C=C \\ \diagup \\ H \end{array}\!\!\!\begin{array}{c} H \\ \diagup \\ \\ \diagdown \\ H \end{array} \xrightarrow{\ H-H\ } \begin{array}{c} H \\ \diagdown \\ H\text{--}C-C\text{---}H \\ \diagup \\ H \end{array}\!\!\!\begin{array}{c} H \\ \diagup \\ \\ \diagdown \\ H \end{array} \tag{4-1}$$

Acetylene is the first member of a general class of unsaturated hydrocarbons with a carbon-carbon triple bond.

4.2 THE TRIPLE BOND

If one of the bonds to a carbon atom is a triple bond, then that carbon atom can only be attached to two other atoms. Since a carbon atom has one $2s$ and three $2p$ orbitals available for bonding, the two equivalent orbitals needed can be formed by blending the $2s$ orbital with a $2p$ orbital to make two sp orbitals (half s, half p). These orbitals lie in a straight line; that is, the angle between them will be 180° to minimize any repulsion between electrons placed in these orbitals. One electron is placed in each hybrid orbital. The remaining two valence electrons occupy the two other $2p$ orbitals, which are mutually perpendicular, and also perpendicular to the hybrid sp orbitals.

The union of two sp carbon atoms to form a triple bond is shown in Figure 4-1. Overlap of two sp orbitals results in one bond, a sigma (σ) bond, analogous to an ordinary single bond. If the two carbons are properly oriented with respect to one another, the p-orbitals also overlap, forming two mutually perpendicular pi (π) bonds (π_1 and π_2 in Figure 4-1).

Acetylenes are linear, as shown in Figures 4-1 and 4-2. The carbon-carbon triple bond distance is 1.21 Å, shorter than the double (1.34 Å) or single (1.54 Å) bonds. This is because three electron pairs draw the nuclei closer together than two or one pair.

4.3 NOMENCLATURE OF ALKYNES

Hydrocarbons which contain a triple bond are known as acetylenes or **alkynes.** The general rules for nomenclature are identical with

Figure 4-1

The triple bond consists of the end-on overlap of two sp hybrid orbitals to form a σ bond plus the edgewise overlap of two parallel oriented p orbitals to form two π bonds.

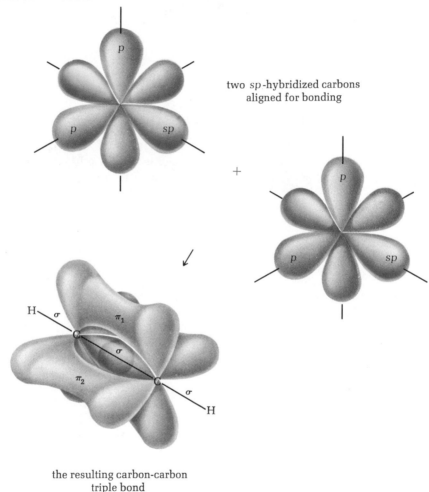

two *sp*-hybridized carbons
aligned for bonding

the resulting carbon-carbon
triple bond

those of the alkenes, except that the ending is -yne. The following examples illustrate the rules.

$$CH_3-C{\equiv}C-H \qquad \overset{1}{CH_3}-\overset{2}{C}{\equiv}\overset{3}{C}-\overset{4}{CH_2}-\overset{5}{CH_3} \qquad \overset{1}{H}-\overset{2}{C}{\equiv}\overset{3}{C}-\overset{4}{CH}-CH_3$$
$$\underset{CH_3}{\vert}$$

propyne   2-pentyne   3-methyl-1-butyne

Figure 4-2
Models of acetylene.

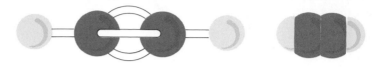

4.4 ADDITION REACTIONS OF ALKYNES

Some reactions of the triple bond are analogous to those of the double bond. The same reagents which add to alkenes also add to alkynes. Addition may be illustrated by the reaction of acetylene with bromine.

$$H—C\equiv C—H \xrightarrow{\text{Br}_2} \underset{\substack{|\\Br}}{\overset{\substack{Br\\|}}{H—C}}=\underset{\substack{|\\Br}}{\overset{\substack{Br\\|}}{C—H}} \xrightarrow{\text{Br}_2} \underset{\substack{|\\Br}}{\overset{\substack{Br\\|}}{H—C}}—\underset{\substack{|\\Br}}{\overset{\substack{Br\\|}}{C—H}} \qquad (4\text{-}2)$$

acetylene   1,2-dibromoethene   1,1,2,2-tetrabromoethane

Tetrachloroethane, prepared in an analogous manner, is used as a solvent and to prepare other chlorinated compounds (Chapter 8).

Electrophilic additions to the triple bond usually proceed more slowly than analogous double bond reactions, because sp carbons, having more s character (50%) than sp^2 carbons ($33\frac{1}{3}$%) attract the bonding electrons closer to the nuclei. Thus the π electrons are more firmly bound and less susceptible to attack by electrophiles.

The triple bond adds hydrogen in the presence of a catalyst (platinum). Usually the product is the corresponding alkane, but with a special palladium catalyst it is possible to stop the reaction at the alkene. This is particularly useful for preparing *cis* alkenes, since both hydrogens add to the triple bond from the same "side" (i.e., the catalyst surface).

$$CH_3—C\equiv C—CH_3 \xrightarrow[\substack{\text{Pd Lindlar}\\\text{catalyst}}]{\text{H—H}} \overset{CH_3}{\underset{H}{}}C=C\overset{CH_3}{\underset{H}{}} \qquad (4\text{-}3)$$

2-butyne   cis-2-butene

Addition of unsymmetrical reagents to the triple bond follows Markownikoff's rule.

$$H-C{\equiv}C-H \xrightarrow{\text{HCl}} \underset{\substack{| \\ H \ \ Cl}}{H-C{=}C-H} \xrightarrow{\text{HCl}} \underset{\substack{| \ \ | \\ Cl \ \ H}}{\overset{\substack{Cl \ \ H \\ | \ \ |}}{H-C-C-H}} \quad (4\text{-}4)$$

vinyl chloride   1,1-dichloroethane

$$CH_3-C{\equiv}CH \xrightarrow{\text{HI}} \underset{\substack{| \ \ | \\ I \ \ H}}{CH_3-C{=}CH} \xrightarrow{\text{HI}} \underset{\substack{| \ \ | \\ I \ \ H}}{\overset{\substack{I \ \ H \\ | \ \ |}}{CH_3-C-CH}} \quad (4\text{-}5)$$

2,2-diiodopropane

With certain acids, addition to acetylene is readily interrupted after one mole has reacted. The products are vinyl compounds

$$HC{\equiv}CH + H-X \longrightarrow \underset{H}{\overset{H}{\diagdown}}C{=}C\underset{X}{\overset{H}{\diagup}} \quad (4\text{-}6)$$

a vinyl compound

which, like ethylene (sec 3.4d), may be polymerized. The resulting polymers are frequently useful commercially.

$$\underset{X}{\overset{|}{CH_2{=}CH}} \longrightarrow -\underset{X}{\overset{|}{CH_2CH}}(\underset{X}{\overset{|}{CH_2-CH}})_n\underset{X}{\overset{|}{CH_2-CH}}- \quad (4\text{-}7)$$

a vinyl polymer

Examples include

$$\underset{\text{acrylonitrile}}{CH_2{=}CHCN} \longrightarrow \text{polyacrylonitrile (Orlon, Acrilan)}$$

$$\underset{\text{vinyl acetate}}{CH_2{=}CH\underset{\overset{\|}{O}}{O}CCH_3} \longrightarrow \text{polyvinyl acetate}$$

$$\underset{\text{vinyl chloride}}{CH_2{=}CHCl} \longrightarrow \text{polyvinyl chloride}$$

$$\underset{\text{vinyl methyl ether}}{CH_2{=}CHOCH_3} \longrightarrow \text{a polyvinyl ether}$$

$$(4\text{-}8)$$

Polyacrylonitrile can be spun into fibers for the production of fabrics and sweaters. Polyvinyl acetate emulsions are used as adhesives and in latex paints. Many consumer products, such as floor tile, raincoats and other rubber-like materials, phonograph records, and fabrics are produced from polyvinyl chloride. Polyvinyl methyl ether, from the addition of methyl alcohol to acetylene, is used as an adhesive. Finally, many co-polymers can be made, starting with mixtures of vinyl monomers. Co-polymers of vinyl acetate and vinyl chloride give vinyl resins useful for floor coverings, shoe soles, and furniture coverings.

4.5 REACTIONS OF THE ACETYLENIC HYDROGEN

The orbitals which bind a hydrogen to a singly-, doubly- or triply-bound carbon have increasing s character, in that order. The more s character, the more firmly will the electrons be pulled in toward the carbon nucleus.

$$sp^3\ (\tfrac{1}{4}s) \qquad\qquad sp^2\ (\tfrac{1}{3}s) \qquad\qquad sp\ (\tfrac{1}{2}s)$$

(Remember, s atomic electrons have lower energy and are closer to the nucleus than p electrons.) For this reason, the ease with which a carbon can give up a proton (that is, its acidity) increases as the s- character of the bonding orbital increases. Hydrocarbons are exceedingly weak acids but, with acetylenes, the $\equiv$C—H proton can be removed with sufficient ease to make the reaction synthetically useful.

Acetylene reacts with molten sodium (110°) to liberate hydrogen, and form an **acetylide.**

$$2H-C\equiv C-H + 2Na \xrightarrow{\text{heat}} 2H-C\equiv C:^-Na^+ + H_2 \qquad (4\text{-}9)$$
$$\text{sodium acetylide}$$

More commonly, acetylides are prepared using the very strong base sodamide in liquid ammonia to remove the acetylenic hydrogen.

$$H-C\equiv C-\!\!|H + Na^+NH_2^- \longrightarrow H-C\equiv C:^-Na^+ + H-NH_2 \qquad (4\text{-}10)$$
$$\text{sodamide} \qquad\qquad\qquad\qquad \text{ammonia}$$

The acetylide ion can displace halide ions from alkyl halides. This reaction can be used to synthesize higher alkynes from acetylene. Thus with ethyl bromide, one obtains 1-butyne.

$$H-C\equiv C:^-Na^+ + CH_3CH_2-Br \longrightarrow$$

$$H-C\equiv C-CH_2CH_3 + Na^+Br^- \qquad (4\text{-}11)$$
$$\text{1-butyne}$$

The sequence can be repeated with the remaining acetylenic hydrogen, to give a dialkyl acetylene.

$$H-C\equiv C-CH_2CH_3 + NaNH_2 \xrightarrow[\text{NH}_3]{\text{liquid}}$$
$$Na^+:^-C\equiv C-CH_2CH_3 + H-NH_2 \quad (4\text{-}12)$$

$$Na^+:^-C\equiv C-CH_2CH_3 + CH_3I$$
$$CH_3-C\equiv C-CH_2CH_3 + Na^+I^- \quad (4\text{-}13)$$
2-pentyne

This reaction is also the basis of a laboratory test for compounds with an acetylenic hydrogen. They react with aqueous ammoniacal silver nitrate or cuprous chloride to give water-insoluble, characteristically colored acetylides.

$$R-C\equiv C-H + Ag(NH_3)_2^+ \longrightarrow$$
$$R-C\equiv C:^-Ag^+\downarrow + [H-NH_3]^+ + NH_3 \quad (4\text{-}14)$$
a silver acetylide
(white precipitate)

$$R-C\equiv C-H + Cu(NH_3)_2^+ \longrightarrow$$
$$R-C\equiv C:^-Cu^+\downarrow + [H-NH_3]^+ + NH_3 \quad (4\text{-}15)$$
a copper acetylide
(red precipitate)

The resulting acetylides are explosive when dry and may detonate on touch.

The test distinguishes compounds of the type $RC\equiv CH$ from $RC\equiv CR'$ and from alkenes. (This is sometimes useful, since all three types will decolorize bromine and give a positive Baeyer test with permanganate.)

4.6 PREPARATION OF ALKYNES

Acetylene is produced commercially from limestone and coke. The first step gives calcium carbide (an acetylide). Being the salt of the very weak acid, acetylene, calcium carbide is readily hydrolyzed by water.

$$3\,C + CaO \longrightarrow CaC_2 + CO \quad (4\text{-}16)$$
coke   lime   calcium carbide

$$CaC_2 + 2\,H-OH \longrightarrow H-C\equiv C-H + Ca(OH)_2 \quad (4\text{-}17)$$
(Ca^{2+} $^-$:$C\equiv C$:$^-$)

Acetylene is also produced by cracking natural gas (mainly methane) at very high temperatures. Acetylene is used not only in welding, but as an important chemical raw material.

Substituted acetylenes can be prepared from acetylene by the sequence of reactions given in section 4.5.

The dehydrohalogenation reaction used for alkenes (sec 3.5b) can be adapted to the synthesis of alkynes. The starting material must be a dihalide with both halogens on the same or on adjacent carbon atoms.

$$
\begin{array}{c}
\text{H} \;\; \text{H} \\
| \;\;\; | \\
-\text{C}-\text{C}- \;+\; \text{OH}^- \;\xrightarrow[-\text{Br}^-]{-\text{HOH}}\; \overset{\text{H}}{\underset{\text{Br}}{\diagdown}}\text{C}{=}\text{C}\diagup \;+\; \text{OH}^- \;\xrightarrow[-\text{Br}^-]{-\text{HOH}} \\
| \;\; | \\
\text{Br} \;\; \text{Br}
\end{array}
$$

$$-\text{C}{\equiv}\text{C}- \quad (4\text{-}18)$$

Most commonly, the reaction is used to convert an alkene to an alkyne.

$$
\text{RCH}{=}\text{CH}_2 \;\xrightarrow{\text{Br}_2}\; \underset{\substack{|\;\;\;\;| \\ \text{Br}\;\;\;\text{Br}}}{\text{RCH}{-}\text{CH}_2} \;\xrightarrow[\text{alcohol}]{\text{KOH in}}
$$

$$\text{RC}{\equiv}\text{CH} + \text{K}^+\text{Br}^- + \text{H}_2\text{O} \quad (4\text{-}19)$$

4.7 DIENES: CLASSES, NOMENCLATURE, AND SYNTHESIS

Dienes (*di*-enes; compounds with *two* carbon-carbon double bonds) fall into three categories which depend on the relative positions of the double bonds. If three consecutive carbons are joined by double bonds, the latter are said to be **cumulative;** the simplest member is **allene** (propadiene).* If double bonds alternate with single bonds, they are said to be **conjugated.** This arrangement is common in many natural products. Double bonds separated by more than one single bond usually react quite independently of one another and are called **isolated.**

$$\text{CH}_2{=}\text{C}{=}\text{CH}_2 \qquad \overset{1}{\text{CH}_2}{=}\overset{2}{\text{CH}}{-}\overset{3}{\text{CH}}{=}\overset{4}{\text{CH}_2}$$

allene   1,3-butadiene

(cumulative)   (conjugated)

$$\overset{1}{\text{CH}_2}{=}\overset{2}{\text{CH}}{-}\overset{3}{\text{CH}_2}{-}\overset{4}{\text{CH}_2}{-}\overset{5}{\text{CH}}{=}\overset{6}{\text{CH}_2}$$

1,5-hexadiene

(isolated)

---

*The chemistry of allenes is too specialized for this text; the interested student should consult a more advanced book.

Cyclic systems with more than one double bond are common.

1,3-cyclopentadiene   1,4-cyclohexadiene
(conjugated)   (isolated)

Isoprene (2-methyl-1,3-butadiene) is a particularly important diene because many compounds in nature have carbon skeletons in which two or more isoprene units are joined together. (For examples, see sec 4.10 and Chapter 18.)

$$CH_2{=}C{-}CH{=}CH_2 \qquad C{-}C{-}C{-}C$$
$$\qquad\quad \underset{\displaystyle CH_3}{|} \qquad\qquad\qquad \underset{\displaystyle C}{|}$$

isoprene   isoprene unit

Dienes can be synthesized by extensions of the methods used for alkenes (sec 3.5). 1,3-Butadiene, considered to be the most important commercial diene because of its use in the manufacture of synthetic rubber, is made most cheaply by catalytic dehydrogenation of butane or butenes. Cyclopentadiene is a component of coal gas

$$C_4H_{10} \xrightarrow[\text{heat}]{\text{catalyst}} C_4H_8 \xrightarrow[\text{heat}]{\text{catalyst}} C_4H_6 \qquad\qquad (4\text{-}20)$$

and can be obtained, together with 1,3-butadiene and isoprene, by the cracking of petroleum.

4.8   CONJUGATED SYSTEMS: RESONANCE

In molecules with conjugated double bonds, some overlap of the π orbitals is possible because of their proximity. This leads to unique

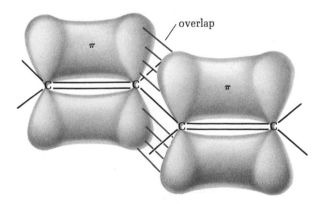

reactions which are not observed when the double bonds are isolated.

Comparison of the addition of bromine to 1,4-pentadiene (*isolated* double bonds) and 1,3-butadiene (*conjugated* double bonds) illustrates the phenomenon. With one mole of bromine, 1,4-pentadiene gives a single product, as expected (the product is identical, regardless of which "end" of the chain reacts).

$$CH_2\!=\!CH\!-\!CH_2\!-\!CH\!=\!CH_2 \xrightarrow{Br_2} CH_2\!-\!CH\!-\!CH_2\!-\!CH\!=\!CH_2 \quad (4\text{-}21)$$

1,4-pentadiene

Br   Br

4,5-dibromo-1-pentene

With 1,3-butadiene, however, two products are observed.

$$\overset{1}{CH_2}\!=\!CH\!-\!CH\!=\!\overset{4}{CH_2} \xrightarrow{Br_2}$$

80% → $\overset{2}{CH_2}\!-\!CH\!=\!CH\!-\!\overset{3}{CH_2}$

Br   Br

1,4-dibromo-2-butene (1,4-addition)

(4-22)

20% → $CH_2\!-\!CH\!-\!CH\!=\!CH_2$

Br   Br

3,4-dibromo-1-butene (1,2-addition)

Only 20% of the expected product, with the bromines on adjacent carbons, is obtained. In the major product, the bromines have added to carbons 1 and 4 of the diene, and a new double bond is present between carbons 2 and 3. The process is called **1,4-addition** in contrast to the usual 1,2-addition.

This unusual result becomes quite reasonable when the reaction mechanism is considered (review sec 3.4a). Addition of bromine to the terminal carbon (according to Markownikoff's rule) gives the intermediate carbonium ion* shown.

$$:\!\overset{..}{\underset{..}{Br}}\!-\!\overset{..}{\underset{..}{Br}}\!: + CH_2\!=\!CH\!-\!CH\!=\!CH_2 \longrightarrow$$

$$:\!\overset{..}{\underset{..}{Br}}\!:^- + Br\!-\!CH_2\!-\!\overset{+}{CH}\!-\!CH\!=\!CH_2 \quad (4\text{-}23)$$

an allylic carbonium ion

This ion contains three adjacent sp^2 carbons, each with a p orbital perpendicular to the molecular plane. As produced in equation 4-23, the carbonium ion would be represented with an empty p orbital at carbon 2, but filled overlapping orbitals at carbons 3 and 4 (structure **A**). But equally possible is structure **B**, where the p orbitals on

---

*Similar arguments apply for a free radical mechanism.

A   B

carbons 2 and 3 are occupied by two electrons, and the p orbital on carbon 4 is vacant. Carbon 4 now becomes the site of the positive charge.

This phenomenon may be represented with more conventional formulas as

$$\text{BrCH}_2\overset{+}{\text{C}}\text{H}-\overset{\frown}{\text{CH}}=\text{CH}_2 \longleftrightarrow \text{BrCH}_2\text{CH}=\overset{\frown}{\text{CH}}-\overset{+}{\text{C}}\text{H}_2$$

A   B

It is clear that the reaction can be completed either at carbon 2 (giving 1,2-addition) or at carbon 4 (giving a 1,4-adduct with a new double bond between carbons 2 and 3).

The explanation required to rationalize 1,4-addition to conjugated dienes is but one example of a much more general concept which pervades organic chemistry, the concept of **resonance**. Consider a carbonium ion* with an adjacent carbon-carbon double bond, the **allyl** carbonium ion: Each carbon is attached to only three other atoms; each is sp^2 hybridized. Perpendicular to the general plane

$$\overset{+}{\text{C}}\text{H}_2-\text{CH}=\text{CH}_2 \qquad \text{or}$$

allyl carbonium ion

*The same ideas apply to free radicals and carbanions.

of the molecule are three p orbitals in which one must place two electrons. It seems quite arbitrary to leave the orbital on carbon 1 vacant and place the electrons in overlapping orbitals on carbons 2 and 3. The alternative choice, shown below, seems equally arbitrary.

$$CH_2=CH-\overset{+}{C}H_2 \qquad \text{or}$$

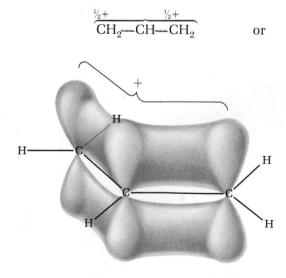

This phenomenon, in which *two or more equivalent structures are possible, with identical positions of the atoms but different arrangements of the electrons is called resonance.* The true structure is not adequately represented by any of the individual classical structures one can write, but is a **resonance hybrid** of all of them. With the allyl cation, the two electrons may be placed in a molecular orbital which encompasses all three carbon atoms.

$$\overset{\frac{1}{2}+}{CH_2}---\overset{}{CH}---\overset{\frac{1}{2}+}{CH_2} \qquad \text{or}$$

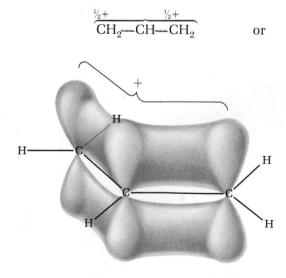

The bonds between carbons 1 and 2, and between carbons 2 and 3 are identical and can be thought of approximately as "one-and-a-half" bonds. The bond length is intermediate between those of single and double bonds.

The charge in the allyl cation is divided equally between carbons 1 and 3. We know from physics that it is better in terms of energy distribution to spread charge out over a large surface than to concentrate it at a point. The same is true with molecules. Thus the allyl cation is much more stable than, say, a propyl cation where charge delocalization by resonance is not possible.

Resonance is a stabilizing force; a resonance hybrid is always more stable than any of the individual structures which contribute to it. This is readily apparent with the allyl cation where each contributing structure has a full positive charge on a single carbon atom, whereas the hybrid has the charge distributed equally between the two terminal carbons. But it is also true of uncharged resonance hybrids as well. We shall see examples in the next chapter.

4.9 OTHER DIENE REACTIONS

One of the most useful methods for synthesizing six-membered rings is the **cycloaddition** of a conjugated diene to an alkene. The simplest example (though not a practical one) is the addition of ethylene to butadiene.

$$\text{diene} \qquad \text{dienophile} \qquad \qquad \text{cyclohexene} \tag{4-24}$$

The reaction type is named after its discoverers, the German chemists Otto Diels and Kurt Alder, who shared the Nobel prize in 1950. The **Diels-Alder reaction** is unusually versatile, since many structural variations are possible in both the **diene** and the **dienophile**. The reactions are often reversible. Some useful examples are:

$$\tag{4-25}$$

1,3-butadiene   4-vinylcyclohexene

butadiene   acrylonitrile   4-cyanocyclohexene   (4-26)

cyclopentadiene   acetylene   norbornadiene   (4-27)

Another thermal reaction of considerable importance is the **Cope rearrangement** (named after the American chemist, Arthur Cope, who was affiliated for many years with MIT). The required structural feature is two π-bonds separated by three single bonds. The simplest case would be 1,5-hexadiene.

1,5-hexadiene   (4-28)

In this example, the product and reactant are identical and can only be distinguished if some isotopic label is introduced. But with other examples, reactant and product are different.

2,6-octadiene   3,4-dimethyl-   (4-29)
                       1,5-hexadiene

1,3-cyclohexadiene   1,3,5-hexatriene   (4-30)

The reaction is quite versatile, since one or more of the carbons may be replaced by other atoms (O, N, S).

4.10   POLYENES: NATURAL AND SYNTHETIC RUBBER

Many polyenes occur in nature, and some are important biological intermediates. **Squalene** ($C_{30}H_{50}$), a hexaene first isolated from shark liver oil, is an intermediate in the biochemical synthesis of cholesterol (sec 18.3).

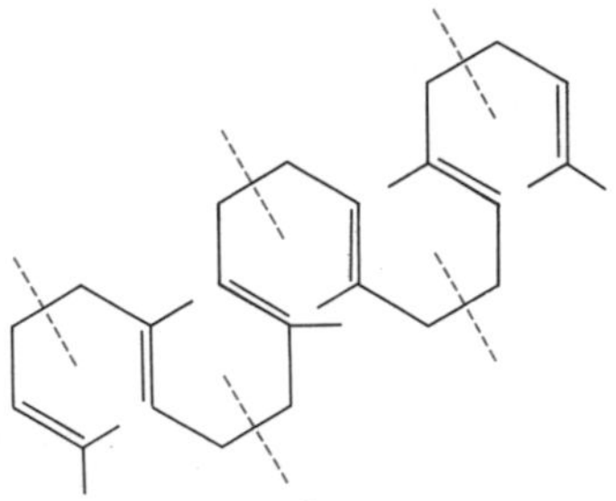

squalene

It is made of six isoprene units (marked off in color). Starting at either end, one sees that the units are joined in head-to-tail fashion, until one comes to the middle of the molecule, where two like ends are joined. Squalene, which has only isolated double bonds, is a colorless, crystalline substance.

Some plant pigments are hydrocarbons with extended conjugated systems. Examples include **lycopene,** the red pigment in ripe tomatoes and watermelon, and the **carotenes,** yellow pigments present in many plants but first isolated from carrots (the β-isomer is most prevalent). Both are made up of eight isoprene units (40 carbons), and have conjugated systems which extend over 22 carbon atoms.

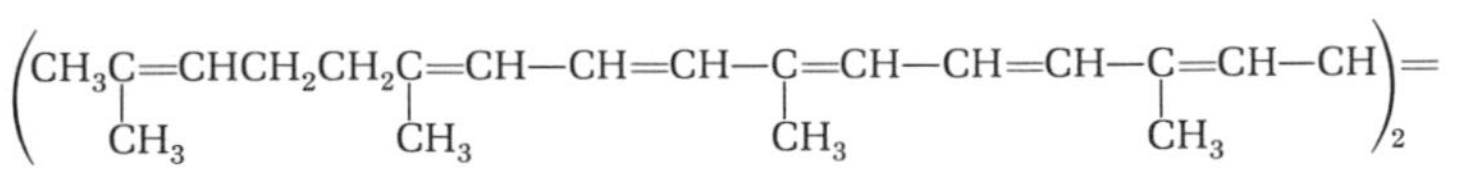

lycopene

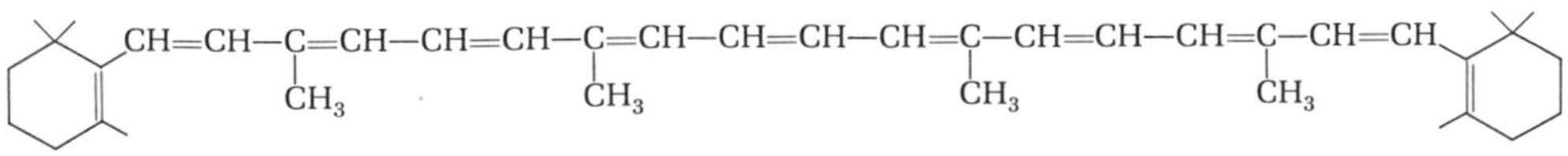

β-carotene

Natural rubber is a complex hydrocarbon polymer. Thermal decomposition of rubber in the absence of air gives mainly the conjugated diene, isoprene. This suggests that rubber is a polymer of isoprene. When isoprene is treated with conventional free radical initiators (review sec 3.4d), it forms a polymer similar to, although not identical with, natural rubber. In this polymer, isoprene units

$$\text{natural rubber} \xrightarrow[\text{no air}]{\text{heat}} \underset{\underset{\displaystyle CH_3}{|}}{CH_2{=}C{-}CH{=}CH_2} \qquad (4\text{-}31)$$

isoprene

are joined to one another either by 1,2- or 1,4-addition. But in natural rubber, the structure is that which would result from almost exclusive 1,4-addition. Furthermore, the remaining double bonds are all *cis*, whereas ordinary free radical polymerization of isoprene led to a polymer with mainly *trans* double bonds.

isoprene

natural rubber

$$(4\text{-}32)$$

It is now known that the same special catalysts which produce linear polyethylene (sec 3.4d) also produce a polymer from isoprene essentially identical with natural rubber by an all *cis*, almost exclusive 1,4-addition. There may be as many as two thousand isoprene units per rubber molecule.

The term **synthetic rubber** or **elastomer** refers to polymers which have *rubber-like* properties, but which may be quite different structurally from natural rubber. All *cis* polybutadiene is now produced in large quantities, and is used mainly blended with natural or other synthetic rubbers. Co-polymers of butadiene and styrene are used in tire manufacture. **Neoprene** is a relatively expensive synthetic rubber which, however, has many special uses because of its resistance to oils, chemicals, air, light, and heat. It is made by the polymerization of **chloroprene** (2-chloro-1,3-butadiene). The monomer is made from acetylene, by the following sequence:

$$HC{\equiv}CH + HC{\equiv}CH \qquad\qquad \underset{\underset{\displaystyle Cl}{|}}{CH_2{=}CH{-}C{=}CH_2}$$

$$\Bigg\downarrow \begin{smallmatrix}Cu_2Cl_2\\ NH_4Cl\end{smallmatrix} \qquad\qquad\qquad \text{chloroprene} \qquad (4\text{-}33)$$

$$\qquad\qquad\qquad\qquad\qquad \Big\uparrow H^+$$

$$\underset{\text{vinylacetylene}}{CH_2{=}CH{-}C{\equiv}CH} \xrightarrow[\text{1,4-addition}]{HCl} \underset{\underset{\displaystyle Cl}{|}}{CH_2{-}CH{=}C{=}CH_2}$$

$$CH_2=CH-\underset{\underset{Cl}{|}}{C}=CH_2 \xrightarrow{\text{catalyst}}$$

(4-34)

chloroprene   neoprene

The gross structure resembles that of natural rubber, but the double bonds are all *trans,* and the methyl substituents are replaced by chlorine, which imparts flame retardant properties to the rubber.

New Concepts, Facts, and Terms

1. Alkynes, acetylenes
2. The triple bond, *sp* hybrid orbitals are linear.
3. Nomenclature
4. Addition to alkynes; Markownikoff's rule is followed; vinyl compounds and vinyl polymers
5. Reactions of the acetylenic hydrogen—its acidity; replacement by metals; synthesis of substituted alkynes; acetylides
6. Preparation of acetylene from coke and lime; conversion of a double bond to a triple bond
7. Dienes; cumulative, conjugated and isolated; isoprene
8. Conjugated systems; 1,4-addition. Resonance, allylic systems, resonance hybrid
9. Diels-Alder reaction; Cope rearrangement
10. Squalene, carotene, natural and synthetic rubber; chloroprene and neoprene

Exercises and Problems

1. Write structural formulas and IUPAC names for all the alkynes and alkadienes with the indicated molecular formulas:
   a. C_5H_8
   b. C_6H_{10}
   c. C_5H_6

2. Name the following compounds by the IUPAC system:
   a. $(CH_3)_2CHC \equiv CH$
   b. $CH_3C \equiv CCH_2CH_3$
   c. $CH_2 = CBr-CH = CH_2$
   d. CH_3—⬡

   e. $(CH_3)_2C = C = C(CH_3)_2$
   f. $CH_2 = CH-C \equiv CH$
   g. $CH_2-C \equiv C-CH_2$ $-(CH_2)_6-$

3. Write structural formulas for each of the following:
   a. 1,4-hexadiene
   b. 3-methyl-1-pentyne
   c. vinylcyclopentane
   d. 1,2-cyclononadiene
   e. 1,4-hexadiyne
   f. 2,3-dimethyl-1,3-cyclopentadiene
   g. 4,4,5-trimethyl-2-hexyne
   h. allyl bromide

4. Starting with calcium carbide and any other inorganic chemicals, write equations for the preparation of:
   a. acetylene
   b. ethane
   c. ethyl iodide
   d. vinylacetylene
   e. 1-butyne
   f. 1,1-diiodoethane
   g. 2,2-dibromobutane
   h. 3-hexyne
   i. 1,1,2,2-tetrabromoethane
   j. 3-hexene

5. Each of the following conversions requires the combination, in the proper sequence, of several reactions described in this and previous chapters. Write equations to show how each could be executed.
   a. 2-butene to 1,3-butadiene (2 steps)
   b. isopropyl alcohol to propyne (3 steps)
   c. n-propyl bromide to 2,2-dichloropropane
   d. 1,3-butadiene to 1,4-dibromobutane
   e. 1-butene to 1-butyne

6. The following facts are known about a compound:
   a. molecular weight: 82
   b. analysis: C, 88%, H, 12%
   c. one mole of the compound takes up two moles of bromine
   d. when shaken with cuprous chloride solution, the compound produces a copper-containing precipitate
   From these data suggest a possible structural formula for the compound.

7. Do you expect two carbons, bound by a triple bond, to rotate freely with respect to one another? Explain. Would any restriction in rotation lead to geometric isomerism, as with alkenes? Explain.

8. Addition of one mole of hydrogen bromide to 1,3-butadiene gives two products, 1-bromo-2-butene and 3-bromo-1-butene. Write all the steps in a reaction mechanism which explains how both products are formed.

9. Predict the products in each of the following reactions:
   a. isoprene + one mole of bromine
   b. 1-vinylcyclohexene + acetylene + heat
   c. sodium acetylide + 2-bromobutane
   d. 1,2-dibromocyclohexane + alcoholic KOH
   e. cyclopentadiene + one mole of hydrogen bromide
   f. 3-methyl-1,5-hexadiene + heat

10. 1,3-Cyclopentadiene is a low boiling liquid (bp 41°) which, on standing at room temperature, produces a low melting solid (mp 33°). If this solid is heated to its boiling point (170°) and distilled slowly, the distillate is found to be cyclopentadiene. Suggest an explanation for these observations.

11. When an electrophile (say, H+) attacks 1,3-butadiene, it always does so at C-1 rather than at C-2. Write the structure for the intermediate carbonium ion in each case, and suggest an explanation for the observed result.

12. Suggest a possible mechanism for the last step, in equation 4-33.

13. Place dashed lines in the appropriate places in the formulas for lycopene and β-carotene to show that they are constituted of isoprene units.

14. Point out, in the formulas in equations 4-29 and 4-30, the structural feature required for the observed Cope rearrangements.

CHAPTER FIVE

AROMATIC HYDROCARBONS

The development of the coal tar industry in Germany during the nineteenth century was a great stimulus to the systematic study of organic chemistry. When coal is converted to coke and coal gas for processing iron and steel, a portion of the coal remains as a tarry residue. Although this unpleasant-looking material is only a small percentage of the original coal, it nevertheless amounts to many tons annually because of the tremendous quantity of coal processed. Distillation of coal tar produces many useful raw materials, all related in structure to the aromatic hydrocarbon, benzene. Aromatics are now synthesized in large quantities from petroleum. The pharmaceutical, dyestuff, and explosives industries are based largely upon these aromatic chemicals.

Early in the development of organic chemistry it became evident that a class of compounds exists which, though highly unsaturated, is uniquely stable. Since the derivatives of these compounds frequently had rather fragrant odors, the compounds were called *aromatic*. We now know that the parent compound of these aromatic substances is *benzene*, first isolated in 1825 by Michael Faraday from illuminating gas. Coal tar is an important source of benzene. Some benzene is present in petroleum, and large amounts are produced from petroleum hydrocarbons by aromatization of alkanes (sec 3.6e).

5.1 THE STRUCTURE OF BENZENE

Analysis and molecular weight determination show that the molecular formula of benzene is C_6H_6. The ratio of carbons to hydrogens suggests that the molecule is unsaturated. However, benzene does not readily decolorize bromine solutions, nor is it easily oxidized by permanganate; both of these reactions are characteristic tests for the double bond. The usual reaction of benzene with bromine, particularly in the presence of certain catalysis, is **substitution.**

$$C_6H_6 + Br_2 \xrightarrow[\text{catalyst}]{\text{FeBr}_3} C_6H_5Br + HBr \qquad (5\text{-}1)$$

bromobenzene

The fact that only *one* monobromo derivative of benzene is obtained implies that all of the hydrogens are equivalent (or that certain of the hydrogens are nonreplaceable, a less likely situation).

When bromine or chlorine reacts with benzene in the absence of oxygen and in the presence of sunlight, an addition product is formed.

$$C_6H_6 + 3\,Cl_2 \xrightarrow[\text{light}]{\text{ultraviolet}} C_6H_6Cl_6 \qquad (5\text{-}2)$$

benzene hexachloride

An important clue to the carbon skeleton of benzene is the observation that it can be catalytically hydrogenated to cyclohexane.

$$C_6H_6 + 3\,H_2 \xrightarrow[\text{heat, pressure}]{\text{Ni catalyst}} \begin{array}{c} CH_2-CH_2 \\ CH_2 \qquad CH_2 \\ CH_2-CH_2 \end{array} \qquad (5\text{-}3)$$

cyclohexane

These experiments suggest a cyclic structure for benzene.

In 1865, Kekulé proposed formula **A** or **B**, a hexagon of carbon atoms with alternating single and double bonds, with one hydrogen attached to each carbon atom.

Figure 5-1
Friedrich August Kekulé (1829–1896) was a pioneer in the development
of structural formulas in organic chemistry. The tetracovalence of
carbon, the importance of chains of carbon atoms, and the structure
of benzene were among his major contributions to chemistry. (The
Bettmann Archive, Inc.)

The Kekulé formula for benzene adequately accounts for its hydrogenation to cyclohexane and also explains why only *one* monosubstitution product is obtained (all six hydrogens are equivalent). But there remains the problem of benzene's resistance to oxidation by permanganate and its lack of facile addition as might be expected from a triene. Kekulé surmounted this difficulty by suggesting that the bonds in **A** and **B** alternated positions so rapidly that each carbon-carbon bond was neither single nor double but something intermediate.

In structures **A** and **B,** corresponding atoms occupy essentially identical positions in space. The formulas differ only in the positions assigned to the π electrons. Once again we are dealing with the phenomenon of **resonance** (review sec 4.8). The structure of benzene is not adequately represented by either of the two Kekulé formulas; instead, *benzene is a resonance hybrid to which the Kekulé structures are contributors.*

Physical measurements support this idea. Benzene is planar, with each carbon at the corner of a regular hexagon. All carbon-carbon bond lengths are identical, unique, and not typical of either single or double bonds. The carbon-carbon bond distance in benzene is 1.40 Å, intermediate between the usual single (1.54 Å) and double (1.34 Å) bond distances. A scale model of benzene is shown in Figure 5.2.

Figure 5-2
Scale model of benzene.

It is useful to consider the structure of benzene in molecular orbital terms. Each carbon in benzene is bound to three other atoms (two carbons, one hydrogen). Bonding is therefore of the sp^2 type, as in ethylene, with angles of 120° between the atoms. Each carbon is bound to a hydrogen through overlap of an sp^2 carbon orbital with the 1s orbital of hydrogen (a σ bond). Each carbon is also bound to its two carbon neighbors by overlap of the sp^2 orbitals, thus providing the σ-bond framework for the hexagon. Finally, each carbon has an additional p orbital at right angles to the plane of the ring, with one electron in it (Figure 5-3). Overlap of these six p orbitals leads to three filled π orbitals, and the net result is a π-electron cloud above and below the plane of the atomic nuclei. This representation will be useful in explaining the chemical behavior of benzene and other aromatic compounds.

Figure 5-3
A molecular orbital representation of the bonding in benzene. Sigma bonds are formed by overlap of sp² orbitals. In addition, each carbon contributes one electron to the π system by overlap of its p orbital with the p orbital of its two neighbors.

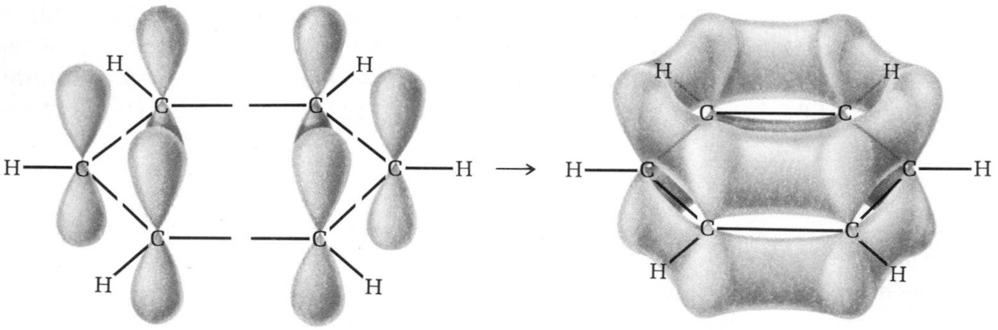

The model of benzene shown in Figure 5-3, though important in understanding its chemistry, is inconvenient to use in chemical equations. In this text we will use the Kekulé contributing structures, keeping in mind always that benzene is a hybrid of these structures and that the double bonds are not fixed in either possible arrangement.* Most commonly, the hydrogen which is attached to each

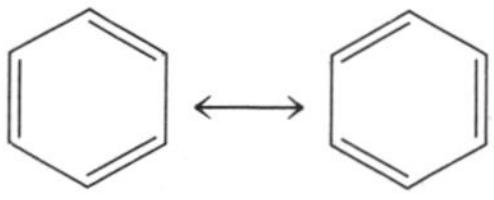

or on its side

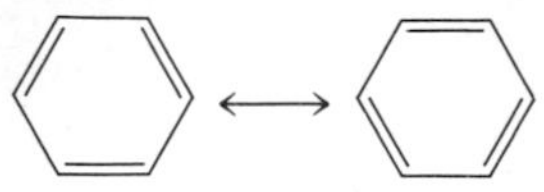

*Another symbol for benzene is a hexagon with a circle in the middle:  . The circle represents the even distribution of the π electrons over all six atoms.

carbon is understood to be there, and is not written explicitly (just as the carbons are understood to occupy the corners of the hexagon, and are not designated by their atomic symbol).

5.2   RESONANCE ENERGY OF BENZENE

We have asserted without experimental proof that a resonance hybrid is more stable than any of its contributing structures. Since the concept of resonance pervades aromatic chemistry, it is fortunate that some experimental support for it, and even some quantitative measure of its importance, can be simply presented.

When a carbon-carbon double bond is hydrogenated, energy is released. The amount of energy is, to a first approximation, nearly independent of the groups attached to the double bond and varies between 26.5–30 kcal/mole (the exact value is usually lower the greater the number of alkyl groups attached to the double bond).

$$\ce{C=C} + \ce{H-H} \longrightarrow \underset{\ce{H}\ \ce{H}}{\ce{-C-C-}} + 26.5\text{--}30 \text{ kcal/mole} \quad (5\text{-}4)$$

When more than one double bond is present in a molecule, multiples of this amount of heat are evolved, depending on the number of double bonds. For example, the heat of hydrogenation of 1,3-cyclohexadiene is about twice that of cyclohexene:

$$\bigcirc + \ce{H-H} \longrightarrow \bigcirc + 28.6 \text{ kcal/mole} \qquad (5\text{-}5)$$

$$\bigcirc + 2\ \ce{H-H} \longrightarrow \bigcirc + 55.4 \text{ kcal/mole} \qquad (5\text{-}6)$$

If benzene were accurately described by one of the Kekulé structures (that is, 1,3,5-cyclohexatriene), one would expect it to have a heat of hydrogenation of 84-86 kcal/mole. (Three separate double bonds like that of cyclohexene gives $3 \times 28.6 = 85.8$ kcal/mole; the sum of cyclohexene and 1,3-cyclohexadiene gives a slightly lower value, $28.6 + 55.4 = 84.0$ kcal/mole).

The experimentally observed value for the heat of hydrogenation of benzene is drastically lower than these predicted values, being only 49.8 kcal/mole. This means that benzene, because of its unique

$$\bigcirc + 3\ \ce{H-H} \longrightarrow \bigcirc \qquad \begin{array}{l} 86 \text{ kcal (predicted)} \\ \underline{50 \text{ kcal}} \text{ (observed)} \end{array} \qquad (5\text{-}7)$$

resonance energy 36 kcal/mole

bonding, is 36 kcal/mole more stable than if it were simply a conjugated triene. This amount of stabilization energy, called the **resonance energy,** is defined as the difference between the actual energy of the resonance hybrid and the calculated energy of the most stable contributing structure. Chemical reactions which would destroy the "aromaticity" or resonance in benzene would sacrifice this stabilization energy. When possible, therefore, benzene (and other aromatic compounds) react in such a way as to retain their aromatic structure and consequently their resonance energy.

5.3 NOMENCLATURE OF BENZENE DERIVATIVES

Since aromatic chemistry developed in a haphazard fashion many years before systematic methods of nomenclature were developed, common names acquired historic respectability and are still frequently used. There are, however, some elements of system. The general class name for aromatic hydrocarbons is **arene.** The benzene ring with one hydrogen atom removed

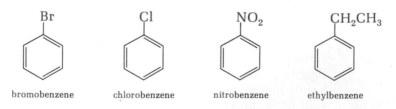

$$C_6H_5- \quad \text{or}$$

is called the **phenyl** group (from the Greek *pheno*, "I bear light," a reference to the original discovery of benzene in illuminating gas). This is an example of an **aryl** group derived from an **ar**omatic compound, as contrasted with alkyl groups derived from aliphatic* compounds. The symbol **Ar-** is used to designate an aryl group.

Monosubstituted benzenes are named as derivatives of benzene.

Br	Cl	NO$_2$	CH$_2$CH$_3$
bromobenzene	chlorobenzene	nitrobenzene	ethylbenzene

---

*The term *aliphatic* originally referred to compounds derived from fats, just as the term *aromatic* referred to natural products with a characteristic aromatic odor. These words have now been generalized in terms of the structures of molecules. Aliphatic compounds are substances related to methane; that is, saturated or unsaturated open-chain or cyclic compounds *not* containing a benzene ring. Aromatic compounds are substances whose structure *is* related to that of benzene. Later in the book, the third large class of organic substances, the *heterocyclic* compounds, will be discussed. These have cyclic structures with at least one element other than carbon as a member of the ring. The hetero atom may be oxygen, nitrogen, sulfur, or some other element.

Some monosubstituted benzenes have common names.

CH₃ benzene structure

CH=CH₂ benzene structure

CH₂Cl benzene structure

toluene
(*not* methylbenzene)

styrene

benzyl chloride
(the group in color is the
benzyl group)

Sometimes it is more convenient to designate the benzene ring as a substituent, as in

$$\overset{1}{C}H_3\overset{2}{C}H\overset{3}{C}H_2\overset{4}{C}H_2\overset{5}{C}H_3$$

2-phenyl pentane

trans-1,2-
diphenyl cyclopropane

When two substituents are present, three isomeric structures are possible. They are usually designated by the prefixes *ortho* (*o*-), *meta* (*m*-) and *para* (*p*-).

ortho-dichloro-
benzene

meta-dichloro-
benzene

para-dichloro-
benzene

The ring may also be numbered; this is particularly useful when three or more substituents are present.

meta-bromoiodobenzene
or 1-bromo-3-iodobenzene

1,2,4-tri-
methylbenzene

3,5-dichlorotoluene

5.4 ELECTROPHILIC AROMATIC SUBSTITUTION

The most common reactions of the aromatic ring involve replacement of one or more of the ring hydrogens by other atoms or groups. Examples include *halogenation, nitration, sulfonation,* and *alkylation.*

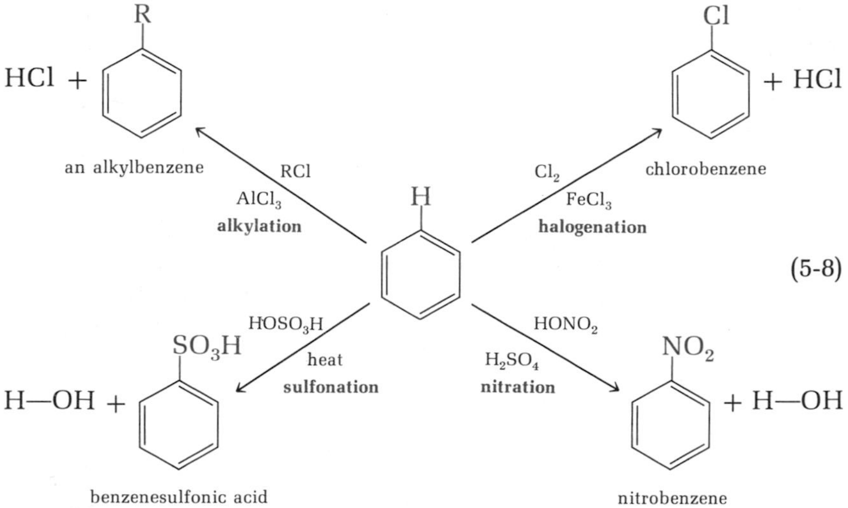

an alkylbenzene   RCl   chlorobenzene
AlCl$_3$ Cl$_2$
alkylation   FeCl$_3$ halogenation

HOSO$_3$H   HONO$_2$
heat   H$_2$SO$_4$
sulfonation   nitration

benzenesulfonic acid   nitrobenzene

$$(5\text{-}8)$$

The conditions and type of catalyst vary depending on the particular reaction and aromatic substrate. The reactions can ordinarily be controlled so as to replace only one hydrogen or more than one, as may be desired. The special problem of the location of a second substituent, after the first one has been introduced, is discussed in section 5.4b. But first we shall consider the reaction mechanism.

5.4a THE MECHANISM OF ELECTROPHILIC AROMATIC SUBSTITUTION

The reactive species in each reaction of equation 5-8 is an electrophile—a seeker of an electron pair. This can be designated by the symbol E$^+$. The structure of the particular electrophile in each reaction will be discussed later.

The aromatic ring, with a π electron cloud above and below the plane of the ring, acts as a supplier of electrons to the electrophile. The first step is analogous to the attack of an electrophile on the π electrons of a carbon-carbon double bond (sec 3.4a).

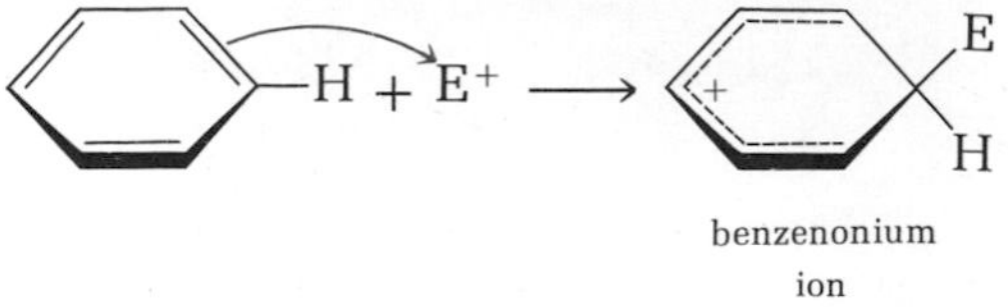

$$(5\text{-}9)$$

benzenonium
ion

The carbonium ion intermediate, called a **benzenonium ion,** has three important contributing resonance structures which distribute

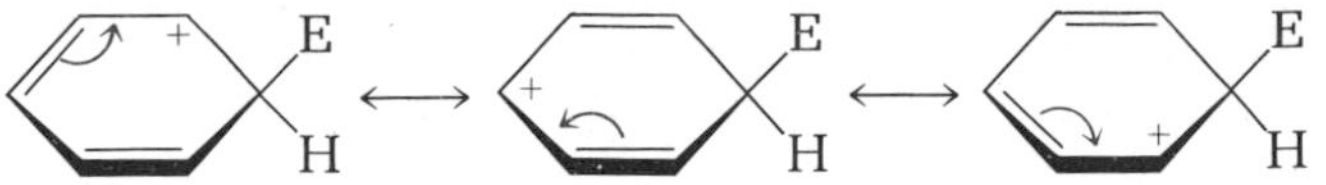

resonance forms of benzenonium ions

the positive charge predominantly over the carbons *ortho* and *para* to the position attacked by the electrophile. -

In additions to alkenes, the reaction is usually completed by combination of the carbonium ion with a nucleophile (eq 3-3). If this were to happen when an aromatic ring is attacked by an electrophile, the product would be a cyclohexadiene, and the aromatic system, with its considerable resonance energy, would be destroyed. Instead, the reaction is completed by loss of a proton, thus restoring the aromatic system.

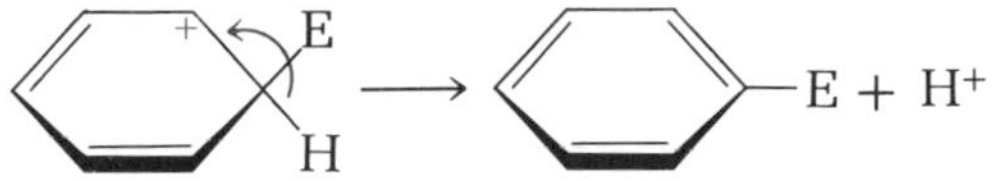

$$-E + H^+ \qquad (5\text{-}10)$$

The net result is the substitution of E for H.

Let us now consider the nature of the electrophile in each part of equation 5-8. The purpose of the iron catalyst in **halogenation** is to produce a positive halogen ion **(halonium ion),** according to equation 5-11.

$$:\overset{..}{\underset{..}{Cl}} \!:\! \overset{..}{\underset{..}{Cl}} \!:\! + FeCl_3 \longrightarrow \quad :\overset{..}{\underset{..}{Cl}}{}^+ \quad + FeCl_4^- \qquad (5\text{-}11)$$

chloronium ion

In **nitration,** sulfuric acid is sometimes added to the nitric acid to facilitate formation of the **nitronium ion,** which is the attacking electrophile.

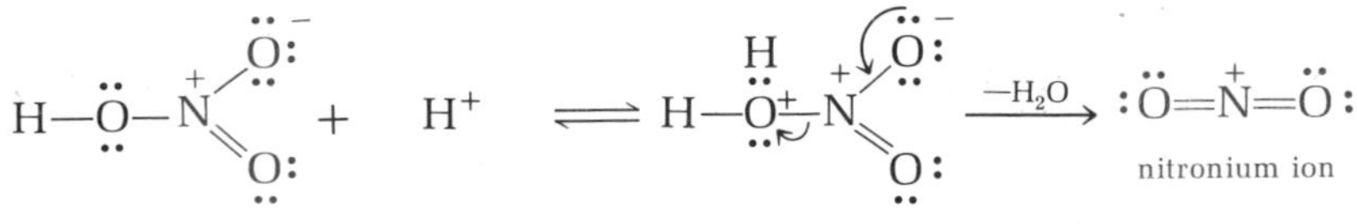

(from
H$_2$SO$_4$
or another
molecule of
HONO$_2$)

$$(5\text{-}12)$$

The active species in **sulfonation** is usually sulfur trioxide or the SO_3H^+ ion.

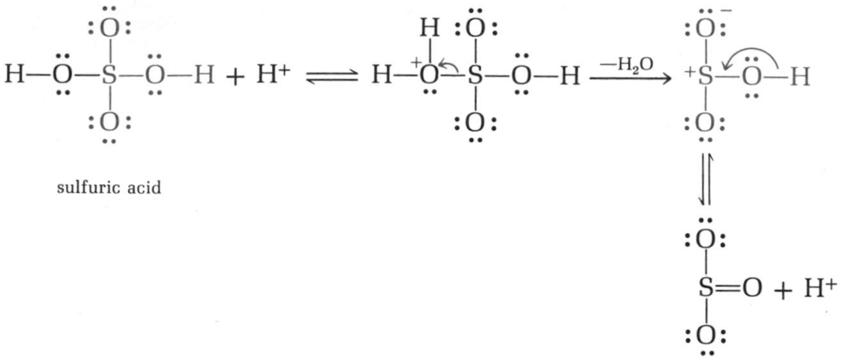

$$H-\overset{\overset{\displaystyle :\ddot{O}:}{|}}{\underset{\underset{\displaystyle :\ddot{O}:}{|}}{S}}-\ddot{O}-H + H^+ \rightleftharpoons H-\overset{\overset{\displaystyle H\;:\ddot{O}:}{|}}{\underset{\underset{\displaystyle :\ddot{O}:}{|}}{\overset{+}{O}}}-\ddot{O}-H \xrightarrow{-H_2O} \quad {}^+S-\ddot{O}-H \qquad (5\text{-}13)$$

sulfuric acid

$$\overset{\overset{\displaystyle :\ddot{O}:}{\|}}{\underset{\underset{\displaystyle :\ddot{O}:}{|}}{S}}{=}O + H^+$$

Finally, an alkyl carbonium ion is the electrophile in **alkylation.***
The aluminum chloride functions as a Lewis acid, to help break the R—Cl bond.

$$R \;\vdots\; -Cl + AlCl_3 \longrightarrow R^+ + AlCl_4^- \qquad (5\text{-}14)$$

Figure 5-4
The electrophiles in common aromatic substitutions.

Cl^+ or Br^+	NO_2^+	SO_3 or SO_3H^+	R^+
halogenation	nitration	sulfonation	alkylation

Each of these electrophiles (Figure 5-4) can function as E^+ in equations 5-9 and 5-10, thus replacing an aromatic proton. Since these groups can, through various reactions, be converted to many others (for example, NO_2 is easily reduced to NH_2), electrophilic aromatic substitution is frequently an important first step in the synthesis of many aromatic compounds.

In special cases, salts of the intermediate carbonium ions in electrophilic aromatic substitutions can actually be isolated. For example, treatment of mesitylene (1,3,5-trimethylbenzene) with deuterium fluoride and boron fluoride at $-30°$ gives, the salt shown in equation 5-15. On warming, the

---

*This reaction (Friedel-Crafts alkylation) is named after Charles Friedel (French) and James Mason Crafts (American), who first discovered it in 1878.

$$(5\text{-}15)$$

yellow salt, mp $-15°$

salt decomposes to mesitylene (40%) or deuteromesitylene (60%); the C—H bond breaks more readily than the C—D bond. Substitution of hydrogen for deuterium (**deuteration**) is another example of an electrophilic aromatic substitution.

5.4b REACTIVITY AND ORIENTATION IN SUBSTITUTION REACTIONS

Substituents affect the reactivity of aromatic rings towards electrophilic substitutions. Compare, for example, the relative rates of nitration of the following compounds:

$k_{rel}^{nitration}$ 24.5   1.0   0.033   0.0000001

These relative rates were measured by allowing a mixture of two compounds (say toluene and benzene) to compete for a limited amount of reagent (say, nitric acid). The rate of consumption of each reactant was then measured by analyzing the mixture from time to time for each component.

Those substituents (Cl, NO_2) which are electron-attracting withdraw electron density from the aromatic ring thus decreasing its reactivity relative to benzene. This is because they decrease the availability of the π electrons to the attacking electrophile. Conversely, those substituents (CH_3) which are electron-releasing relative to hydrogen increase the susceptibility of the aromatic ring to electrophilic attack. Substituents can therefore be classified as **ring-activating** or **ring-deactivating,** depending on whether they increase or decrease ring reactivity toward electrophiles.

Ring-Activating Substituents

Alkyl groups, OH, NH$_2$

Ring-Deactivating Substituents

Halogens, NO$_2$, SO$_3$H, CN, CCl$_3$

In addition to their effect on reactivity, substituents also control the position taken by a *second* substituent relative to themselves. If the process were statistically controlled, one would expect 40% *ortho*, 40% *meta* and 20% *para* substitution, since there are 2 *ortho* and 2 *meta* positions and only 1 *para* position.

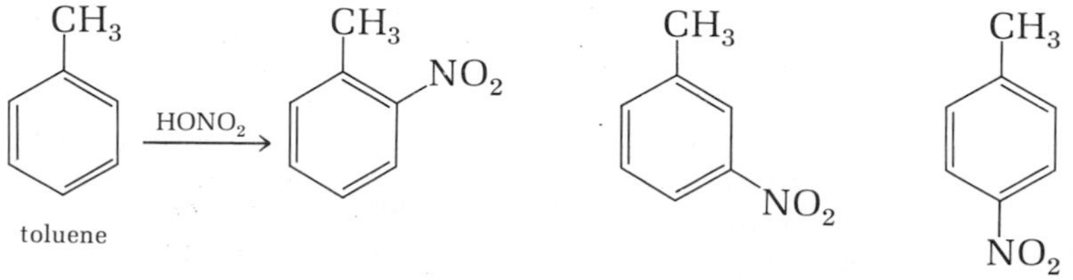

toluene			
observed	59%	4%	37%

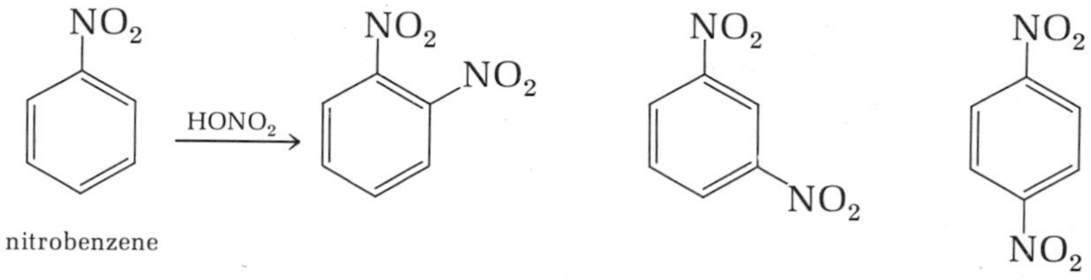

nitrobenzene			
observed	7%	93%	trace
statistical	40%	40%	20%
	ortho	*meta*	*para*

However, experiment shows that the process is *not* statistical. With toluene, 96% of the product is *ortho*- or *para*-substituted, whereas with nitrobenzene 93% of the product is *meta*-substituted. These differences can be rationalized by examining the reaction mechanism.

Consider electrophilic substitution, *ortho, meta,* or *para,* to a methyl group (in toluene). The intermediate benzenonium ions will be

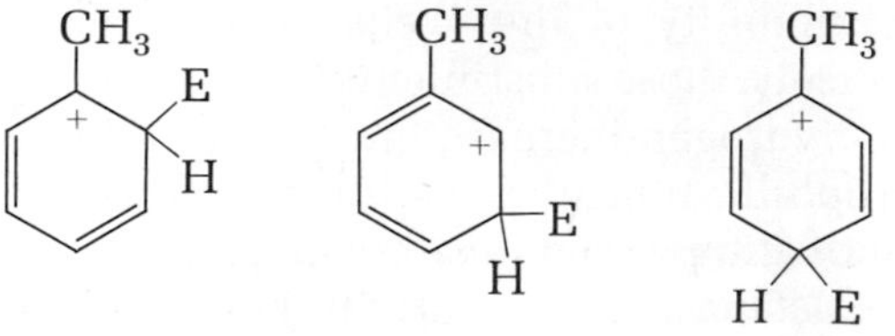

ortho   meta   para

In *ortho-* or *para*-substitution, the positive charge can be placed on the methyl-bearing ring carbon, whereas in *meta*-substitution, this is not possible. It will be recalled (sec 3-4c) that alkyl groups stabilize carbonium ions. Therefore the intermediate benzenonium ions for *o,p*-substitution will be more stable than the intermediate for *m*-substitution. As a consequence, the CH_3 group is said to be **ortho-para-directing.** This is true not only for nitration but for chlorination, sulfonation, alkylation, and any other electrophilic substitution reaction. In all of these reactions, only a small amount of *meta*-product is obtained; most of the product is either the *ortho-* or *para*-isomer.*

With nitrobenzene, the situation is different. The nitrogen atom in a nitro group carries a formal positive charge. Intermediates whose structures require that a second positive charge be placed adjacent to the nitrogen will be unfavorable. Of the three possible benzenonium ions, the one in which the electrophile attacks the meta

| ortho | meta | para |

position keeps the positive charges furthest apart and is therefore preferred. Consequently, the nitro group is said to be **meta-directing.**

In general, *ring-activating (or electron-donating) substituents are also o,p-directing, whereas ring-deactivating (or electron-withdrawing) substituents are m-directing.* The only major exceptions are the halogens. They are ring-deactivating because they are electronegative and withdraw electron density from the aromatic ring. However, they have unshared electron pairs which can stabilize an adjacent positive charge and are therefore *ortho-para*-directing.

---

*The distribution of the product between *ortho-* and *para*-isomers is a more complex problem which will not be treated here.

The importance of the orienting effect of substituents for synthesis can be seen in the following example. If benzene is first nitrated, then chlorinated, the product is almost exclusively *m*-chloronitrobenzene. If the sequence is reversed, a mixture of *o*- and *p*-chloro-

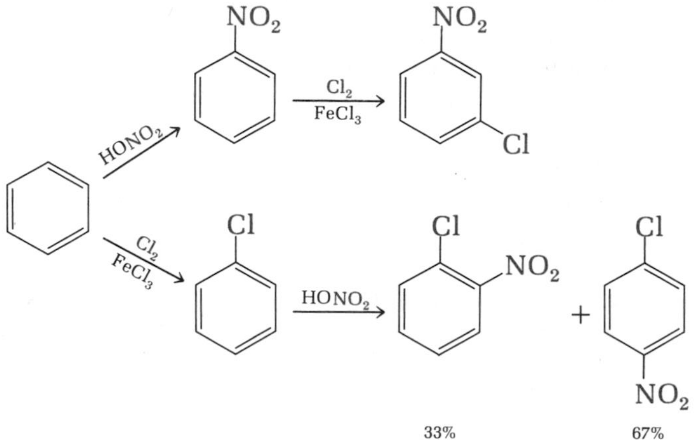

(5-16)

33%   67%

nitrobenzenes is obtained. The group which is introduced first determines the position taken by the second substituent. NO_2 is a *m*-directing group, whereas Cl is an *o,p*-directing group.

When two substituents are already present, the orienting influence of each must be considered in predicting the position to be taken by a third substituent. In general, ring-activating substituents predominate over deactivating ones, when both types are present.

5.5 OTHER REACTIONS OF AROMATIC COMPOUNDS

Many aromatic compounds undergo **catalytic hydrogenation** to cyclohexanes. This is often the best way of making alkylcyclohexanes.

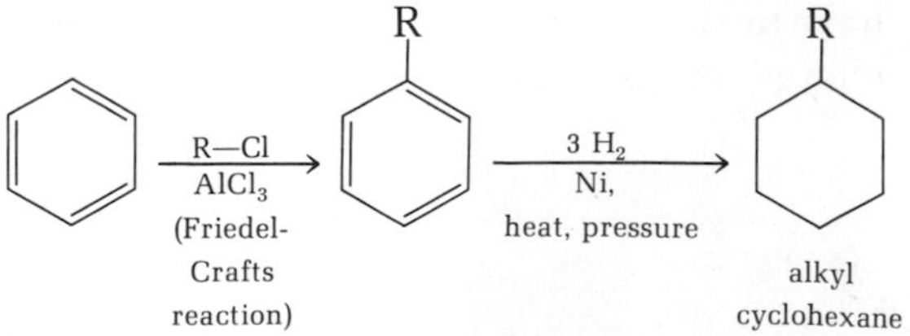

(5-17)

The reaction cannot be stopped with the uptake of one or two moles of hydrogen, since the intermediate cyclohexadienes or cyclohexenes are more easily hydrogenated than the aromatic hydrocarbon.

Alkyl **side chains** on an aromatic ring undergo **free radical substitution** (compare sec 2.7a) very readily. Examples are

$$\underset{\text{toluene}}{\text{CH}_3\text{—}} \quad \xrightarrow[\text{sunlight}]{\text{Cl}_2} \quad \text{CH}_2\text{Cl} + \text{HCl} \qquad (5\text{-}18)$$

benzyl chloride

$$\text{CH}_2\text{CH}_3 \quad \xrightarrow[\text{sunlight}]{\text{Br}_2} \quad \text{CHBrCH}_3 + \text{HBr} \qquad (5\text{-}19)$$

1-bromo-1-phenylethane

The reason for almost exclusive substitution on the carbon adjacent to the aromatic ring in equation 5-19 is that the intermediate benzyl-type free radical can be stabilized by resonance.

$$\cdot\text{CHCH}_3 \quad \longleftrightarrow \quad \text{CHCH}_3 \quad \longleftrightarrow \quad \text{CHCH}_3 \qquad\qquad \text{CH}_2\text{—CH}_2\cdot$$

resonance stabilization by the aromatic ring   the odd electron
cannot be delocalized

The behavior of alkylbenzenes toward strong oxidizing agents illustrates strikingly the remarkable stability of the aromatic ring. Toluene is converted to benzoic acid almost quantitatively, when heated with alkaline potassium permanganate.

$$\text{—CH}_3 \quad \xrightarrow[\text{heat}]{\text{KMnO}_4} \quad \text{—}\overset{\text{O}}{\underset{}{\text{C}}}\text{—OH} \qquad (5\text{-}20)$$

benzoic acid

Despite the customary ease with which unsaturated compounds are attacked by permanganate, and despite the stability of alkanes toward the same reagent, *the aliphatic side chain is oxidized and the aromatic ring is left intact.* Rings with longer side chains or with more than one alkyl group are similarly oxidized.

$$\text{—CH}_2\text{CH}_3 \quad \longrightarrow \quad \text{—}\overset{\text{O}}{\underset{}{\text{C}}}\text{—OH} + \text{CO}_2 \qquad (5\text{-}21)$$

$$\text{CH}_3\text{—}\text{—CH}_3 \quad \longrightarrow \quad \text{HO—}\overset{\text{O}}{\underset{}{\text{C}}}\text{—}\text{—}\overset{\text{O}}{\underset{}{\text{C}}}\text{—OH} \qquad (5\text{-}22)$$

p-xylene   terephthalic acid

A variety of oxidants may be used. Terephthalic acid, one of the two raw materials for the synthesis of Dacron and other polyesters (sec 10.6d), is manufactured in large quantities by the air oxidation of *p*-xylene in the presence of certain metal ion catalysts (eq 5-22).

5.6 SIMPLE BENZENE DERIVATIVES

Several well-known substances are synthesized directly from benzene or toluene by electrophilic aromatic substitution. **Trinitrotoluene (TNT),** the important explosive, is manufactured by nitration of toluene, usually in several steps.

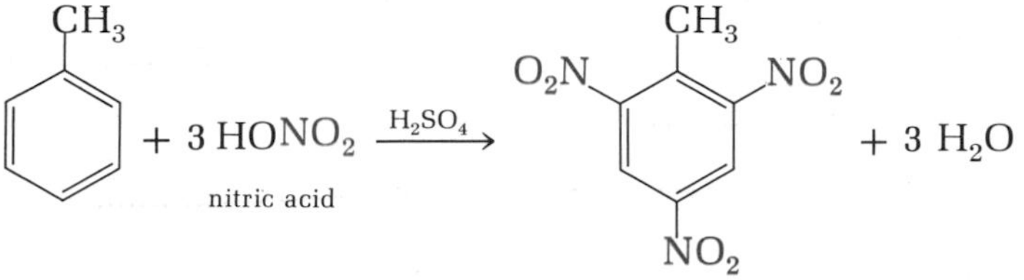

$$\text{(benzene ring with }CH_3) + 3\ HONO_2 \xrightarrow{H_2SO_4} \text{(TNT structure: }O_2N,\ CH_3,\ NO_2,\ NO_2) + 3\ H_2O \qquad (5\text{-}23)$$

nitric acid

2,4,6-trinitrotoluene (TNT)

Dichlorination of benzene gives ***p*-dichlorobenzene,** a crystalline substance used as a moth repellant.

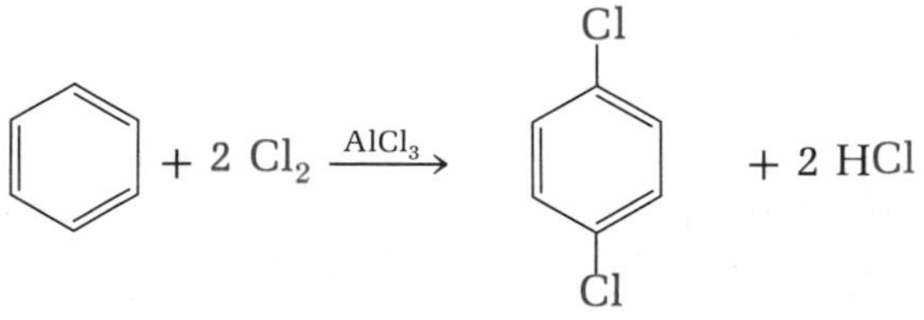

$$\text{(benzene)} + 2\ Cl_2 \xrightarrow{AlCl_3} \text{(Cl}\text{—benzene—}Cl) + 2\ HCl \qquad (5\text{-}24)$$

p-dichlorobenzene

The ortho isomer, obtained simultaneously, is useful as a special solvent.

Styrene is produced from benzene and ethylene. The first step is a Friedel-Crafts type alkylation, to give ethylbenzene.

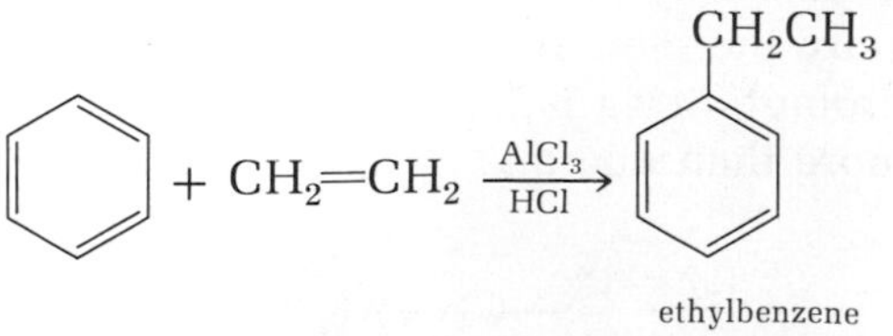

$$\text{(benzene)} + CH_2{=}CH_2 \xrightarrow[HCl]{AlCl_3} \text{(benzene—}CH_2CH_3) \qquad (5\text{-}25)$$

ethylbenzene

Addition of a proton (from HCl) to ethylene gives the ethyl cation which is the electrophile that attacks the benzene ring. It is quite common to use an alkene in place of an alkyl halide in the Friedel-Crafts reaction.

The ethylbenzene is catalytically dehydrogenated to styrene.

$$\overset{CH_2CH_3}{\underset{}{\bigcirc}} \quad \xrightarrow[650°]{Fe_2O_3 \text{ or ZnO}} \quad \overset{CH{=}CH_2}{\underset{\text{styrene}}{\bigcirc}} \quad + H_2 \qquad (5\text{-}26)$$

Styrene, a liquid which boils at 145°, can be readily polymerized (review sec 3.4d). Polystyrene is chemically inert, light weight,

$$n \quad \overset{CH{=}CH_2}{\underset{\text{styrene}}{\bigcirc}} \quad \xrightarrow[\text{heat}]{\text{peroxide catalyst}} \quad \left[\overset{CH{-}CH_2}{\underset{}{\bigcirc}} \right]_n \qquad (5\text{-}27)$$

strong, transparent, and useful as a structural material and insulator (Figure 5-5). The most important synthetic rubbers are co-polymers of styrene and butadiene.

Figure 5-5
The structure below, 80 feet in diameter, is a dome-shaped ceiling roof for a theater-convention hall in Traverse City, Michigan. Styrofoam insulation is used as the sole structural component in the roof, a process that was developed by Dow Chemical Company. (Courtesy of Dow Chemical Company.)

5.7 ADDITIONAL AROMATIC COMPOUNDS

Naphthalene, anthracene, and phenanthrene are aromatic hydro-carbons with "fused" benzene rings—rings which share two common carbon atoms. These hydrocarbons, all of which can be obtained

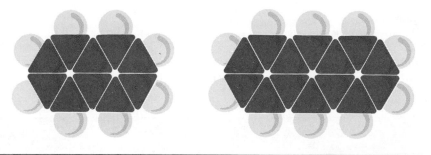

naphthalene   anthracene   phenanthrene

from coal tar, are resonance-stabilized aromatic substances which undergo electrophilic substitution reactions.

Figure 5-6
Molecular models of naphthalene (left) and anthracene.

It was quite natural for chemists to inquire whether the lower and higher homologs of benzene, cyclobutadiene and cyclooctatetraene, were also aromatic (that is, peculiarly stable, resistant to oxidation, and capable of undergoing substitution rather than addition reac-

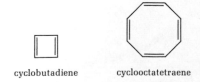

cyclobutadiene   cyclooctatetraene

tions). Cyclobutadiene proved extraordinarily difficult to prepare and extremely reactive—certainly not aromatic. Cyclooctatetraene was easier to synthesize, but turned out to be tub-shaped rather than planar so that overlap of all eight p-orbitals was not possible. It

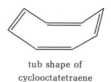

tub shape of
cyclooctatetraene

readily adds four moles of bromine, is easily oxidized by permanganate, and behaves like a tetra-ene, not an aromatic compound.

Some non-benzenoid aromatic compounds are known, however. Molecular orbital theory predicts that planar cyclic conjugated systems with 2, 6, 10, 14 . . . π electrons [$(4n + 2)\pi$ electrons, where n is an integer] will be aromatic, whereas those with 4, 8, 12 . . . will not. In accord with theory, the cyclopentadiene anion (6 π electrons) is particularly stable, whereas the corresponding cation (4 π electrons) is not. However in the seven-membered ring analogs, the cation (6 π electrons) is stable, but the anion (8 π electrons) is not.

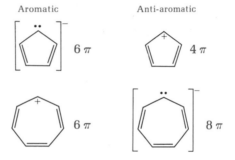

The extension of the aromaticity concept to non-benzenoid compounds is a recent development in organic chemistry, and has proved to be one of the more spectacular successes of molecular orbital theory.

New Concepts, Facts, and Terms

1. Structure of benzene; Kekulé formula; resonance hybrid. Aromatic C—C bond distance is 1.40 Å.
2. Bonding in benzene; σ-framework with π electrons above and below the ring plane
3. Resonance energy; the difference between the actual energy of the resonance hybrid and the calculated energy of the most stable contributing structure
4. Nomenclature; phenyl and benzyl groups; *ortho, meta, para*
5. Electrophilic aromatic substitutions; halogenation, nitration, sulfonation, alkylation (Friedel-Crafts reaction), deuteration
6. Benzenonium ion; halonium ion, nitronium ion as electrophiles

7. Ring-activating, *ortho-para*-directing groups: alkyl, OH, NH$_2$. Ring-deactivating, *meta*-directing groups: NO$_2$, SO$_3$H, CN, CCl$_3$. Ring-deactivating, *ortho-para*-directing groups: the halogens (F, Cl, Br, I).
8. Other reactions of aromatic compounds:
    a. catalytic hydrogenation ⟶ cyclohexanes
    b. side-chain oxidation ⟶ aromatic acids
    c. side-chain halogenation: a free-radical chain reaction
9. TNT, *p*-dichlorobenzene, styrene, polystyrene
10. Naphthalene, anthracene, phenanthrene
11. Non-benzenoid aromatic compounds

Exercises and Problems

1. Write structural formulas for the following compounds:
    a. 1,3,5-tribromobenzene   f. 2,3-diphenylbutane
    b. *m*-chlorotoluene   g. *p*-bromostyrene
    c. *o*-diethylbenzene   h. 4-ethyl-2-chloro-3,5-dinitrotoluene
    d. isopropylbenzene   i. *m*-chlorobenzenesulfonic acid
    e. benzyl bromide   j. *p*-bromobenzoic acid

2. Name the following compounds:

a. 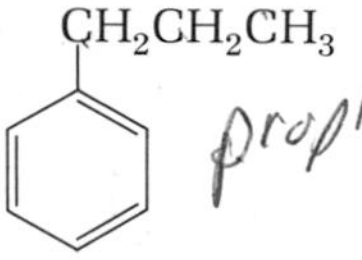 CH$_2$CH$_2$CH$_3$ *propylbenzene*

e. 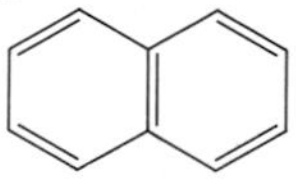 *naphthalene*

b. 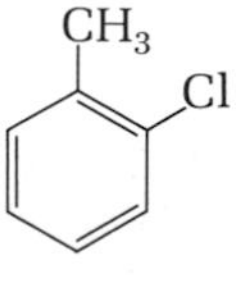 CH$_3$, Cl *ortho Chlorotoluene*

f. 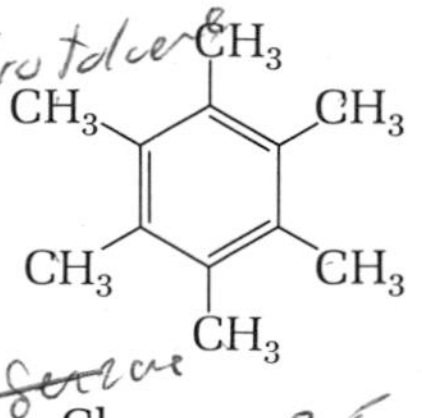 CH$_3$ ... CH$_3$ *prepare this for z*

c. 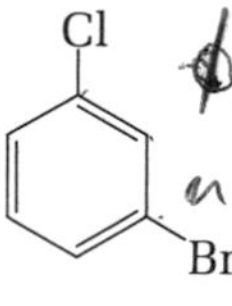 Cl, Br *Bromo 3 chloro benzene* / *m Bromo Chloro benzene*

g. 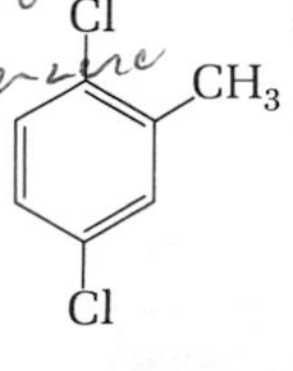 Cl, CH$_3$, Cl *2,5 Dichloro toluene*

d. 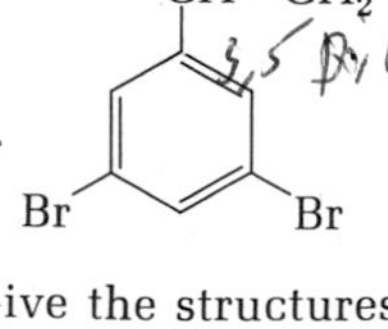 CH=CH$_2$, Br, Br *3,5 Dibromo-styrene*

h. 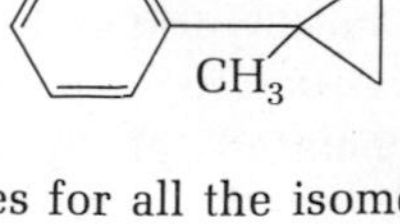 CH$_3$

3. Give the structures and names for all the isomers of the
    a. trimethylbenzenes
    b. dichloronitrobenzenes

4. The following compounds are isomers of benzene (C_6H_6). Tell how each could be distinguished from benzene by a simple chemical test.

a. $CH_2{=}CH{-}CH{=}CH{-}C{\equiv}CH$ *addition to Br₂/Pt*

b.

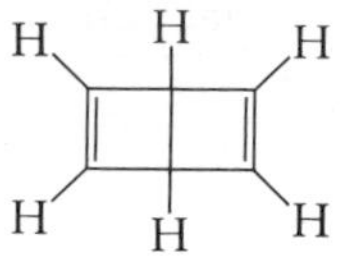

c.

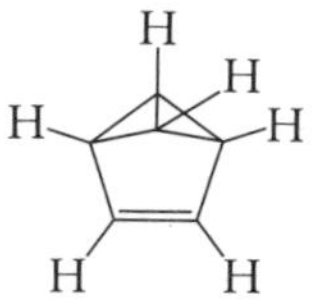

d.

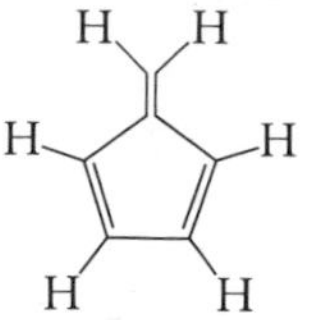

e.

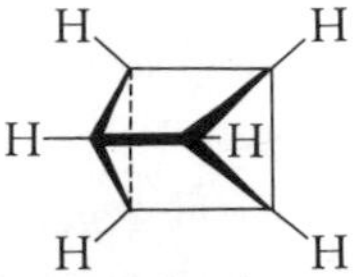

5. The observed amount of heat evolved when 1,3,5,7-cyclooctatetraene is hydrogenated to cyclooctane is 110.0 kcal/mole. Does the compound have any significant resonance energy? Explain.

6. There are three isomeric xylenes (dimethylbenzenes). Each is treated with bromine (one mole) and $FeBr_3$, and the monobrominated products are isolated. One of the xylenes gives a single bromoxylene, another gives two bromoxylenes, and the third isomer gives three. Deduce the structures of each xylene and write equations for its bromination.

7. Give the structures and names of the following aromatic hydrocarbons.
   a. C_8H_{10}; has three possible monobromo derivatives
   b. C_9H_{12}; has one possible mononitro derivative
   c. C_9H_{12}; has four possible mononitro derivatives

8. Using structural formulas, illustrate with equations the halogenation, nitration, and sulfonation of benzene and toluene, introducing only a single substituent in each case.

9. Write out each step in the mechanism of the following reactions:
   a. benzene + ethyl bromide + $AlBr_3 \longrightarrow$ ethylbenzene
   b. toluene + nitric acid + sulfuric acid $\longrightarrow$ p-nitrotoluene
   c. toluene + chlorine + uv light $\longrightarrow$ benzyl chloride

10. Indicate the main monosubstitution products in each of the following reactions, keeping in mind that certain substituents are *meta*-directing and others are *ortho-para*-directing.
   a. toluene + chlorine (Fe catalyst)
   b. nitrobenzene + concentrated sulfuric acid (heat)
   c. bromobenzene + chlorine (Fe catalyst)
   d. benzenesulfonic acid + concentrated nitric acid (heat)
   e. chlorobenzene + bromine (Fe catalyst)
   f. ethylbenzene + bromine (Fe catalyst)
   g. nitrobenzene + concentrated nitric acid (heat)
   h. iodobenzene + bromine (Fe catalyst)

11. Explain why, in equation 5-15, the D^+ becomes attached to a hydrogen-bearing carbon rather than one of the methyl-bearing carbons, in the intermediate benzenonium ion.

12. Three isomeric tribromobenzenes are known, with melting points at 44°, 87°, and 119°. When each is nitrated (under conditions which introduce a single nitro group), they give, respectively, three, two, and one mono-nitro-tribromobenzenes. Assign structures to the tribromobenzenes and their nitration products.

13. Draw all possible contributing structures to the intermediate benzenonium ions when toluene is brominated (*o*, *m* and *p*). Explain why *o,p* products predominate. Repeat for the bromination of nitrobenzene, and explain why the *meta* product predominates.

14. Using benzene or toluene as the only organic starting materials, devise syntheses for each of the following:
    a. *m*-bromonitrobenzene
    b. *p*-toluenesulfonic acid
    c. *p*-nitroethylbenzene
    d. methylcyclohexane
    e. *p*-bromobenzoic acid
    f. styrene
    g. *p*-bromonitrobenzene
    h. 2-chloro-4-nitrotoluene
    i. 3,5-dinitrochlorobenzene
    j. 1-methyl-4-*t*-butylcyclohexane

15. When propylene is substituted for ethylene in equation 5-25, the product is isopropylbenzene, *not* n-propylbenzene. Explain.

16. The structure of the nitro group, $-NO_2$, is usually shown as

Yet experiments show that the two N—O bonds have the same length, 1.21 Å, intermediate between 1.36 Å for the N—O single bond and 1.18 Å for the N=O double bond. Draw structural formulas which explain this observation.

17. a. An acid $C_7H_6O_2$ results from the oxidation of an aromatic hydrocarbon whose formula is C_9H_{12}. Suggest two possible structures for the original hydrocarbon.
    b. A hydrocarbon with the molecular formula C_9H_{12} yields o-phthalic acid as the only organic product of oxidation. What is the structural formula of the hydrocarbon?

18. Draw a mechanism for the free radical polymerization of styrene (eq 5-27).

19. There are two possible monobromonaphthalenes. Draw all resonance contributors to the benzenonium ion-type intermediates for the formation of each isomer, preserving in each case the benzenoid ring which is not attacked by the electrophile. Predict which bromonaphthalene will be produced in larger yield.

20. Draw all the "Kekulé" structures of anthracene. Are any of the three rings not aromatic in these structures? Anthracene, on reduction, readily forms a dihydro derivative, after which further hydrogenation becomes more difficult. What is the structure of the dihydroanthracene, and why is further reduction more difficult?

CHAPTER SIX

ALCOHOLS AND PHENOLS

Alcohols and phenols contain the hydroxyl unit as their functional group. Ethyl alcohol has been known since antiquity, although its chemical structure was not understood until the nineteenth century. It was first produced by the fermentation of fruits. All possible alcohols with four carbons or less are commercially available, as are many selected higher alcohols. Phenols are important industrial chemicals, used as antiseptics, and in the synthesis of medicinals, plastics, and dyestuffs.

Alcohols may be thought of as being derived from saturated or unsaturated hydrocarbons by replacing a hydrogen atom by a hydroxyl group (—OH). If a hydrogen atom attached to an aromatic ring is replaced by a hydroxyl group, the compound is known as a **phenol.** The chemistry of alcohols and phenols, then, involves the

properties of a hydroxyl group attached to carbon (—$\overset{\displaystyle |}{\underset{\displaystyle |}{C}}$—OH).

Alcohols and phenols also may be considered as organic analogs of water in which one of the hydrogen atoms has been replaced by an organic group.

H—OH   R—OH   Ar—OH

water   alcohol   phenol

It will not be too surprising, therefore, to find some similarity in the chemical behavior of water and its organic counterparts.

6.1 NOMENCLATURE AND CLASSIFICATION

For the lower members of the alcohol series, common names, the alkyl group names followed by *alcohol,* are often used.

CH_3OH methyl alcohol $CH_3CH_2CH_2OH$ n-propyl alcohol

CH_3CH_2OH ethyl alcohol $CH_3\underset{\displaystyle OH}{\overset{\displaystyle |}{C}}HCH_3$ isopropyl alcohol

The four four-carbon alcohols (derived from n- and isobutane) are also generally known by their common names.

$CH_3CH_2CH_2CH_2OH$ $CH_3—\underset{\displaystyle CH_3}{\overset{\displaystyle |}{C}}H—CH_2OH$

normal butyl alcohol   isobutyl alcohol

$CH_3CH_2\underset{\displaystyle OH}{\overset{\displaystyle |}{C}}HCH_3$ $CH_3—\underset{\displaystyle CH_3}{\overset{\displaystyle |}{\underset{\displaystyle |}{C}}}—OH$

secondary butyl alcohol   tertiary butyl alcohol

The IUPAC system is used for naming more complex alcohols. The longest chain of carbon atoms which includes the hydroxyl group is the basis for the name, and the ending **-ol** is used to indicate the alcohol function. The position of the hydroxyl group is indicated by the number of the carbon atom to which it is attached.

$$CH_3CH_2OH$$

ethanol

(ethyl alcohol)

$$\overset{1}{C}H_2\overset{2}{C}H_2\overset{3}{C}H_2\overset{4}{C}H_2$$
$$\;\;|\quad\quad\quad\quad\;\;|$$
$$OH\quad\quad\quad OH$$

1,4-butanediol

$$\overset{1}{}\;\;\overset{2}{\;}-\overset{}{CHCH_3}$$
$$|$$
$$OH$$

1-phenylethanol

$$\overset{}{CH_3}$$
$$\overset{1}{C}H_3-\overset{2}{\underset{|}{\overset{|}{C}}}-\overset{3}{C}H_3$$
$$OH$$

2-methyl-2-propanol

(*t*-butyl alcohol)

$$\overset{5}{C}H_3\overset{4}{C}H=\overset{3}{C}H-\overset{2}{\underset{|}{C}H}-\overset{1}{C}H_3$$
$$OH$$

3-pentene-2-ol

Figure 6-1

Models of methyl alcohol (left) and isopropyl alcohol.

The chemistry of alcohols sometimes depends upon whether the hydroxyl group is attached to a carbon that is bound to one, two, or three other carbon atoms. The compounds are known, respectively, as *primary, secondary,* or *tertiary* alcohols.

$$\begin{array}{ccc} H & H & R \\ | & | & | \\ R-C-OH & R-C-OH & R-C-OH \\ | & | & | \\ H & R & R \end{array}$$

primary   secondary   tertiary

Some specific examples of these three classifications are

$$CH_3CH_2-OH \qquad CH_3CH_2\overset{\overset{\displaystyle CH_3}{|}}{CH}-OH \qquad CH_3CH_2\overset{\overset{\displaystyle CH_3}{|}}{\underset{\underset{\displaystyle CH_3}{|}}{C}}-OH$$

primary   secondary   tertiary

Phenols are usually named as derivatives of the parent compound

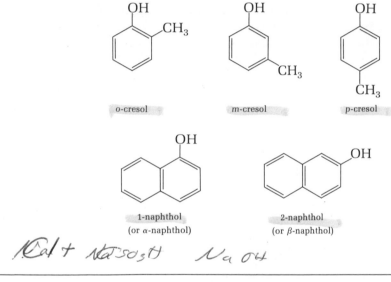

phenol   p-chlorophenol   2,4,6-tribromophenol

or by common names. Some of the more important of these are indicated below.

o-cresol   m-cresol   p-cresol

1-naphthol   2-naphthol
(or α-naphthol)   (or β-naphthol)

Cal + NaSO₃H   Na OH

Figure 6-2
Scale model of phenol.

6.2   **PHYSICAL PROPERTIES**

The physical properties of some of the lower members of the alcohol series are given in Table 6-1. The similarity between the lower

C—C—C—C —x   pain
         |
         x

C—C—C tert
   |
   C

Table 6-1
Physical Properties of Normal Primary Alcohols

C—C—C—C sec
       |
       x

C
‖
C—C—C—x iso

NAME	FORMULA	BP, °C	SOLUBILITY IN H₂O, G./100 G. AT 20°
Methyl	CH_3OH	65	Completely miscible
Ethyl	CH_3CH_2OH	78.5	Completely miscible
n-Propyl	$CH_3CH_2CH_2OH$	97	Completely miscible
n-Butyl	$CH_3CH_2CH_2CH_2OH$	117.7	7.9
n-Amyl	$CH_3CH_2CH_2CH_2CH_2OH$	137.9	2.7
n-Hexyl	$CH_3CH_2CH_2CH_2CH_2CH_2OH$	155.8	0.59

members of the series and water is indicated by their miscibility with water in all proportions. This is true when the carbon chain is short; that is, when the hydroxyl group comprises a significant percentage by weight of the alcohol molecule. As the size of the alkyl group is increased, the molecule gradually becomes more like a hydrocarbon and less like water. The solubility in water decreases with increasing chain length. The solubility of the lower alcohols in water and the relatively high boiling points of alcohols (compared to hydrocarbons of equal molecular weight) are primarily a consequence of hydrogen bonding, a phenomenon which is discussed more fully in section 6.3.

Phenol, a colorless crystalline low-melting solid with a medicinal odor, is moderately soluble in cold water. It causes burns when in contact with the skin and must be handled with care. In dilute solutions, it is used as an antiseptic and disinfectant. Hexylresorcinol is an effective antiseptic used in mouth washes and throat lozenges (Sucrets).

hexylresorcinol

6.3 THE HYDROGEN BOND

Ethanol, CH_3CH_2OH, boils at 78°, whereas its isomer, the gas dimethyl ether, $CH_3—O—CH_3$, boils at −24°, and propane, $CH_3CH_2CH_3$, with nearly the same molecular weight, boils at an even lower temperature (−42°). Indeed the boiling point of ethanol is a little above that of hexane, which has nearly twice the molecular weight. Thus alcohols behave as if their molecular weights were appreciably higher than their formulas indicate. This peculiarity is related to the presence of the hydroxyl group in alcohols.

When a hydrogen atom is covalently bound to a strongly electronegative element (O, N, F) the σ bond is distorted toward that atom, giving the hydrogen a partial positive charge.

$$\overset{\delta+\ \ \delta-}{H-F} \qquad \overset{\delta+\ \ \delta-}{H-O} \qquad \overset{\delta+\ \ \delta-}{H-N}$$

This polarization causes an electrostatic attraction between molecules. Because of the small size of the hydrogen atom, close intermolecular approach is possible, and the **hydrogen bond** (represented by dots) between two electronegative atoms can be quite strong.

$$\overset{R}{\underset{\delta-\ \ \ \delta+}{O-H}} \cdots \overset{R}{\underset{\delta-\ \ \ \delta+}{O-H}} \cdots \overset{R}{\underset{\delta-\ \ \ \delta+}{O-H}}$$

The energy required to break a hydrogen bond varies with the particular electronegative atom and the group(s) attached to it. The strength of such bonds is only about ten per cent of that of most covalent bonds, but is still appreciable (five to ten kcal/mole).

The reason for the anomalously high boiling point of alcohols, then, is that energy is required not only to vaporize each molecule, but to break the hydrogen bonds between them. Because of intermolecular association, hydrogen-bonded liquids have effective molecular weights appreciably higher than their formula weights.

Lower alcohols can readily replace water molecules in the hydrogen-bonded network. This accounts for the complete miscibility or high solubility of alcohols in water.

$$\begin{array}{c} H-O \\ H-O \diagup \diagdown H \\ H-O \diagup \diagdown R \\ \diagdown H \end{array}$$

The infrared spectra of alcohols provide direct experimental evidence for hydrogen bonding (sec 13.3a).

Hydrogen bonds are exceedingly important in connection with the structures of proteins and nucleic acids, as we shall see later (Chapters 17 and 18).

6.4 ACIDITY AND BASICITY OF ALCOHOLS AND PHENOLS

Like water, alcohols are amphoteric. The unshared electrons on oxygen can accept a proton from a stronger acid.

$$H-\overset{..}{\underset{..}{O}}-H + H^+ \rightleftharpoons \left[H-\overset{H}{\underset{..}{\overset{|}{O}}}-H \right]^+ \qquad (6\text{-}1)$$

oxonium ion

$$R-\overset{..}{\underset{..}{O}}-H + H^+ \rightleftharpoons \left[R-\overset{H}{\underset{..}{\overset{|}{O}}}-H \right]^+ \qquad (6\text{-}2)$$

alcohol acting   alkyloxonium ion
as a base

This is the first step in the acid-catalyzed dehydration of alcohols to alkenes (sec 3.5a) and is also important in the conversion of alcohols to ethers (sec 7.2).

Alcohols may also serve as proton sources (acids). When treated with reactive metals such as sodium, hydrogen gas is liberated. The reaction is completely analogous to that of water.

$$2\, HO-H + 2\, Na \longrightarrow 2\, Na^+\, OH^- + H_2 \qquad (6\text{-}3)$$

sodium hydroxide

$$2\, RO-H + 2\, Na \longrightarrow 2\, Na^+\, OR^- + H_2 \qquad (6\text{-}4)$$

sodium alkoxide

The product is a **metal alkoxide** which, in alcohol solution, is a strong base analogous to sodium hydroxide in water. Specific alkoxides are named by using the root of the alkyl group.

$$2\, CH_3O-H + 2\, Na \longrightarrow 2\, Na^+\, OCH_3^- + H_2 \qquad (6\text{-}5)$$

methanol   sodium methoxide

Alkoxides are useful as basic catalysts for organic reactions.
Autoprotolysis occurs to a somewhat lesser extent with alcohols than with water.

$$HOH + HOH \rightleftharpoons \left[\overset{H}{\underset{}{\overset{|}{HOH}}} \right]^+ + OH^- \qquad K = 1 \times 10^{-14} \quad (6\text{-}6)$$

$$ROH + ROH \rightleftharpoons \left[\overset{H}{\underset{}{\overset{|}{ROH}}} \right]^+ + OR^- \qquad K \cong 1 \times 10^{-16} \quad (6\text{-}7)$$

Thus alcohols are weaker acids than water, or alkoxides are stronger bases than hydroxides. Alkoxides are therefore hydrolyzed, if dissolved in an excess of water.

$$Na^+OR^- + H{-}OH \longrightarrow ROH + Na^+OH^- \qquad (6\text{-}8)$$

The reaction shown in equation 6-8 proceeds largely to the right. Stated another way, alkoxides are not ordinarily prepared in the laboratory from alcohols and sodium hydroxide.

Phenols are much stronger acids than alcohols or water. The principal reason is that the negative charge in the phenoxide ion can be delocalized over the aromatic ring through resonance, whereas in alkoxides or hydroxide, the negative charge is confined to the oxygen atom (Figure 6-3).

Figure 6-3
The negative charge in phenoxide ion can be delocalized from the oxygen to the ortho *and* para *positions of the aromatic ring, whereas it is confined to the oxygen in the alkoxide or hydroxide ion.*

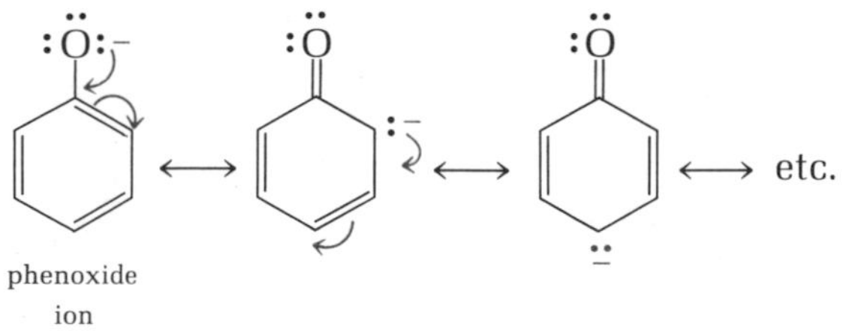

phenoxide
ion

alkoxide
ion

hydroxide
ion

The approximate ionization constants of ethanol, water, and phenol are 10^{-20}, 10^{-14}, and 10^{-10} respectively; phenol is 10,000 (10^4) times as strong an acid as water. Conversely, phenoxide ion is a much weaker base than hydroxide ion. For this reason, phenols can be converted to phenoxides by treatment with metal hydroxides.

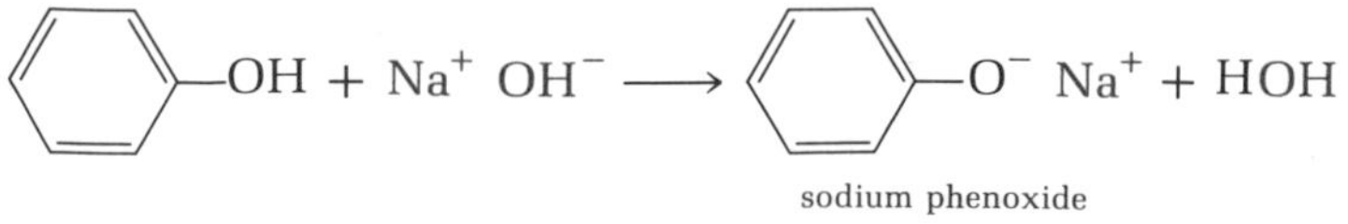

$$\text{C}_6\text{H}_5{-}OH + Na^+\ OH^- \longrightarrow \text{C}_6\text{H}_5{-}O^-\ Na^+ + HOH \qquad (6\text{-}9)$$

sodium phenoxide

Phenols can therefore be extracted from organic mixtures by aqueous base. The free phenol can then be obtained by acidifying the basic extract with a strong mineral acid:

$$\text{C}_6\text{H}_5\!-\!\text{O}^-\,\text{Na}^+ + \text{H}^+\text{Cl}^- \longrightarrow \text{C}_6\text{H}_5\!-\!\text{OH} + \text{Na}^+\,\text{Cl}^- \quad (6\text{-}10)$$

The acidity of phenols can be enhanced by the introduction of electron-withdrawing groups in the aromatic ring. Picric acid (2,4,6-trinitrophenol) is comparable in strength to hydrochloric acid.

picric acid

Ammonium picrate is a strong explosive used in the warheads of naval shells.

6.5 PREPARATION OF ALCOHOLS

Three general methods for preparing alcohols will be described.* The precursors are alkenes or alkyl halides.

6.5a HYDRATION OF ALKENES

Hydration of a double bond can be accomplished by addition of cold concentrated sulfuric acid to an alkene and hydrolysis of the resulting alkyl hydrogen sulfate.

$$\text{H}\!-\!\text{C}\!=\!\text{C}\!-\!\text{H} + \text{H}^{+-}\text{OSO}_3\text{H} \xrightarrow{\text{cold}} \text{H}\!-\!\text{C}\!-\!\text{C}\!-\!\text{OSO}_3\text{H} \quad (6\text{-}11)$$

ethylene   sulfuric acid   ethyl hydrogen sulfate

$$\text{CH}_3\text{CH}_2\text{OSO}_3\text{H} + \text{H}\!-\!\text{OH} \longrightarrow \text{CH}_3\text{CH}_2\text{OH} + \text{HOSO}_3\text{H} \quad (6\text{-}12)$$

steam   ethyl alcohol

*For another general laboratory method, see section 9.6f. In addition, some special industrial methods are employed for lower members of the series.

The reactions are the inverse of the dehydration of an alcohol to an alkene (sec 3.5a). The net result of the sequence is the addition of water to a double bond in accord with Markownikoff's rule. This method is used commercially for the manufacture of ethyl, isopropyl, secondary butyl, and tertiary butyl alcohols. The alkenes are readily available from petroleum sources, and the sulfuric acid is recovered in the process.

6.5b HYDROBORATION

Diborane (B_2H_6), generated from sodium borohydride ($NaBH_4$) and boron trifluoride, reacts with alkenes to form trialkylboranes. These may be oxidized with hydrogen peroxide to trialkylborates which, on alkaline hydrolysis, give the alcohol.

$$6\ RCH{=}CH_2 + (BH_3)_2 \longrightarrow 2\ (\overset{\overset{\displaystyle H}{|}}{R}CH{-}CH_2)_3B \qquad (6\text{-}13)$$

$$(RCH_2CH_2)_3B \xrightarrow{H_2O_2} (RCH_2CH_2O)_3B \xrightarrow[3\ HOH]{OH^-}$$
$$3\ RCH_2CH_2OH + H_3BO_3 \qquad (6\text{-}14)$$

The overall result of equations 6-13 and 6-14 is the addition of water to the alkene, *but in the reverse direction prescribed by Markownikoff's rule.* This is because the diborane adds in the sense of hydride ion (H^-) and positive boron.

$$\begin{array}{ccc} R{-}CH{=}CH_2 & & RCH{-}CH_2 \\ \curvearrowright & \longrightarrow & |\quad\ \ | \\ H{-}BH_2 & & H\quad\ BH_2 \end{array} \qquad (6\text{-}15)$$

This method, therefore, complements the more conventional hydration of alkenes with sulfuric acid (sec 6.5a).

6.5c HYDROLYSIS OF ALKYL HALIDES: NUCLEOPHILIC
DISPLACEMENTS

Alkyl halides can be hydrolyzed to the corresponding alcohols. The reaction conditions are determined by the structure of the particular halide.

Primary alcohols are prepared by heating primary halides with aqueous base.

$$RCH_2X + OH^- \xrightarrow{heat} RCH_2OH + X^- \qquad (6\text{-}16)$$
$$\text{(X=Cl, Br, I)}$$

The reaction rate depends upon the concentration of both reactants (alkyl halide and hydroxide ion) and the rate increases in direct proportion to the base concentration. The hydroxide ion is said to *displace* the bromide ion in an S_N2 **nucleophilic displacement.** (S = substitution; N = nucleophilic—hydroxide ion seeks a nucleus to which it can donate an electron pair; and 2 = bimolecular—two molecular species are involved in the rate-determining step).

It can be shown that the hydroxide ion (nucleophile) approaches the sp^3 carbon from the side opposite that to which the halogen is attached. As the halogen departs, the bonds to carbon invert, much like an umbrella in a strong wind.

$$HO^- + \underset{\underset{R}{|}}{\overset{\overset{H}{|}}{H-C-X}} \xrightarrow[\text{determining}]{\text{rate-}} \left[HO\cdots\overset{\overset{H}{|}}{\underset{\underset{R}{|}}{\overset{\delta-}{C}}}\cdots X \right] \longrightarrow HO-\overset{\overset{H}{|}}{\underset{\underset{R}{|}}{C}}H + X^-$$

$$\text{primary alkyl} \qquad\qquad \text{transition state} \qquad\qquad\qquad (6\text{-}17)$$
$$\text{halide}$$

The reaction rate increases in the order $X = Cl < Br < I$.

S_N2 displacements are common in organic chemistry, and we shall see many examples with nucleophiles other than hydroxide ion, and with other "leaving groups" besides the halogens.

The hydrolysis of tertiary halides illustrates an alternate mechanism for displacement reactions. One might anticipate that if the hydrogens on the halogen-bearing carbon were replaced by alkyl groups, approach of the hydroxide ion to the backside of the C—X bond would become hindered. Hydrolysis by the S_N2 mechanism should therefore decrease sharply in rate in the order primary > secondary > tertiary.

HO$^=$ - - - - -

approach to the back side of the
C—X bond is sterically hindered
in a tertiary halide

In contrast to expectation, one finds that tertiary halides hydrolyze rapidly to alcohols, *but that the reaction rate is independent of hydroxide ion concentration.* The rate-determining step, therefore, cannot involve hydroxide ion. The mechanism involves a two-step process (illustrated for *t*-butyl bromide in eq 6-18), the rate-determining step being ionization to a carbonium ion. The intermediate carbonium ion, in a second step, reacts rapidly with the solvent.

$$\underset{\substack{\text{\textit{t}-butyl bromide}}}{\overset{\displaystyle CH_3}{\underset{\displaystyle CH_3}{CH_3\text{-}\overset{\displaystyle |}{\underset{\displaystyle |}{C}}\text{-}Br}}} \xrightarrow[\text{slow step}]{} \underset{\substack{\text{\textit{t}-butyl carbonium ion}\\ \text{(planar)}}}{\overset{\displaystyle CH_3}{\underset{\displaystyle CH_3\quad CH_3}{C^+}}} + Br^- \qquad (6\text{-}18a)$$

$$\underset{\substack{CH_3\quad CH_3}}{\overset{\displaystyle CH_3}{C^+}} + :\overset{..}{\underset{..}{O}}H \xrightarrow{\text{fast}} \left[(CH_3)_3C:\overset{..}{\underset{..}{O}}:H\right]^+ \longrightarrow (CH_3)_3COH + H^+$$

$$\underset{\text{\textit{t}-butyloxonium ion}}{} \qquad\qquad (6\text{-}18b)$$

This mechanism is called an $\mathbf{S_N1}$ **nucleophilic displacement**, the one referring to the unimolecular nature of the rate-determining step. Halides which can produce a stable carbonium ion react by this mechanism. The rate by this mechanism therefore increases sharply in the order primary $\ll$ secondary $<$ tertiary (review sec 3.4c).

Secondary alcohols may be produced from secondary halides by either mechanism; strongly alkaline conditions and less stable carbonium ions favor the S_N2 path.

A competing side reaction in all these hydrolyses is the dehydrohalogenation of the alkyl halide to an alkene (sec 3.5b). Displacement is favored by aqueous solvents, whereas elimination is enhanced by using less polar, alcoholic solvents.

6.5d INDUSTRIAL METHODS

Methanol is sometimes known as **wood alcohol**, because at one time it was produced by the destructive distillation of wood. At present, however, very little methanol is produced in this way. It is now manufactured synthetically from carbon monoxide and hydrogen. High temperatures (400°) and pressures (about 150 atmospheres) are employed. The catalyst may be a mixture of the oxides of zinc, copper, and chromium. Normal propyl alcohol and isobutyl alcohol are by-products of this synthesis.

$$CO + 2\,H_2 \xrightarrow[400°,\ 150\ atm.]{ZnO-Cr_2O_3} H-\overset{\displaystyle H}{\underset{\displaystyle H}{\overset{\displaystyle |}{\underset{\displaystyle |}{C}}}}-OH \qquad (6\text{-}19)$$

Methanol is the starting material in the manufacture of formaldehyde and is also used as an antifreeze and as a solvent for shellacs and varnishes.

It is highly toxic, causing permanent blindness if taken internally. It is added to grain alcohol to make the latter unfit for human consumption, when the grain alcohol is to be used for industrial purposes rather than in beverages. It has also been used as a fuel in experimental internal combustion engines. The exhaust gases are low in air pollutants.

Ethanol is the most important industrial alcohol. It can be prepared by the fermentation of blackstrap molasses, the residues which result from the purification of cane sugar. The molasses is fermented with yeast, an organic catalyst which transforms the sugar into alcohol and carbon dioxide. The starch in potatoes, grain, and similar substances can be converted by malt into sugar which, when fermented, gives ethyl alcohol. Because of its source, ethyl alcohol is sometimes known as **grain alcohol.**

$$C_{12}H_{22}O_{11} + H_2O \xrightarrow{\text{yeast}} 4\ CH_3CH_2OH + 4\ CO_2 \qquad (6\text{-}20)$$

$\qquad$ cane sugar $\qquad\qquad\qquad\qquad$ ethyl alcohol

The most important method for producing ethyl alcohol industrially involves the hydration of ethylene (eqs 6-11 and 6-12).

Ordinary commercial alcohol is a constant-boiling mixture of alcohol (95%) and water (5%) which cannot be further purified by distillation. **Absolute alcohol** can be prepared from 95% alcohol by chemical methods as, for example, with quicklime (CaO) which reacts with the water but not with the alcohol.

Ethanol has been known since earliest times, particularly as an ingredient in all fermented beverages. The term *proof,* in reference to alcoholic beverages, is approximately twice the volume percentage of alcohol present. Ethanol is a raw material for several industrial syntheses, including that of the anesthetic ether. The solvent properties of ethanol are used in making resins and varnishes and in certain medicines (for example, tincture of iodine is a solution of iodine in ethanol).

Another alcohol produced by a fermentation process is *1-butanol.* It is used to manufacture quick-drying solvents for automobile lacquers and varnishes.

6.6 PREPARATION OF PHENOLS

Annual production of phenol in the United States approximates one billion pounds, making it one of the most important industrial organic chemicals. Much of it is used to make plastics (sec 9.8), but it is also a raw material for certain medicinals, weed killers, detergents, and miscellaneous products. At least five commercial processes for its synthesis are in use; three will be described here. The first of these is also suitable for the laboratory, and another laboratory method is deferred to Chapter 12.

6.6a SULFONATION AND ALKALI FUSION

The sodium salts of aromatic sulfonic acids, when fused with alkali, give phenols.

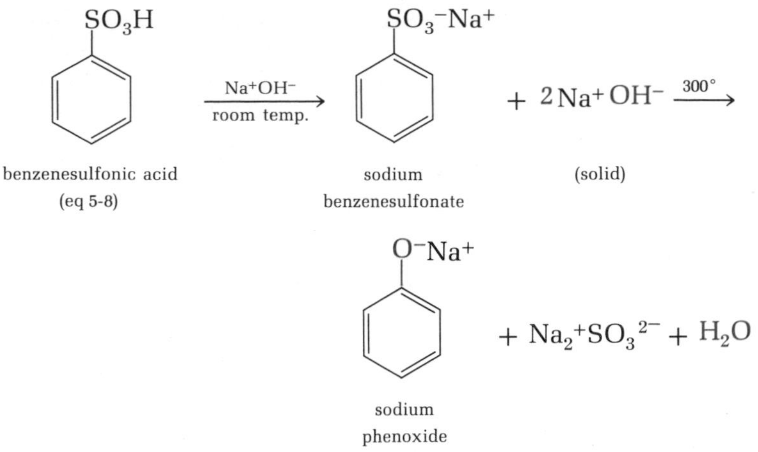

benzenesulfonic acid
(eq 5-8)

sodium
benzenesulfonate

(solid)

sodium
phenoxide

$$+ \ Na_2{}^+SO_3{}^{2-} + H_2O \quad (6\text{-}21)$$

The free phenol may be liberated from the sodium phenoxide by acid (eq 6-10).

Even carbon dioxide is a sufficiently strong acid for this purpose.

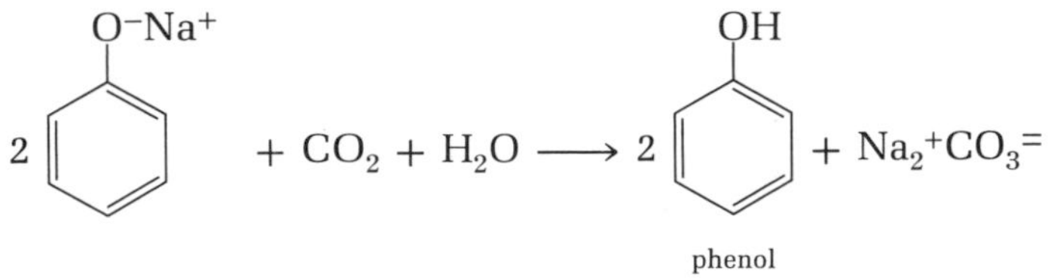

phenol

$$2 \quad \cdots \quad + \ CO_2 + H_2O \longrightarrow 2 \quad \cdots \quad + \ Na_2{}^+CO_3{}^= \quad (6\text{-}22)$$

The final step in equation 6-21 is a **nucleophilic aromatic substitution.** Hydroxide ion attacks the aromatic ring at the carbon which holds the sulfonate group, and in a second step, sulfite ion is lost.

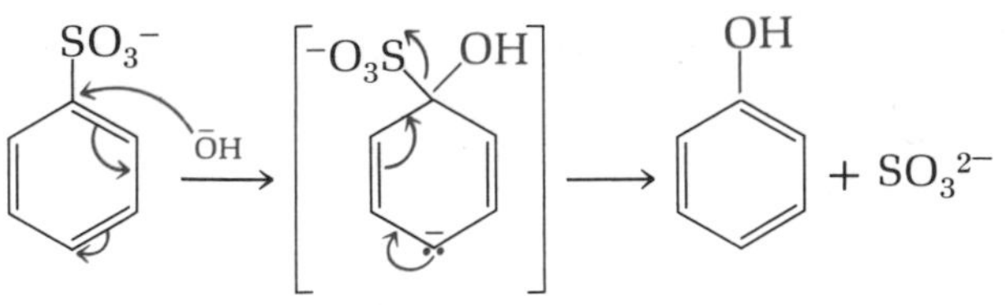

$$+ \ SO_3{}^{2-} \quad (6\text{-}23)$$

In this example, strenuous conditions are required (300°), but such displacements can be facilitated if the negative charge on the aromatic ring, in the intermediate shown in brackets, is stabilized by electron-withdrawing groups in the *ortho* or *para* positions (sec 8.5).

6.6b   THE DOW PROCESS

In the Dow Chemical Company process, chlorobenzene is converted to phenol with alkali at high temperatures and pressures. An emulsion of the reactants is continuously passed through the reactor at such a rate that conversion is essentially complete when the mixture emerges.

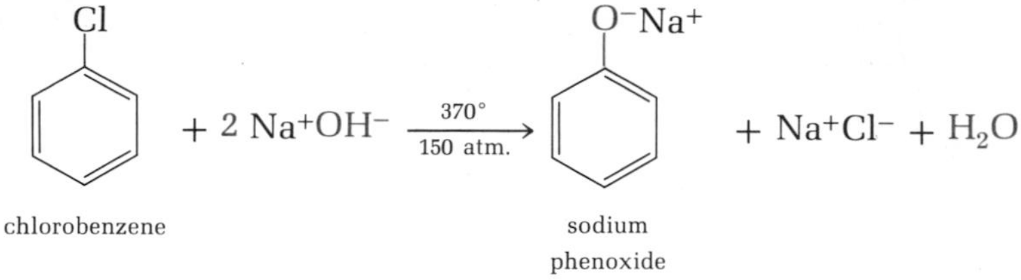

$$+ \, 2 \, Na^+OH^- \xrightarrow[150 \text{ atm.}]{370°} \qquad + \, Na^+Cl^- + H_2O \quad (6\text{-}24)$$

chlorobenzene

sodium
phenoxide

Acidification of the mixture converts the sodium phenoxide to free phenol.

6.6c   FROM CUMENE (ISOPROPYLBENZENE)

Cumene is available commercially through the Friedel-Crafts alkylation of benzene with propylene:

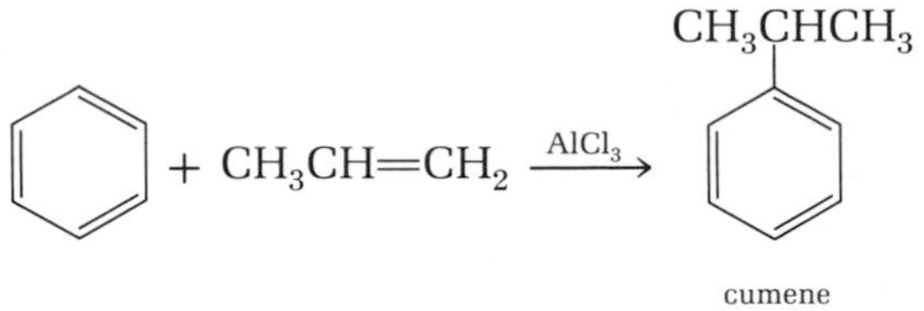

$$+ \, CH_3CH{=}CH_2 \xrightarrow{AlCl_3} \qquad\qquad (6\text{-}25)$$

cumene

Air is blown into liquid cumene, converting it to a hydroperoxide (by a free-radical mechanism). This, on reaction with strong acid, gives phenol and acetone, by an interesting ionic rearrangement (problem 12b).

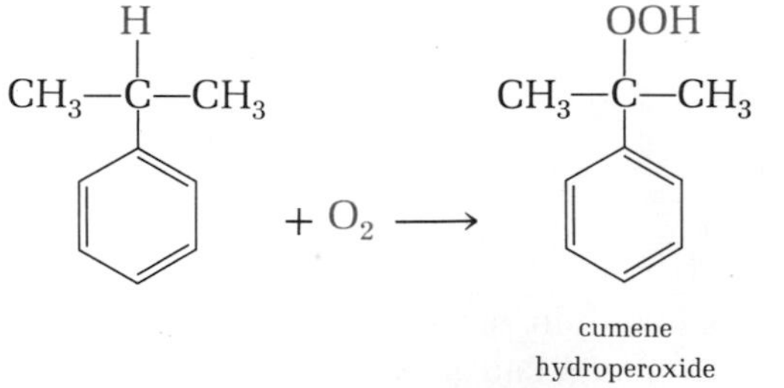

$$+ \, O_2 \longrightarrow \qquad\qquad (6\text{-}26a)$$

cumene
hydroperoxide

$$\xrightarrow[H_2O]{H^+} \qquad + \, CH_3{-}\overset{\displaystyle O}{\underset{\displaystyle \|}{C}}{-}CH_3 \quad (6\text{-}26b)$$

acetone

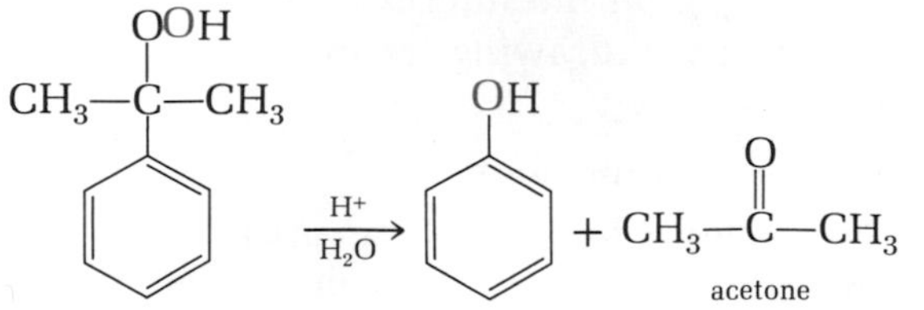

6.7 REACTIONS OF ALCOHOLS AND PHENOLS

6.7a ALKYL HALIDES FROM ALCOHOLS

The hydroxyl group of alcohols can be replaced by halogens. Hydrogen halides may be used for this purpose.

$$R—OH + H—X \longrightarrow R—X + H—OH \qquad (6\text{-}27)$$

$$\underset{\text{alcohol}}{} \qquad \underset{\text{alkyl halide}}{}$$

The reaction is the reverse of the hydrolysis of alkyl halides to alcohols (sec 6.5c) and proceeds by the same two mechanistic paths (S_N1 or S_N2), depending on the structure of R. In the first step, the alcohol is protonated by the acid (eq 6-2). If R is primary, the halide ion displaces a water molecule from the oxonium ion, in an S_N2 process.

$$X^- + R—\overset{\overset{\displaystyle H}{|}}{\underset{\displaystyle \cdot\cdot}{O}}{}^{\pm}H \longrightarrow R—X + H_2O \qquad S_N2 \qquad (6\text{-}28)$$

If R is tertiary, ionization precedes reaction with X^-.

$$R—\overset{\overset{\displaystyle H}{|}}{O}{}^{\pm}H \underset{}{\overset{-H_2O}{\rightleftharpoons}} R^+ \overset{X^-}{\longrightarrow} R—X \qquad S_N1 \qquad (6\text{-}29)$$

The reaction proceeds under increasingly mild conditions as one goes from primary to secondary to tertiary alcohols. Tertiary butyl alcohol reacts with concentrated hydrochloric acid at room temperature in a few minutes to form tertiary butyl chloride by an S_N1 mechanism. Normal butyl alcohol, which reacts by an S_N2 mechanism, requires zinc chloride as a catalyst and heat for several hours. The structure of the nucleophile is also important, iodide reacting the fastest, chloride the slowest, and bromide at an intermediate rate.

The reaction of an alcohol with a hydrogen halide is reversible. Several irreversible methods for bringing about the replacement of —OH by —X have been devised. The reaction of alcohols with *thionyl chloride* is driven to completion because two of the products are gases.

$$ROH + \underset{\substack{\text{thionyl} \\ \text{chloride}}}{Cl—\overset{\overset{\displaystyle O}{\|}}{S}—Cl} \overset{-HCl\uparrow}{\longrightarrow} RO—\overset{\overset{\displaystyle O}{\|}}{S}—Cl \overset{\text{heat}}{\longrightarrow} RCl + SO_2\uparrow \qquad (6\text{-}30)$$

Phosphorus trihalides react, giving high-boiling phosphorous acid as one of the products. The alkyl halide is usually the lowest boiling component of the reaction mixture and can be removed by distillation.

$$3 \text{ ROH} + \text{PX}_3 \longrightarrow 3 \text{ RX} + \text{H}_3\text{PO}_3 \qquad (6\text{-}31)$$

Alkyl halides are important intermediates in organic synthesis (Chapter 8), and they are frequently obtained from the corresponding alcohols by one of the above methods.

6.7b ESTERS OF INORGANIC ACIDS

Alkyl halides can be considered as organic derivatives of the hydrogen halides, through replacement of the acidic hydrogen by an organic group. Corresponding derivatives of other mineral acids are known; the compounds are called *esters* of inorganic acids.

	Acid		Ester
HX	Hydrogen halide	RX	alkyl halide
HONO_2	Nitric acid	RONO_2	alkyl nitrate
HOSO_3H	Sulfuric acid	ROSO_3H	alkyl hydrogen sulfate
		ROSO_3R	dialkyl sulfate
H_3PO_4	Phosphoric acid	$\overset{\displaystyle O}{\overset{\|}{\text{ROP(OH)}_2}}$	alkyl dihydrogen phosphate
		$\overset{\displaystyle O}{\overset{\|}{\text{(RO)}_2\text{POH}}}$	dialkyl hydrogen phosphate
		$\text{(RO)}_3\text{P}{=}\text{O}$	trialkyl phosphate

Organic nitrates may be prepared from alcohols and nitric acid. The products are explosive.

$$\text{ROH} + \text{HONO}_2 \longrightarrow \text{RONO}_2 + \text{H}_2\text{O} \qquad (6\text{-}32)$$

When sulfuric acid reacts *in the cold* with alcohols, particularly primary ones, the products are alkyl hydrogen sulfates. With excess alcohol, sulfates are obtained.

$$\text{ROH} + \text{HOSO}_3\text{H} \xrightarrow{-\text{H}_2\text{O}} \text{ROSO}_3\text{H} \xrightarrow[-\text{HOH}]{\text{R—OH}} \text{ROSO}_3\text{R} \quad (6\text{-}33)$$

Alkyl hydrogen sulfates are intermediates in the synthesis of alcohols (eq 6-11) and ethers (eq 7-4).

Phosphate esters are particularly important in many biochemical processes. They are synthesized enzymatically in the cell. All three types of esters are known, as well as esters of di- and tri-phosphoric acids. Phosphates are important in many enzymatic reactions and in the structure of nucleic acids (Chapter 18).

$$RO-\underset{\underset{OH}{|}}{\overset{\overset{O}{||}}{P}}-O-\underset{\underset{OH}{|}}{\overset{\overset{O}{||}}{P}}-OH \qquad\qquad RO-\underset{\underset{OH}{|}}{\overset{\overset{O}{||}}{P}}-O-\underset{\underset{OH}{|}}{\overset{\overset{O}{||}}{P}}-O-\underset{\underset{OH}{|}}{\overset{\overset{O}{||}}{P}}-OH$$

alkyl diphosphate   alkyl triphosphate

6.7c ESTERS OF ORGANIC ACIDS

Alcohols react with organic acids to form *esters*.

$$ROH + R'-\overset{\overset{O}{||}}{C}-OH \xrightarrow{H^+} R'-\overset{\overset{O}{||}}{C}-OR + H_2O \qquad (6\text{-}34)$$

alcohol   carboxylic   ester
        acid

This reaction and class of compounds will be discussed in detail in Chapter 10.

6.7d OXIDATION OF ALCOHOLS

Alcohols are the most important precursors of carbonyl compounds (Chapter 9). Oxidation of primary alcohols gives *aldehydes*, whereas oxidation of secondary alcohols gives *ketones*, and tertiary alcohols are not easily oxidized. For a detailed discussion of the oxidation reaction, see Chapter 9.

$$RCH_2OH \longrightarrow RCH{=}O \qquad\qquad (6\text{-}35)$$

aldehyde

$$R_2CHOH \longrightarrow R_2C{=}O \qquad\qquad (6\text{-}36)$$

ketone

6.7e REACTIONS OF THE AROMATIC RING IN PHENOLS

The hydroxyl group on an aromatic ring is *ortho-para*-directing and ring-activating in electrophilic substitutions. This is because the unshared electron pair on oxygen can stabilize a positive charge adjacent to it.

intermediate in electrophilic substitution *para* to a phenolic hydroxyl group

Phenol can be nitrated with *dilute aqueous* nitric acid to yield
p-nitrophenol.

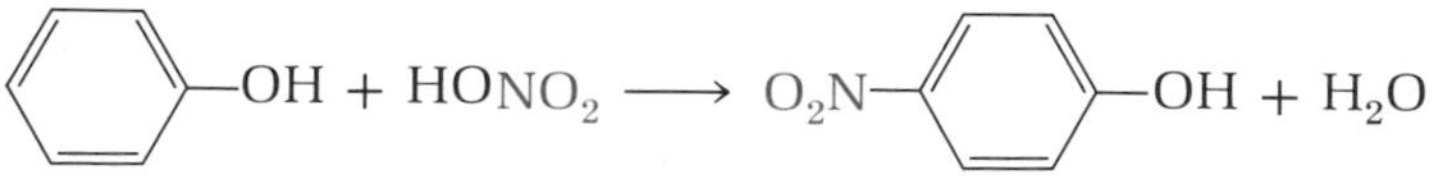

$$\text{C}_6\text{H}_5\text{—OH} + \text{HONO}_2 \longrightarrow \text{O}_2\text{N—C}_6\text{H}_4\text{—OH} + \text{H}_2\text{O} \qquad (6\text{-}37)$$

p-nitrophenol

It can also be brominated rapidly with *bromine water* producing
2,4,6-tribromophenol.

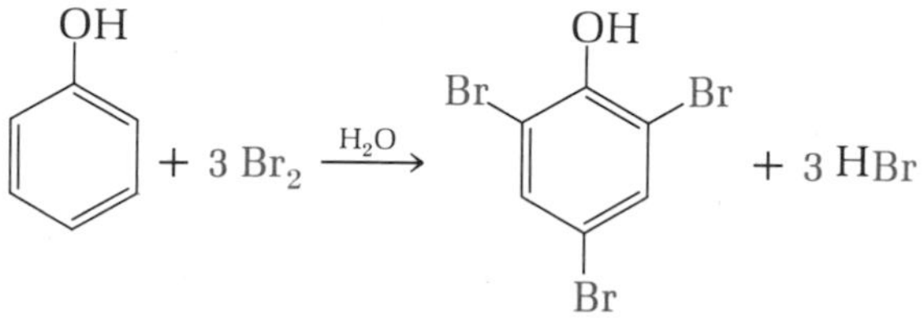

$$\text{C}_6\text{H}_5\text{OH} + 3\,\text{Br}_2 \xrightarrow{\text{H}_2\text{O}} \text{tribromophenol} + 3\,\text{HBr} \qquad (6\text{-}38)$$

2,4,6-tribromophenol

But typical phenolic properties are destroyed when the aromatic ring
is hydrogenated.

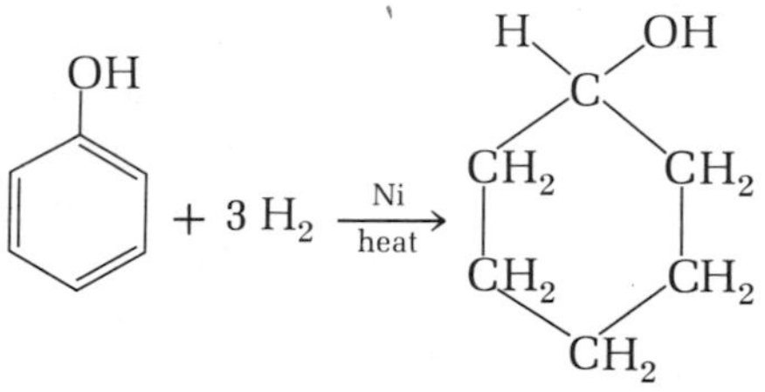

$$\text{C}_6\text{H}_5\text{OH} + 3\,\text{H}_2 \xrightarrow[\text{heat}]{\text{Ni}} \text{Cyclohexanol} \qquad (6\text{-}39)$$

Cyclohexanol

Cyclohexanol behaves like a secondary aliphatic alcohol, not a
phenol.

6.8 POLYHYDRIC ALCOHOLS AND PHENOLS

Compounds with several hydroxyl groups attached to different
carbon atoms in the same molecule are well known, and some are
commercially important.

Alcohols with more than one hydroxyl group per molecule are
prepared by reactions that are modifications of the general methods
described in section 6.5. **Ethylene glycol** (compounds with two —OH
groups are called glycols) may be prepared from ethylene by the
following sequence of reactions:

$$CH_2{=}CH_2 \; + \quad HO^-Cl^+ \quad \longrightarrow \quad \underset{\underset{\displaystyle OH \quad Cl}{|\quad\quad|}}{CH_2{-}CH_2} \xrightarrow[\text{in } H_2O]{\underset{Na_2{}^+CO_3{}^{2-}}{OH^-}}$$

hypochlorous acid   ethylene

(chlorine + water)   chlorohydrin

$$\underset{\underset{\displaystyle OH \quad\; OH}{|\quad\quad\;|}}{CH_2{-}CH_2} \qquad (6\text{-}40)$$

ethylene

glycol

(1,2-ethanediol)

Aqueous sodium carbonate is basic, and the hydroxide ions displace chloride ions from ethylene chlorohydrin.

Ethylene glycol, a permanent antifreeze, is completely miscible with water, lowers its freezing point, and is not as volatile (bp 197°) as methanol (bp 65°). Glycol is also one of the two raw materials used to manufacture Dacron (sec 10.6d).

Glycerol, an alcohol with three hydroxyl groups, is manufactured synthetically from propylene.

$$CH_3CH{=}CH_2 \xrightarrow[400\text{-}500°]{Cl_2} \underset{\underset{\displaystyle Cl}{|}}{CH_2{-}CH{=}CH_2} \xrightarrow[H_2O]{Na^+OH^-}$$

allyl chloride

$$\underset{\underset{\displaystyle OH}{|}}{CH_2{-}CH{=}CH_2} \qquad (6\text{-}41)$$

allyl alcohol

(2-propene-1-ol)

This first step involves the substitution of halogen, rather than addition (sec 3.4f). The allyl alcohol is subsequently treated with hypochlorous acid followed by hydrolysis.

$$\underset{\underset{\displaystyle OH}{|}}{CH_2{-}CH{=}CH_2} \xrightarrow{HO^-Cl^+} \underset{\underset{\displaystyle OH \quad\; OH \quad Cl}{|\quad\quad\;|\quad\;\;|}}{CH_2{-}CH{-}CH_2} \xrightarrow[H_2O]{Na^+OH^-}$$

$$\underset{\underset{\displaystyle OH \quad\; OH \quad\; OH}{|\quad\quad\;|\quad\quad\;|}}{CH_2{-}CH{-}CH_2} \qquad (6\text{-}42)$$

glycerol

(1,2,3-propanetriol)

Glycerol is also a major by-product in the manufacture of soap (Chapter 11).

Glycerol (glycerine) is a syrupy, colorless, high-boiling liquid with a distinctly sweet taste. It mixes with water and alcohol in all proportions. Glycerol is used as a moistening agent in tobacco, for its soothing qualities in shaving and toilet soaps, and in the manufacture of plastics, explosives, cellophane, etc. (Figure 6-4). *Glyceryl trinitrate* (eq 6-43) is the principal explosive ingredient in dynamite. It is also used as a vasodilator to reduce arterial tension in heart disease.

$$
\begin{array}{l}
CH_2OH \\
| \\
CHOH \\
| \\
CH_2OH
\end{array}
+ 3\ HONO_2 \xrightarrow{H_2SO_4}
\begin{array}{l}
CH_2ONO_2 \\
| \\
CHONO_2 \\
| \\
CH_2ONO_2
\end{array}
+ 3\ H_2O \qquad (6\text{-}43)
$$

glycerol   glyceryl trinitrate
(nitroglycerine)

Polyhydric phenols can be made by suitable modifications of the methods in section 6.6, and by special methods. *Catechol,* is used as an antioxidant in paints. *Resorcinol* is used in the manufacture of adhesives, dyes and antiseptics. The chief use of *hydroquinone* is as a photographic developer.

catechol   resorcinol   hydroquinone

6.9 THIOLS: SULFUR ANALOGS OF ALCOHOLS AND PHENOLS

Sulfur is in the same group as oxygen in the periodic table, and forms similar compounds. The sulfur analogs of alcohols are called *thiols* or *mercaptans*. The —SH group is known as the *sulfhydryl* group. Perhaps the most distinguishing feature of thiols is their disagreeable odor; *n*-butyl mercaptan ($CH_3CH_2CH_2CH_2SH$; 1-butanethiol) is the compound responsible for the odor of the skunk's defense mechanism.

The chemistry of thiols is in some respects like that of alcohols. They can be prepared from alkyl halides by an S_N2 displacement with sodium hydrosulfide, Na^+SH^-.

$$
R—X + Na^+SH^- \longrightarrow R—SH + Na^+X^- \qquad (6\text{-}44)
$$

Mercaptans are more acidic than alcohols and form insoluble salts

Figure 6-4

Glycerol serves many industries. It is used in the manufacture of alkyd resins for protective coatings, as a plasticizer for cellophane and plastics, and as a moistening agent in tobacco products. Its unique combination of properties has made it an important ingredient in drugs, cosmetics, and toilet goods. It has also found increasing use in foods and beverages.

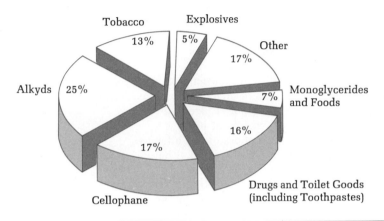

with heavy metal ions, such as mercury (hence the origin of their name, which means "to capture mercury").

$$2\,\text{RSH} + \text{Hg}^{2+} \longrightarrow (\text{RS})_2\text{Hg} + 2\,\text{H}^+ \qquad (6\text{-}45)$$

mercaptan   mercaptide

Mercaptans are readily oxidized to disulfides.

$$2\,\text{RS—H} + \text{H}_2\text{O}_2 \longrightarrow \text{RS—SR} + 2\,\text{H}_2\text{O} \qquad (6\text{-}46)$$

mercaptan   disulfide

These reactions are important in protein chemistry, since the amino acid cysteine contains a sulfhydryl group (Chapter 17).

A SUGGESTION

Although only a few specific examples have been cited for each reaction in this chapter, it should be recalled that these reactions are due to the hydroxyl group, and that, in general, most organic compounds that contain the alcohol or phenol function will behave in a similar fashion chemically. The student should learn these reactions whenever possible in terms of the general formulas R—OH and Ar—OH and apply them to specific cases by substituting the appropriate group for R— or Ar— in the compounds.

New Concepts, Facts, and Terms

1. Alcohols, phenols; primary, secondary, and tertiary alcohols
2. Nomenclature, sec 6.1. The IUPAC ending for alcohols is *-ol*. The four butyl groups (normal, secondary, iso, and tertiary). Common names for phenols (cresols, naphthols).
3. Physical properties
4. Hydrogen bond
5. Acidity and basicity of alcohols and phenols; alkyloxonium ions; alkoxides; resonance in phenoxide ions
6. Preparation of alcohols
   a. hydration of alkenes
   b. hydroboration of alkenes
   c. hydrolysis of alkyl halides; S_N2 and S_N1 displacements
   d. industrial methods
      i. methanol from $CO + H_2$
      ii. fermentation (ethanol, 1-butanol)
7. Preparation of phenols
   a. alkali fusion of sodium arylsulfonates; nucleophilic aromatic substitution
   b. the Dow process
   c. from cumene hydroperoxide
8. Reactions of alcohols and phenols
   a. preparation of alkyl halides, using HX, $SOCl_2$, or PX_3
   b. esters of inorganic acids; alkyl nitrates, sulfates and phosphates
   c. oxidation to aldehydes and ketones
   d. electrophilic substitution in phenols; OH is a ring-activating, *ortho-para*-directing substituent.
9. Polyhydric alcohols and phenols; ethylene glycol, glycerol, glyceryl trinitrate
10. Sulfur analogs; thiols or mercaptans

Exercises and Problems

1. Write structural formulas for each of the following compounds:
   a. 2,2-dimethyl-1-butanol
   b. o-bromophenol
   c. 2,3-pentanediol
   d. 1,3-diphenyl-2-propanol
   e. n-propyl hydrogen sulfate
   f. 3-pentene-2-ol
   g. sodium p-bromophenoxide
   h. 3-methylcyclopentanol
   i. 1-phenylethanol
   j. 2-bromo-1-naphthol

2. Draw the structures of the eight isomeric pentanols, $C_5H_{11}OH$. Name

each by the IUPAC system, and classify each as primary, secondary, or tertiary.

3. Name each of the following compounds:

a. $CH_3CH_2CH(OH)CH_3$

b.

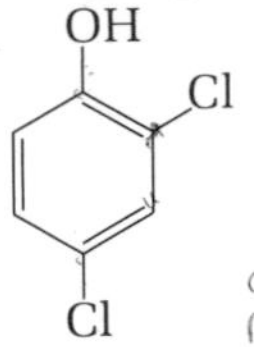

c. $CH_3CHBrC(CH_3)_2OH$

d.

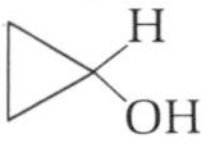

e.

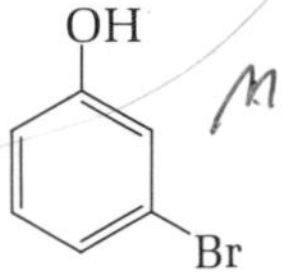

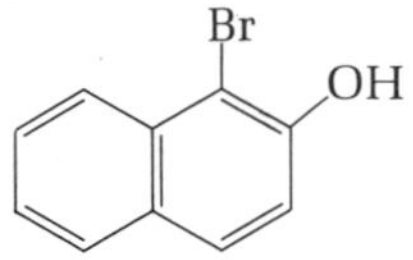

g. $CH_3CH=CHCH_2OH$

h. $CH_3CH(SH)CH_3$

i. $CH_2(OH)CH(OH)CH(OH)CH_2OH$

j. 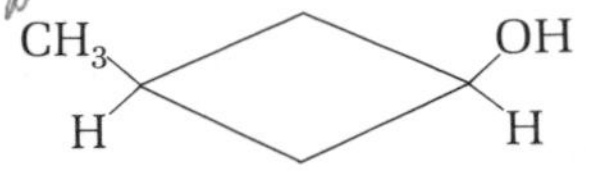 CH_3

4. Explain why each of the following names is unsatisfactory, and give a correct name:

a. isopropanol

b. 2,2-dimethyl-3-butanol

c. 2-ethyl-1-propanol

d. 1-propene-3-ol

e. 5-chlorocyclohexanol

f. phenyl alcohol

g. 6-bromo-p-cresol

h. 2,3-propanediol

5. Arrange the compounds in each of the following groups in order of increasing solubility in water, and explain briefly the principles on which your answer is based:

a. ethanol; ethyl chloride; 1-hexanol

b. 1-pentanol; 2,3-pentanediol; $CH_2OH(CHOH)_3CH_2OH$

c. phenol; benzene; sodium phenoxide; hydroquinone

6. Arrange methanol, water, and dimethyl ether, first in order of increasing boiling points and then in order of increasing formula weight. Explain why these series might have been expected to parallel one another and why, in fact, they do not.

7. Do you expect methyl mercaptan to have a higher or lower boiling point than methyl alcohol? Explain.

8. What is meant by the statement, "Alcohols are amphoteric"? Illustrate with equations.

9. a. Arrange the following compounds in order of increasing acidity, and explain the reasons for your choices: phenol, p-chlorophenol, cyclohexanol, p-cresol.

b. Draw the contributing resonance structures for the p-nitrophenoxide ion. Do you expect p-nitrophenol to be a stronger or weaker acid than phenol? Why?

10. Write equations for the preparation of
    a. *t*-butyl alcohol from isobutylene
    b. 1-phenylethanol from styrene
    c. 2-phenylethanol from styrene
    d. 1-butanol from 1-bromobutane
    e. allyl alcohol from propene
    f. cyclohexanol from benzene
    g. resorcinol from benzene
    h. *m*-nitrophenol from benzene
    i. 2,3-butanediol from 2-butene
    j. benzyl alcohol from toluene

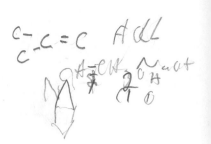

11. Write equations for each of the following reactions:
    a. t-butyl alcohol + hydrochloric acid
    b. 1-pentanol + sodium metal
    c. benzyl chloride + aqueous sodium hydroxide
    d. *p*-cresol + hydrogen (Ni catalyst)
    e. allyl alcohol + hypochlorous acid
    f. cyclopentanol + phosphorus tribromide
    g. 1-phenylethanol + thionyl chloride
    h. 1-butanol + cold concentrated sulfuric acid
    i. cyclohexanol + aqueous sodium hydroxide
    j. 1-naphthol + aqueous sodium hydroxide

12. a. Write a free radical mechanism for the oxidation of cumene to
       cumene hydroperoxide (eq 6-26a).
    b. Write an ionic mechanism for the conversion of cumene hydro-
       peroxide to phenol and acetone (eq 6-26b). Your mechanism must
       explain how the benzene ring becomes attached to the "colored"
       oxygen. (HINT: Begin by protonating the oxygen to which hydrogen
       is attached, in the peroxide.)

13. Indicate how the following mixtures could be separated *without* the
    use of distillation.
    a. benzene and phenol
    b. phenol and cyclohexanol
    c. 1-propanol and 1-heptanol

14. Describe a simple test (reagent, equation, and readily observable change)
    which will distinguish
    a. 3-methylpentane from 3-pentanol
    b. allyl alcohol from isopropyl alcohol
    c. *p*-cresol from benzyl alcohol
    d. 1-butanethiol from 1-butanol

15. A pure liquid **A** reacts with metallic sodium to evolve hydrogen. With
    concentrated HBr it gives **B** which, on treatment with alcoholic sodium
    hydroxide, gives only cyclohexene. Give the structural formulas for **A**
    and **B**, and write equations for all reactions described.

16. A crystalline solid C_7H_8O is only slightly soluble in water, but readily
    dissolves in dilute aqueous sodium hydroxide. Treating the compound

with bromine water gives a product $C_7H_6Br_2O$. Give a structural formula for the original compound, and write equations for each reaction.

17. a. 1-Bromo-2,2-dimethylpropane (neopentyl bromide) reacts *very slowly* in comparison with 1-bromo-3,3-dimethylbutane in an S_N2 displacement with hydroxide ion. Draw models of the transition states which explain this observation.
    b. The chloride

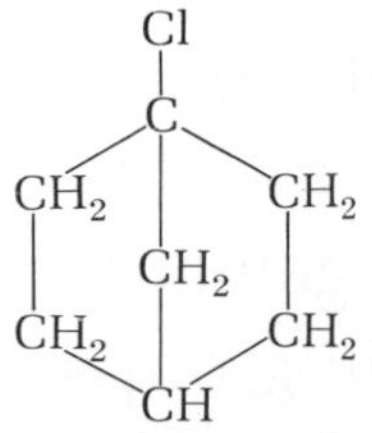

or

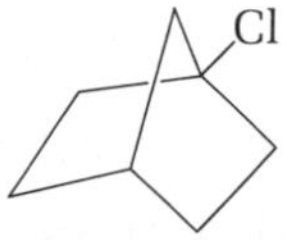

cannot readily be hydrolyzed to the corresponding alcohol. Explain.

18. In the reaction of an alcohol with an acid to form an ester, water can be eliminated in two possible ways. These are illustrated by appropriate use of colors in the two parts of equation 6-34. Write equations for each step which clearly illustrate the two different mechanisms.

19. Give equations to show the steps required in each of the following conversions:
    a. 1-butanol to 2-bromobutane
    b. 1-butene to sodium 2-butoxide
    c. isobutyl alcohol to isobutyl bromide
    d. *p*-cresol to 4-methylcyclohexene
    e. *t*-butyl alcohol to isobutane
    f. cyclohexanol to 1,2-cyclohexanediol
    g. 2-methylnaphthalene to 2-methylnaphthol
    h. ethanol to ethanethiol
    i. toluene to 2,6-dibromo-*p*-cresol
    j. 1-methylcyclohexene to 2-methylcyclohexanol

20. What is the mechanism for the final steps in equations 6-40 and 6-42?

CHAPTER SEVEN

ETHERS

To the layman, the word *ether* is usually synonymous with the well-known anesthetic. To the organic chemist, however, ethers are a general class of compounds whose characteristic arrangement involves an oxygen atom to which two organic groups are attached. When the two groups are ethyl, the substance is the common anesthetic. But other organic groups impart different properties so diverse that the materials can be used as artificial flavors, solvents, and refrigerants. Ethers are relatively inert chemically, and consequently see use as solvents in organic chemical reactions.

We have seen that alcohols can be considered as derivatives of water in which one of the hydrogen atoms is replaced by an alkyl group. Although unrelated chemically to water, **ethers** are structurally related to it with *both* hydrogens replaced by organic groups. Ethers are isomers of alcohols with the same number of carbon atoms.

Figure 7-1
Models showing the structural relationship between water, alcohols, and ethers.

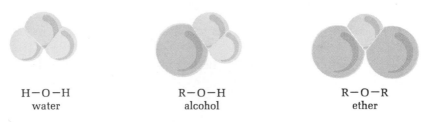

H—O—H   R—O—H   R—O—R
water   alcohol   ether

7.1 NOMENCLATURE

Ethers may be either aliphatic, aromatic, or mixed. The organic groups may be identical or different. Frequently, common names are used. Usually the name of each organic group is given, followed by the word *ether*.

$CH_3—O—CH_2CH_3$ $CH_3CH_2—O—CH_2CH_3$ $CH_3—O—CH(CH_3)_2$
methyl ethyl ether   diethyl ether   methyl isopropyl ether

diphenyl ether $CH_3—O—$ methyl phenyl ether
                          (anisole)

The RO— group is known as the alkoxyl group ($CH_3O—$ is methoxyl). In the IUPAC system, ethers are named as alkoxy-substituted hydrocarbons. For example:

$$\overset{1}{C}H_3\overset{2}{C}H\overset{3}{C}H_2\overset{4}{C}H_2\overset{5}{C}H_3$$
$$OCH_3$$
2-methoxypentane

7.2   PREPARATION OF ETHERS

Ethers are usually prepared from alcohols by one of the following methods.

7.2a   INTERMOLECULAR DEHYDRATION OF ALCOHOLS

When an alcohol is dehydrated, water may split out either intramolecularly (that is, within the molecule itself) to yield an alkene,

$$
\begin{array}{c}
\text{H H} \\
\ \ |\ \ \ | \\
\text{H}-\text{C}-\text{C}-\text{OH} \\
\ \ |\ \ \ | \\
\text{H H}
\end{array}
\longrightarrow
\begin{array}{c}
\text{H H} \\
\ \ |\ \ \ | \\
\text{H}-\text{C}=\text{C}-\text{H} + \text{H}_2\text{O} \\
\ \ \ \ \ \ \text{ethylene}
\end{array}
\qquad (7\text{-}1)
$$

or intermolecularly (between two molecules) to form an ether.

$$
\begin{array}{c}
\text{H H} \\
\ \ |\ \ \ | \\
\text{H}-\text{C}-\text{C}-\text{OH} \\
\ \ |\ \ \ | \\
\text{H H}
\end{array}
\ \
\begin{array}{c}
\text{H H} \\
\ \ |\ \ \ | \\
\text{HO}-\text{C}-\text{C}-\text{H} \\
\ \ |\ \ \ | \\
\text{H H}
\end{array}
\longrightarrow
$$

$$
\begin{array}{c}
\text{H H}\ \ \ \ \ \text{H H} \\
\ \ |\ \ \ |\ \ \ \ \ \ |\ \ \ | \\
\text{H}-\text{C}-\text{C}-\text{O}-\text{C}-\text{C}-\text{H} + \text{H}_2\text{O} \\
\ \ |\ \ \ |\ \ \ \ \ \ |\ \ \ | \\
\text{H H}\ \ \ \ \ \text{H H} \\
\text{ethyl ether}
\end{array}
\qquad (7\text{-}2)
$$

The reaction course depends upon the conditions, which may be controlled to prepare either the alkene or the ether.

   In practice, sulfuric acid is frequently used as the dehydrating agent. In the manufacture of the common anesthetic ethyl ether, ethanol is dissolved in sulfuric acid to form the intermediate ethyl hydrogen sulfate.

$$
\text{CH}_3\text{CH}_2\text{OH} + \text{HOSO}_3\text{H} \longrightarrow \underset{\text{ethyl hydrogen sulfate}}{\text{CH}_3\text{CH}_2\text{OSO}_3\text{H}} + \text{H}_2\text{O} \quad (7\text{-}3)
$$

This solution is then heated to about 140°, and more alcohol is added, whereupon the ether distils.

$$
\text{CH}_3\text{CH}_2\text{OSO}_3\text{H} + \text{HOCH}_2\text{CH}_3 \xrightarrow{140°}
$$

$$
\text{CH}_3\text{CH}_2\text{OCH}_2\text{CH}_3 + \text{HOSO}_3\text{H} \qquad (7\text{-}4)
$$

The reaction can be regarded as an S_N2 displacement either of bisulfate ion or water, alcohol acting as the nucleophile.

$$CH_3CH_2\overset{..}{\underset{..}{O}}H + CH_3CH_2-OSO_3H \xrightarrow{-OSO_3H^-}$$

$$CH_3CH_2\overset{..}{\underset{..}{O}}H + CH_3CH_2-\overset{+}{O}H_2 \xrightarrow{-H_2O}$$

$$CH_3CH_2\overset{+}{\underset{\underset{H}{|}}{O}}:CH_2CH_3 \underset{+H^+}{\overset{-H^+}{\rightleftharpoons}} CH_3CH_2\overset{..}{\underset{..}{O}}CH_2CH_3 \qquad (7\text{-}5)$$

At 140°, the main product is ether, whereas at higher temperatures (say, 180°) ethylene is produced by an elimination reaction.

If one alkyl group is tertiary, the reaction may proceed by a carbonium ion (S_N1) mechanism.

$$(CH_3)_3C\overset{..}{\underset{..}{O}}H \overset{H^+}{\rightleftharpoons} \left[(CH_3)_3C\overset{\overset{H}{|}}{\underset{..}{O}}H\right]^+ \overset{-H_2O}{\rightleftharpoons} (CH_3)_3C^+$$

t-butyl alcohol $\qquad\qquad\qquad\qquad$ t-butyl cation

$$(CH_3)_3COCH_3 \overset{-H^+}{\rightleftharpoons} \left[(CH_3)_3C:\overset{+}{\underset{\underset{H}{|}}{O}}CH_3\right]^+ \overset{CH_3\overset{..}{O}H}{\longleftarrow} \qquad (7\text{-}6)$$

t-butyl methyl
ether

Such reactions are usually run cold, since tertiary alcohols are so readily dehydrated to alkenes.

7.2b THE WILLIAMSON SYNTHESIS

In 1851, the British chemist Alexander William Williamson devised a general method for the synthesis of pure ethers. The reactants are the sodium salt of an alcohol or phenol and an alkyl halide or sulfate.

$$RO^-Na^+ + R'X \longrightarrow ROR' + Na^+X^- \qquad (7\text{-}7)$$

The reaction is a typical S_N2 displacement and works well only when R' is a primary or secondary alkyl group. The nucleophile is an alkoxide or phenoxide ion.

$$R\overset{..}{\underset{..}{O}}:\longrightarrow R'-X \longrightarrow R\overset{..}{\underset{..}{O}}:R' + X^- \qquad (7\text{-}8)$$

The R and R' groups may be identical or different. Both reactants can be prepared from alcohols. One sequence for making methyl n-propyl ether is:

$$2\ CH_3OH + 2\ Na \longrightarrow 2\ CH_3O^-Na^+ + H_2 \qquad (7\text{-}9a)$$

$$CH_3CH_2CH_2OH + HBr \longrightarrow CH_3CH_2CH_2Br + H_2O \qquad (7\text{-}9b)$$

$$\underset{\substack{\text{sodium}\\\text{methoxide}}}{CH_3O^-Na^+} + \underset{\substack{n\text{-propyl}\\\text{bromide}}}{CH_3CH_2CH_2Br} \longrightarrow \underset{\substack{\text{methyl }n\text{-propyl}\\\text{ether}}}{CH_3OCH_2CH_2CH_3} + Na^+Br^- \qquad (7\text{-}9c)$$

Alkyl sulfates are often used to prepare ethers of phenols.

$$\underset{\text{phenol}}{\bigcirc\!\!-OH} + \underset{\text{methyl sulfate}}{(CH_3)_2SO_4} \xrightarrow{\text{NaOH}}$$

$$\underset{\text{anisole}}{\bigcirc\!\!-OCH_3} + CH_3OSO_3^-Na^+ \qquad (7\text{-}10)$$

7.3 PROPERTIES OF ETHERS

Dimethyl and methyl ethyl ether are gases, whereas ethyl ether and higher homologs in the aliphatic series are liquids. Aromatic ethers are liquids or solids. Ethers are only slightly soluble in water. Ordinary (ethyl) ether is an unusually good solvent for many organic compounds, and is frequently used to extract desirable organic products from plants or other sources. It dissolves the organic matter, but leaves salts and other inorganic substances behind. Ether is only slightly soluble in water, and separates from it as an upper layer. Its low boiling point (35°) makes it easy to remove from an extract and easy to recover. Ethyl ether is particularly flammable, however, and one must exercise caution in handling it. It is also easily oxidized by air to a nonvolatile, explosive peroxide; for this reason, distillations of ether extracts must not be carried to dryness with excessive heating.

The most notable chemical property of ethers is their lack of reactivity. The ether linkage is particularly stable to dilute acids or alkalies, and to many oxidizing and reducing agents. Sodium does not react with ethers, a property that distinguishes ethers from their isomers, the alcohols. In general, ethers resemble saturated hydrocarbons in their inert chemical behavior.

Ethers are weak bases, by virtue of the unshared electron pairs on oxygen. They are therefore soluble in concentrated sulfuric acid, a property that may be used to distinguish them from saturated hydrocarbons. Solubility depends on protonation, to form a dialkyl-oxonium ion.

$$R\!-\!\ddot{O}\!-\!R' + H^+X^- \rightleftharpoons R\!-\!\overset{\overset{\displaystyle H}{|}}{\ddot{O}^+}\!-\!R' + X^- \qquad (7\text{-}11)$$

dialkyloxonium ion

When they are heated with concentrated mineral acids, ethers are cleaved. Hydrobromic or hydriodic acids are frequently used for this purpose. The reaction involves protonation of the ether (eq 7-11) followed either by an S_N2 displacement by halide ion or, if one of the R groups is tertiary, an S_N1 reaction.

$$R\!-\!X + R'\!-\!\ddot{O}H \xleftarrow[\substack{S_N2 \\ \text{(R is primary} \\ \text{or secondary)}}]{X^-} R\!-\!\overset{\overset{\displaystyle H}{|}}{\ddot{O}^+}\!-\!R' \xrightarrow[\substack{S_N1 \\ \text{R is} \\ \text{(tertiary)}}]{} R^+ + R'\!-\!OH \quad (7\text{-}12)$$

$$\big\downarrow \xrightarrow{X^-} R\!-\!X$$

The alcohol (R'—OH) may react further if excess HX is used to give a second mole of halide (R'X). For example, methyl n-propyl ether reacts with excess hydriodic acid as shown.

$$CH_3OCH_2CH_2CH_3 + 2\ HI \xrightarrow{100°} CH_3I + CH_3CH_2CH_2I + H_2O \quad (7\text{-}13)$$

Aromatic ethers readily yield phenols by the same type of reaction.

phenetole   phenol

$$+ \text{HBr} \longrightarrow \qquad \qquad -OH + CH_3CH_2Br \quad (7\text{-}14)$$

The reaction stops at the phenol because, except for special cases, phenols cannot be converted directly to aryl halides (sec 8.5).

7.4 SOME IMPORTANT ETHERS

Dimethyl ether is manufactured industrially for use as a refrigerant. It may also be used as an anesthetic, but its effect is short-lived. The most important ether is **diethyl ether,** the common anesthetic. Like other anesthetics, ether is toxic in high concentrations, but the spread between the concentration necessary to produce anesthesia and its lethal action is sufficient to permit safe usage. **Diphenyl ether**

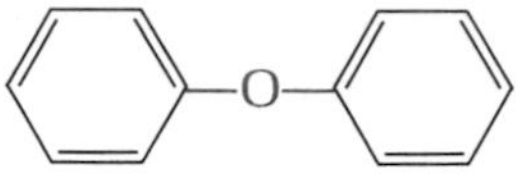

diphenyl ether

is obtained in large quantities as a by-product in the commercial production of phenol by the Dow process (sec 6.6b). It has a high boiling point (259°) and is a rather inert liquid which is used as a medium for the transfer of heat.

Certain important naturally occurring aromatic compounds contain the ether linkage as, for example, eugenol (oil of cloves) and anethole (oil of anise).

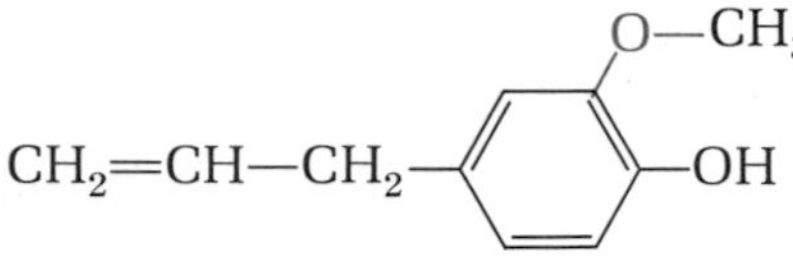

eugenol 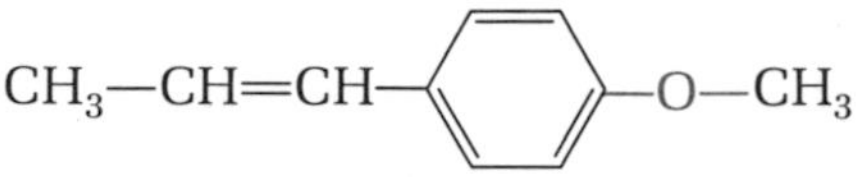   anethole

7.5 EPOXIDES (OXIRANES): ETHYLENE OXIDE

Alkenes can be converted by certain oxidizing agents to cyclic ethers called **epoxides** or **oxiranes.**

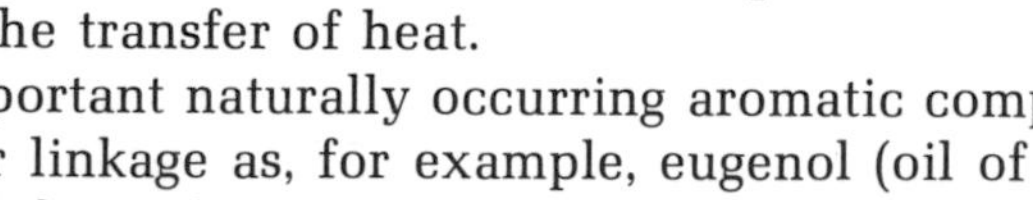

$$R-CH=CH-R' \xrightarrow{[O]} R-\underset{\underset{O}{\diagdown\diagup}}{CH-CH}-R' \qquad (7\text{-}15)$$

an epoxide
or oxirane

Ethylene oxide, the simplest member of this series, was first synthesized by Wurtz in 1859. It is an important commercial chemical, being second in production only to ethanol, as a derivative of ethylene. About 30% of ethylene production goes into its manufacture. Two commercial processes are used, the most important involving direct oxidation with air:

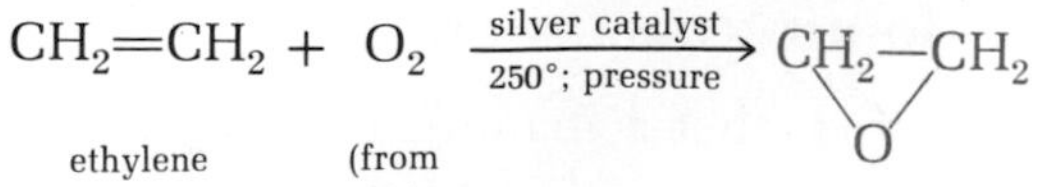

$$CH_2=CH_2 + O_2 \xrightarrow[\text{250°; pressure}]{\text{silver catalyst}} CH_2-CH_2 \qquad (7\text{-}16)$$

ethylene   (from   ethylene
                air)   oxide

Ethylene oxide is also prepared from ethylene via the intermediate ethylene chlorohydrin.

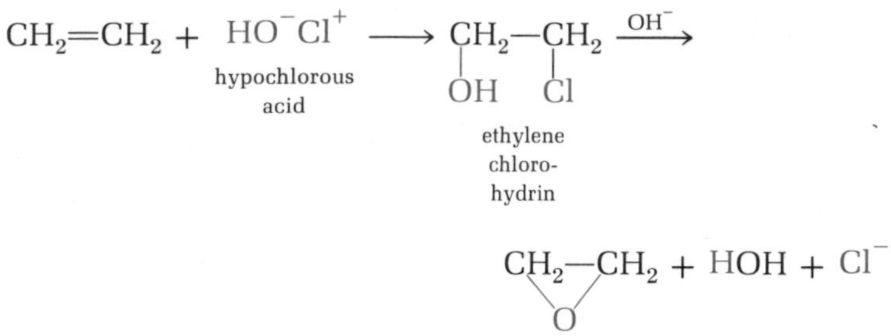

$$CH_2{=}CH_2 + HO\overset{-}{}Cl\overset{+}{} \longrightarrow \underset{\underset{\displaystyle OH}{|}}{CH_2}{-}\underset{\underset{\displaystyle Cl}{|}}{CH_2} \xrightarrow{\ OH^-\ }$$

hypochlorous
acid

ethylene
chloro-
hydrin

$$CH_2{-}CH_2 + HOH + Cl^- \qquad (7\text{-}17)$$

The first step in this synthesis has been discussed (eq 3-14). The second step is an intramolecular example of the Williamson synthesis (sec 7-2b), as shown by the following mechanism:

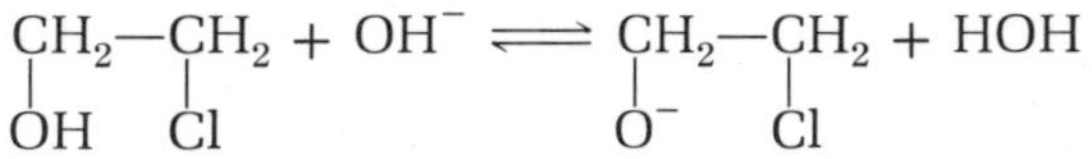

$$\underset{\underset{\displaystyle OH}{|}}{CH_2}{-}\underset{\underset{\displaystyle Cl}{|}}{CH_2} + OH^- \;\rightleftharpoons\; \underset{\underset{\displaystyle O^-}{|}}{CH_2}{-}\underset{\underset{\displaystyle Cl}{|}}{CH_2} + HOH \qquad (7\text{-}18a)$$

$$\underset{\underset{\displaystyle O^-}{|}}{CH_2}{-}\overset{\overset{\displaystyle Cl}{|}}{CH_2} \xrightarrow{\ S_N2\ } CH_2{-}CH_2 + Cl^- \qquad (7\text{-}18b)$$

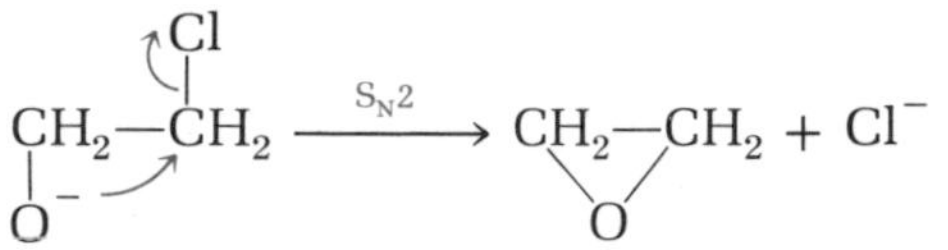

Annual United States production capacity for ethylene oxide exceeds 2.2 billion pounds. The commercial importance of ethylene oxide rests on its remarkable reactivity, enabling its conversion to many useful products. The carbon-oxygen bond is broken far more readily than the same link in an ordinary ether. The reason for this is the strain due to the abnormally small angles necessitated by the compound's structure.

Ring opening of epoxides is usually acid-catalyzed. The mechanism for the acid-catalyzed addition of water to ethylene oxide to make ethylene glycol is typical.

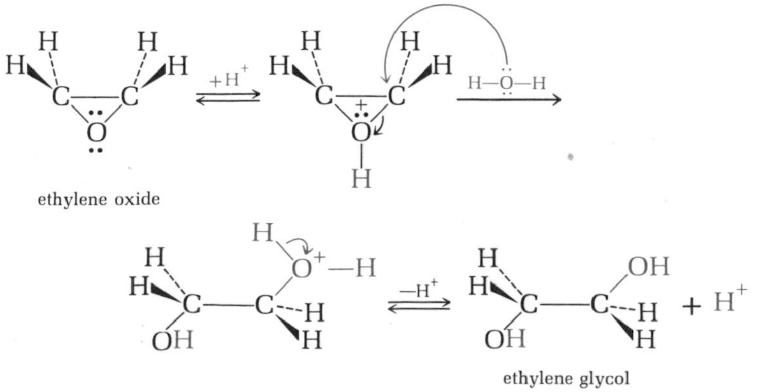

ethylene oxide

$$(7\text{-}19)$$

ethylene glycol

The net result is the addition of a molecule of water and the opening of the ring.

$$\underset{O}{\overset{\displaystyle CH_2-CH_2}{\diagdown \diagup}} + H-OH \longrightarrow \underset{\underset{OH}{|} \quad \underset{OH}{|}}{CH_2-CH_2} \qquad (7\text{-}20)$$

More than half the ethylene oxide produced goes into the manufacture of ethylene glycol which is consumed primarily as permanent antifreeze, but is also used in the manufacture of fibers and films.

Ethylene oxide reacts with many other nucleophiles besides water. Typical of these are alcohols, glycols and ammonia.

$$\underset{O}{\overset{\displaystyle CH_2-CH_2}{\diagdown \diagup}} \begin{cases} \xrightarrow{CH_3OH} HOCH_2CH_2OCH_3 \\ \qquad\qquad \text{2-methoxyethanol} \\ \xrightarrow{HOCH_2CH_2OH} HOCH_2CH_2OCH_2CH_2OH \qquad (7\text{-}21) \\ \qquad\qquad \text{diethylene glycol} \\ \xrightarrow{NH_3} HOCH_2CH_2NH_2 \\ \qquad\qquad \text{ethanolamine} \end{cases}$$

2-Methoxyethanol is added to jet fuels to prevent the formation of ice crystals. Being both an alcohol and an ether, the substance is soluble in water and in organic solvents. Over 50 million pounds of **diethylene glycol** are produced annually. It is used as a plasticizer in the resin used to bind cork granules in gaskets and tile, as a solvent for dyes, and as a selective solvent in petroleum refining. **Ethanolamine,** a water-soluble organic base, is used to absorb acid gases from refineries and in dry-ice manufacture, for the recovery and concentration of carbon dioxide.

Epichlorhydrin is another important commercial epoxide, being one of the two raw materials for the manufacture of **epoxy resins.** Addition of hypochlorous acid to allyl chloride (sec 3-4f) produces 1,3-dichloro-2-propanol which, with base, cyclizes to epichlorhydrin.

$$CH_2{=}CH{-}CH_2Cl \xrightarrow[H_2O]{Cl_2} \underset{\underset{Cl}{|} \quad \underset{OH}{|} \quad \underset{Cl}{|}}{CH_2-CH-CH_2} \xrightarrow{OH^-}$$

1,3-dichloro-2-propanol

$$\underset{O}{\overset{\displaystyle CH_2-CH_2-CH_2Cl}{\diagdown \diagup}} \qquad (7\text{-}22)$$

epichlorhydrin

The other raw material is bisphenol-A (a dihydric phenol, readily synthesized from phenol and acetone). The first step in the base-catalyzed polymerization of the two reactants is believed to be an S_N2 attack by phenoxide ion on the epoxide ring.

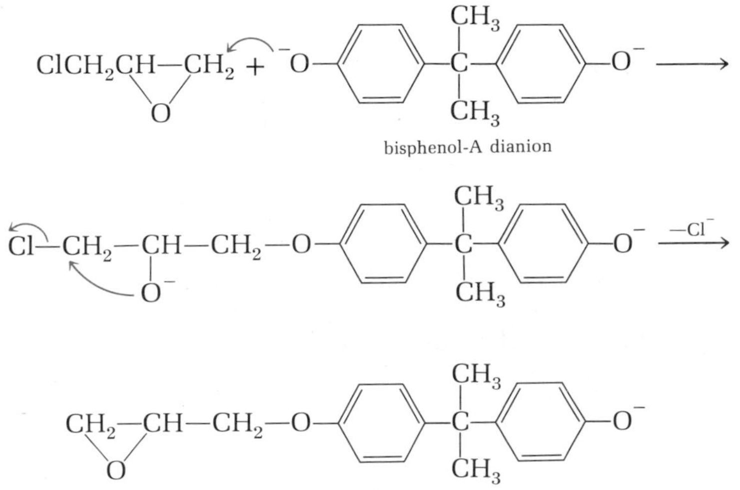

bisphenol-A dianion

(7-23)

Another epoxide ring is generated by displacement of chloride ion. Since bisphenol-A is bifunctional, the same reaction can occur at the other "end" of the phenol, and since a new epoxide ring is produced each time, the process can be repeated to build a polymer with alternating units.

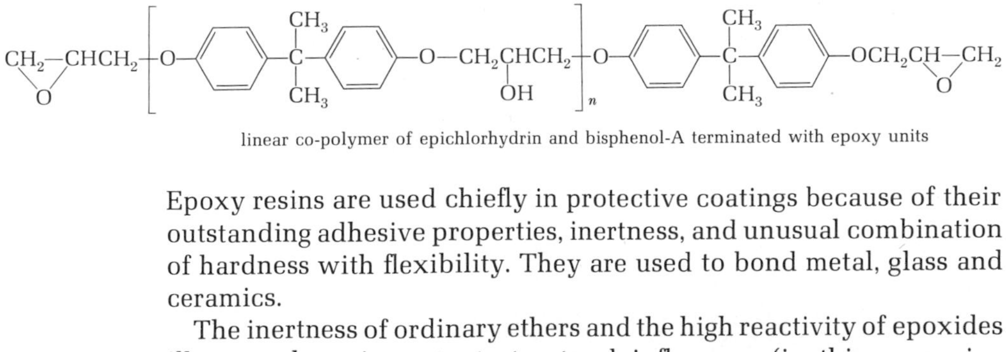

linear co-polymer of epichlorhydrin and bisphenol-A terminated with epoxy units

Epoxy resins are used chiefly in protective coatings because of their outstanding adhesive properties, inertness, and unusual combination of hardness with flexibility. They are used to bond metal, glass and ceramics.

The inertness of ordinary ethers and the high reactivity of epoxides illustrate how important structural influences (in this case, ring strain) can be on the chemical behavior of a single functional group.

7.6 SULFUR ANALOGS OF ETHERS

Thioethers, or **sulfides,** can be prepared from mercaptans by a method analogous to the Williamson synthesis.

$$RS^-Na^+ + R'X \longrightarrow RSR' + Na^+X^-$$

a sodium   thioether
mercaptide   (sulfide)

(7-24)

$$CH_3S^-Na^+ + CH_3CH_2Br \longrightarrow CH_3SCH_2CH_3 + Na^+Br^-$$

methyl ethyl sulfide

Sulfides may be oxidized to sulfoxides or sulfones.

$$R\overset{\cdot\cdot}{\underset{\cdot\cdot}{S}}R \xrightarrow{H_2O_2} R-\overset{:\overset{\cdot\cdot}{O}:^-}{\underset{\cdot\cdot}{S^+}}-R \xrightarrow{HNO_3} R-\overset{:\overset{\cdot\cdot}{O}:^-}{\underset{:\overset{\cdot\cdot}{O}:^-}{S^{++}}}-R$$

sulfide   sulfoxide   sulfone

(7-25)

Dimethyl sulfoxide (bp 189°) is an excellent solvent for polar substances, even some salts, and is also used as an intermediate in organic synthesis.

Allyl sulfide is a constituent of garlic and onions. **Mustard gas** is a powerful vesicant which has been used in chemical warfare.

$CH_2=CHCH_2-S-CH_2CH=CH_2$ $ClCH_2CH_2-S-CH_2CH_2Cl$

allyl sulfide   2-chloroethyl sulfide
(mustard gas)

New Concepts, Facts, and Terms

1. Structure; relation of ethers to water and alcohols
2. Nomenclature; alkoxy groups
3. Preparation of ethers
   a. intermolecular dehydration of alcohols; may be S_N2 or S_N1
   b. Williamson synthesis; S_N2 displacement by alkoxides or phenoxides on alkyl halides or sulfates
4. Properties of ethers
   a. inert; used as solvents
   b. basic to strong acids
   c. cleaved by mineral acids; may be S_N2 or S_N1
5. Important ethers; dimethyl (refrigerant), diethyl (anesthetic, solvent), diphenyl (inert heat exchanger), eugenol and anethole (flavors and fragrances).
6. Epoxides; ethylene oxide
   a. preparation by intramolecular Williamson method
   b. reaction with nucleophiles; commercial source of ethylene glycol
   c. epichlorhydrin; epoxy resins
7. Sulfur analogs; sulfides, sulfoxides, sulfones

Exercises and Problems

1. Write structural formulas for each of the following:
   a. n-propyl ether
   b. methyl t-butyl ether
   c. 3-methoxyhexane
   d. allyl ether
   e. p-bromophenyl ethyl ether
   f. diethylene glycol
   ~~g. ethylene glycol dimethyl ether~~
   h. isopropyl sulfide
   i. chloromethyl ether
   j. cis-2-butene oxide

2. Name each of the following: *Isobutyl phenyl ether*
   1 methyl ethleer ether

   a. $(CH_3)_2CHOCH(CH_3)_2$ *ethyl*
   b. $(CH_3)_2CHCH_2OCH_3$ *isobutene methyl ether*
   c. $CH_3CH\!-\!CH_2$ *propylene oxide*
   $\underset{O}{}$
   p Bromo phenyl methyl ether

   f. [benzene ring]$-OC(CH_3)_3$ *2 Ethlyoxy Pentave*
   g. $CH_3CH(OCH_2CH_3)CH_2CH_2CH_3$ *3 cloro ethyl oxide*
   h. $CH_2\!-\!CH\!-\!CH_2Cl$ *epichlorohydrine*
   $\underset{O}{}$

   d. Br$-$[benzene ring]$-OCH_3$

   e. $CH_3OCH_2CH_2OH$
   ethanol methly ether
   2 methoxy ethanol

   i. $CH_3SCH_2CH_2CH_3$ *methly a propyl sulfide*

   j. [cyclopentane ring]$-OCH_3$

3. Ethyl n-propyl ether may be prepared by heating a mixture of ethyl and n-propyl alcohols with sulfuric acid. What other ethers might be simultaneously produced? Illustrate by writing equations.

4. Give equations for two different combinations of reagents for preparing the following ethers by the Williamson method:
   a. methyl ethyl ether   b. methyl sec-butyl ether

5. Only one combination of reagents (phenoxide ion and alkyl halide) is effective in the Williamson synthesis of an aryl ether (say, phenyl n-butyl ether). Write the equation for this reaction. Why cannot the alkoxide and aryl halide be used?

6. Write equations for each of the following. If no reaction occurs, so indicate.

   a. ethanol + H_2SO_4 $\xrightarrow{140°}$
   b. n-butyl ether + boiling aqueous NaOH $\longrightarrow$
   c. methyl n-propyl ether + excess HBr (hot) $\longrightarrow$
   d. n-propyl ether + Na $\longrightarrow$
   e. potassium t-butoxide + ethyl iodide $\longrightarrow$
   f. ethyl ether + cold concentrated H_2SO_4 $\longrightarrow$

7. Treatment of cis-1-bromo-2-methylcyclopentane with sodium methoxide gives, as the only ether, trans-1-methoxy-2-methylcyclopentane. Draw a transition state which explains this result. Recalling that sodium methoxide is a strong base, what other reaction products might be expected?

8. Explain why, when a mixture of methanol and *t*-butyl alcohol is treated with acid, a high yield of *t*-butyl methyl ether results (eq 7-6), and little methyl ether and no *t*-butyl ether are formed.

9. Starting with ethylene oxide, give equations for synthesizing the following compounds:
   a. $HOCH_2CH_2Br$
   b. $HOCH_2CH_2OC_6H_5$
   c. $CH_3OCH_2CH_2OCH_3$
   d. $N(CH_2CH_2OH)_3$

10. An unknown ether, when it is heated with excess HBr, gives 1,4-dibromobutane as the only product. Write a structure for the ether and equations for the reactions.

11. What chemical test would distinguish the compounds in each pair? Indicate what is visually observed in each test.
    a. *n*-propyl ether and hexane
    b. ethyl phenyl ether and allyl phenyl ether
    c. 2-butanol and methyl *n*-propyl ether
    d. phenol and anisole

12. Write equations for each of the following (some may involve several steps):
    a. *n*-butyl alcohol to *n*-butyl ether
    b. benzene and ethylene to ethyl phenyl ether
    c. *n*-propyl ether to propene
    d. ethylene to 2-ethoxyethanol
    e. phenol and ethanol to ethyl cyclohexyl ether
    f. propene to allyl ether
    g. ethyl bromide to ethyl sulfide
    h. propylene to propylene oxide

13. Write equations which show the mechanisms of each step in equation 7-22. Explain why the intermediate is 1,3-dichloro-2-propanol, not 2,3-dichloro-1-propanol.

14. Write equations which show the mechanism of each part of equation 7-21.

15. Give a series of equations which explain how an epoxy resin with the structure shown in section 7.5 is obtained from the product of equation 7-23.

16. Write equations for the reactions of $CH_2{=}CHCH(OCH_3)CH_2OH$ with the following reagents:
    a. dilute $KMnO_4$
    b. potassium
    c. aqueous sodium hydroxide
    d. excess HBr and heat
    e. bromine
    If no reaction would be expected, indicate this.

17. An organic compound with the molecular formula $C_4H_{10}O_3$ shows the properties of both an alcohol and an ether. When treated with an excess of hydrogen bromide, it yields only one organic compound, ethylene dibromide. Draw a structural formula for the original compound.

18. Compound **A**, $C_8H_{10}O_2$, absorbs three moles of H_2 on catalytic hydrogenation to give **B**. Treatment of **B** with excess hot HBr gives methyl bromide and **C**, $C_6H_{10}Br_2$. Bromination of **A** (with Br_2 and Fe) gives a single monobromo derivative, $C_8H_9BrO_2$. Deduce structures for all the compounds, and write equations for each reaction.

CHAPTER EIGHT

ORGANIC HALOGEN COMPOUNDS

Halogen-containing organic compounds are rare in nature, but many of these substances have been synthesized in the laboratory or prepared commercially. Alkyl and aryl halides are the workhorses of synthetic organic chemistry, since they are easily converted to many other classes of compounds. On the other hand, polyhalogen compounds often have practical uses in their own right. For example, they are used as dry-cleaning agents, anesthetics, insecticides, fire extinguishers, aerosol propellants, and refrigerants.

It is convenient to consider separately those compounds with only one halogen atom (alkyl and aryl halides) and those with more than one halogen atom per molecule (polyhalides). The most common and synthetically useful monohalides are chlorides, bromides, or iodides; fluorides are rarely used, though polyhalides with fluorine are quite common.

Organic halogen compounds have already been mentioned many times in this text (especially in sec 2.7a, 2.8a and b, 3.4, 3.5b, 4.4, 4.8, 5.4, 5.5, 6.5c, 6.7a and 7.2b). The student might find it useful to review these sections. In this chapter we shall correlate and extend this material, review certain reaction mechanisms, consider organometallic compounds, and discuss several commercially useful polyhalogen compounds.

8.1 PREPARATION OF ALKYL AND ARYL HALIDES

8.1a HALOGENATION OF HYDROCARBONS

Direct halogenation of alkanes frequently gives a mixture of products (sec 2.7a) unless the alkane structure is symmetric, as in the vapor phase chlorination of cyclopropane (eq 8-1); cf. cyclopentane, eq 2-11.

$$\underset{\substack{\diagup\diagdown\\ CH_2}}{CH_2-CH_2} + Cl_2 \xrightarrow[\text{light}]{\text{ultraviolet}} \underset{\substack{\diagup\diagdown\\ CH_2}}{CH_2-CHCl} + HCl \;(+ \text{ some dichloro compounds})$$

$$(8\text{-}1)$$

The reaction is also useful if one type of hydrogen in the molecule is much more easily replaced than the others. This is true of hydrogens on a carbon adjacent to a double bond (eq 3-28a) or an aromatic ring (eq 5-19). In all cases, the reaction proceeds by a free radical mechanism (sec 2.7a). Radicals adjacent to a double bond or aromatic ring are resonance-stabilized; hydrogens at these positions are therefore most easily removed and replaced by halogens.

$$\dot{C}H_2-CH{=}CH_2 \longleftrightarrow CH_2{=}CH-\dot{C}H_2$$
allylic free radical

benzylic free radical

It is not surprising, therefore, that bromination of cumene proceeds as shown in equation 8-2, despite the presence of eleven other replaceable hydrogens.

$$CH_3-CH-CH_3 + Br_2 \xrightarrow{\text{sunlight}} CH_3-\overset{\overset{\displaystyle Br}{|}}{C}-CH_3 + HBr \qquad (8\text{-}2)$$

cumene   α-bromocumene

Aromatic hydrogens are replaced by an ionic mechanism (sec 5.4). Mixtures may be obtained, but the composition of these mixtures is governed by the orientation rules (sec 5.4b). Thus toluene gives mainly o- and p-chlorotoluenes. These melt at $-34°$ and $+7.5°$ respectively, and are readily separated by fractional crystallization.

$$\overset{CH_3}{\bigcirc} + 2\,Cl_2 \xrightarrow{\text{FeCl}_3} \overset{CH_3}{\overset{}{\bigcirc}}{}^{Cl} \quad \text{and} \quad \overset{CH_3}{\underset{Cl}{\bigcirc}} + 2\,HCl \qquad (8\text{-}3)$$

58%   42%

Naphthalene, under similar conditions, produces predominantly the 1-halogenated isomer.

$$\text{(naphthalene)} + Br_2 \xrightarrow[\text{reflux}]{CCl_4} \text{(1-bromonaphthalene)} + HBr \qquad (8\text{-}4)$$

8.1b ALKYL HALIDES FROM ALCOHOLS

Alcohols may be converted to alkyl halides by the use of such reagents as hydrogen halides, phosphorus trihalides, or thionyl chloride (sec 6.7a). These are perhaps the most common laboratory methods for making alkyl halides. Reactive unsaturated alcohols can be converted to unsaturated halides; in aqueous acid, the oxygen is protonated more rapidly than is the carbon-carbon double bond.

$$CH_2{=}CH-CH_2OH + \text{conc. HCl} \longrightarrow CH_2{=}CH-CH_2Cl + H_2O$$

allyl alcohol   allyl chloride

$$(8\text{-}5)$$

8.1c ADDITION TO ALKENES

Alkenes add hydrogen halides to give alkyl halides. The reaction follows Markownikoff's rule (sec 3.4c). Even alkyl fluorides can be made this way.

$$CH_3CH{=}CH_2 + HF \xrightarrow{0°} CH_3\underset{\underset{F}{|}}{C}HCH_3 \qquad (8\text{-}6)$$

isopropyl fluoride

8.1d   HALOGEN EXCHANGE

Iodides and fluorides are frequently made from chlorides or bromides by exchange of one halogen atom for another.

Organic iodides can be prepared from organic chlorides and sodium iodide when acetone is used as the solvent. The equilibrium is shifted to the right because sodium chloride is considerably less soluble in acetone than is sodium iodide. The reaction is an S_N2 displacement which proceeds best when R is a primary group.

$$R{-}Cl + Na^+I^- \xrightarrow{acetone} R{-}I + {\downarrow}NaCl \qquad (8\text{-}7)$$

Simple alkyl fluorides can be prepared from the corresponding chlorides by reaction with metallic fluorides. Mercurous fluoride is

$$2\ CH_3CH_2CH_2CH_2Cl + Hg_2F_2 \xrightarrow{heat}$$

$$2\ CH_3CH_2CH_2CH_2F + Hg_2Cl_2 \quad (8\text{-}8)$$

most commonly used. This method supplements the addition of HF to alkenes.

8.2   USE OF ALKYL HALIDES IN SYNTHESIS BY DISPLACEMENT

The utility of alkyl halides in organic synthesis is due in part to the ease with which a halogen atom may be displaced by nucleophiles (review sec 6.5c). The reaction can be used to introduce a variety of functional groups. Some common examples are given in equations 8-9 through 8-16.

		Nucleophile		*Product (type)*			
RX	+	OH⁻	⟶	ROH (alcohol)	+	X⁻	(8-9)
RX	+	OR⁻	⟶	ROR (ether)	+	X⁻	(8-10)
RX	+	CN⁻	⟶	RCN (nitrile)	+	X⁻	(8-11)
RX	+	NH₂⁻	⟶	RNH₂ (amine)	+	X⁻	(8-12)
RX	+	SH⁻	⟶	RSH (thioalcohol)	+	X⁻	(8-13)
RX	+	SR⁻	⟶	RSR (thioether)	+	X⁻	(8-14)
RX	+	⁻C≡CR	⟶	RC≡CR (alkyne)	+	X⁻	(8-15)
RX	+	N₃⁻	⟶	RN₃ (azide)	+	X⁻	(8-16)

The reagent is usually the sodium or potassium salt of the nucleophile. The solvent is most commonly alcohol, though water (eq 8-9) or liquid ammonia (eq 8-12 and 8-15) may be used.

The student will recognize several of these reactions as having already been discussed. Equation 8-9 represents a general alcohol synthesis (sec 6.5c). Equation 8-10 is the key step in the Williamson ether synthesis (sec 7.2b). Organic sulfur compounds are frequently prepared via equations 8-13 and 8-14 (sec 6.9 and 7.6), and equation 8-15 represents a general alkyne synthesis (sec 4.5).

The success or failure of these reactions depends on the structure of R and X, and on the reaction conditions. Elimination (sec 3.5) can compete with substitution. Some discussion of the mechanisms of these reactions may be helpful in understanding these problems.

8.3 SUBSTITUTION AND ELIMINATION MECHANISMS

When an alkyl halide with a hydrogen on the carbon adjacent to the halogen-bearing carbon is treated with a nucleophile, two reactions are possible—substitution or elimination.

$$\text{substitution(S)} \quad -\overset{\text{H}}{\underset{|}{C}}-\overset{|}{\underset{|}{C}}-Nu + X:^- \qquad (8\text{-}17a)$$

$$\text{elimination(E)} \quad \overset{}{\underset{}{>}}C=C\overset{}{\underset{}{<}} + NuH + X:^- \qquad (8\text{-}17b)$$

Each of these reactions can occur by two mechanisms—bimolecular (S_N2 and E2) or unimolecular (S_N1 or E1).

$$Nu:^- + \quad \overset{S_N2}{\longrightarrow} \quad + X^- \qquad (8\text{-}18a)$$

$$Nu:^- + \quad \overset{E2}{\longrightarrow} \quad NuH + \; >C=C< \; + X^- \qquad (8\text{-}18b)$$

$$\overset{-X^-}{\longrightarrow} \quad \text{carbonium ion}$$

$$\overset{S_N1}{\underset{+Nu:^-}{\longrightarrow}} \quad -C-C-Nu \qquad (8\text{-}19a)$$

$$\overset{E1}{\underset{-H^+}{\longrightarrow}} \quad >C=C< \; + H^+ \qquad (8\text{-}19b)$$

Predictions can be made as to which of the four possible mechanisms will be operative with a particular organic halide. Consider, for example, variations in the structure of R (in RX). Bulky substituents on C-1 will sterically hinder S_N2 attack at this carbon. Thus S_N2 substitutions are facile for primary halides (RCH_2X) where these substituents are hydrogen but very slow for tertiary halides (R_3CX) where these substituents are alkyl groups. In contrast, the E2 reaction involves initial attack on a hydrogen at C-2 and is relatively unaffected by substituents at C-1. It is not surprising, therefore, that when *t*-butyl chloride is treated with a strong base (Nu:$^-$ = OH$^-$), the product is mainly isobutylene (elimination) and not *t*-butyl alcohol (substitution).

$$OH^- + \quad \underset{\substack{t\text{-butyl chloride}}}{\overset{\substack{H \\ CH_3 \\ CH_3}}{C-C}} \quad \xrightarrow{E2} \quad H_2O + \underset{\substack{\text{isobutylene} \\ (\text{no } t\text{-butyl alcohol})}}{CH_2=C(CH_3)_2} + Cl^- \qquad (8\text{-}20)$$

Substituents which stabilize carbonium ions enhance the rates of unimolecular (S_N1 and E1) reactions. The effect is perhaps best seen in a case (eq 8-21) where the substituent (Y) is quite remote from the reaction site.

$$(8\text{-}21)$$

The relative reaction rates of several compounds are shown in Table 8-1. The rate difference between compounds with the electron-donating methoxy (OCH_3) and the electron-withdrawing nitro (NO_2) group is a spectacular 13,000,000 !

Table 8-1
Effect of Substituents on Hydrolysis Rates (eq. 8-21; 90% aqueous acetone at 25°)

Y =	H	OCH$_3$	CH$_3$	Cl	NO$_2$
Relative Rate =	1.0	3500	26	0.32	0.00026

These mechanisms have been subjected to many experimental tests, some of which are described in problems nine and ten at the end of this chapter.

8.4 STEREOCHEMISTRY OF SUBSTITUTION AND ELIMINATION

Each of the four mechanisms in equations 8-18 and 8-19 has certain stereochemical consequences which have been tested by experiment.

In S_N2 displacements, the nucleophile attacks carbon at the rear of the leaving group. For example, if *cis*-3-methyl-1-chlorocyclopentane (eq 8-22) is treated with sodium methoxide (Williamson synthesis), the resulting ether has the methyl and methoxyl groups *trans*.

$$(8\text{-}22)$$

The reaction is said to take place with *inversion* at the carbon where displacement occurred. This is because bonding to the leaving group begins at the back lobe of the orbital as depicted in eq 8-23. In the

transition state

$$(8\text{-}23)$$

transition state, the leaving group (Cl) and the attacking nucleophile (OCH_3) are both partially bonded to the carbon and are aligned perpendicular to a plane containing the remaining three groups.

The stereochemical result in S_N1 displacements is quite different. If we make the cyclopentyl chloride tertiary instead of secondary by adding a methyl group, and if we use the less nucleophilic (neutral) methanol in place of the methoxide ion, the experimental result is as shown in equation 8-24. Both possible isomeric ethers are obtained, though not in exactly equal amounts. The *cis* ether corresponds to *retention* (i.e., the OCH_3 retains the same position held by the Cl), whereas the *trans* ether corresponds to *inversion* of configuration at the reaction center.

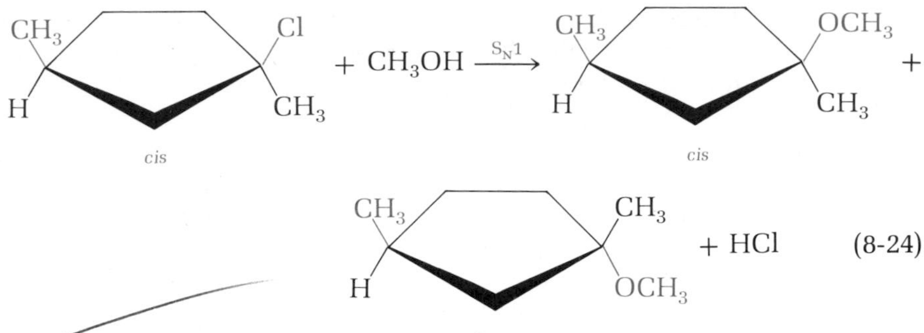

$$+ \; HCl \qquad (8\text{-}24)$$

trans

The result can be explained by a carbonium ion mechanism (eq 8-25). Loss of chloride ion gives a planar carbonium ion, with

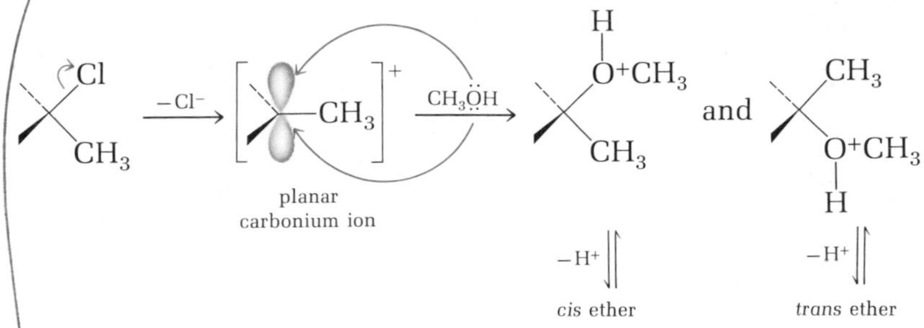

$$(8\text{-}25)$$

planar
carbonium ion

cis ether   *trans* ether

a vacant *p*-orbital perpendicular to the plane of the atoms. The nucleophile may then react at either face to give both isomeric ethers. If the chloride ion is still close to the carbonium ion when the solvent attacks, it may partially shield the face from which it departed. The observed slight preference for inversion over retention in the product can be explained in this way.

Stereochemistry is also important in elimination reactions. The preferred geometry for E2 reactions seems to be one in which the two leaving groups and the carbons to which they are attached lie in a single plane, with the leaving groups in a *transoid* arrangement. This is shown for dehydrohalogenation in equation 8-26.

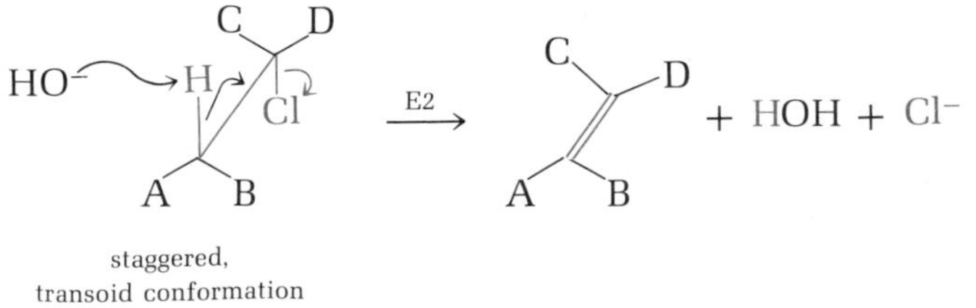

$$+ \; HOH + Cl^- \qquad (8\text{-}26)$$

staggered,
transoid conformation

The reason for this preference is that the bonds to the departing groups (i.e., the C—H and C—Cl bonds in the example shown) eventually become the π-bond in the resulting alkene. They must be properly aligned in the same plane for this to occur. Of course the *cisoid* coplanar arrangement (eq 8-27) would also satisfy this requirement. Though sometimes observed, it is less common

$$+ \; HOH + Cl^- \qquad\qquad (8\text{-}27)$$

eclipsed, cisoid conformation

because it requires the less stable eclipsed rather than staggered conformation of the alkyl halide. The two paths are easily distinguished, since they give different (isomeric) alkenes. In problem 11 at the end of this chapter, experiments are described which support the conclusion that the transoid coplanar geometry is preferred for E2 eliminations.

E1 reactions proceed via the same carbonium ion intermediate as S_N1 reactions, and like them are not stereospecific (eq 8-25).

The stereochemical requirements of substitution and elimination reactions discussed here are not restricted to the chemistry of alkyl halides. They are important whenever these reaction mechanisms are involved and frequently apply to reactions in biological systems.

8.5 NUCLEOPHILIC SUBSTITUTION IN ARYL HALIDES

Halogens attached to an aromatic ring or a double bond do not ordinarily undergo displacement by either an S_N1 or an S_N2 mechanism. Because of resonance contributors such as those shown in sec 5.4b which place a positive charge on halogen and give some double-bond character to the carbon-halogen bond, ionization of these halides by an S_N1 mechanism is considerably more difficult than with alkyl halides. The usual backside approach of a nucleophile for S_N2 displacement is blocked by the aromatic ring in aryl halides.

Simple aryl and vinyl halides are therefore usually inert to most nucleophiles. Nevertheless, displacements are possible, particularly when strongly electron-withdrawing groups are located *ortho* and/or *para* to the halogen (compare eq 8-28 and 8-29).

$$+ \; CH_3O^- \longrightarrow \text{ no reaction} \qquad\qquad (8\text{-}28)$$

$$+ \text{CH}_3\text{O}^- \xrightarrow[\text{temperature}]{\text{room}} \qquad + \text{Cl}^- \qquad (8\text{-}29)$$

Reactions such as that depicted by equation 8-29 are bimolecular; their rates depend on the concentrations both of aryl halide and of nucleophile. Any postulated mechanism must explain why electron-withdrawing groups in *ortho/para* positions are so effective in promoting the reaction. The established mechanism parallels that of electrophilic aromatic substitution (sec 5.4a) except that the attacking reagent is a nucleophile rather than an electrophile and the intermediate is, consequently, a carbanion rather than a carbonium ion. Equation 8-30 describes a general mechanism for this type of reaction.

carbanion intermediate

The carbanion intermediate has a negative charge concentrated primarily at the *ortho* and *para* positions. Any group in those positions which can stabilize (or delocalize) that charge should facilitate the reaction. Nitro groups are particularly effective in this regard, since the negative charge can be delocalized to the oxygens through resonance.

Thus, electron-withdrawing groups facilitate nucleophilic aromatic substitution, just as electron-donating substituents facilitate electrophilic aromatic substitution. Without these stabilizing groups, the reaction conditions must be quite severe (eq 6-21 and 6-23). Even then, the reaction may not occur. Nucleophilic aromatic substitution has been useful in determining protein structures (sec 17.7).

8.6 ORGANOMETALLIC COMPOUNDS

In organometallic compounds, carbon is directly linked to a metallic element. Such compounds have been prepared from virtually every metallic element in the periodic table. Many (those derived from magnesium, lithium, mercury, tin, boron, and aluminum) are reactive and useful as synthetic intermediates. Others (for example *tetraethyl lead,* the well-known gasoline anti-knock, *ethylmercuric chloride,* an important fungicide, and the antiseptic dye *mercurochrome*) are useful in their own right.

$(CH_3CH_2)_4Pb$ CH_3CH_2HgCl

tetraethyl lead   ethylmercuric
                                      chloride

mercurochrome

8.6a PREPARATION OF GRIGNARD REAGENTS

By far the most useful organometallic reagents in synthesis were discovered by the French organic chemist, Victor Grignard (pronounced *green-yard*). He found that magnesium turnings gradually dissolved when stirred with an ether solution of an alkyl or aryl halide. The resulting solutions are known as **Grignard reagents.** The solvent is important in stabilizing the organomagnesium compound.

The exact structure of Grignard reagents is still a subject of research, but they can be represented as alkyl- or arylmagnesium halides. They most frequently behave chemically as if the alkyl or aryl group were *negative* (a carbanion or a nucleophile), although the carbon-magnesium bond is only partially ionic.

$$R\!-\!X + Mg \xrightarrow{\text{ether}} \overset{\delta- \quad \delta+}{R\,Mg\,X} \qquad (8\text{-}31)$$

Grignard reagent

$$CH_3\!-\!I + Mg \xrightarrow{\text{ether}} CH_3MgI \qquad (8\text{-}32)$$

methyl iodide   methylmagnesium iodide

$$\text{⬡}\!-\!Br + Mg \xrightarrow{\text{ether}} \text{⬡}\!-\!MgBr \qquad (8\text{-}33)$$

phenylmagnesium bromide

Figure 8-1
Victor Grignard (1871–1935) is noted primarily as the discoverer of the reagent which bears his name and as the editor of a comprehensive French treatise on organic chemistry. He received the Nobel prize in 1912 for his work in synthetic chemistry. (Courtesy of the Journal of Chemical Education.)

8.6b REACTIONS OF GRIGNARD REAGENTS

Grignard reagents must be prepared in scrupulously dry apparatus, for they react vigorously with water (or any other O—H compound). The reaction can be used to prepare hydrocarbons from alkyl or aryl halides.

$$\overset{\delta-}{R}\overset{\delta+}{MgX} + \overset{\delta+}{H}\overset{\delta-}{-OH} \longrightarrow RH + Mg^{2+}OH^-X^- \qquad (8\text{-}34)$$

If D_2O is used in place of H_2O, deuterium can be introduced in a particular position, as in the following example:

$$CH_3-\!\!\!\!\bigcirc\!\!\!\!-Br \xrightarrow[\text{ether}]{Mg} CH_3-\!\!\!\!\bigcirc\!\!\!\!-MgBr \xrightarrow{D_2O}$$

$$CH_3-\!\!\!\!\bigcirc\!\!\!\!-D \qquad (8\text{-}35)$$

p-deuterotoluene

This is a useful way of isotopically labeling organic compounds.

Grignard reagents may act as nucleophiles in S_N2 displacements. For example, in reactions with ethylene oxide, ring-opening occurs (compare with eq 7-23).

$$R \overset{\delta-}{\overbrace{}} \overset{\delta+}{MgX} + CH_2\!\!-\!\!CH_2 \xrightarrow{S_N2} R\!-\!CH_2CH_2O\!-\!\overset{+}{MgX} \xrightarrow{H\!-\!OH}$$

$$R\!-\!CH_2CH_2OH + Mg^{2+}OH^-X^- \quad (8\text{-}36)$$

Hydrolysis of the resulting alkoxymagnesium halide gives a primary alcohol with two more carbons than the original alkyl halide from which the Grignard reagent was prepared. The perfume constituent, oil of roses (2-phenylethanol), can be prepared using this reaction sequence.

$$\text{⟨⟩}\!-\!MgBr \xrightarrow[\text{O}]{CH_2\!-\!CH_2} \text{⟨⟩}\!-\!CH_2CH_2O^-\overset{+}{MgBr} \xrightarrow{H_2O}$$

$$(8\text{-}37)$$

$$\text{⟨⟩}\!-\!CH_2CH_2OH$$

2-phenylethanol
(oil of roses)

Grignard reagents may displace halogens from certain inorganic halides. Silicon tetrachloride reacts to give products with one to four organic groups. By regulating the ratio of reagents and reaction

$$SiCl_4 \xrightarrow{RMgX} RSiCl_3 \xrightarrow{RMgX} \underset{\substack{\text{dialkyldichloro-}\\\text{silane}}}{R_2SiCl_2} \xrightarrow{RMgX}$$

$$R_3SiCl \xrightarrow{RMgX} \underset{\text{tetraalkylsilane}}{R_4Si} \quad (8\text{-}38)$$

conditions, any of the four types of organosilicon compounds can be obtained. The di-substituted compounds are particularly useful, for on hydrolysis they give **silicone polymers.** By varying R, the properties of the polymer can be modified over a wide range.

$$R_2SiCl_2 + H_2O \longrightarrow \underset{\substack{|\\R}}{\overset{\substack{R\\|}}{-\!Si\!-}}O\!-\underset{\substack{|\\R}}{\overset{\substack{R\\|}}{Si\!-}}O\!- + HCl \quad (8\text{-}39)$$

silicone polymer

Silicone polymers are non-wettable and are resistant to high temperatures. They are used in automobile polishes, waterproof films, and high-temperature lubricants.

The use of Grignard reagents in other syntheses will be discussed in later sections (9.6f and 10.4c).

8.7   POLYHALOGEN COMPOUNDS

Many polyhalogen compounds are produced commercially, either for use as such or in the form of polymers. It is convenient to consider separately those with one carbon and those with more than one carbon.

8.7a   POLYHALOMETHANES

Direct chlorination (sec 2.7a) is used to make each of the chlorinated methanes. However, **carbon tetrachloride** can also be manufactured by chlorination of carbon disulfide (obtained by heating sulfur with coke in an electric furnace). The products, CCl_4 and S_2Cl_2, are separated by fractional distillation.

$$CS_2 \; + \; 3\,Cl_2 \; \xrightarrow{\;Fe\;} \; CCl_4 \; + \; S_2Cl_2 \qquad (8\text{-}40)$$

carbon   carbon   sulfur
disulfide   tetrachloride   monochloride

Carbon tetrachloride is a colorless, dense liquid with a mild, somewhat pleasant odor. Although some carbon tetrachloride is used as a dry-cleaner or fire extinguisher, most of it is converted to tetrachloroethylene (eq 8-43) and freons (eq 8-41). **Chloroform** ($CHCl_3$) and **methylene chloride** (CH_2Cl_2) are useful, relatively inert, low-boiling solvents.

The mixed chlorofluoromethanes are known as **freons**. These are colorless gases which are nontoxic, nonflammable, and noncorrosive. They are used extensively as industrial and domestic refrigerants in air conditioning and deep-freeze units. Freons are produced commercially from chlorocarbons, by fluorination.

$$CCl_4 \; \xrightarrow[\;SbF_5\;]{\;HF\;} \; CCl_3F \; \xrightarrow[\;SbF_5\;]{\;HF\;} \; CCl_2F_2 \qquad (8\text{-}41)$$

trichlorofluoromethane   dichlorodifluoromethane
(freon-11)   (freon-12)

Freon production has soared during the last few years, owing to freon's use as the inert propellant in aerosol bombs (bombs used for insecticides, hair sprays, paints, suntan lotions, shave creams, and the like).

8.7b OTHER POLYHALIDES

Trichloroethylene and **tetrachloroethylene** are the major solvents used commercially to degrease metals and to dry-clean fabrics. They can be made from acetylene (eq 8-42) by chlorine addition and dehydrochlorination steps.

$$HC{\equiv}CH \xrightarrow[\text{FeCl}_3]{2\ \text{Cl}_2} \underset{\text{tetrachloroethane}}{CHCl_2CHCl_2} \xrightarrow[-\ \text{HCl}]{\text{Ca(OH)}_2} \underset{\text{trichloroethylene}}{CHCl{=}CCl_2} \xrightarrow{\text{Cl}_2} \quad (8\text{-}42)$$

$$\underset{\text{pentachloroethane}}{CHCl_2CCl_3} \xrightarrow[-\ \text{HCl}]{\text{Ca(OH)}_2} \underset{\text{tetrachloroethylene}}{CCl_2{=}CCl_2}$$

Tetrachloroethylene is also obtained by pyrolysis of carbon tetrachloride.

$$2\ CCl_4 \xrightarrow{800\text{--}900^\circ} CCl_2{=}CCl_2 + 2\ Cl_2 \qquad (8\text{-}43)$$

Tetrafluoroethylene is made in two steps from chloroform.

$$CHCl_3 \xrightarrow[\text{SbF}_5]{\text{HF}} \underset{\text{chlorodifluoromethane}}{CHClF_2} \xrightarrow[-\ \text{HCl}]{700^\circ} \underset{\text{tetrafluoroethylene}}{CF_2{=}CF_2} \qquad (8\text{-}44)$$

Figure 8-2
Teflon coatings withstand the elevated temperatures used in cooking and prevent food from sticking to the utensils, making them easier to clean. (Courtesy of E. I. DuPont DeNemours & Company, Inc.)

When treated with peroxides it produces an extremely inert polymer
Teflon, which is resistant to almost all chemical reagents. Fluorine-

$$n \quad CF_2{=}CF_2 \xrightarrow[\text{catalyst}]{\text{peroxide}} {+}CF_2CF_2{\,}_{\overline{n}} \qquad (8\text{-}45)$$

Teflon

containing polymers are being increasingly used where protective
coatings are needed (Figure 8-2).

   Many more complex polyhalogen compounds are known. A few
of these are the well-known insecticides **DDT** (dichlorodiphenyl-
trichloroethane), **chlordane, aldrin,** and **dieldrin.** Although these

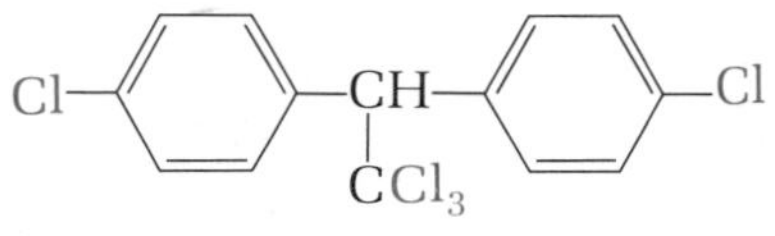

DDT

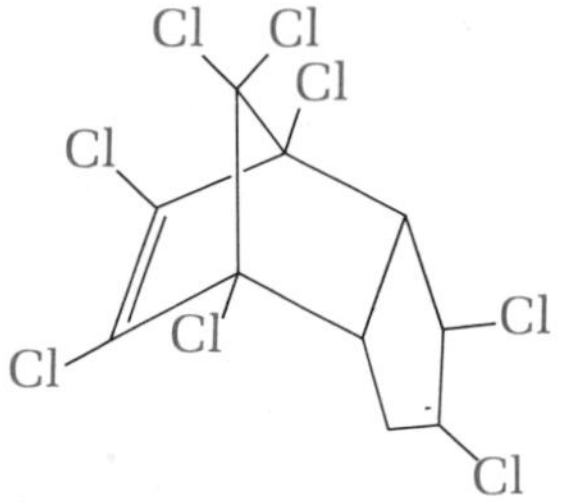

chlordane

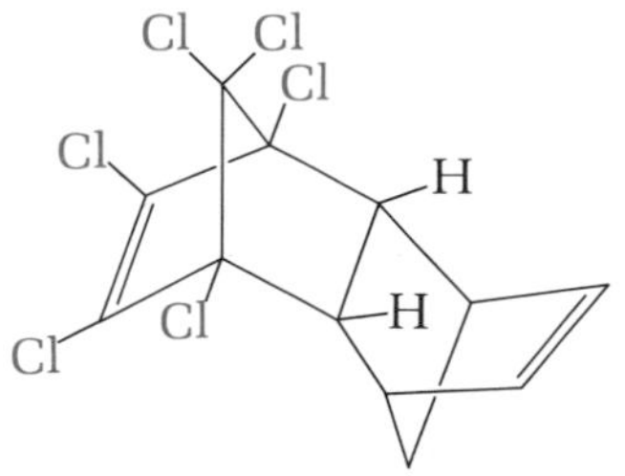

aldrin

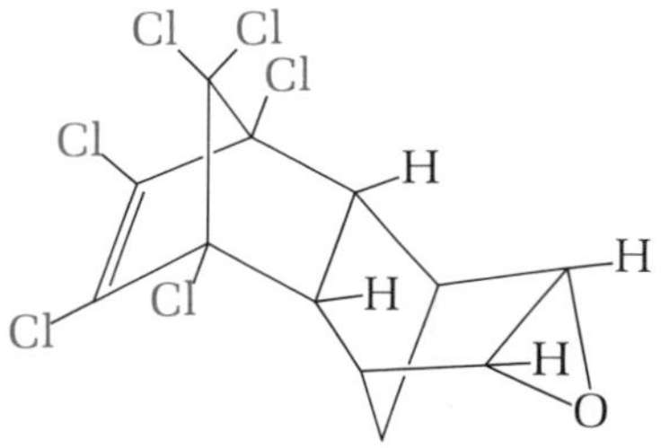

dieldrin

insecticides have been invaluable in controlling diseases such as
malaria and in improving agricultural yields, their indiscriminate use
can be severely criticized because of the insecticides' persistence and
high toxicity to animal life.

New Concepts, Facts, and Terms

1. Preparation of alkyl and aryl halides
   a. halogenation of hydrocarbons; free radical mechanism for alkanes,
      ionic mechanism for arenes
   b. replacement of OH in alcohols
   c. addition of HX to alkenes
   d. replacement of one halogen by another, especially for iodides and
      fluorides

2. Displacement of halides by nucleophiles; used to prepare alcohols, ethers, nitriles, amines, thioalcohols, thioethers, acetylenes, and azides
3. Mechanisms for substitution (S_N1, S_N2) and elimination (E1, E2)
4. Stereochemistry: S_N2 inversion; S_N1 inversion and retention; E2-transoid coplanar
5. Nucleophilic aromatic substitution—aided by electron-withdrawing groups in *ortho* and/or *para* positions
6. Organometallic compounds; Grignard reagents (RMgX). Used to make hydrocarbons, alcohols, silanes. Silicone polymers.
7. Polyhalogen compounds; carbon tetrachloride, freons, tri- and tetra-chloroethylene, Teflon, DDT and other insecticides

Exercises and Problems

1. Write structural formulas for the following compounds:
   a. *m*-chlorotoluene
   b. 2,3-dibromobutane
   c. tribromomethane
   d. *cis*-1,3-dichlorocyclobutane
   e. *p*-bromobenzyl chloride
   f. allyl iodide
   g. trichlorofluoromethane
   h. *o*-chloroiodobenzene
   i. vinyl bromide
   j. 1-fluorobutane

2. Name the following compounds:
   a. $(CH_3)_3CCH_2Br$

   b. $Br-\langle\bigcirc\rangle-Cl$

   f.

   g. $BrCH_2C\equiv CCH_2Br$

   c. $CH_3CHBrCH(CH_3)CH_2CH_2Cl$
   d. $C_6H_5CH_2CH_2CH_2Cl$
   e. $CH_3CH_2CF_2CH_3$

   h. $CH_3CH_2CH_2MgCl$

   i. $Cl-\langle\bigcirc\rangle-OCH_3$

   j. $CH_2{=}CBrCH_3$

3. Write out all the steps in the free radical mechanism for equation 8-2. Explain why substitution occurs on the carbon adjacent to the benzene ring.

4. Write out the steps in the mechanism for equation 8-4. Repeat, for substitution in the 2-position on the naphthalene ring. Compare the intermediate carbonium ions, and suggest an explanation for the fact that substitution at C-1 predominates.

5. Write equations which show how 2-propanol could be converted to each of the following:
   a. isopropyl chloride
   b. allyl chloride
   c. 1,2-dibromopropane
   d. 2-methoxypropane
   e. 1-chloro-2-propanol
   f. isopropylbenzene

6. Write equations for the following reactions:
   a. 1-bromobutane + sodium iodide in acetone
   b. sec-butyl chloride + sodium ethoxide
   c. *t*-butyl bromide + water
   d. benzyl bromide + sodium cyanide
   e. *n*-propyl iodide + sodium acetylide
   f. 2-chloropropane + sodium hydrosulfide
   g. allyl chloride + sodium amide (NaNH$_2$)
   h. 1,4-dibromobutane + sodium cyanide
   i. 1-methyl-1-bromocyclohexane + *n*-propyl alcohol (reflux)
   j. *p*-bromotoluene + sodium ethoxide (reflux)

7. 2-Bromobutane can be obtained in one step from each of the following precursors: butane, 2-butanol, 1-butene, and 2-butene. Write an equation for each method. Describe the advantages or disadvantages of each.

8. Starting with an unsaturated hydrocarbon, show how each of the following could be prepared:
   a. 1,2-dibromobutane
   b. 1,1-dichloroethane
   c. 1,2,3,4-tetrabromobutane
   d. cyclohexyl iodide
   e. 1,4-dibromo-2-butene
   f. 1,1,2,2-tetrachloropropane
   g. 1-bromo-1-phenylethane
   h. 1,2,5,6-tetrabromocyclooctane

9. The same ratio of *t*-butyl alcohol to isobutylene is obtained whether one hydrolyzes *t*-butyl chloride, bromide, or iodide. Explain (see equation 8-19).

10. Hydrolysis of 3-bromo-1-butene in water gives a mixture of two alcohols. However, if 20% aqueous sodium hydroxide is used in place of water, only one alcohol is obtained. Explain these results in terms of nucleophilic displacement mechanisms.

11. Menthyl chloride (derived from the natural product menthol) and neomenthyl chloride differ only in the stereochemistry of the C—Cl bond, as shown:

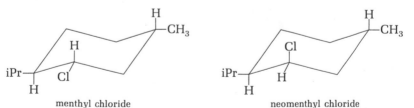

menthyl chloride   neomenthyl chloride

When each is treated with a strong base (Na$^+$$^-OC_2H_5$ in C$_2$H$_5$OH), menthyl chloride gives 100% 2-menthene whereas neomenthyl chloride gives mainly (75%) 3-menthene.

2-menthene   3-menthene

Show how these results are consistent with the transoid coplanar geometry for the transition state in E2 eliminations.

12. Write out the steps in the mechanism for equation 8-29. Show by appropriate resonance structure how each nitro group stabilizes the carbanion intermediate.

13. When 2,4,6-trinitroanisole is treated with potassium ethoxide in ethanol, a crystalline salt can be isolated. If the salt is warmed, it gives a mixture of nearly equal amounts of 2,4,6-trinitroanisole and 2,4,6-trinitrophenyl ethyl ether. What is the structure of the salt? Explain the observed reactions.

14. Which aromatic bromide do you expect to be more reactive toward displacement by methoxide ion, 4-nitrobromobenzene or 3,5-dimethyl-4-nitrobromobenzene? Consider the geometry of the intermediate carbanion in your decision, and explain your choice.

15. A Grignard reagent can be used in the following conversions. Give equations for each.
    a. sec-butyl bromide to butane
    b. allyl bromide to 4-pentene-1-ol
    c. 1-propanol to $CH_3CH_2CH_2D$
    d. 2-phenylethanol to 4-phenyl-1-butanol
    e. cyclohexene to monodeuterocyclohexane

16. Write out the mechanisms for the various steps in equation 8-42.

17. When chloroform is treated with a strong base in the presence of an alkene, a dichlorocyclopropane is obtained, frequently in good yield. For example,

Suggest a mechanism for this reaction.

CHAPTER NINE

ALDEHYDES AND KETONES

Formaldehyde, long known as a disinfectant and as a preservative for biological specimens, is the simplest of a series of compounds built around a carbonyl group (a carbon atom linked to an oxygen by a double bond). Like other unsaturated compounds, aldehydes and ketones are highly reactive and undergo addition reactions at the carbonyl group. Many natural substances are aldehydes and ketones. Among them are the various flavors from almonds, cinnamon, and the vanilla bean; perfumes; camphor; certain vitamins; and sex hormones.

Aldehydes and ketones contain the carbonyl group, $\mathord{>}C{=}O$. If one of the two groups attached to the carbonyl atom is a hydrogen atom,

$$\overset{\displaystyle O}{\underset{}{\overset{\|}{-C}}}\!-H,$$

the compounds are **aldehydes.** The aldehyde group is $-\overset{O}{\overset{\|}{C}}-H$, sometimes abbreviated $-CHO$. The remaining group may be another hydrogen atom or an alkyl or aryl group.

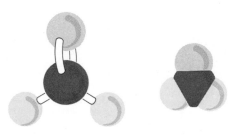

$$H-\overset{\displaystyle O}{\overset{\|}{C}}-H \qquad R-\overset{\displaystyle O}{\overset{\|}{C}}-H \qquad Ar-\overset{\displaystyle O}{\overset{\|}{C}}-H$$

formaldehyde   aliphatic aldehyde   aromatic aldehyde

Figure 9-1
Models of formaldehyde.

Those compounds in which both groups attached to the carbonyl carbon are organic (alkyl or aryl) are known as **ketones.** The groups need not be identical.

Figure 9-2
Models of acetone.

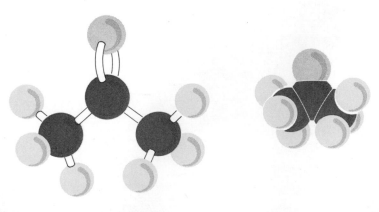

$$R\overset{\overset{\displaystyle O}{\|}}{-}C-R \qquad R\overset{\overset{\displaystyle O}{\|}}{-}C-Ar \qquad Ar\overset{\overset{\displaystyle O}{\|}}{-}C-Ar$$

  aliphatic ketone   mixed ketone   aromatic ketone

9.1 NOMENCLATURE

The simplest aldehydes are known by common names which are closely related to the names of the acids formed from them by oxidation. *Formaldehyde*, for example, gives *formic acid* (HCO_2H) when oxidized; *acetaldehyde* gives *acetic acid*. Table 9-1 gives the formulas, names, and boiling points of the simplest aldehydes.

Common names are also used for certain ketones. These are derived by naming the alkyl or aryl groups attached to the carbonyl, though in certain cases special names, which end in *-one*, are employed.

$$CH_3\overset{\overset{\displaystyle O}{\|}}{C}CH_3 \qquad CH_3\overset{\overset{\displaystyle O}{\|}}{C}CH_2CH_3 \qquad CH_3CH_2\overset{\overset{\displaystyle O}{\|}}{C}CH_2CH_3$$

   acetone   methyl ethyl ketone   diethyl ketone

   cyclohexanone   acetophenone   benzophenone
                    (methyl phenyl ketone)   (diphenyl ketone)

Table 9-1
Boiling Points of Some Simple Aldehydes

NAME	FORMULA	BP
Formaldehyde (methanal)	$H-\overset{\overset{\displaystyle O}{\|}}{C}-H$	$-21°$
Acetaldehyde (ethanal)	$CH_3-\overset{\overset{\displaystyle O}{\|}}{C}-H$	$20.2°$
Propionaldehyde (propanal)	$CH_3CH_2\overset{\overset{\displaystyle O}{\|}}{C}-H$	$48.8°$
n-Butyraldehyde (butanal)	$CH_3CH_2CH_2\overset{\overset{\displaystyle O}{\|}}{C}-H$	$75.7°$
Isobutyraldehyde (2-methylpropanal)	$CH_3-\overset{\displaystyle CH}{\underset{\displaystyle CH_3}{|}}-\overset{\overset{\displaystyle O}{\|}}{CH}$	$61°$
Benzaldehyde	$\overset{\overset{\displaystyle O}{\|}}{C}-H$	$179.5°$

In the IUPAC system, the characteristic ending for the aldehydes is *-al,* derived from the first syllable of **al**dehyde, and that for the ketones is *-one,* derived from the last syllable of ket**one.** The examples below illustrate this method of naming.

$$\overset{4}{C}H_3-\overset{3}{C}H_2-\overset{2}{C}H_2-\overset{1}{\overset{\overset{\displaystyle O}{\|}}{C}}H$$

butanal

$$\overset{4}{C}H_3-\overset{3}{\underset{\underset{\displaystyle CH_3}{|}}{C}}H-\overset{2}{C}H_2-\overset{1}{\overset{\overset{\displaystyle O}{\|}}{C}}H$$

3-methylbutanal

$$\overset{1}{C}H_3-\overset{2}{\overset{\overset{\displaystyle O}{\|}}{C}}-\overset{3}{C}H_2-\overset{4}{C}H_2-\overset{5}{C}H_3$$

2-pentanone

$$\overset{1}{C}H_3-\overset{2}{\overset{\overset{\displaystyle O}{\|}}{C}}-\overset{3}{C}H_2-\overset{4}{\underset{\underset{\displaystyle CH_3}{|}}{C}}H-\overset{5}{C}H_3$$

4-methyl-2-pentanone

Since the aldehyde function must terminate a carbon chain, it is not necessary to designate its position by a numerical prefix. When more than one functional group is present, several endings may be necessary, as in the following examples:

$$\overset{4}{C}H_2=\overset{3}{C}H-\overset{2}{C}H_2-\overset{1}{\overset{\overset{\displaystyle O}{\|}}{C}}-H$$

3-butenal

$$\overset{1}{C}H_3-\overset{2}{\overset{\overset{\displaystyle O}{\|}}{C}}-\overset{3}{C}H_2-\overset{4}{\overset{\overset{\displaystyle OH}{|}}{C}}H-\overset{5}{C}H_3$$

4-pentanol-2-one

9.2 OCCURRENCE, PROPERTIES, AND USES

Except for formaldehyde, which is a gas, the simpler aliphatic aldehydes are colorless liquids. The lower members of the series have a penetrating odor, but with increasing molecular weight, the odor becomes more fragrant. This is particularly true in the aromatic series, where many of the aldehydes are used for their flavor or perfume. Some ketones also have pleasant odors and are used extensively in the blending of perfumes. Acetone, the simplest ketone, is particularly useful as a solvent, since it dissolves most organic compounds yet is completely miscible with water. It is a solvent for coatings—which range from nail polish to exterior enamel paints.

Some of the common natural substances in which important carbonyl compounds occur are almonds (benzaldehyde), cinnamon (cinnamaldehyde), the vanilla bean (vanillin), the sex hormones (testosterone, progesterone), camphor (from the camphor tree), oil of citronella (citronellal), the anti-hemorrhagic vitamin K, the perfume fixative derived from the musk deer (muscone, a cyclic ketone with a 15-membered ring), and innumerable others (Figure 9-4).

Figure 9-3
Model of benzaldehyde.

Figure 9-4
Some naturally occurring carbonyl compounds.

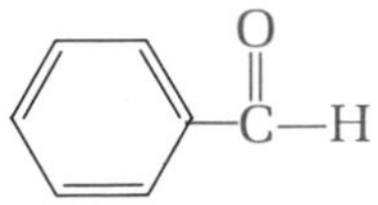

benzaldehyde
(oil of almonds)

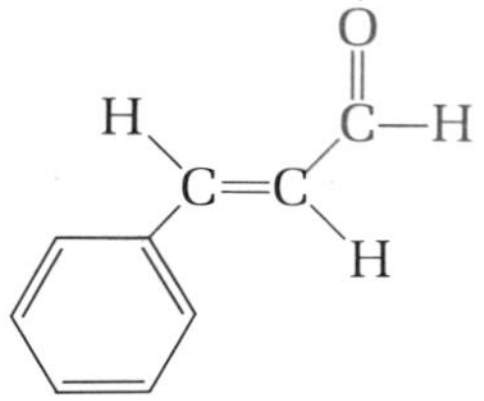

cinnamaldehyde
(cinnamon)

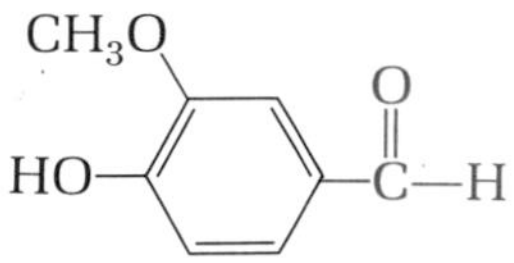

vanillin
(vanilla bean)

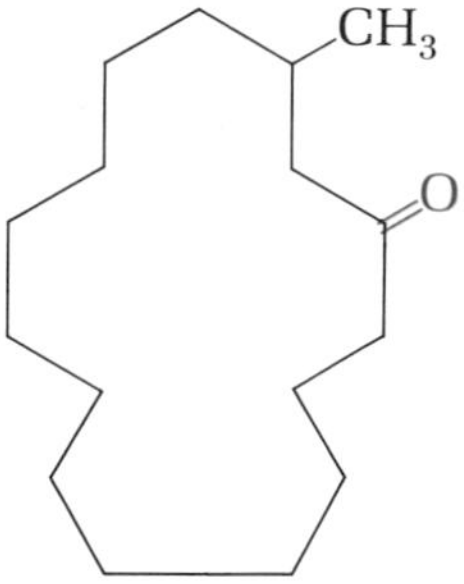

muscone
(musk deer)

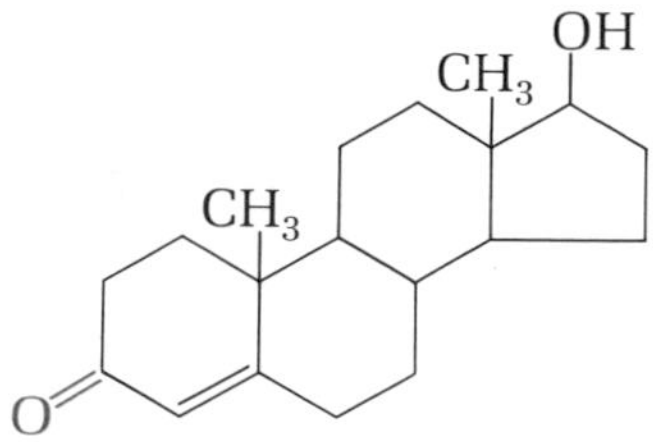

testosterone
(male sex hormone)

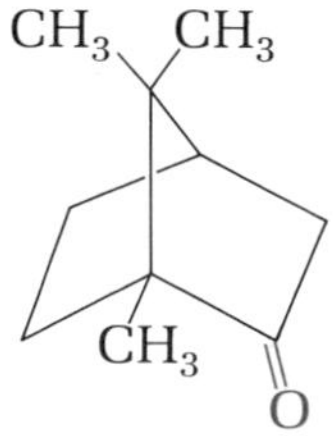

camphor

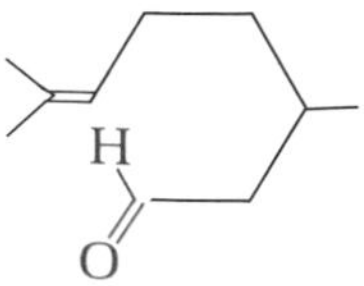

citronellal

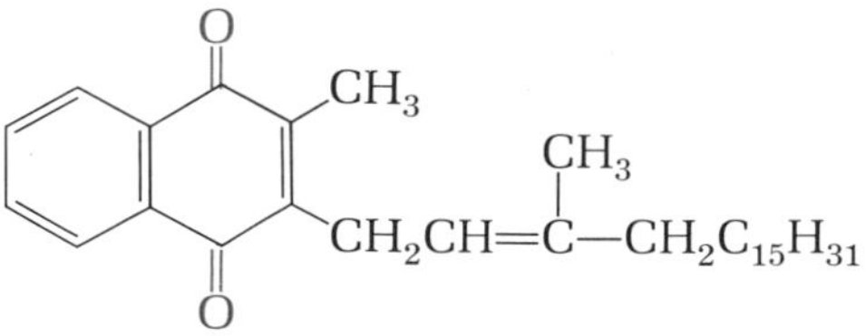

vitamin K

9.3 OXIDATION STATES OF OXYGEN-CONTAINING FUNCTIONAL GROUPS

A carbon atom is in its most reduced state when it is completely bonded to hydrogen atoms or to other saturated carbon atoms, as in the alkanes. As C—H bonds (or C—C bonds) are replaced by C—O bonds, the carbon atom attains higher oxidation states. In carbon dioxide, all four bonds are to oxygen atoms, and no further oxidation is possible. For this reason, carbon dioxide is the ultimate oxidation product of all organic compounds, whether by combustion (sec 2.7c) or by metabolism (sec 16.8).

Various oxygen-containing functional groups occur at oxidation states intermediate between those of the alkanes and carbon dioxide. These are summarized in Table 9-2. In this table methane and its oxygenated derivatives are used as examples.

Compounds at the same oxidation level are interconvertible without using oxidants or reductants, provided that the proper reagent and reaction conditions can be found. Alcohols and ethers provide a good example (eq 7-2, 7-6, 7-12). Compounds at different oxidation levels should also be interconvertible, provided the appropriate oxidizing or reducing agents can be found. Table 9-2 permits us to predict that organic acids might be prepared by oxidizing alkanes, alcohols, or aldehydes; that aldehydes might be prepared by oxidizing alcohols or by reducing acids (or their derivatives); and that aldehydes or ketones may be reduced to alcohols or to alkanes. We

Table 9-2

The Oxidation States of Oxygen-Containing Functional Groups

BONDS TO OXYGEN	COMPOUNDS AT EACH OXIDATION LEVEL		
0	CH_4 alkanes		
1	CH_3—OH alcohols	CH_3—OR ethers	
2	CH_2=O aldehydes (and ketones)	HO—CH_2—OR hemiacetal (and hemiketals)	RO—CH_2—OR acetals (and ketals)
3	$HC\overset{O}{\underset{OH}{}}$ acids	$HC\overset{O}{\underset{OR}{}}$ esters	$HC{-}OR$ with OR / OR ortho esters
4	O=C=O carbon dioxide	$RO{-}\overset{O}{\overset{\|}{C}}{-}OR$ carbonates	

increasing oxidation levels (vertical arrow, downward along left side) shall see that these and other predictions derived from Table 9-2 are valid.

Functional groups which do not contain oxygen may also be classified as to oxidation level. Alkyl halides, for example, are readily converted to alcohols (sec 6.5c) by hydrolytic reactions which do not involve oxidation or reduction. Therefore, *carbon with one carbon-halogen (C—X) bond is at the same oxidation level as carbon with one C—O bond.* A logical extension would be that a di-halogenated carbon $\left(\,\diagdown\!\!\diagup\!\!\text{CX}_2\right)$ is at the same oxidation level as alde-hydes or ketones $\left(\,\diagdown\!\!\diagup\!\!\text{C=O}\,\right)$. We shall see shortly that this prediction is useful in devising a synthetic method for carbonyl compounds (sec 9-4b).

9.4   PREPARATION OF ALDEHYDES AND KETONES

The ubiquity of aldehydes and ketones in natural products has challenged organic chemists to devise many methods for their synthesis. Only a few of the more important ones will be presented here; others will be described later, in conjunction with the reactions of other functional groups.

9.4a   OXIDATION OF ALCOHOLS

Aldehydes may be prepared by the oxidation of primary alcohols; ketones, by the oxidation of secondary alcohols (sec 6.7d). The reaction can be achieved catalytically by passing the alcohol vapor over copper gauze or powder at about 300°. When performed in this

$$\text{H}\!-\!\!\underset{\diagup}{\overset{\diagdown}{\text{C}}}\!-\!\text{OH} \xrightarrow[300°]{\text{Cu}} \underset{\diagup}{\overset{\diagdown}{\text{C}}}\!=\!\text{O} + \text{H}_2 \tag{9-1}$$

primary or
secondary
alcohol   aldehyde or
ketone

way, the reaction is called **dehydrogenation** (indeed, the word *al-dehyde* arises as an abridgement of the phrase *alcohol dehydro-genation*).

Laboratory oxidation of alcohols is most frequently accomplished using chromic acid (H_2CrO_4) as the oxidant. This is prepared from either sodium dichromate ($\text{Na}_2\text{Cr}_2\text{O}_7$) or chromic oxide ($\text{CrO}_3$) and acid (usually sulfuric or acetic). The reaction involves the formation of a chromate ester (review sec 6.7b) which is thermally unstable, and undergoes an elimination reaction in which the chromium, originally Cr^{VI}, is reduced to Cr^{IV} when the Cr—O bond cleaves.

$$-\underset{\underset{H}{|}}{C}-OH + H_2CrO_4 + H^+ \;\rightleftharpoons\; -\underset{\underset{H}{|}}{C}-O-\underset{\underset{O}{\overset{O}{||}}}{Cr}-OH \;+\; H_3O^+ \qquad (9\text{-}2)$$

alcohol   an alkyl hydrogen chromate (chromate ester)

$$-\underset{\underset{H}{|}}{C}-O-\underset{\underset{O}{\overset{O}{||}}}{Cr}-OH \;\longrightarrow\; \underset{\text{aldehyde or}\atop\text{ketone}}{\diagdown C=O} \;+\; H^+ \;+\; {}^{-}CrO_3H \qquad (9\text{-}3)$$

The chromous acid (H_2CrO_3) is unstable and quickly disproportionates to Cr^{VI} and the chromic ion, Cr^{3+}. Since chromic acid is orange and chromic ion is green, the oxidation can be followed by the color change.

A typical example would be the oxidation of the secondary alcohol cyclohexanol to the ketone cyclohexanone.

$$\begin{array}{c}
CH_2\text{—}CH_2 \\
\diagup \qquad \diagdown \\
CH_2 \qquad CHOH \\
\diagdown \qquad \diagup \\
CH_2\text{—}CH_2
\end{array}
\xrightarrow[\text{H}^+,\ \text{heat}]{\text{CrO}_3}
\begin{array}{c}
CH_2\text{—}CH_2 \\
\diagup \qquad \diagdown \\
CH_2 \qquad C=O \\
\diagdown \qquad \diagup \\
CH_2\text{—}CH_2
\end{array}
\qquad (9\text{-}4)$$

When aldehydes are prepared this way, they must be removed from the reaction mixture as they are formed; otherwise, they will be further oxidized to acids (sec 9.6a). Fortunately aldehydes have lower boiling points than their corresponding alcohols. By carrying out the oxidation at a temperature slightly above the boiling point of the aldehyde, it may be removed by distillation as it is formed.

$$CH_3CH_2CH_2OH \xrightarrow[50\text{-}60°]{\text{CrO}_3,\ \text{H}^+} CH_3CH_2CHO \qquad (9\text{-}5)$$

1-propanol, bp 97°   propanal, bp 49°

9.4b HYDROLYSIS OF DIHALIDES

Compounds with two halogens on the same carbon atom are readily hydrolyzed by base to aldehydes or ketones. For example, benzaldehyde is prepared commercially from toluene by chlorination of the side chain (sec 5.5) followed by hydrolysis.

$$
\begin{array}{c}
\text{CH}_3 \\

\end{array}
\xrightarrow[\text{UV light}]{2\ \text{Cl}_2}
\begin{array}{c}
\text{CHCl}_2 \\

\end{array}
\xrightarrow[\text{H}_2\text{O}]{\text{Na}^+\text{OH}^-}
\left[
\begin{array}{c}
\text{H}-\overset{\overset{\text{Cl}}{|}}{\text{C}}-\text{O}-\text{H} \\

\end{array}
\right]
\xrightarrow{-\text{HCl}}
\begin{array}{c}
\text{H}\diagdown\ \diagup\text{O} \\
\text{C} \\

\end{array}
\quad (9\text{-}6)
$$

toluene   benzal
chloride   benzaldehyde

After displacement of the first halogen by hydroxyl, a base-catalyzed dehydrohalogenation of the intermediate, shown in brackets, produces the carbonyl compound.

If the di-halogenated carbon is not at the end of the carbon chain, the product is a ketone rather than an aldehyde. A general ketone synthesis from alkynes is possible via this route.

$$
\text{R}-\text{C}\equiv\text{CH}
\xrightarrow[\text{(sec 4.4)}]{2\ \text{HX}}
\text{R}-\overset{\overset{\text{X}}{|}}{\underset{\underset{\text{X}}{|}}{\text{C}}}-\text{CH}_3
\xrightarrow[\text{H}_2\text{O}]{\text{OH}^-}
\text{R}-\overset{\overset{\text{O}}{\|}}{\text{C}}-\text{CH}_3
\quad (9\text{-}7)
$$

9.4c SYNTHESIS WITH CARBON MONOXIDE

Aromatic aldehydes may be prepared by the **Gatterman-Koch reaction,** which involves passing a mixture of carbon monoxide and hydrogen chloride into an aromatic hydrocarbon in which aluminum chloride and cuprous chloride are suspended as catalysts.

$$
\text{CH}_3\text{---}\hexagon + \text{CO} \xrightarrow[\text{Cu}_2\text{Cl}_2]{\overset{\text{HCl}}{\text{AlCl}_3}} \text{CH}_3\text{---}\hexagon\text{---}\overset{\overset{\text{O}}{\|}}{\text{C}}\text{---H} \quad (9\text{-}8)
$$

toluene   *p*-tolualdehyde

In the **Oxo process,** an alkene is treated with carbon monoxide and hydrogen, in the presence of a catalyst such as dicobalt octacarbonyl $[\text{Co(CO)}_4]_2$. The reaction is equivalent to adding formaldehyde, H—CHO, to the double bond.

$$
\text{RCH}=\text{CH}_2 + \text{CO} + \text{H}_2 \xrightarrow[\text{heat, pressure}]{\text{catalyst}}
$$

$$
\underset{\underset{\text{H}}{|}}{\text{RCHCH}_2\text{CHO}} + \underset{\underset{\text{CHO}}{|}}{\text{RCHCH}_2\text{---H}} \quad (9\text{-}9)
$$

Of the two aldehyde products, the straight-chain isomer usually predominates.

9.4d SPECIAL METHODS

The most important aldehydes and ketones from the standpoint of large scale industrial production are formaldehyde, acetaldehyde, and acetone.

Formaldehyde is a gaseous substance which is very soluble in water. It is produced when a mixture of air and vaporized methanol is passed over a copper or silver gauze at 250°–300°.

$$2 \ CH_3OH + O_2 \xrightarrow[250°-300°]{Ag \ or \ Cu} 2 \ HCHO + 2 \ H_2O \qquad (9\text{-}10)$$

The gaseous product may then be dissolved in water to prepare **formalin,** a 40 per cent aqueous solution. Formaldehyde is used for the silvering of mirrors, as a disinfectant and fumigant, as a preservative for biological specimens, and in the synthesis of various resins, plastics, and synthetic glue.

Acetaldehyde may be prepared by a similar oxidation of ethanol. A better commercial synthesis of acetaldehyde, however, involves hydration of acetylene. Dilute sulfuric acid and mercuric sulfate are the catalysts for the addition of water to the triple bond.

$$HC\equiv CH + H-OH \xrightarrow[H^+]{HgSO_4} \left[\begin{array}{c} H \ \ OH \\ | \ \ \ | \\ H-C=C-H \end{array} \right] \longrightarrow CH_3CHO \qquad (9\text{-}11)$$

enol intermediate   acetaldehyde

The **enol** (unsaturated alcohol) intermediate—in this case, vinyl alcohol—rearranges to acetaldehyde by a mechanism which is discussed in detail in section 9.7. Hydration of other alkynes follows Markownikoff's rule, and gives ketones.

The preparation of acetaldehyde from acetylene makes it a very inexpensive industrial chemical, since the original raw materials are limestone, coke, and water or methane. Numerous important chemicals, including acetic acid, DDT, and butadiene, are manufactured from acetaldehyde.

Acetone is an important solvent used in the manufacture of explosives, lacquers, paint removers, plastics, drugs, and disinfectants. There are several commercial sources of this compound. It is prepared by the catalytic dehydrogenation of isopropyl alcohol (readily available by hydration of propylene, sec 6.5a). Large quantities are also made by a fermentation process discovered during the first World War by Chaim Weizmann, British chemist and later first president of Israel. Sugar derived from cornstarch or molasses, when fermented with the bacterium *Clostridium acetobutylicum*, yields about 30 per cent acetone, 60 per cent n-butyl alcohol, and 10 per cent ethyl alcohol. Acetone is also a by-product from the manufacture of phenol from cumene hydroperoxide (eq 6-26b). Finally, acetone is produced from acetic acid, using a metallic oxide as catalyst.

$$CH_3-\overset{\displaystyle O}{\overset{\|}{C}}-OH \quad \xrightarrow[\substack{ThO_2, \\ heat}]{MnO \ or} \quad CH_3\overset{\displaystyle O}{\overset{\|}{C}}CH_3 + CO_2 + H_2O \qquad (9\text{-}12)$$

$$CH_3-\overset{\displaystyle O}{\overset{\|}{C}}-OH$$

acetone

acetic acid

skip

The mechanism of this reaction is not well understood.

9.5   THE CARBONYL GROUP

The electronic structure of a carbon-oxygen double bond resembles that of a carbon-carbon double bond in that it consists of a σ and a π component (Figure 9-5). The carbonyl carbon atom is trigonal, being attached to oxygen and two other atoms. Its orbitals are therefore of the sp^2 type (Figure 3-1). The σ bond results from overlap of a carbon sp^2 orbital with an oxygen p orbital. It is symmetric about the axis which joins the carbon and oxygen nuclei. The π bond is formed by overlap of the remaining p orbital on carbon with another oxygen p orbital. As a consequence of the π bond, the carbon, the oxygen, and the two other atoms attached to the carbon lie in a single plane. The electrons in the π bond lie above and below that plane.

Figure 9-5
Bonding in the carbonyl group.

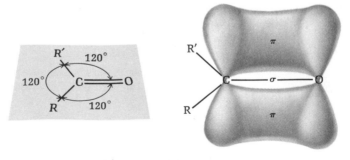

There is, however, an important respect in which the carbon-oxygen and carbon-carbon double bonds differ. The carbon-oxygen double bond involves two atoms with widely different electronegativities. Consequently, the bonding electrons are not shared equally, but are displaced toward the more electronegative oxygen atom. The carbonyl group may be described as a resonance hybrid of two extreme structures, one covalent and the other dipolar, with one electron pair displaced entirely toward the oxygen.

$$\overset{\backslash}{\underset{/}{C}}\!\!=\!\!\overset{..}{\underset{..}{O}}: \longleftrightarrow \overset{\backslash}{\underset{/}{C}}\!\!\overset{+}{-}\!\!\overset{..}{\underset{..}{O}}:^{-}$$

resonance contributors to the carbonyl group

The C=O bond distance is only 1.24 Å, whereas the C—O distance (in ethers or alcohols) is about 1.43 Å. Because of the polar carbonyl group, aldehydes and ketones have higher boiling points than hydrocarbons of equal molecular weight. Hydrogen bonding in pure aldehydes and ketones is unimportant; for this reason, they usually boil below alcohols with an equal number of carbon atoms. However, the lower members of the series probably owe their water solubility in part to hydrogen bonding between the carbonyl oxygen and water molecules.

$$\overset{\backslash}{\underset{/}{C}}\!\!\overset{\delta+}{=}\!\!\overset{\delta-}{\underset{..}{O}}:\cdots\text{H}\!\!-\!\!\text{O}\overset{\diagup\text{H}}{}$$

9.6   REACTIONS OF ALDEHYDES AND KETONES

Our discussion of aldehyde and ketone reactions will be divided into three major groups—oxidations, additions to the carbon-oxygen double bond, and base-catalyzed reactions which are a consequence of the acidity (though weak) of some carbonyl compounds.

9.6a   OXIDATION

Aldehydes and ketones behave differently toward oxidizing agents. Aldehydes are *easily oxidized* to acids containing the same number of carbon atoms.

$$R\!-\!\overset{\overset{\text{O}}{\|}}{C}\!-\!H \xrightarrow{[O]} R\!-\!\overset{\overset{\text{O}}{\|}}{C}\!-\!OH \tag{9-13}$$

aldehyde   acid

Ketones, however, have no hydrogen attached to the partially oxidized carbon atom. They may be oxidized *only under more severe conditions,* and the oxidation cleaves the carbon chain.

$$\underset{\substack{\text{one of these carbon-carbon} \\ \text{bonds must be broken in oxidation}}}{R\!-\!\overset{\displaystyle O}{\overset{\|}{C}}\!\diagdown\!\diagup\!R'}$$

Several laboratory tests to distinguish aldehydes from ketones take advantage of the different susceptibilities of these compounds toward oxidation.

For example, in **Tollens' test** the silver ammonia complex ion* is reduced by aldehydes to metallic silver within a few minutes at room temperature. The equation for the reaction may be written as:

$$\underset{\text{aldehyde}}{R\overset{\displaystyle O}{\overset{\|}{C}}H} + \underset{\substack{\text{silver ammonia} \\ \text{complex ion} \\ \text{(colorless)}}}{2\ Ag(NH_3)_2{}^+} + 3\ OH^- \longrightarrow$$

$$\underset{\text{acid anion}}{R\overset{\displaystyle O}{\overset{\|}{C}}\!-\!O^-} + \underset{\substack{\text{silver} \\ \text{mirror}}}{2\ Ag\!\downarrow} + 4\ NH_3\!\uparrow + 2\ H_2O \quad (9\text{-}14)$$

or in the less precise, but more easily remembered form, illustrated for formaldehyde.

$$\underset{\text{formaldehyde}}{H\overset{\displaystyle O}{\overset{\|}{C}}H} + 2\ AgOH \xrightarrow{\ NH_4OH\ } \underset{\text{formic acid}}{H\overset{\displaystyle O}{\overset{\|}{C}}\!-\!OH} + 2\ Ag\!\downarrow + H_2O \qquad (9\text{-}15)$$

If the vessel in which the test is performed is thoroughly cleaned, the silver will deposit as a mirror on the glass surface. This reaction is commonly used in silvering glass, formaldehyde usually being the aldehyde employed. Ketones do not react with Tollens' reagent.

9.6b   NUCLEOPHILIC ADDITION REACTIONS—MECHANISTIC CONSIDERATIONS

Because of electron withdrawal by the oxygen (sec 9.5), the carbonyl carbon is susceptible to attack by reagents which can supply an electron pair, i.e., nucleophiles. The nucleophile approaches the carbonyl carbon from a direction perpendicular to the carbonyl plane, and the π electrons are displaced to the oxygen (eq 9-16). The carbonyl carbon, previously trigonal, becomes tetrahedral. If the

---

*As silver hydroxide is insoluble, the silver ion must be complexed to keep it in solution in a basic medium.

$$\text{Nu:}^- + \text{ }C\!\!=\!\!\ddot{O}\text{:} \rightleftharpoons \overset{\text{Nu}}{\underset{}{\underset{}{C}}}\!\!-\!\!\ddot{O}\text{:}^- \tag{9-16}$$

nucleophile is an anion, the product will also be an anion. If the nucleophile is a neutral molecule, the product will be a dipolar ion. In either case, the product has a negative charge on the oxygen; the oxygen atom then acquires a proton from the solvent, which is usually hydroxylic (alcohol, water, or an acid).

$$\overset{\text{Nu}}{\underset{}{C}}\!\!-\!\!\ddot{O}\text{:}^- + \text{ SOH} \rightleftharpoons \overset{\text{Nu}}{\underset{}{C}}\!\!-\!\!\text{OH} + \text{SO}^- \tag{9-17}$$

(solvent)

The net result is the addition of Nu—H to the C=O bond.

$$C\!\!=\!\!O + \text{Nu}\!-\!\text{H} \longrightarrow \overset{\text{Nu}}{\underset{}{C}}\!\!-\!\!\text{OH} \tag{9-18}$$

Although these reactions are reversible, further transformations of the initial adduct frequently drive the process to completion (sec 9.6d).

Nucleophilic additions to carbonyl groups may be catalyzed by bases or acids. Bases may transform a weak nucleophile to a strong one by removing a proton.

$$\text{Nu}\!-\!\text{H} + \text{B} \rightleftharpoons \text{Nu:}^- + \text{BH}^+ \tag{9-19}$$

weak neutral   base   strong anionic
nucleophile   nucleophile

An example is the conversion of an alcohol (weak nucleophile) to an alkoxide ion (strong nucleophile).

$$\text{R}\ddot{O}\text{H} + \text{B} \rightleftharpoons \text{R}\ddot{O}\text{:}^- + \text{BH}^+ \tag{9-20}$$

Acids catalyze the addition of rather weak nucleophiles to carbonyl compounds by protonating the carbonyl oxygen atom. This enhances the electron deficiency of the carbonyl carbon by making it positively charged (a carbonium ion).

$$C\!\!=\!\!\ddot{O}\text{:} + \text{H}^+ \longrightarrow \left[C\!\!=\!\!\overset{+}{\ddot{O}}\text{H} \longleftrightarrow \overset{+}{C}\!\!-\!\!\ddot{O}\text{H} \right] \tag{9-21}$$

The choice of acid or base as a catalyst will depend on the reactivity of the particular nucleophile and carbonyl compound.

Ketones are generally less reactive than aldehydes in nucleophilic addition reactions. This is because (*a*) two R groups around the carbonyl carbon are bulkier than one R and an H and (*b*) R groups are electron-releasing compared with hydrogen, and this tends to neutralize the partial positive charge on carbon more in ketones than in aldehydes.

9.6c   ACETALS AND KETALS

In the presence of an acid catalyst, alcohols, using the unshared electron pair on oxygen for attack, are sufficiently nucleophilic to react with most aldehydes.

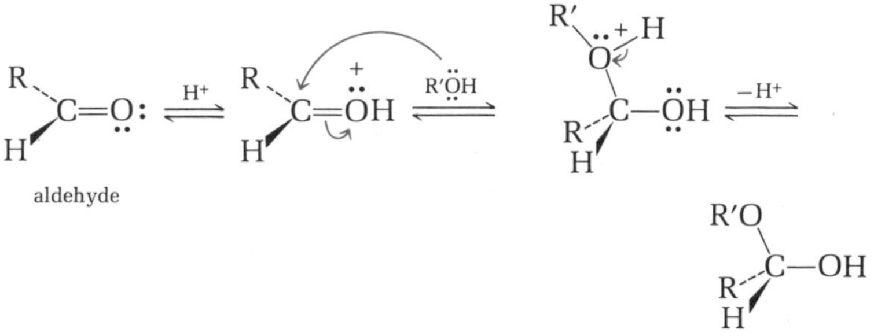

$$\text{(9-22)}$$

hemiacetal

The product, a **hemiacetal,** contains both an alcohol and an ether function at what was originally the carbonyl carbon. The addition is reversible.

In the presence of excess alcohol, further reaction occurs with the formation of water and an acetal.

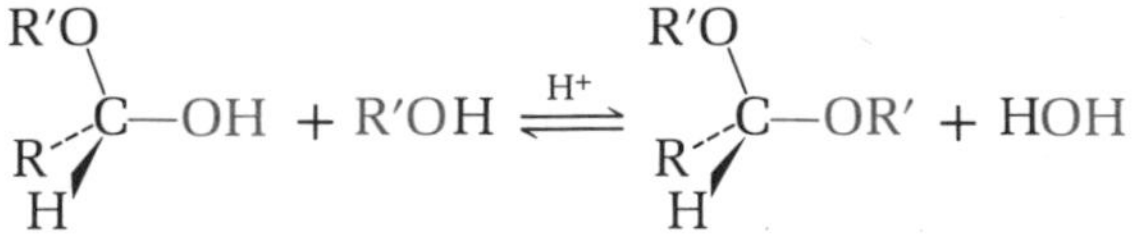

$$\text{(9-23)}$$

hemiacetal   acetal

This reaction involves protonation of the hemiacetal hydroxyl group followed by loss of water to form a resonance-stabilized carbonium ion. Reaction of this ion with the alcohol, which is usually the solvent, gives the final product.

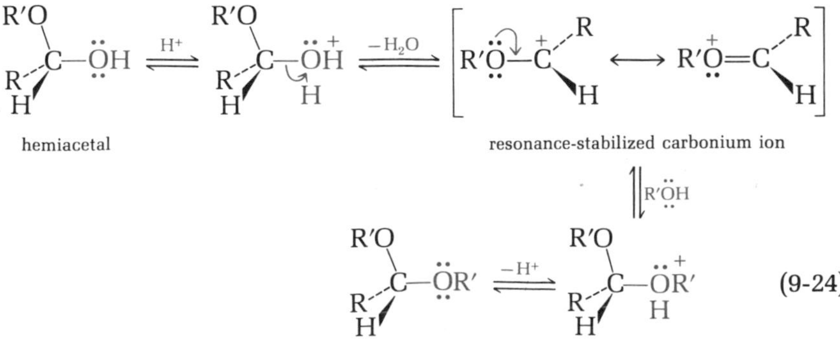

hemiacetal   resonance-stabilized carbonium ion

$$\text{(9-24)}$$

acetal

A specific example, using acetaldehyde and methanol, is shown in equation 9-25.

$$CH_3\overset{\displaystyle O}{\overset{\|}{C}}H + CH_3OH \rightleftharpoons CH_3\underset{OCH_3}{\overset{OH}{\underset{|}{C}}}H \quad \underset{\text{dry HCl}}{\overset{CH_3OH}{\rightleftharpoons}}$$

acetaldehyde   hemiacetal

$$CH_3\underset{OCH_3}{\overset{OCH_3}{\underset{|}{C}}}H + H_2O \quad (9\text{-}25)$$

acetaldehyde
dimethyl acetal

The acetal structure involves two ether linkages at a single carbon atom. Like ethers, acetals are stable toward alkali. However, since all the steps in equations 9-24 and 9-22 are reversible, acetals are readily cleaved by acids to their aldehyde and alcohol components.

Hemiacetal and acetal groups are important structural features of carbohydrates (Chapter 16).

Ketones react with alcohols in an analogous manner to form **ketals.** If a glycol is used as the alcohol component, the product will have a cyclic structure.

$$\underset{CH_3}{\overset{CH_3}{>}}C=O + \overset{HO-CH_2}{\underset{HO-CH_2}{\overset{|}{|}}} \overset{H^+}{\rightleftharpoons} \underset{CH_3}{\overset{CH_3}{>}}C\underset{O-CH_2}{\overset{O-CH_2}{<}}\overset{|}{\underset{|}{}} + H_2O \quad (9\text{-}26)$$

acetone   ethylene glycol   acetone-ethylene
glycol ketal

9.6d NITROGEN DERIVATIVES OF ALDEHYDES AND KETONES

The unshared electron pair on nitrogen enables ammonia and its derivatives to act as nucleophiles toward carbonyl compounds.

$$>C=\ddot{O}: + :\ddot{N}H_3 \rightleftharpoons \underset{\overset{+}{N}H_3}{\overset{|}{>}}C-\ddot{O}:^- \xrightarrow[\text{the more basic O}^-]{\substack{\text{remove H}^+ \text{ from} \\ \text{the ammonium-} \\ \text{type ion and} \\ \text{place it on}}} \underset{NH_2}{\overset{|}{>}}C-\ddot{O}H \quad (9\text{-}27)$$

The reaction is reversible, and the products from ammonia itself are relatively unimportant. However, the reaction is important with certain ammonia derivatives in which one hydrogen is replaced by various other groups. In these cases, initial nucleophilic addition to the carbonyl group is followed by dehydration.

$$\ce{C=O} + \overset{\cdot\cdot}{N}H_2\!-\!Z \;\rightleftharpoons\; \left[\begin{array}{c} \ce{H\!\!\diagdown N \diagup Z} \\ \ce{C{-}OH} \end{array} \right] \xrightarrow{-HOH} \ce{C=N\diagup Z} \qquad (9\text{-}28)$$

The net result is conversion of the carbon-oxygen double bond to a carbon-nitrogen double bond. Table 9-3 shows several examples of ammonia derivatives used in this reaction, together with the structures and names of the products obtained.

Oximes, hydrazones and their phenyl and 2,4-dinitrophenyl derivatives are frequently crystalline solids whose characteristic melting points can be used to identify a particular carbonyl compound. Phenylhydrazine has been an important reagent in elucidating the structures of sugars (sec 16.5b). Imines (sometimes called Schiff's bases) are important in many biological reactions, particularly in the interconversion of amino acids and carbonyl compounds derived from protein and carbohydrate metabolic pathways respectively.

Two typical examples of the formation of nitrogen derivatives of carbonyl compounds are shown in equations 9-29 and 9-30.

$$\begin{array}{c} CH_3 \\ \diagdown \\ CH_3 \diagup \end{array} C=O + H_2N\!-\!OH \longrightarrow \begin{array}{c} CH_3 \\ \diagdown \\ CH_3 \diagup \end{array} C=NOH + H_2O \qquad (9\text{-}29)$$

| acetone | hydroxylamine | acetone oxime |

$$\text{benzaldehyde} \quad \ce{-CH=O} + H_2NNH\!-\!\text{phenylhydrazine} \longrightarrow$$

$$\ce{-CH=NNH-} + H_2O \quad (9\text{-}30)$$

benzaldehyde phenylhydrazone

9.6e CYANOHYDRINS

Addition of hydrogen cyanide to the carbonyl group leads to a class of compounds known as **cyanohydrins.** These compounds have a

$$\ce{C=O} + \ce{CN-} \rightleftharpoons \overset{CN}{\underset{}{C}}\!-\!O^- \underset{}{\overset{H^+}{\rightleftharpoons}} \overset{CN}{\underset{}{C}}\!-\!OH \qquad (9\text{-}31)$$

a cyanohydrin

Table 9-3
Nitrogen Derivatives of Carbonyl Compounds

Z (eq 9-28)	FORMULA OF AMMONIA DERIVATIVE	NAME	FORMULA OF CARBONYL DERIVATIVE	NAME
$-OH$	NH_2OH	hydroxylamine	C=NOH	oxime
$-NH_2$	NH_2NH_2	hydrazine	C=NNH_2	hydrazone
$-NHC_6H_5$	$NH_2NHC_6H_5$	phenylhydrazine	C=NNHC_6H_5	phenylhydrazone
$-NH$ (2,4-dinitrophenyl, NO_2 groups)	NH_2NH (2,4-dinitrophenyl, NO_2 groups)	2,4-dinitrophenylhydrazine	C=N-NH (2,4-dinitrophenyl, NO_2 groups)	2,4-dinitrophenylhydrazone
$-R$ or Ar	RNH_2 or $ArNH_2$	primary amine	C=NR or C=NAr	imine

cyano and a hydroxyl group attached to the same carbon atom. Usually the reaction is carried out by mixing the carbonyl compound with aqueous sodium or potassium cyanide, after which mineral acid is slowly added. The steps are reversible; the equilibrium favors the product when the groups attached to the carbonyl carbon are fairly small, as in aldehydes or methyl ketones (eq 9-32).

$$CH_3-\overset{\overset{\displaystyle O}{\|}}{C}-CH_3 + HCN \longrightarrow CH_3-\overset{\overset{\displaystyle OH}{|}}{\underset{\underset{\displaystyle CN}{|}}{C}}-CH_3 \qquad (9\text{-}32)$$

acetone

acetone cyanohydrin

Cyanohydrins are useful intermediates in the synthesis of several classes of natural products such as hydroxy acids, amino acids, and sugars.

9.6f ADDITION OF GRIGNARD REAGENTS

Grignard reagents (sec 8.6) usually react with carbonyl compounds by nucleophilic addition.

$$\overset{\displaystyle}{C}{=}O + \bar{R}\overset{+}{MgX} \longrightarrow \overset{\displaystyle R}{\underset{\displaystyle}{\overset{|}{C}}}-\bar{O}\overset{+}{MgX} \qquad (9\text{-}33)$$

intermediate addition
product (a magnesium alkoxide)

The reaction is carried out by slowly adding an ether solution of the aldehyde or ketone to the Grignard reagent. Addition is frequently exothermic, and complete even at room temperature. After all the carbonyl compound is added, the intermediate addition product (a magnesium alkoxide, sec 6.4) is hydrolyzed, usually with acid, to obtain an alcohol.

$$\overset{\displaystyle R}{\underset{\displaystyle}{\overset{|}{C}}}-\bar{O}\overset{+}{MgX} + H^+Cl^- \xrightarrow{H_2O} \overset{\displaystyle R}{\underset{\displaystyle}{\overset{|}{C}}}-OH + Mg^{2+}X^-Cl^- \qquad (9\text{-}34)$$

an alcohol

This reaction is exceedingly versatile. By the appropriate choice of Grignard reagent and aldehyde or ketone, virtually any alcohol can be synthesized. The particular class of alcohol produced depends upon the choice of carbonyl compound, as is apparent from equation 9-34. Formaldehyde, in which both of the undesignated bonds are to hydrogen atoms, gives primary alcohols. Other aldehydes give secondary alcohols, whereas ketones produce tertiary alcohols.

$$H-\overset{\overset{\displaystyle O}{\|}}{C}-H \xrightarrow[\text{2. H}_2\text{O, H}^+]{\text{1. RMgX}} H-\overset{\overset{\displaystyle OH}{|}}{\underset{\underset{\displaystyle R}{|}}{C}}-H \text{ (or RCH}_2\text{OH)} \qquad (9\text{-}35)$$

formaldehyde   a primary alcohol

$$R-\overset{\overset{\displaystyle O}{\|}}{C}-H \xrightarrow[\text{2. H}_2\text{O, H}^+]{\text{1. RMgX}} R-\overset{\overset{\displaystyle OH}{|}}{\underset{\underset{\displaystyle R}{|}}{C}}-H \left(\text{or } \overset{\displaystyle R}{\underset{\displaystyle R}{\diagup}}\hspace{-0.3em}CH-OH \right) \qquad (9\text{-}36)$$

all other aldehydes   a secondary alcohol

$$R-\overset{\overset{\displaystyle O}{\|}}{C}-R \xrightarrow[\text{2. H}_2\text{O, H}^+]{\text{1. RMgX}} R-\overset{\overset{\displaystyle OH}{|}}{\underset{\underset{\displaystyle R}{|}}{C}}-R \left(\text{or } R\overset{\displaystyle R}{\underset{\displaystyle R}{\diagup}}\hspace{-0.3em}C-OH \right) \qquad (9\text{-}37)$$

a ketone   a tertiary alcohol

Equations 9-35 to 9-37 show that one of the R groups (in black) attached to the hydroxyl-bearing carbon of the alcohol must come from the Grignard reagent; those parts of the alcohol shown in color originate in the carbonyl compound.

When the R groups are not identical, more than one combination of Grignard reagent and carbonyl compound may give the desired alcohol. For example, 1-phenylethanol may be prepared from acetaldehyde and phenylmagnesium bromide (eq 9-38) or from benzaldehyde and methylmagnesium bromide (eq 9-39).

$$CH_3-\overset{\overset{\displaystyle O}{\|}}{C}-H + \text{〈 〉}-MgBr \xrightarrow[\text{2. H}_2\text{O, H}^+]{\text{1. add}} CH_3-\overset{\overset{\displaystyle OH}{|}}{C}H-\text{〈 〉} \qquad (9\text{-}38)$$

acetaldehyde   phenylmagnesium   1-phenylethanol
                        bromide

$$CH_3MgBr + H-\overset{\overset{\displaystyle O}{\|}}{C}-\text{〈 〉} \xrightarrow[\text{2. H}_2\text{O, H}^+]{\text{1. add}} CH_3-\overset{\overset{\displaystyle OH}{|}}{C}H-\text{〈 〉} \qquad (9\text{-}39)$$

methylmagnesium   benzaldehyde   1-phenylethanol
bromide

The intermediate addition compound, formed in the first step, is of course identical from each pair of reactants.

$$CH_3-\overset{\overset{\displaystyle \overset{-}{O}\ \overset{+}{MgBr}}{|}}{\underset{\underset{\displaystyle H}{|}}{C}}-\text{〈 〉}$$

Hydrolysis, the second step in the reaction sequence, gives the desired alcohol. The choice between the possible sets of reactants may be based on their availability and cost, or it may be made on chemical grounds. For example, since the reaction involves nucleophilic attack at the carbonyl carbon, the yield may be enhanced by selecting the carbonyl compound which has the smallest R groups.

9.6g BISULFITE ADDITION COMPOUNDS

Aldehydes and a few ketones which are not sterically hindered by having large groups attached to the carbonyl carbon atom react with saturated aqueous sodium bisulfite to form white crystalline addition products. Excess bisulfite shifts the equilibrium to the right.

$$\underset{H}{\overset{R}{>}}C{=}O + HOSO_2^-Na^+ \rightleftharpoons \underset{\underset{H}{R}}{\overset{SO_3^-Na^+}{\underset{|}{\overset{|}{C}}}}{-}OH \qquad (9\text{-}40)$$

sodium bisulfite   bisulfite addition compound

These addition compounds may readily be reconverted to carbonyl compounds by treatment with acid or base. Thus, bisulfite addition compounds may be used to separate aldehydes from mixtures with other compounds. The solid formaldehyde compound is used as a way of transporting or storing the gaseous aldehyde.

A qualitative test for aldehydes is based upon the addition of bisulfite. Fuchsin (magenta) is a dyestuff which, on treatment with sulfur dioxide, forms a colorless substance. When this reagent is treated with an aldehyde, a new colored compound is formed, and the solution changes from colorless to pink. This is known as **Schiff's test for aldehydes.**

9.6h REDUCTION

As Table 9-2 implies, it should be possible to reduce aldehydes and ketones to alcohols or to hydrocarbons. Both reactions are known.

Aldehydes and ketones can be reduced to the corresponding alcohols either by catalytic hydrogenation under pressure or by chemical reducing agents. Hydrogenation requires a catalyst (nickel, copper chromite) and may be pictured as the addition of hydrogen to the carbonyl group.

$$>C{=}O + H{-}H \xrightarrow{\text{catalyst}} \underset{}{\overset{H}{>}}C{-}OH \qquad (9\text{-}41)$$

For example:

$$\text{acetophenone} \quad \underset{O}{\overset{||}{C}}{-}CH_3 + H_2 \xrightarrow[\substack{\text{heat,} \\ \text{pressure}}]{Ni} \underset{OH}{\overset{|}{C}H}{-}CH_3 \qquad (9\text{-}42)$$

acetophenone   1-phenylethanol

Recently, metallic hydrides have been used extensively as reducing agents in organic chemistry. The most common of these are lithium aluminum hydride ($LiAlH_4$) and sodium borohydride ($NaBH_4$). Reduction occurs by nucleophilic attack of hydride on the carbonyl carbon atom. Since the carbon-carbon double bond is not readily attacked by nucleophiles, these reagents may reduce a carbon-oxygen double bond to the corresponding alcohol without affecting a carbon-carbon double bond.

$$CH_3-CH=CH-\overset{\overset{\displaystyle O}{\|}}{C}H \xrightarrow{\text{LiAlH}_4} CH_3CH=CH-CH_2OH \qquad (9\text{-}43)$$

crotonaldehyde   crotyl alcohol

The carbonyl group of aldehydes or ketones may be reduced to a methylene ($-CH_2-$) group with amalgamated zinc and concentrated hydrochloric acid. The reaction is known as the **Clemmensen reduction**.

$$\xrightarrow[\text{HCl}]{\text{Zn—Hg}} \qquad (9\text{-}44)$$

propiophenone   n-propylbenzene

9.7 THE EFFECT OF THE CARBONYL GROUP ON HYDROGENS ATTACHED TO THE α-CARBON*

The C—H bond is normally only very weakly acidic, in contrast to the O—H bond, which readily gives up protons to a base. The difference is apparent by comparing the acidities of phenol and toluene. The 10^{25} difference in acidity is due mainly to the much

$K_a = 10^{-10}$ $K_a = 10^{-35}$

greater electronegativity of oxygen, and its concomitant ability to carry a negative charge.

*The carbon atom adjacent to the carbonyl group is known as the α (alpha) carbon atom:

$$\overset{\gamma}{R}CH_2\overset{\beta}{C}H_2\overset{\alpha}{C}H_2\overset{\overset{\displaystyle O}{\|}}{C}H$$

The acidity of C—H bonds α to a carbonyl group is enhanced because the negative charge in the carbanion which results from proton removal can be delocalized, by resonance, to the carbonyl oxygen.

$$\text{Base} \longrightarrow \text{H} \underset{}{\overset{}{\underset{}{\Big\backslash}}} \text{C}-\text{C} \overset{\ddot{\text{O}}\!:}{\underset{\text{R}}{\Big\langle}} \xrightarrow{-\text{BH}^+} \left[\underset{\text{R}}{\overset{\ddot{\text{O}}\!:}{\text{C}-\text{C}}} \longleftrightarrow \underset{\text{R}}{\overset{:\ddot{\text{O}}\!:}{\text{C}=\text{C}}} \right] \qquad (9\text{-}45)$$

enolate anion

The **enolate anion,** as it is called, has most of the negative charge on oxygen, though it is a resonance hybrid with a small percentage of the charge on the α-carbon. When a solution of an enolate anion is acidified, the ion can accept a proton either at the α-carbon (giving back the original carbonyl compound) or at the oxygen, giving an *enol* (which has a double bond *-ene* and alcohol *-ol* on the same carbon atom).

$$\underset{\text{keto form}}{\text{H}\,\underset{}{\overset{}{\Big\backslash}}\,\text{C}-\text{C}\overset{\text{O}}{\underset{\text{R}}{\Big\langle}}} \underset{+\text{H}^+}{\overset{-\text{H}^+}{\rightleftharpoons}} \left[\begin{array}{c} \underset{}{\overset{\ddot{\cdot}\cdot}{\text{C}-\text{C}}}\overset{\text{O}}{\underset{\text{R}}{\Big\langle}} \\ \updownarrow \\ \underset{\text{R}}{\overset{\text{O}^-}{\text{C}=\text{C}}} \end{array} \right] \underset{-\text{H}^+}{\overset{+\text{H}^+}{\rightleftharpoons}} \underset{\text{R}}{\overset{\text{O}\,\text{H}}{\text{C}=\text{C}}} \qquad (9\text{-}46)$$

enolate anion   enol form

Carbonyl compounds with an α-hydrogen are in equilibrium, via the enolate anion, with the corresponding enol. The carbonyl form is referred to as the **keto** form, and the phenomenon is known as **keto-enol tautomerism.** *It is different from resonance, because keto-enol forms differ in the location of a hydrogen atom as well as in electron distribution.*

Because of the O—H bond, enols are, in general, more acidic than the corresponding keto forms. Reactions which should lead to an enol (as, for example, the hydration of acetylenes—equation 9-11) give the carbonyl product instead. Most aldehydes and ketones exist predominantly in the keto form (for example, only 0.0003% enol is present in ordinary acetone). However, some stable enols are known, the most familiar being the *phenols* (Chapter 6). Here the keto structure would disrupt the resonance stabilization of the aromatic ring.

enol form   keto form   (9-47)

Aldehydes and ketones which have an α-hydrogen undergo several important reactions via their enolate anions. The reactions are usually initiated by a base.

9.7a THE ALDOL CONDENSATION

Enolate anions can function as nucleophiles in additions to the carbon-oxygen double bond. The reaction is useful in synthesis because a new carbon-carbon bond is formed, thus permitting one to construct large molecules from smaller ones. For example, when acetaldehyde is treated with base, the four-carbon product **aldol*** (3-hydroxybutanal) is obtained.

$$2\ CH_3\overset{\overset{\displaystyle O}{\|}}{C}H \xrightarrow{\ OH^-\ } CH_3\overset{\overset{\displaystyle OH}{|}}{C}H-CH_2\overset{\overset{\displaystyle O}{\|}}{C}H \qquad (9\text{-}48)$$

aldol
(3-hydroxybutanal)

The steps in the reaction mechanism involve base-catalyzed formation of the enolate anion (eq 9-48a), nucleophilic attack of this anion on the carbonyl group of another aldehyde molecule (eq 9-48b), and stabilization of the resulting alkoxide by abstraction of a proton from the solvent (eq 9-48c).

$$\overset{\alpha}{C}H_3-\overset{\overset{\displaystyle O}{\|}}{C}-H + OH^- \rightleftharpoons \overset{..}{\overset{..}{C}}H_2-\overset{\overset{\displaystyle O}{\|}}{C}-H + HOH \qquad (9\text{-}48a)$$

$$CH_3-\overset{\overset{\displaystyle O}{\|}}{C}H + \overset{..}{C}H_2-\overset{\overset{\displaystyle O}{\|}}{C}H \rightleftharpoons CH_3\overset{\overset{\displaystyle O^-}{|}}{C}H-CH_2\overset{\overset{\displaystyle O}{\|}}{C}H \qquad (9\text{-}48b)$$

$$CH_3\overset{\overset{\displaystyle O^-}{|}}{C}H-CH_2\overset{\overset{\displaystyle O}{\|}}{C}H + HOH \rightleftharpoons CH_3\overset{\overset{\displaystyle OH}{|}}{C}H\overset{\alpha}{|}CH_2\overset{\overset{\displaystyle O}{\|}}{C}H + OH^- \qquad (9\text{-}48c)$$

The dashed line in the final formula (eq 9-48c) shows that the α-

---

* The word *aldol* comes from the structure of the product, which is both an *aldehyde* and an *alcohol*.

carbon of one aldehyde molecule has become bonded to what was originally the carbonyl carbon of the second aldehyde molecule. The product is a β-hydroxyaldehyde.

When a higher aldehyde is used, branching results at the α-carbon of the resulting aldol.

$$
\underset{}{RCH_2\overset{O}{\overset{\|}{C}}H} \;+\; \underset{}{RCH_2\overset{O}{\overset{\|}{\underset{\alpha}{C}}}H} \;\xrightarrow{\;OH^-\;}\; \underset{}{RCH_2\overset{OH}{\overset{|}{C}}H}-\underset{\underset{R}{|}}{\overset{}{C}H}\overset{O}{\overset{\|}{\underset{\alpha}{C}}}H
\tag{9-49}
$$

an aldol
(a β-hydroxyaldehyde)

Aldols are useful in synthesis. They are more easily dehydrated than most alcohols because the double bond in the resulting unsaturated aldehyde is conjugated with the carbonyl group. Acetaldehyde is converted to crotonaldehyde, 1-butanol, and butadiene commercially via the aldol condensation.

$$
\underset{\text{aldol}}{CH_3\overset{OH}{\overset{|}{C}}HCH_2\overset{O}{\overset{\|}{C}}H} \xrightarrow[-H_2O]{H_2SO_4} \underset{\text{crotonaldehyde}}{CH_3CH{=}CH\overset{O}{\overset{\|}{C}}H} \xrightarrow{4[H]}
$$

$$
\underset{n\text{-butyl alcohol}}{CH_3CH_2CH_2CH_2OH} \xrightarrow[-H_2]{-H_2O} \underset{\text{1,3-butadiene}}{CH_2{=}CH{-}CH{=}CH_2} \tag{9-50}
$$

Aldols can be reduced to 1,3-diols. The mosquito repellant **6-12** (2-ethylhexane-1,3-diol) is synthesized from n-butyraldehyde via its

$$
CH_3CH_2CH_2\overset{OH}{\overset{|}{C}}H-\underset{\underset{CH_2CH_3}{|}}{CH}-CH_2OH
$$

2-ethylhexane-1,3-diol (6-12)

aldol (eq 9-49, R = CH₃CH₂—) which is catalytically reduced using hydrogen and nickel.

Aldehydes that do not have an α-hydrogen, such as formaldehyde (which has no α-carbon), or aromatic aldehydes cannot produce enolate anions. They can, however, react with enolate anions produced by another carbonyl compound giving rise to a *crossed* aldol product. For example, benzaldehyde and acetaldehyde react, in the presence of base, to produce cinnamaldehyde (a flavor constituent of cinnamon).

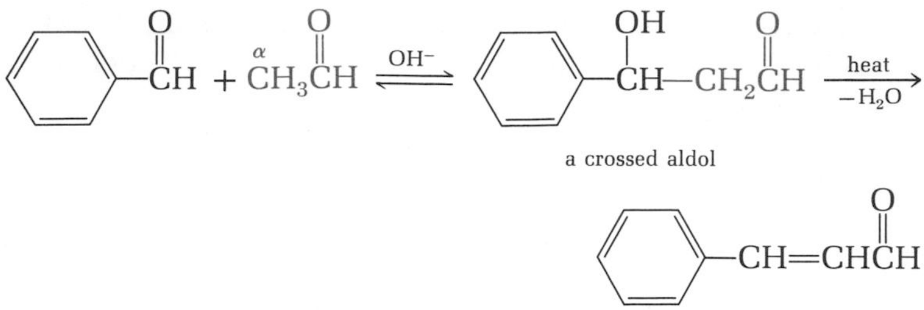

a crossed aldol

(9-51)

cinnamaldehyde

9.7b THE HALOFORM REACTION

Enolate anions react rapidly with halogens (chlorine, bromine, iodine) to yield an α-halocarbonyl compound and halide ion.

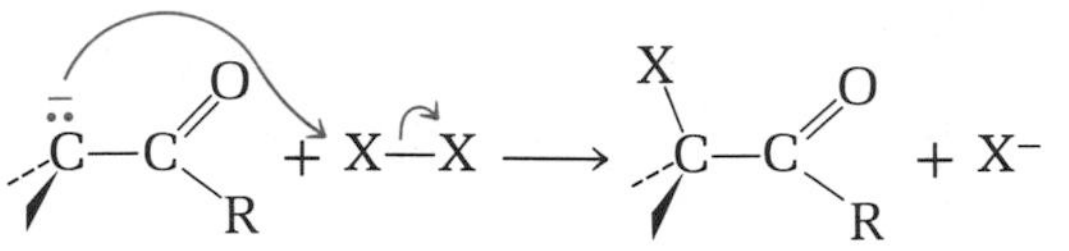

enolate anion

(9-52)

X = Cl, Br, or I

Thus in the presence of base, the α-hydrogens of an aldehyde or ketone can be readily replaced by halogens. The electron-withdrawing effect of the first α-halogen enhances the acidity of any remaining hydrogens on that α-carbon, so that further replacement of the remaining hydrogens by halogen is even more facile than the first. For acetaldehyde, the sequence is

$$CH_3CHO \xrightarrow[OH^-]{X_2} CH_2X\,CHO \xrightarrow[OH^-]{X_2} CHX_2\,CHO \xrightarrow[OH^-]{X_2} CX_3\,CHO \quad (9\text{-}53)$$

acetaldehyde   a trihalo-acetaldehyde

where each substitution consists of two steps, formation of the enolate anion and subsequent reaction of that anion with a halogen molecule.

$$CH_3CHO + OH^- \rightleftharpoons \bar{C}H_2CHO + H_2O \quad (9\text{-}53a)$$

$$\bar{C}H_2CHO + X_2 \longrightarrow CH_2X\,CHO + X^- \quad (9\text{-}53b)$$

The bond between a carbonyl carbon and a trihalogenated α-carbon is weakened by the opposing inductive effects of the halogens and oxygen atom attached to these carbons, and is readily cleaved by strong base. One of the products is a **haloform** (a trihalomethane); the other organic product is the anion of an organic acid. The **haloform reaction** is summarized in equation 9-54.

$$CH_3 \overset{\overset{\displaystyle O}{\|}}{C}R + 3\,X_2 + 4\,OH^- \longrightarrow$$

$$CX_3H + RCO_2^- + 3\,X^- + 3\,H_2O \quad (9\text{-}54)$$

a haloform

X = Cl, Br or I

The first steps in the reaction are the replacement of the three α-hydrogens in the methyl group by halogens, according to equation 9-53. The final step, carbon-carbon bond cleavage at the dashed colored line, occurs by the following mechanism:

$$CX_3 - \overset{\overset{\displaystyle O}{\|}}{C} - R + OH^- \longrightarrow CX_3 - \overset{\overset{\displaystyle O^-}{|}}{\underset{\displaystyle OH}{C}} - R \longrightarrow$$

$$CX_3^- + \overset{\overset{\displaystyle O}{\|}}{R C} - OH \quad (9\text{-}54a)$$

a trihalomethyl   an acid
anion

$$CX_3^- + RCO_2H \xrightarrow[\substack{\text{proton} \\ \text{transfer} \\ \text{to the anion} \\ \text{of the} \\ \text{weaker acid}}]{} CX_3H + RCO_2^- \quad (9\text{-}54b)$$

a haloform

The haloform reaction is most commonly employed as the **iodoform test** for the $CH_3 \overset{\overset{\displaystyle}{}}{\underset{\underset{\displaystyle O}{\|}}{C}}\!-\!R$ group. Here, the carbonyl compound to be tested is treated with iodine and base. If the compound is a *methyl ketone* (or acetaldehyde, R=H in eq 9-54), a yellow crystalline precipitate of iodoform, CI_3H, deposits. This can be easily identified by its medicinal odor and melting point (119°). Since alkaline solutions of halogens are good oxidizing agents, alcohols with the structure $CH_3 \overset{\overset{\displaystyle}{}}{\underset{\underset{\displaystyle OH}{|}}{C}}H\!-\!R$ also give a positive test; they are first oxidized to methyl ketones (sec 9.4a) which then react further in the usual way.

By methods which avoid using a base, it is possible to make tri-α-halogenated carbonyl compounds. The most important of these commercially is **chloral** (trichloroacetaldehyde), which is manufactured industrially from ethanol and chlorine.

$$CH_3CH_2OH \xrightarrow[-2\ HCl]{Cl_2} CH_3CHO \xrightarrow[-3\ HCl]{3\ Cl_2} CCl_3CHO \quad (9\text{-}55)$$

chloral

Chloral forms a hydrate, $CCl_3CH(OH)_2$, which is used in medicine as a sleep-producer (knock-out drops). Most chloral is used as one of the two raw materials for DDT synthesis.*

$$CCl_3CH=O \quad \xrightarrow[-H_2O]{H_2SO_4} \quad (9\text{-}56)$$

chloral   chlorobenzene   *p,p'*-dichlorodiphenyltrichloroethane
(DDT)

9.8   USEFUL POLYMERS FROM ALDEHYDES

The lower aldehydes, with dilute acid, form cyclic trimers which exhibit no aldehydic properties, but which may be decomposed by heat or acids to furnish the corresponding aldehyde. Like bisulfite addition compounds, these polymers are sometimes convenient ways of storing or transporting the aldehydes.

$$\xrightarrow{H^+} \qquad (9\text{-}57a)$$

formaldehyde   trioxane

$$\xrightarrow{H^+} \qquad (9\text{-}57b)$$

acetaldehyde   paraldehyde

Formaldehyde also forms a linear, low-molecular weight polymer known as **polyoxymethylene,** or **paraformaldehyde,** by slow evaporation of an aqueous solution of the aldehyde. Complete removal of

$$(n+2)\ H\overset{O}{\overset{\|}{C}}H \xrightarrow{H_2O} HOCH_2(OCH_2)_nOCH_2OH \qquad (9\text{-}58)$$

paraformaldehyde

* See the last sentence of Chapter 8 regarding the use of DDT as an insecticide.

water permits further polymerization to a high-molecular weight polymer. This can "unzip" to formaldehyde on heating. (Note that although all internal carbons in the chain have the acetal structure, the terminal carbons have a hemiacetal structure.) If the terminal carbons are protected by forming an ester, the resulting polymer is stable and can be extruded or molded into many useful articles. This polymer is marketed by the DuPont Company under the trade name **Delrin.**

$$CH_3 \overset{O}{\overset{\|}{C}} OCH_2(OCH_2)_n OCH_2 O \overset{O}{\overset{\|}{C}} CH_3$$

Delrin

Formaldehyde undergoes a base-catalyzed reaction with phenols, which leads to a useful class of polymers called **Bakelites.** In the first stage of the reaction, the enolate anion of a phenol undergoes nucleophilic addition to the formaldehyde carbon-oxygen double bond.

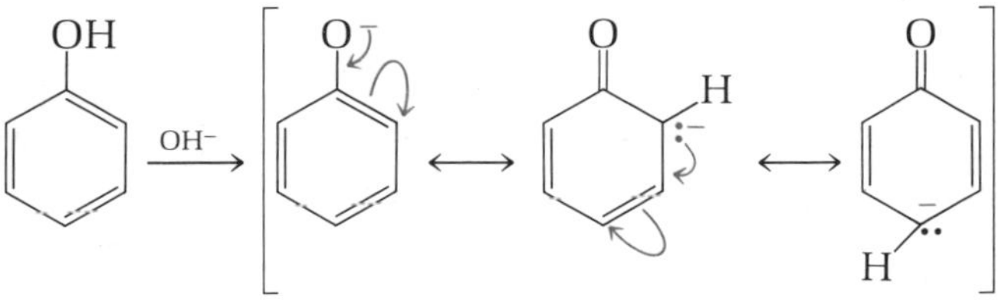

(9-59)

phenolate anion

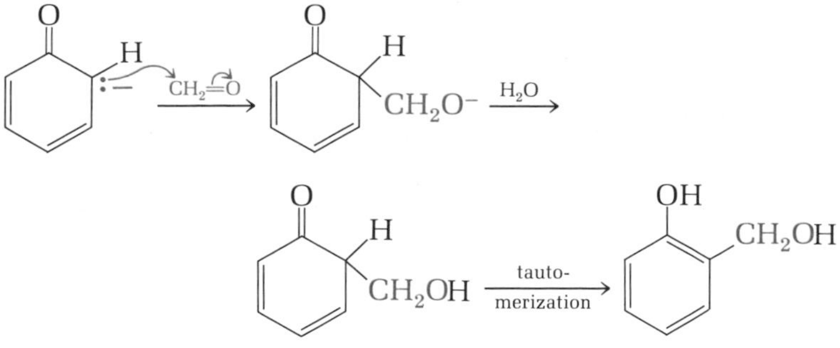

(9-60)

o-hydroxybenzyl
alcohol

The first product is an o- or p-hydroxybenzyl alcohol which reacts further with phenol, eliminating water to form a methylene (—CH₂—) bridge between two aromatic rings.

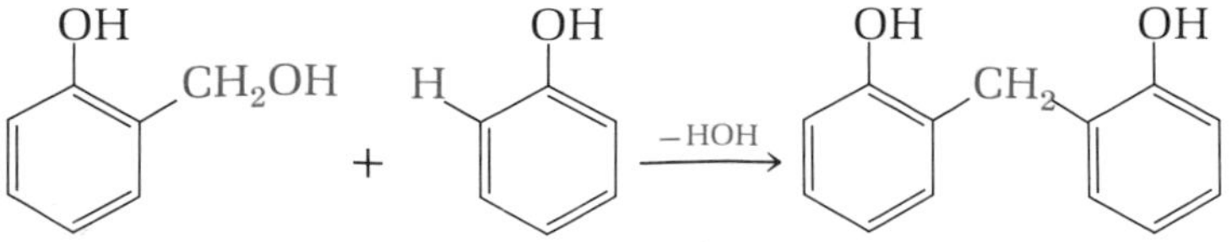

(9-61)

Such condensations continue, *ortho* and *para* to the phenolic hydroxyl groups (sec 6-7e), to build long-chain and cross-linked polymers, as is shown in the following partial formula (the methylene groups are derived from the formaldehyde, the remainder of the polymer from the phenol).

partial formula of a Bakelite

Various degrees of polymerization and cross-linking may be obtained by adjusting reaction conditions and by the addition of substituted phenols. Phenol-formaldehyde resins are used commercially in protective coatings, laminated plywood, and in molded products.

New Concepts, Facts, and Terms

1. Nomenclature: *al* ending for aldehydes, *one* for ketones. Some common names to remember are formaldehyde, acetaldehyde, benzaldehyde, acetone, acetophenone, and benzophenone.
2. Oxidation states of carbon; count the number of C—H and C—O bonds.
3. Preparation
   a. oxidation of alcohols: primary alcohols ⟶ aldehydes; secondary alcohols ⟶ ketones. Dehydrogenation.
   b. hydrolysis of dihalides in which both halogens are attached to the same carbon atom
   c. syntheses with CO; Gatterman-Koch reaction; Oxo process
   d. special methods; formaldehyde from methanol, acetaldehyde from acetylene, acetone from acetic acid
4. Nature of the carbonyl group; planar, sp^2 bonding, polarized with the oxygen end negative, susceptible to attack mainly by electron-donating reagents (nucleophiles).

5. Reactions
   a. oxidation; aldehydes are more easily oxidized than ketones; Tollens' silver mirror test.
   b. nucleophilic additions; may be acid or base catalyzed.
   c. acetals and ketals—from carbonyl compounds and alcohols; hemi-acetals
   d. nitrogen derivatives; nucleophilic addition followed by dehydration; see Table 9-3 for summary
   e. cyanohydrins from HCN addition
   f. Grignard reagents; used to make alcohols; formaldehyde ⟶ primary alcohols, any other aldehyde ⟶ secondary alcohols, ketones ⟶ tertiary alcohols
   g. bisulfite addition; Schiff's aldehyde test
   h. reduction to alcohols; lithium aluminum hydride, sodium boro-hydride; Clemmensen reduction of ⟍C=O to ⟍CH$_2$
6. Reactions involving α-hydrogens.
   a. enol, enolate anion, keto-enol tautomerism
   b. aldol condensation; gives a β-hydroxyaldehyde; crossed aldol con-densations
   c. haloform reaction; iodoform test for $CH_3\overset{\displaystyle O}{\overset{\|}{C}}-$ and $CH_3CH(OH)-$ groups
   d. Chloral, DDT
7. Polymers from aldehydes; trioxane, paraldehyde, Delrin, Bakelite, phenol-formaldehyde resins

Exercises and Problems

1. Name each of the following compounds:
   a. $CH_3CH_2COCH_2CH_3$
   b. $CH_3(CH_2)_4CHO$
   c. $(C_6H_5)_2CO$

   d. Br—⟨benzene ring⟩—CHO

   e.

   f. $(CH_3)_3CCHO$
   g. $CH_3CH{=}CHCHO$
   h. $CH_3CH{=}CHCOCH_3$
   i. $CH_2BrCOCH_3$
   j. $CH_3CH(OH)COCH_3$

2. Write structural formulas for each of the following:
   a. 2-octanone
   b. 4-methylpentanal
   c. m-chlorobenzaldehyde
   d. 3-methylcyclohexanone
   e. 2-butenal
   f. benzyl phenyl ketone
   g. p-tolualdehyde
   h. p,p'-dihydroxybenzophenone
   i. 2,2-dibromohexanal
   j. 1-phenyl-2-butanone

3. Give a definition or example of the following terms:
   a. acetal
   b. hemiacetal
   c. ketal
   d. hemiketal
   e. cyanohydrin
   f. dehydrogenation
   g. oxime
   h. phenylhydrazine
   i. phenylhydrazone
   j. enol

4. Write an equation for the reaction, if any, of benzaldehyde with each of the following, and name the organic product.
   a. Tollens' reagent
   b. hydroxylamine
   c. H_2, nickel
   d. ethylmagnesium bromide, then H_3O^+
   e. phenylhydrazine
   f. sodium bisulfite
   g. cyanide ion, H^+
   h. methanol, dry HCl
   i. ethylene glycol, H^+
   j. lithium aluminum hydride

5. Give equations for the preparation of the following carbonyl compounds:
   a. 2-pentanone from an alcohol
   b. pentanal from an alcohol
   c. cyclohexanone from phenol (two steps)
   d. acetone from propyne
   e. p-chlorobenzaldehyde from toluene
   f. octanal from 1-heptene

6. What simple chemical test can distinguish between the compounds in the following pairs:
   a. 2-pentanone and 3-pentanone
   b. hexanal and 3-hexanone
   c. benzyl alcohol and benzaldehyde
   d. cyclopentanone and 2-cyclopentenone
   e. methanol and ethanol

7. Each of the following questions pertains to section 9.6c.
   a. Following the patterns set in equations 9-22 and 9-24, write out all the steps in the mechanism of the reaction in equation 9-26.
   b. Acetaldehyde dimethylacetal (eq 9-25), when heated with aqueous acid, is readily cleaved to acetaldehyde and methanol. Write out the steps in the reaction mechanism.
   c. Acetaldehyde dimethyl acetal is readily cleaved by acid, whereas the geometrically similar methyl sec-butyl ether is much more difficult to cleave. Explain why.
   d. Acetaldehyde reacts with ethyl mercaptan (section 6.9) and acid to give a mercaptal (analogous to an acetal). Write out the equations for the reaction.

8. a. Write out the steps in the mechanisms of the reactions in equations 9-29 and 9-30.
   b. Explain how the reaction between acetone and hydroxylamine (eq 9-29) can be catalyzed by acid, but retarded or stopped if the reaction medium is too acidic.

9. Write equations for the reaction of each of the following with methyl-magnesium bromide, followed by hydrolysis with aqueous acid.
   a. acetaldehyde
   b. acetophenone
   c. formaldehyde
   d. cyclohexanone

10. Using a Grignard reagent and the appropriate aldehyde or ketone, show how each of the following could be prepared.
    a. 1-pentanol
    b. 3-pentanol
    c. 2-methyl-2-butanol
    d. 1-cyclopentylcyclopentanol
    e. 1-phenyl-1-propanol
    f. 3-butene-2-ol

11. Show, by means of equations, that in the synthesis of an alcohol from a Grignard reagent and an aldehyde or ketone, the following statements are true:
    a. 1-butanol (a primary alcohol) can be made from only one combination of reagents.
    b. 2-butanol (a secondary alcohol) can be made from two different combinations of reagents.
    c. 3-methyl-3-hexanol (a tertiary alcohol) can be made from three different combinations of reagents.

12. Give a specific example of (a) a secondary alcohol which can be made from only one combination of Grignard reagent and aldehyde; (b) a tertiary alcohol from one combination of Grignard reagent and ketone.

13. Equation 9-40 is more complicated than it first appears, since the attacking nucleophile is the sulfur atom. Write out the steps in the mechanism—in particular, those which show the proton exchanges.

14. Write equations for each of the following reactions:
    a. propionaldehyde, OH^-, heat
    b. ethyl phenyl ketone, bromine, base
    c. cyclopentanone, 2 moles of bromine, base
    d. 3-cyclopentenone, lithium aluminum hydride
    e. phenyl *n*-propyl ketone, zinc amalgam, HCl

15. Cyclohexanone is treated with sodium methoxide in a large excess of deuteromethanol (CH_3OD) as the solvent. The cyclohexanone which is recovered has a molecular weight four atomic mass units higher than ordinary cyclohexanone. What is the structure of the recovered cyclohexanone, and explain with equations how it is formed.

16. a. Write out all of the steps in the mechanism of the reaction in equation 9-49.
    b. Write out the steps in the synthesis of the mosquito repellant **6-12** from butanal.

17. An alkyl bromide reacts with magnesium in ether to form a Grignard reagent which reacts with water to give butane, and with acetaldehyde to give, after hydrolysis, 3-methyl-2-pentanol. Give the formula of the alkyl bromide and equations for all reactions mentioned.

18. Compound **A** reacts with methylmagnesium bromide to give, after hydrolysis, compound **B**. Chromic acid oxidation of **B** gives **C**, $C_5H_{10}O$, which gives a crystalline product with 2,4-dinitrophenylhydrazine and a positive iodoform test. Give the formulas of **A–C** and equations for all reactions mentioned.

19. Using the aldol condensation as the first step, followed by any other necessary reactions, show with equations how the following compounds can be prepared:
    a. 2-ethyl-2-hexenal
    b. 2-methyl-1,3-pentanediol
    c. 3-phenyl-1-propanol
    d. 4-methyl-4-pentanol-2-one
    e. 2-methyl-3-phenylpropenal

20. Crotonaldehyde (eq 9-50), when treated with a dilute solution of sodium methoxide in excess deuteromethanol (CH_3OD), gives $CD_3CH{=}CHCHO$. When a mixture of benzaldehyde and crotonaldehyde is treated with strong base, a good yield of the aldehyde $C_6H_5CH{=}CH{-}CH{=}CH{-}CHO$ can be obtained. Write equations which explain these observations.

CHAPTER TEN

CARBOXYLIC ACIDS AND THEIR DERIVATIVES

It was known to the ancients that a sour taste developed in wine left exposed to the air, and this process is still used in the manufacture of vinegar. Ethyl alcohol in the wine is oxidized by bacteria to a dilute solution of acetic acid, one of the simplest members of a series of compounds known as the carboxylic acids. Many occur in nature, either free or in combination with other organic substances. Acids combine with alcohols to form esters, which occur in fruit flavors, perfumes, and vegetable and animal fats and oils. Modification of the carboxyl group leads to related compounds known as acid derivatives, many of which have important uses.

Carboxylic acids, which constitute the most important class of organic acids, contain the **carboxyl group** (the name is a contraction of the two components of the group, carbonyl and hydroxyl). When

carboxyl group	aliphatic carboxylic acid	aromatic carboxylic acid

the formula is written on one line, it may be abbreviated to RCOOH or RCO_2H. Typical examples are acetic ($R = CH_3$) and benzoic ($Ar = C_6H_5$) acids.

$$CH_3C \overset{O}{\underset{OH}{\diagup}} \quad \text{or } CH_3CO_2H \qquad \qquad \overset{O}{\underset{OH}{\diagup}} \quad \text{or } C_6H_5CO_2H$$

acetic acid (an aliphatic acid)	benzoic acid (an aromatic acid)

Figure 10-1
Models of acetic acid.

Figure 10-2
Model of benzoic acid.

Several classes of compounds closely related in structure to carboxylic acids are known as **acid derivatives.** The most important of these are derived from acids by replacing the hydroxyl by various other groups.

$$R—C \overset{O}{\underset{OH}{\big\langle}}$$

$R—C\overset{O}{\underset{X}{\big\langle}}$	$R—C\overset{O}{\underset{O—C—R}{\big\langle}}$ $\overset{}{\underset{O}{\|}}$	$R—C\overset{O}{\underset{OR}{\big\langle}}$	$RC\overset{O}{\underset{NH_2}{\big\langle}}$
acid halide (X = any halogen)	acid anhydride	ester	amide*

10.1   NOMENCLATURE OF ACIDS

Because of their wide distribution and abundance in natural products, the carboxylic acids were among the first organic compounds studied. Consequently, many of them are known by common names, often derived from a Latin or Greek name indicating the original source of the acid. Formic acid (L., *formica*, ant) was obtained by distillation of ants. Others are acetic acid (L., *acetum*, vinegar), butyric acid (L., *butyrum*, butter), and stearic acid (Gr., *stear*, beef suet).

The aliphatic acids are sometimes known as **fatty acids** because many of them were first isolated from natural fats. The common names in Table 10-1 should be memorized, because they often provide the root of the names of other compounds.

The IUPAC system employs the suffix **-oic acid,** and applies all the other rules. Because of its structure, the carboxyl group (analogous to the aldehyde group) must come at the end of a carbon chain; consequently, no number is used to locate this group. Table 10-1 indicates the names of the first few members in the aliphatic series.

In the IUPAC system the chain is numbered from the carboxyl carbon to designate substituent positions. However, if common names are used, substituents are located by Greek letters, beginning with the α carbon.

$$\overset{\beta}{\underset{3}{}}CH_3—\overset{\alpha}{\underset{2}{}}CH—\overset{O}{\underset{OH}{C\big\langle}} \qquad \overset{\gamma}{\underset{4}{}}CH_3—\overset{\beta}{\underset{3}{}}CH—\overset{\alpha}{\underset{2}{}}CH_2—\overset{O}{\underset{OH}{C\big\langle}}$$
$$\underset{Br}{\big|} \qquad\qquad\qquad \underset{CH_3}{\big|}$$

2-bromopropanoic acid   3-methylbutanoic acid
(α-bromopropionic acid)   (β-methylbutyric acid)

*The hydrogens on the nitrogen may be substituted by organic groups; the term "amide" also applies to these compounds.

Table 10-1
Aliphatic Carboxylic Acids

FORMULA	COMMON NAME	IUPAC NAME
$HCOOH$	formic acid	methanoic acid
CH_3COOH	acetic acid	ethanoic acid
CH_3CH_2COOH	propionic acid	propanoic acid
$CH_3CH_2CH_2COOH$	n-butyric acid	butanoic acid
$CH_3CH_2CH_2CH_2COOH$	valeric acid	pentanoic acid
$CH_3CH_2CH_2CH_2CH_2COOH$	n-caproic acid	hexanoic acid
$CH_3(CH_2)_{10}COOH$	lauric acid	dodecanoic acid
$CH_3(CH_2)_{14}COOH$	palmitic acid	hexadecanoic acid
$CH_3(CH_2)_{16}COOH$	stearic acid	octadecanoic acid

The following examples illustrate the nomenclature of acids with cyclic R groups:

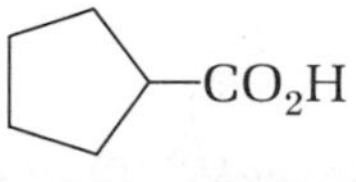

cyclopentanecarboxylic
acid

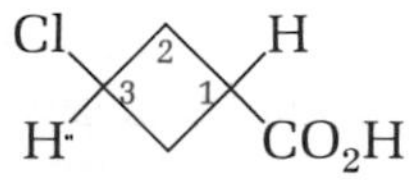

trans-3-chlorocyclo-
butanecarboxylic acid

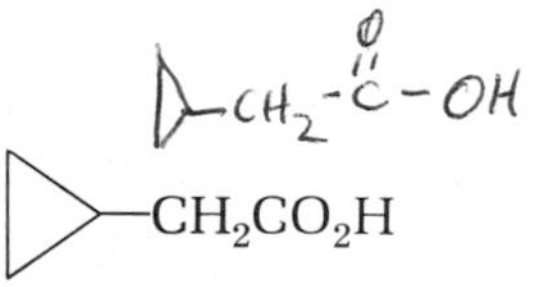

cyclopropaneacetic
acid

Aliphatic dicarboxylic acids are known almost exclusively by their common names (Table 10-2). However, the IUPAC system can be used, in which case the suffix **dioic acid** is used. Thus the IUPAC name for glutaric acid is pentanedioic acid.

Aromatic acids are given either common names or ones derived from the parent compound, benzoic acid, as in the following examples:

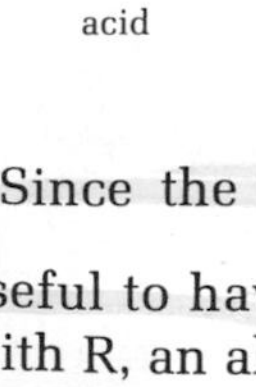

p-chlorobenzoic
acid

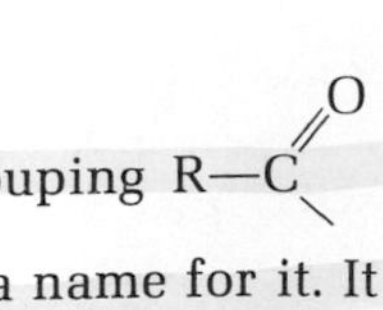

m-toluic acid

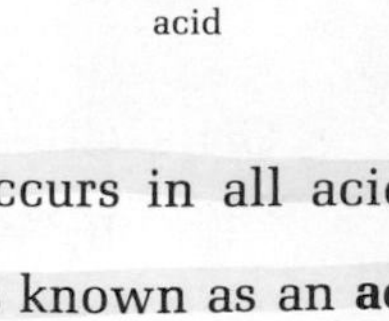

1-naphthoic
acid

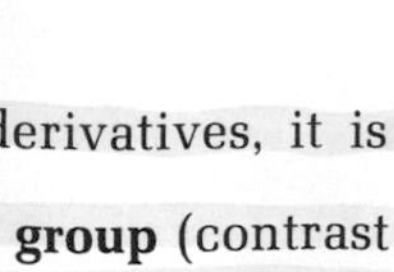

phthalic
acid

Since the grouping R—C occurs in all acid derivatives, it is useful to have a name for it. It is known as an **acyl group** (contrast with R, an alkyl group). Particular acyl groups are named by changing the -*ic* ending of the acid to -*yl*.

Table 10-2
Aliphatic Dicarboxylic Acids

FORMULA	NAME
HOOC—COOH	oxalic acid
HOOC—CH_2—COOH	malonic acid
HOOC—CH_2CH_2—COOH	succinic acid
HOOC—$CH_2CH_2CH_2$—COOH	glutaric acid
HOOC—$CH_2CH_2CH_2CH_2$—COOH	adipic acid

acetyl group   propionyl group   benzoyl group

10.2 ACIDITY OF CARBOXYLIC ACIDS

Compounds containing a carboxyl group are acidic by virtue of their ability to donate a proton to a more basic substance. In water, carboxylic acids dissociate to a carboxylate ion and an oxonium ion (eq 10-1).

carboxylate ion   oxonium ion

The degree of dissociation depends on the structure of R. Generally, carboxylic acids are weak so that dissociation is far from complete. They may, however, be neutralized by strong bases (eq 10-2). The salts thus produced are partially hydrolyzed in aqueous solution.

carboxylic acid   strong base   a sodium salt

10.2a IONIZATION CONSTANTS OF ACIDS

The value of the equilibrium constant for equation 10-1 is a measure of the strength of the acid. The equilibrium constant is

$$K_a = \frac{[RCO_2^-][H_3O^+]}{[RCO_2H]}$$

The greater its numerical value, the stronger the acid (i.e., the greater the concentration of hydronium ions, H_3O^+, in the solution).*

The determination of K_a is outlined briefly for acetic acid.

$$K_a = \frac{[H_3O^+][CH_3COO^-]}{[CH_3COOH]}$$

The concentration of hydrogen ions, *determined experimentally,* in a 0.1 M aqueous solution of acetic acid is 0.00135 gram-ions per liter. Since one CH_3COO^- is formed every time an H_3O^+ is formed, the concentration of acetate ions must also be 0.00135 gram-ions per liter. The amount of undissociated acetic acid is 0.1 − 0.00135 moles per liter. Substitution of these values in the equation gives the numerical value of K_a for acetic acid.

$$K_a = \frac{[0.00135][0.00135]}{[0.1 - 0.00135]} = 1.8 \times 10^{-5}$$

Alternatively, an acid solution may be half neutralized by base. At this point, $[RCO_2H] = [RCO_2^-]$ and K_a is simply equal to the concentration of H_3O^+ present at the half neutralization point.

The ionization constants of several carboxylic acids and some reference compounds are given in Table 10-3.

Table 10-3
Ionization Constants of Some Carboxylic Acids

NAME	FORMULA	K_a
Acetic acid	CH_3COOH	1.8×10^{-5}
Propionic acid	CH_3CH_2COOH	1.4×10^{-5}
n-Butyric acid	$CH_3CH_2CH_2COOH$	1.6×10^{-5}
Formic acid	$HCOOH$	2.1×10^{-4}
Chloroacetic acid	$ClCH_2COOH$	1.5×10^{-3}
Dichloroacetic acid	$Cl_2CHCOOH$	5.0×10^{-2}
Trichloroacetic acid	Cl_3CCOOH	2.0×10^{-1}
2-Chlorobutyric acid	$CH_3CH_2CHClCOOH$	1.4×10^{-3}
3-Chlorobutyric acid	$CH_3CHClCH_2COOH$	8.9×10^{-5}
Benzoic acid	C_6H_5COOH	6.6×10^{-5}
o-Chlorobenzoic acid	$o\text{-}ClC_6H_4COOH$	12.5×10^{-4}
m-Chlorobenzoic acid	$m\text{-}ClC_6H_4COOH$	1.6×10^{-4}
p-Chlorobenzoic acid	$p\text{-}ClC_6H_4COOH$	1.0×10^{-4}
p-Nitrobenzoic acid	$p\text{-}NO_2C_6H_4COOH$	4.0×10^{-4}
Phenol	C_6H_5OH	1.0×10^{-10}
Ethanol	CH_3CH_2OH	about 10^{-20}

*[ ] stands for activity, or, in dilute solutions, molar concentration of the species designated within the brackets. The concentration of water is omitted from the denominator, since it is used in large excess and remains relatively constant. It is incorporated into the value of K_a.

ci
① H
O R
NH₂

There are two major factors which influence the strengths of carboxylic acids, the **resonance effect** and the **inductive effect.**

10.2b THE RESONANCE EFFECT

One may wonder why the —OH of the carboxyl group should give up protons more easily than the —OH of alcohols or phenols. The relative strengths of ethyl alcohol, phenol, and acetic acid are given in Table 10-3. Acetic acid is about one hundred thousand times as strong as phenol, and the latter is 10^{10} or ten thousand million times as strong an acid as ethanol. Delocalization of charge through resonance is the major factor in these differences (review Figure 6-3).

In the carboxylate ion, the negative charge is distributed equally between the two oxygen atoms, since the two contributors to the

$$\left[R-C\!\!\begin{array}{c}\ddot{O}\\ \ddot{\underline{O}}{:}^-\end{array} \longleftrightarrow R-C\!\!\begin{array}{c}\ddot{\underline{O}}{:}^-\\ \ddot{O}\end{array}\right] \quad \text{or} \quad R-C\!\!\begin{array}{c}O\\ O\end{array}^-$$

contributors to the resonance hybrid carboxylate
anion

resonance hybrid are identical. Physical evidence regarding carbon-oxygen bond lengths bears out this contention. Whereas in formic acid the two carbon-oxygen bonds have different lengths, its sodium salt has two identical carbon-oxygen bonds. Furthermore, the bond

$$\begin{array}{cc} 1.23\text{Å}\quad H-C\!\!\begin{array}{c}O\\ O-H\end{array} & 1.27\text{Å}\quad H-C\!\!\begin{array}{c}O\\ O\end{array}^- Na^+ \\ 1.36\text{Å} & 1.27\text{Å} \end{array}$$

formic acid   sodium formate

lengths in the formate ion are of intermediate length between those of the normal double and single carbon-oxygen bonds.

In molecular orbital terms, the carboxyl carbon is trigonal, using sp^2 bonding (sec 3.2) for attachment to R and two oxygens. The system is planar, with 120° bond angles. The remaining carbon p orbital, perpendicular to this plane, overlaps with similar p orbitals on both oxygen atoms (Figure 10-3). This results in a dispersal of electron density over all three atoms, though the negative charge is located primarily on the two oxygens.

Charge delocalization is greater in carboxylate ions than in phenoxide or alkoxide ions, and accounts for the observed order of acidities: carboxylic acids > phenols > alcohols.

10.2c THE INDUCTIVE EFFECT

Electron-withdrawing groups increase the acidity of carboxylic acids. Compare, for example, the K_a of acetic acid with those of

Figure 10-3
*Carboxylate ion, showing over-
lap of the carbon p orbital
with similar orbitals on both
oxygens, leading to charge delocalization.*

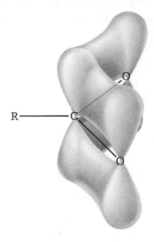

mono-, di-, and trichloroacetic acids (Table 10-3). Chloroacetic acid
is 100 times stronger than acetic acid, and trichloroacetic acid is yet
another 100 times stronger than chloroacetic acid. These data can
be explained by the inductive effect of the chlorines, which tends
to remove electron density from the carboxylate group and disperse
the negative charge.

chloroacetate ion   acetate ion

Looked at another way, the carbon-chlorine dipole places a positive
charge near the negative charge on the carboxylate group in the
chloroacetate ion. This partial neutralization of charge stabilizes the
anion.

The effect falls off with distance (compare the K_a's of 2- and
3-chlorobutyric acids, Table 10-3), but can be effectively relayed
through an aromatic ring (compare the three chlorobenzoic acids
with benzoic acid, Table 10-3).

Formic acid is about ten times stronger than acetic acid. This
suggests that hydrogen is electron-withdrawing compared to methyl,
or conversely, that a methyl group is electron-donating compared
to hydrogen. This is consistent with the observation that methyl (and
other alkyl) groups stabilize carbonium ions (which have a positive
carbon, and are electron-demanding, sec 3.4c) and are ring-activating

and *ortho-para* directing in electrophilic aromatic substitution (sec 5.4b).

10.3 PHYSICAL PROPERTIES OF ACIDS

The lower aliphatic acids are liquids with sharp or unpleasant odors. Butyric acid is responsible for the disagreeable odor of rancid butter; the C_6, C_8, and C_{10} straight-chain acids have the unpleasant odor associated with goats. Higher fatty acids (greater than ten carbons) are waxlike solids. Stearic acid, for example, is mixed with paraffin in the manufacture of wax candles. The dicarboxylic and aromatic acids are crystalline solids.

Table 10-4
Physical Properties of Some Acids

NAME OF ACID	FORMULA	BP, °C	MP, °C	SOLUBILITY, GRAMS/100g H₂O AT 25°
Formic	HCOOH	101	8	miscible
Acetic	CH_3COOH	118	17	miscible
Propionic	CH_3CH_2COOH	141	−22	miscible
n-Butyric	$CH_3CH_2CH_2COOH$	164	−8	miscible
Caproic	$CH_3(CH_2)_4COOH$	202	−1.5	0.4
Capric	$CH_3(CH_2)_8COOH$	268	31	almost insoluble
Palmitic	$CH_3(CH_2)_{14}COOH$	356	64	almost insoluble
Stearic	$CH_3(CH_2)_{16}COOH$	383	69	almost insoluble
Benzoic	C_6H_5COOH	249	122	about 0.4
Oxalic	$(COOH)_2$	—	189	about 15

Carboxylic acids are polar and, like the alcohols (sec 6.3), form hydrogen bonds with themselves or with other molecules. Molecular weight determination shows that formic acid and acetic acid consist largely of dimers (two formula weights per molecule) even in the vapor state.

$$CH_3-C\overset{O\cdots H-O}{\underset{O-H\cdots O}{}}C-CH_3$$

The two units are probably held together by hydrogen bonds as indicated in the above formula. This explains the high boiling points of the carboxylic acids in comparison with their formula weights. The solubility of the lower molecular weight acids in water is probably due to hydrogen bonding with the solvent.

10.4 PREPARATIONS OF ACIDS

Three methods will be described here—oxidation of other organic compounds, hydrolysis of organic cyanides (nitriles), and the reaction of a Grignard reagent with carbon dioxide. The choice amongst these alternatives may depend on whether the synthesis is commercial or to be done in the laboratory, on availability of starting materials, and on the presence of other groups which might interfere with a particular method.

10.4a OXIDATION

Table 9-12 indicates that one might hope to obtain acids by oxidizing other organic compounds (alkanes, alkenes, alcohols, aldehydes).

Oxidation of alkanes to acids is not readily controlled. However, when an alkyl group is a side-chain on an aromatic ring, oxidation occurs readily no matter how long the chain.

toluene   benzoic acid   n-propylbenzene   (10-3)

Carbons not directly attached to the aromatic ring are oxidized to carbon dioxide. The oxidizing agent may be dilute nitric acid or potassium permanganate. There may be more than one side chain, or the two side chains may be joined in the form of a ring as in the case of naphthalene.

naphthalene   phthalic acid   (10-4)

Oxidation of an alkyl side-chain on an aromatic ring occasionally may be accomplished indirectly, as in the chlorination of toluene to benzotrichloride, followed by hydrolysis to benzoic acid (compare with sec 9-13 and eq 9-6).

benzo-   benzoic
trichloride   acid   (10-5)

Vigorous oxidation of an alkene with potassium permanganate cleaves the double bond. The type of product obtained depends on the number of alkyl groups attached to each end of the double bond. An unsubstituted end ($=CH_2$) is converted to carbon dioxide, a disubstituted end ($=CR_2$) yields a ketone, and a monosubstituted end ($=CHR$) forms a carboxyl group. Cycloalkenes give dicarboxylic acid if no substituents are attached to the double bond.

$$\text{cyclohexene} \xrightarrow[\text{heat}]{KMnO_4} \text{adipic acid} \quad (10\text{-}6)$$

Primary alcohols and aldehydes, when oxidized, give acids with the same number of carbon atoms per molecule.

$$RCH_2OH \xrightarrow{Na_2Cr_2O_7} RCO_2H \xleftarrow{KMnO_4} RCHO \quad (10\text{-}7)$$

Terminal diols can be oxidized to dicarboxylic acids.

$$\underset{\text{1,4-butanediol}}{HOCH_2CH_2CH_2CH_2OH} \xrightarrow{[O]} \underset{\text{succinic acid}}{HO_2CCH_2CH_2CO_2H} \quad (10\text{-}8)$$

Acetic acid is made commercially by the catalytic oxidation of ethanol or acetaldehyde. The acetaldehyde is usually prepared from acetylene (eq 9-11).

$$CH_3CHO \xrightarrow[\text{MnO}]{\text{air}} CH_3CO_2H \quad (10\text{-}9)$$

$$CH_2CH_3OH$$

10.4b HYDROLYSIS OF ORGANIC CYANIDES (NITRILES)

Organic cyanides (also called nitriles) can be prepared from primary or secondary alkyl halides by S_N2 displacement with cyanide ion (eq 8-11). Hydrolysis, either in acid or base, converts the cyanide to a carboxylic acid. The nitrogen of the cyano group is converted in base to ammonia (or in acid, to NH_4^+). The overall equation for the sequence is:

$$\underset{\substack{\text{alkyl}\\\text{halide}}}{RX} \xrightarrow{CN^-} \underset{\text{nitrile}}{RC\equiv N} \xrightarrow[H^+]{2\,H_2O} \underset{\text{acid}}{RCO_2H} + NH_4^+ \quad (10\text{-}10)$$

The acid formed has one more carbon atom than the alkyl halide; this is sometimes a useful way of lengthening a carbon chain by one carbon atom. Equations 10-11 to 10-13 illustrate with a specific example.

$$CH_3CH_2CH_2OH \xrightarrow[ZnCl_2]{HCl} CH_3CH_2CH_2Cl + H_2O \qquad (10\text{-}11)$$
1- propanol

$$CH_3CH_2CH_2Cl \xrightarrow{Na^+CN^-} CH_3CH_2CH_2CN + Na^+Cl^- \qquad (10\text{-}12)$$
n-propyl cyanide,
or butyronitrile

$$CH_3CH_2CH_2CN \xrightarrow[H^+]{H_2O} CH_3CH_2CH_2CO_2H + NH_4^+ \qquad (10\text{-}13)$$
butanoic acid

The mechanism of nitrile hydrolysis involves nucleophilic attack on the carbon-nitrogen triple bond. The reaction is mechanistically similar to nucleophilic attack on the carbonyl group of aldehydes or ketones (sec 9.6b).

$$(10\text{-}14)$$

acid   amide

The first mole of water converts the nitrile to an **amide,** which frequently can be isolated. The step leading to the amide involves tautomerism similar to keto-enol tautomerism in carbonyl compounds (sec 9.7). A second mole of water is required to hydrolyze the amide to an acid (sec 10.8).

10.4c   THE GRIGNARD METHOD

Grignard reagents (sec 8.6) react with carbon dioxide in a manner analogous to their reaction with other carbonyl compounds (sec 9.6f). The initial adduct, upon hydrolysis with acid, yields a carboxylic acid.

$$R-MgX + O{=}C{=}O \longrightarrow R-\overset{\overset{\displaystyle O}{\|}}{C}-OMgX \xrightarrow{HX}$$

$$R-\overset{\overset{\displaystyle O}{\|}}{C}-OH + Mg^{2+}X_2^- \qquad (10\text{-}15)$$

The reaction gives good yields and is widely applicable; R may be aliphatic or aromatic. Like the nitrile method (sec 10.4b), this procedure can be used to lengthen a carbon chain by one carbon atom, as in the following examples:

$$CH_3CH_2Br \xrightarrow[\text{ether}]{Mg} CH_3CH_2MgBr \xrightarrow{CO_2}$$

ethyl bromide
(2-carbon chain)

$$CH_3CH_2CO_2MgBr \xrightarrow[H_2O]{H^+} CH_3CH_2CO_2H \quad (10\text{-}16)$$

propionic acid
(3-carbon chain)

p-bromotoluene   p-toluic acid

10.5 SALTS OF CARBOXYLIC ACIDS

Carboxylic acids react with bases to form salts (eq 10-2). The crystalline salt may be obtained by using equivalent amounts of acid and base and evaporating the water.

Salts of carboxylic acids are named as one would name the salt of an inorganic acid. The cation is named first, followed by the name of the carboxylate anion. The latter is named by changing the *ic* ending of the acid to *ate*, as in these examples:

sodium acetate   potassium benzoate   calcium propionate

Sodium and potassium salts of long-chain fatty acids are known as **soaps** (Chapter 11). They are water soluble, and their solutions

sodium palmitate
(a soap)

are usually weakly alkaline due to hydrolysis (the reverse of eq 10-2). Carboxylates of bivalent cations such as calcium, magnesium, and barium are sparingly soluble in water. Calcium propionate is added to bread to prevent mold growth; sodium benzoate is another common food preservative. Lithium stearate and other heavy metal salts, when blended with lubricating oils, form automotive greases.

10.6 ESTERS

When an organic acid is heated with a primary or secondary alcohol, especially in the presence of a little mineral acid as a catalyst, equilibrium is established with the ester and water.

$$\underset{\text{acid}}{R-\overset{\overset{\displaystyle O}{\|}}{C}-OH} + \underset{\text{alcohol}}{R'OH} \overset{H^+}{\rightleftharpoons} \underset{\text{ester}}{R-\overset{\overset{\displaystyle O}{\|}}{C}-OR'} + H_2O \qquad (10\text{-}18)$$

The process is known as **esterification.** The equilibrium can be shifted toward ester formation by using an excess of the alcohol* or acid, whichever is least expensive. Alternatively, the water may be removed as it is formed (by distillation with a partially miscible solvent such as benzene) to drive the reaction forward.

10.6a NOMENCLATURE AND PROPERTIES

Esters are named in a manner analogous to salts. The alkyl group from the alcohol is named first, followed by the name of the acid, with the *ic* ending changed to *ate*.

$$\overset{\overset{\displaystyle O}{\|}}{CH_3C}-OCH_3 \qquad \overset{\overset{\displaystyle O}{\|}}{CH_3C}-OCH_2CH_3 \qquad \overset{\overset{\displaystyle O}{\|}}{CH_3CH_2CH_2C}-OCH_3$$

methyl acetate ethyl acetate methyl butanoate

$$CH_3\overset{\overset{\displaystyle O}{\|}}{C}-O-\bigcirc \qquad \bigcirc-\overset{\overset{\displaystyle O}{\|}}{C}-OCH_3$$

phenyl acetate methyl benzoate

Esters are usually rather pleasant-smelling substances which are responsible for the flavor and fragrance of many fruits and flowers. Some of the more common are ethyl formate (artificial rum flavor),

---

*This is the most common procedure and is called *Fischer esterification,* after the great nineteenth century German organic chemist who developed the method.

Figure 10-4
Models of the ester, ethyl acetate.

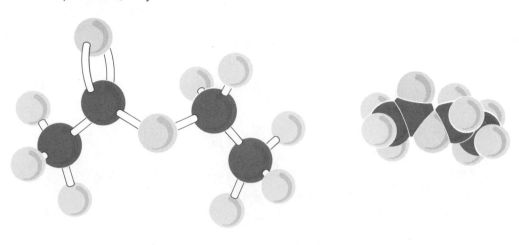

n-pentyl acetate (bananas), octyl acetate (oranges), ethyl butyrate (pineapples), and pentyl butyrate (apricots). Artificial flavors of strawberry, cherry, raspberry, apple, etc., are made largely from mixtures of esters.

10.6b   MECHANISM OF ESTERIFICATION

In equation 10-18, the ester is depicted as having been constructed from the *acyl* group of the acid and the *alkoxy* group of the alcohol (and *not* from the acyloxy group of the acid and the alkyl group of the alcohol). This fact has been established experimentally, using isotopic labeling. For example, reaction of benzoic acid with methanol enriched in the O^{18} isotope of oxygen gave labeled methyl benzoate; the water contained none of the enrichment O^{18}. This

$$\langle\!\!\!\bigcirc\!\!\!\rangle\!\!-\!\!\overset{\displaystyle O}{\overset{\|}{C}}\!\!-\!OH + HO^{18}CH_3 \overset{H^+}{\longrightarrow} \langle\!\!\!\bigcirc\!\!\!\rangle\!\!-\!\!\overset{\displaystyle O}{\overset{\|}{C}}\!\!-\!O^{18}CH_3 + HOH$$

methyl benzoate   (10-19)

result can be explained by the mechanism shown in equation 10-20. The purpose of the acid catalyst is to protonate the carboxyl group (step 1), thus enhancing the susceptibility of the carboxyl carbon to nucleophilic attack by the alcohol (step 2).

$$(10\text{-}20)$$

Step 3 is a proton transfer to one of the hydroxyl groups, which then departs as H_2O (step 4) leaving the protonated ester which is in equilibrium with the ester itself by proton loss (step 5). The net result of this scheme is to replace the OH of the acid by the OR′ group of the alcohol (eq 10-18). The entire scheme is reversible and applies not only to most esterifications, but also to the reverse process, the acid-catalyzed hydrolysis of esters.

In step 2 a trigonal (sp^2) carbon becomes a tetrahedral (sp^3) carbon. This step involves some compression of groups (the angles decrease from 120° to 109.5°) and is usually rate-determining. Bulky R groups in the acid or R′ groups in the alcohol can sharply decrease the esterification rate because of steric hindrance in this step.

10.6c REACTIONS OF ESTERS

Esters may be converted to their component alcohols and acids by boiling with aqueous sodium hydroxide. The reaction is known as **saponification,** since it involves the same process as the making of soap from fats (Chapter 11).

$$(10\text{-}21)$$

ester   alkali   salt of an acid   alcohol

For most esters, the reaction rate depends on the concentration of ester and hydroxide ion; the rate-determining step is therefore probably nucleophilic attack by hydroxide ion on the carbonyl carbon of the ester.

$$R-\overset{:\ddot{O}:^-}{\underset{R'\ddot{O}}{C}}=O: + :\ddot{O}H^- \overset{slow}{\rightleftharpoons} R-\overset{|}{\underset{R'\ddot{O}}{C}}-OH \rightleftharpoons \overset{R}{\underset{HO}{C}}=O + R'\ddot{O}:^-$$

(10-22)

Since alkoxide ions are strong bases (sec 6.4), the final step is rapid proton transfer to produce the carboxylate anion and the alcohol.

$$R-\overset{O}{\underset{OH}{C}} + R'O^- \overset{fast}{\longrightarrow} R-\overset{O}{\underset{O^-}{C}} + R'OH$$

(10-23)

Saponification of esters can be used to prepare acids and alcohols from naturally occurring esters. The reaction is also usually the first step in the structure determination of an unknown ester isolated from nature; if the acid and alcohol components can be identified, the structure of the original ester can be deduced. Nucleophiles other than hydroxide ion can be used in this reaction. For example, esters react with ammonia to form amides.

$$\underset{\text{ester}}{R-\overset{O}{\underset{OR'}{C}}} + :NH_3 \longrightarrow \underset{\text{amide}}{R-\overset{O}{\underset{NH_2}{C}}} + R'OH$$

(10-24)

One ester can be converted to another if the nucleophile is an alcohol.

$$R-\overset{O}{\underset{OR'}{C}} + R''OH \overset{basic}{\underset{catalyst}{\rightleftharpoons}} R-\overset{O}{\underset{OR''}{C}} + R'OH$$

(10-25)

This process, called **transesterification**, is reversible. It can be driven forward by using excess R''OH, but most commonly the reaction is used when R'OH has a much lower boiling point than R''OH, in which case the former can simply be removed by distillation as the reaction proceeds. This technique is used in the production of Dacron (sec 10.6d).

Esters (and thus indirectly, acids) can be reduced either chemically or catalytically. One obtains not only a mole of the same alcohol as would be produced by saponification (R'OH) but also the primary alcohol (RCH$_2$OH) corresponding to reduction of the acid portion of the ester (review Table 9-2).

$$R-C \underset{OR'}{\overset{O}{\big\|}} \xrightarrow[\substack{\text{Na in alcohol,} \\ \text{or } H_2 \text{ and} \\ \text{copper chromite} \\ \text{catalyst}}]{\text{LiAlH}_4, \text{ or}} RCH_2OH + R'OH \qquad (10\text{-}26)$$

This reaction is used commercially to obtain long-chain primary alcohols from fats (sec. 11.2c).

10.6d POLYESTERS

Dicarboxylic acids react with dihydric alcohols to form **polyesters.** The synthetic fiber **Dacron** is the polyester of terephthalic acid and ethylene glycol. The terephthalic acid must be extremely pure, since small amounts of the *meta* or *ortho* isomer would drastically alter the geometry of the polymer molecule. The dimethyl ester is more easily purified than the acid itself. In the first step, pure dimethyl terephthalate is transesterified with ethylene glycol; methanol, the

$$\text{dimethyl terephthalate} + HOCH_2CH_2OH \underset{}{\overset{-OCH_3}{\rightleftharpoons}} \qquad (10\text{-}27)$$

dimethyl terephthalate   ethylene glycol

$$2\ CH_3OH + \text{(diester)} \xrightarrow[n \text{ moles}]{\text{heat}}$$

$$(n-1)\ HOCH_2CH_2OH + \text{(Dacron polyester)}$$

Dacron polyester

lowest boiling component of the mixture is distilled as the reaction proceeds. The temperature is then raised, and a second transesterification occurs as ethylene glycol is distilled from the mixture, leaving polyester ($n \cong 100 \pm 20$). The fiber produced from this polyester has exceptional elastic recovery properties which contribute to its crease resistance. It also absorbs little moisture, its properties when wet being almost the same as when dry. The same polyester can be converted to a film (Mylar) of unusual strength.

If a trihydroxy alcohol, such as glycerol, is employed, the possibility for cross-linking appears. The **glyptal** resins which are used chiefly as coatings in the lacquer and paint industry are polyesters of **glycerol** and **phthalic** acid.

Figure 10-5
From tiny holes in a spinneret, fine filaments extrude which are twisted together to make a single end of continuous-filament yarn. The principle applies to nylon, Orlon acrylic fiber, Dacron polyester fiber, and many other man-made yarns. The number of filaments in a single thread of man-made yarn ranges from one, as in sheer hosiery, to 2,934 in tire cord. (Courtesy of E. I. DuPont DeNemours & Company, Inc.)

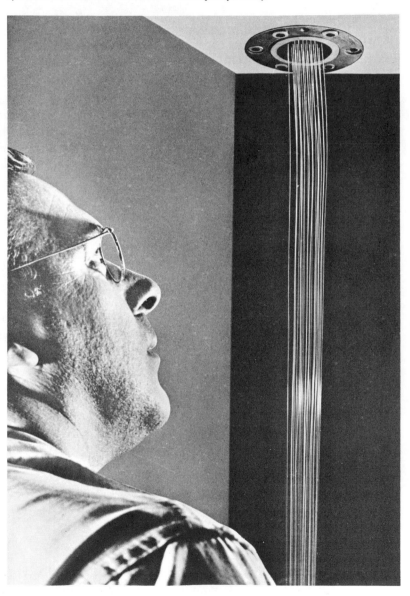

10.7 ACID HALIDES AND ANHYDRIDES

Many reactions of acids and their derivatives involve attack of a nucleophile on the carbonyl carbon. Specific examples already discussed are illustrated in equations 10-20, 10-22, 10-24, and 10-25. These reactions may be generalized as

$$
\begin{matrix}
R \\
 \\
L
\end{matrix}
C\!\!=\!\!\overset{..}{\underset{..}{O}}\!: + \overset{..}{N}u^- \rightleftharpoons
R-\overset{\overset{:\overset{..}{O}:^-}{|}}{\underset{L}{C}}-Nu \rightleftharpoons
\begin{matrix}
R \\
 \\
Nu
\end{matrix}
C\!\!=\!\!\overset{..}{\underset{..}{O}}\!: + :L^- \qquad (10\text{-}28)
$$

wherein some nucleophile Nu replaces a leaving group L through a tetrahedral intermediate. As discussed in detail for aldehydes and ketones (sec 9.6b), the process may be catalyzed by acid or base. The reactions usually proceed more slowly than additions to aldehydes or ketones because the group L has an unshared electron pair which can release electrons to the carbonyl carbon, thus decreasing its susceptibility to nucleophilic attack.

$$
R-C\!\!\!\begin{matrix}\overset{\displaystyle \overset{..}{O}:}{} \\ \underset{\displaystyle \overset{..}{L}}{}\end{matrix} \longleftrightarrow R-C\!\!\!\begin{matrix}\overset{\displaystyle \overset{..}{O}:^-}{} \\ \underset{\displaystyle L^+}{}\end{matrix}
$$

One way in which to enhance the reactivity of acid derivatives is to make L an electron-withdrawing group. This will remove electron density from the carbonyl carbon and make it more susceptible to nucleophilic attack. The **acyl halides** and **acid anhydrides** are the two most common types of **acid derivatives** used for this purpose.

10.7a STRUCTURE AND NOMENCLATURE

Acid halides are named using the name of the particular acyl group (sec 10.1), as in these examples.

$$
CH_3-C\!\!\!\begin{matrix}\overset{\displaystyle O}{\diagup} \\ \diagdown Cl\end{matrix}
\qquad
CH_3CH_2CH_2C\!\!\!\begin{matrix}\overset{\displaystyle O}{\diagup} \\ \diagdown Br\end{matrix}
\qquad
\bigcirc\!\!-C\!\!\!\begin{matrix}\overset{\displaystyle O}{\diagup} \\ \diagdown Cl\end{matrix}
$$

acetyl chloride   butanoyl bromide   benzoyl chloride

Although all types of halides (F, Cl, Br, I) are known, the acyl chlorides are the least expensive and most common.

Figure 10-6
Models of the acyl halide, acetyl chloride.

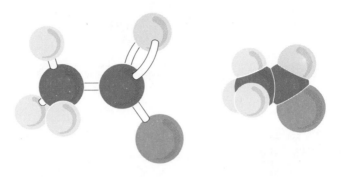

Acid anhydrides are derived from acids by removal of the elements of water. This relationship is shown below for two familiar inorganic acids.

HO
   C=O   O=C=O   HO—S—OH   S=O
HO

carbonic   carbon dioxide   sulfuric   sulfur trioxide
acid   (carbonic anhydride)   acid   (sulfuric anhydride)

The same relationship exists between carboxylic acids and their anhydrides.

two molecules of a   a carboxylic
carboxylic acid   anhydride

Anhydrides are named from the corresponding acid, replacing the word *acid* with anhydride.

$$CH_3-\overset{\overset{O}{\|}}{C}-OH$$

$$+ \longrightarrow$$

$$CH_3-\overset{\overset{O}{\|}}{C}-OH$$

$$\overset{\overset{O}{\|}}{CH_3C}-O-\overset{\overset{O}{\|}}{CCH_3}$$

acetic anhydride

$$\begin{array}{c} CH_3CH_2\overset{O}{\underset{\diagdown}{C}} \\ \overset{\diagup}{O} \\ CH_3CH_2\overset{\diagup}{\underset{\diagdown}{C}} \\ \overset{O}{} \end{array}$$

propionic (or
propanoic) anhydride

benzoic anhydride

Figure 10-7
Models of acetic anhydride.

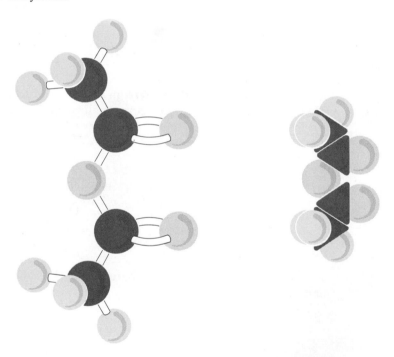

10.7b   PREPARATION

Acyl halides are prepared from acids by reaction with phosphorus
halides (PCl$_3$, PCl$_5$) or thionyl chloride (SOCl$_2$). The reactions are
analogous to the preparation of alkyl halides from alcohols (eq 6-30
and 6-31).

$$3\ R-\overset{O}{\underset{OH}{C}} + PCl_3 \longrightarrow 3\ R-\overset{O}{\underset{Cl}{C}} + \underset{\text{phosphorous acid}}{P(OH)_3} \qquad (10\text{-}29)$$

$$R-C\overset{O}{\underset{OH}{\diagdown}} + PCl_5 \longrightarrow R-C\overset{O}{\underset{Cl}{\diagdown}} + HCl + \underset{\substack{\text{phosphorus} \\ \text{oxychloride}}}{POCl_3} \qquad (10\text{-}30)$$

$$R-C\overset{O}{\underset{OH}{\diagdown}} + SOCl_2 \longrightarrow R-C\overset{O}{\underset{Cl}{\diagdown}} + HCl\uparrow + SO_2\uparrow \qquad (10\text{-}31)$$

Although detailed mechanisms have not been established, it is reasonable to expect that the first step involves formation of HCl and an unstable mixed anhydride which then collapses to the products. For thionyl chloride, the scheme would be

$$R-C\overset{O}{\underset{OH}{\diagdown}} + SOCl_2 \xrightarrow{-HCl} \left[R-C\overset{O}{\underset{\substack{O \\ Cl-S \diagdown O}}{\diagdown}} \right] \longrightarrow R-C\overset{O}{\underset{Cl}{\diagdown}} + SO_2 \quad (10\text{-}32)$$

mixed anhydride

Phosphorus halides are used to prepare low-boiling acyl halides, which can be removed from the reaction mixture by distillation (the phosphorus-containing reaction products are fairly high boiling). Thionyl chloride is especially useful for preparing higher-boiling acyl halides, because two of the reaction products are gases (SO_2, HCl) and any excess thionyl chloride (a low-boiling liquid) is readily removed by distillation.*

Acid anhydrides may be prepared from acyl halides by reaction with salts. The reaction proceeds by nucleophilic attack of the carboxylate ion on the carbonyl carbon of the acyl halide (eq 10-28).

$$\underset{O}{R-\overset{\overset{\displaystyle O}{\|}}{C}-Cl} + \underset{O}{R-\overset{\overset{\displaystyle O}{\|}}{C}-O^-Na^+} \xrightarrow{\text{heat}} \underset{O}{R-\overset{\overset{\displaystyle O}{\|}}{C}-O-\overset{\overset{\displaystyle O}{\|}}{C}-R} + Na^+Cl^- \quad (10\text{-}33)$$

Some anhydrides of dicarboxylic acids can be prepared simply by heating the acid; this is particularly so if the carboxyl groups are favorably situated geometrically, as with phthalic acid.

---

*Acyl halides *cannot* be prepared by the action of a hydrohalic acid (HX) on the carboxylic acid. Although the hydroxyl group of alcohols can be replaced by halogen in this way (eq 6-27), the reaction, which is a reversible one in any event, proceeds entirely in the opposite direction with acids. The acyl halides are very rapidly hydrolyzed to the corresponding acid and HX (sec 10.7c).

phthalic acid   phthalic anhydride

$$\text{(10-34)}$$

Phthalic anhydride is used commercially; over 500 million pounds are produced annually for use in plasticizers and polyester resins for surface coatings (sec 10.6d).

   Acetic anhydride is the major aliphatic anhydride of commercial importance. Over a billion pounds are produced annually and are used to make acetates of alcohols, the most common being cellulose acetate and aspirin. A special procedure is used for acetic anhydride's manufacture. Acetic acid is added to **ketene,** an unusually reactive compound which can be made by pyrolysis of acetone or acetic acid.

$$\text{(10-35)}$$

$$CH_3-\overset{O}{\overset{\|}{C}}-OH + HO-\overset{O}{\overset{\|}{C}}-CH_3$$

10.7c   REACTIONS

Acyl halides and anhydrides are readily hydrolyzed by water to their corresponding acids. They react with alcohols to form esters, and with ammonia or amines to form amides. These reactions are all analogous in mechanism to the similar reactions of esters (sec 10.6c, and eq 10-28). Some specific examples are the following processes:

Hydrolysis (produces acids):

$$CH_3\overset{O}{\overset{\|}{C}}Cl + HOH \xrightarrow[\text{rapid}]{0^\circ} CH_3\overset{O}{\overset{\|}{C}}OH + HCl \qquad \text{(10-36)}$$

$$CH_3\overset{O}{\overset{\|}{C}}-O-\overset{O}{\overset{\|}{C}}CH_3 + HOH \xrightarrow[\text{temp.}]{\text{room}} 2\,CH_3\overset{O}{\overset{\|}{C}}-OH \qquad \text{(10-37)}$$

Alcoholysis (produces esters):

$$\text{C}_6\text{H}_5-\overset{\overset{\displaystyle O}{\|}}{\text{C}}-\text{Cl} + \text{CH}_3\text{OH} \xrightarrow[\text{temp.}]{\text{room}} \text{C}_6\text{H}_5-\overset{\overset{\displaystyle O}{\|}}{\text{C}}-\text{OCH}_3 + \text{HCl} \quad (10\text{-}38)$$

methyl benzoate

$$\text{CH}_3\overset{\overset{\displaystyle O}{\|}}{\text{C}}-\text{O}-\overset{\overset{\displaystyle O}{\|}}{\text{C}}\text{CH}_3 + \text{CH}_3\text{CH}_2\text{CH}_2\text{OH} \xrightarrow{\text{warm}}$$

$$\text{CH}_3\overset{\overset{\displaystyle O}{\|}}{\text{C}}-\text{OCH}_2\text{CH}_2\text{CH}_3 + \text{CH}_3\text{CO}_2\text{H} \quad (10\text{-}39)$$

n-propyl acetate

Ammonolysis (produces amides):

$$\text{CH}_3\overset{\overset{\displaystyle O}{\|}}{\text{C}}-\text{Cl} + 2\,\text{NH}_3 \longrightarrow \text{CH}_3\overset{\overset{\displaystyle O}{\|}}{\text{C}}-\text{NH}_2 + \text{NH}_4{}^+\text{Cl}^- \quad (10\text{-}40)$$

acetamide

$$(10\text{-}41)$$

phthalic anhydride   phthalamic acid

In general, the halides are more reactive than the anhydrides, and aliphatic acyl halides or anhydrides are more reactive than the aromatic acid derivatives. These observations can be rationalized, since reactivity is a function of electron density at the carbonyl carbon (the lower the electron-density, the more reactive the compound will be towards nucleophiles—i.e., electron suppliers).

Acyl halides and acid anhydrides are often used to synthesize aromatic ketones in a reaction analogous to the Friedel-Crafts alkylation reaction (sec 5.4a). The reaction, known as **acylation,** is an electrophilic aromatic substitution.

$$\left.\begin{array}{c} \text{R}-\overset{\overset{\displaystyle O}{\|}}{\text{C}}-\text{Cl} \\ \text{or} \\ \text{R}-\overset{\overset{\displaystyle O}{\|}}{\text{C}}-\text{O}-\overset{\overset{\displaystyle O}{\|}}{\text{C}}-\text{R} \end{array}\right\} + \text{C}_6\text{H}_6 \xrightarrow{\text{AlCl}_3} \text{R}-\overset{\overset{\displaystyle O}{\|}}{\text{C}}-\text{C}_6\text{H}_5 + \begin{array}{c}\text{HCl} \\ \text{or} \\ \text{RCO}_2\text{H}\end{array} \quad (10\text{-}42)$$

The electrophile is an **acyl cation** (or acylium ion), as illustrated for acyl halides in the mechanism:

$$R-\overset{\overset{\displaystyle O}{\|}}{C}-Cl + AlCl_3 \;\rightleftharpoons\; R-\overset{+}{C}=O + AlCl_4^- \qquad (10\text{-}43a)$$

acyl cation

$$+ H^+ \quad (10\text{-}43b)$$

$$H^+ + AlCl_4^- \;\rightleftharpoons\; HCl + AlCl_3 \qquad (10\text{-}43c)$$

The orientation rules for electrophilic substitution (sec 6.4b) are followed, as illustrated in this example.

$$CH_3-\overset{\overset{\displaystyle O}{\|}}{C}-\text{—}-CH_3 + HCl \quad (10\text{-}44)$$

p-methylacetophenone

10.8 AMIDES

Simple amides are obtained by the reaction of ammonia with a variety of acid derivatives, such as esters (eq 10-24), acyl halides (eq 10-40) and acid anhydrides (eq 10-41). They are also intermediates in the hydrolysis of nitriles (eq 10-14) and can also be obtained from ammonium salts of carboxylic acids, by heat.

$$R-\overset{\overset{\displaystyle O}{\|}}{C}-O^- \; NH_4^+ \;\xrightarrow{\text{heat}}\; R-\overset{\overset{\displaystyle O}{\|}}{C}-NH_2 + H_2O \qquad (10\text{-}45)$$

amide

$$CH_3C\overset{O}{\underset{OH}{\Big\langle}} + NH_3 \xrightarrow{\text{neutralization}} CH_3C\overset{O}{\underset{O^-NH_4^+}{\Big\langle}} \xrightarrow{\text{heat}}$$

ammonium acetate

$$CH_3C\overset{O}{\underset{NH_2}{\Big\langle}} + H_2O \quad \text{(10-46)}$$

acetamide

Figure 10-8
Models of the amide, acetamide.

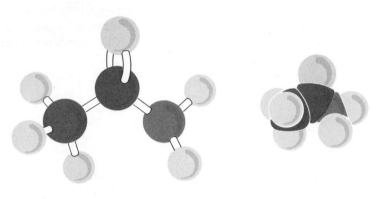

Amides are named from the corresponding acids, by replacing the *ic* or *oic* acid ending by *amide.*

$$H-C\overset{O}{\underset{NH_2}{\Big\langle}} \qquad CH_3-C\overset{O}{\underset{NH_2}{\Big\langle}} \qquad \bigcirc\!\!-C\overset{O}{\underset{NH_2}{\Big\langle}}$$

   formamide   acetamide   benzamide

Amides in which one or both of the —NH₂ hydrogens are replaced by organic groups are discussed in Chapter 12.

With the exception of formamide (liquid, bp 210°), most amides are colorless, crystalline solids which, because of their characteristic melting points, may be used to identify the corresponding acid.

Amides are the least susceptible of the various acid derivatives to nucleophilic attack at the carbonyl carbon. They hydrolyze slowly even in boiling water. In moderately concentrated alkali or acid, however, hydrolysis to the acid does occur. In base, the carboxylate anion and ammonia gas are formed, whereas in acid, the products are the carboxylic acid and ammonium ion.

$$R-C\overset{O}{\underset{NH_2}{\diagdown}} + OH^- \xrightarrow{\text{heat}} R-C\overset{O}{\underset{O^-}{\diagdown}} + NH_3 \qquad (10\text{-}47)$$

$$R-C\overset{O}{\underset{NH_2}{\diagdown}} + H_3O^+ \xrightarrow{\text{heat}} R-C\overset{O}{\underset{OH}{\diagdown}} + NH_4^+ \qquad (10\text{-}48)$$

Amides are sluggish in their reactivity toward nucleophiles because the unshared electron pair on nitrogen can supply electron density to the carbonyl group. This phenomenon can be represented by the two resonance contributors.

$$R-C\overset{\ddot{O}:}{\underset{\underset{H}{|}}{\diagdown}} \longleftrightarrow R-C\overset{:\ddot{O}:^-}{\underset{\underset{H}{|}}{\diagdown}}$$

The dipolar (salt-like) structure helps explain the high melting and boiling points of amides. Consistent with this postulate is the unusually short C—N bond length (1.32 Å, compared with the usual C—N single bond distance of 1.47 Å). The partial double bond character of the C—N bond in amides tends to restrict rotation about the bond and causes the four atoms attached to the carbon and nitrogen to lie in one plane (compare with the C=C bond, sec 3.2). This has important consequences for the structures of proteins (Chapter 17) and other molecules with amide linkages.

Amides may be dehydrated to nitriles, thus providing an alternate route to the synthesis of this class of compounds (compare with eq 8-11).

$$R-C\overset{O}{\underset{NH_2}{\diagdown}} \xrightarrow[\text{heat}]{P_2O_5} R-C{\equiv}N + H_2O \qquad (10\text{-}49)$$

10.9 UREA, A SPECIAL AMIDE

Urea is the most important derivative of carbonic acid, being its diamide (the monoamide, **carbamic acid,** is unstable). It is a colorless,

$$HO-\overset{\displaystyle O}{\overset{\|}{C}}-OH \qquad \left[HO-\overset{\displaystyle O}{\overset{\|}{C}}-NH_2\right] \qquad H_2N-\overset{\displaystyle O}{\overset{\|}{C}}-NH_2$$

carbonic acid   carbamic acid   urea
(unstable)

water-soluble, crystalline solid (mp 133°). The normal end-product of human metabolism of nitrogen-containing foods (proteins), urea is excreted to the extent of 30g daily in the urine of an average adult. Friedrich Wöhler is credited with the first laboratory synthesis (in 1828) of urea, which he obtained by boiling an aqueous solution of ammonium cyanate.

$$NH_4^+{}^-OCN \xrightarrow{\text{heat}} NH_2CONH_2 \qquad (10\text{-}50)$$

ammonium cyanate   urea

The result was important as an early example of the phenomenon of isomerism. It also provided a bridge across the gap of inorganic and organic chemistry which eventually helped overthrow the theory that a "vital force" was uniquely present in organic (but not inorganic) compounds.

Urea is now produced commercially (over two billion pounds annually) primarily for use as a fertilizer, but also for the manufacture of urea-formaldehyde plastics and certain drugs. Ammonia and carbon dioxide in aqueous solution are treated under pressure (2500 psi) at about 150–200°. The last step is an example of the general reaction given in equation 10-45.

$$O=C{\overset{\curvearrowright}{=}}\ddot{\underset{\cdot\cdot}{O}}: + \overset{\cdot}{\cdot}NH_3 \rightleftharpoons \left[O=C\overset{:\ddot{O}:^-}{\underset{\overset{+}{N}H_3}{\Big\langle}}\right] \overset{NH_3}{\rightleftharpoons}$$

carbamic acid

$$O=C\overset{:\ddot{O}:^-\ NH_4^+}{\underset{NH_2}{\Big\langle}} \xrightarrow{175°} O=C\overset{NH_2}{\underset{NH_2}{\Big\langle}} + H_2O \quad (10\text{-}51)$$

ammonium carbamate   urea

When urea is heated gently, ammonia is evolved and the triamide **biuret** is produced. An alkaline solution of biuret gives a violet-pink

$$2\ H_2N-\overset{\displaystyle O}{\overset{\|}{C}}-NH_2 \xrightarrow[\text{gently}]{\text{heat}} H_2N-\overset{\displaystyle O}{\overset{\|}{C}}-NH-\overset{\displaystyle O}{\overset{\|}{C}}-NH_2 + NH_3\uparrow \qquad (10\text{-}52)$$

biuret

color when copper sulfate is added, due to the formation of a complex ion. Since a similar geometric arrangement of amide groups is present in peptides and proteins (Chapter 17), the biuret test can be used to detect their presence.

10.10 TWO UNIQUE ACIDS

Formic acid is unique since the carboxyl group is attached to a hydrogen atom rather than a carbon atom, as in all other members of the series. For this reason it undergoes certain special reactions.

Unlike other carboxylic acids, formic acid is easily oxidized; indeed, viewed in one way, formic acid has the same functional group as aldehydes.

$$\underset{\text{H}-\overset{\displaystyle O}{\overset{\|}{\text{C}}}-\text{OH}}{} \qquad \underset{\text{H}-\overset{\displaystyle O}{\overset{\|}{\text{C}}}-\text{OH}}{}$$

two ways of viewing formic acid

Even such a mild oxidizing agent as Tollens' reagent (sec 9.6a) converts formic acid to carbon dioxide and water (thus giving a positive silver mirror test, similar to aldehydes).

$$\text{H}-\overset{\displaystyle O}{\underset{\displaystyle OH}{\text{C}}} + \text{Ag}_2\text{O} \longrightarrow \text{H}_2\text{O} + \text{CO}_2 + 2\,\underset{\substack{\text{silver}\\\text{mirror}}}{\text{Ag}} \qquad (10\text{-}53)$$

If one tries to prepare the acid chloride of formic acid (formyl chloride), carbon monoxide and hydrogen chloride are obtained instead.

$$3\,\text{H}-\overset{\displaystyle O}{\underset{\displaystyle OH}{\text{C}}} + \text{PCl}_3 \longrightarrow \text{P(OH)}_3 + 3\left[\underset{\text{formyl chloride}}{\text{H}-\overset{\displaystyle O}{\underset{\displaystyle Cl}{\text{C}}}}\right] \longrightarrow 3\,\text{CO} + 3\,\text{HCl}$$

$$(10\text{-}54)$$

Formyl chloride spontaneously decomposes at room temperature, though it can be prepared from carbon monoxide and hydrogen chloride at liquid nitrogen temperatures ($-190°$).

Formic acid also decomposes to carbon monoxide when it is warmed with strong acids such as sulfuric acid. The reaction probably proceeds via a formyl cation.

$$H-C\overset{O}{\underset{O-H}{\lesseqqgtr}} \underset{H_2SO_4}{\overset{H^+}{\rightleftharpoons}} H-C\overset{O}{\underset{\underset{H}{\overset{|}{O}-H}}{\lessgtr}} \overset{-H_2O}{\rightleftharpoons} H-\overset{+}{\underset{}{C}}=O \overset{-H^+}{\rightleftharpoons} CO \qquad (10\text{-}55)$$

formyl
cation

Thus formic acid can be used as a means of storing and generating carbon monoxide.

Because of its acidity and reducing properties, formic acid is used industrially as an inexpensive organic acid. It is produced commercially from carbon monoxide and alkali; the acid is liberated from its sodium salt by careful acidification.

$$CO + Na^+OH^- \xrightarrow[\text{pressure}]{200^\circ} H-C\overset{O}{\underset{O^-Na^+}{\lessgtr}} \xrightarrow{H^+} H-C\overset{O}{\underset{OH}{\lessgtr}} \qquad (10\text{-}56)$$

sodium formate   formic acid

The blistering action caused by the stings of bees and ants is due to formic acid.

Oxalic acid, the first member of the dicarboxylic acid series, also has certain distinctive properties. A crystalline solid, it decomposes at its melting point to carbon dioxide and formic acid, which further breaks down to carbon monoxide and water.

$$\underset{HO}{\overset{O}{\lessgtr}}C-C\underset{OH}{\overset{O}{\lessgtr}} \xrightarrow[-CO_2]{189^\circ} \underset{HO}{\overset{O}{\lessgtr}}C-H \longrightarrow CO + H_2O \quad (10\text{-}57)$$

Though it does not have any C—H bonds, oxalic acid is nevertheless readily oxidized (for example, by permanganate) to carbon dioxide and water. Use is made of its reducing properties to remove stains and ink spots. It reduces the colored, insoluble ferric salts of ink to the colorless, soluble ferrous salts. Rust stains may be removed from porcelain in the same way. It is also used in the bleaching of straw and wood, and in the printing and dying of cloth.

Oxalic acid is produced commercially by heating sodium formate; the mechanism of this rather old reaction is still not well understood.

$$2\ H-C\overset{O}{\underset{O^-Na^+}{\lessgtr}} \xrightarrow{400^\circ} H_2 + \underset{O}{\overset{O}{\underset{\underset{O^-Na^+}{\overset{|}{C}}}{\overset{|}{\underset{C}{\overset{\lessgtr}{}}}}}\overset{O^-Na^+}{} \qquad (10\text{-}58)$$

sodium formate   sodium oxalate

The free acid is obtained from the salt by acidification.

Oxalic acid also occurs in nature in the cell sap of many plants, usually as a salt. The free acid is toxic, but it is usually removed from foods (such as rhubarb) which contain it, by the cooking process.

New Concepts, Facts and Terms

1. The carboxyl group (from *carbonyl* and *hydroxyl*). Acid derivatives: halides, anhydrides, esters, amides
2. Nomenclature
   a. many common names
   b. IUPAC: *oic acid*
   c. the acyl group, $R-C\overset{O}{\diagup}$
3. Acidity; K_a is influenced by resonance and inductive effects. Electron-withdrawing substituents enhance acidity.
4. Acids are polar; frequently dimers.
5. Preparation
   a. oxidation of aromatic side-chains, alkenes, primary alcohols or aldehydes
   b. hydrolysis of cyanides (nitriles)
   c. Grignard reagent + CO_2; methods (b) and (c) provide a way of increasing the carbon chain length by one carbon atom.
6. Salts; named as metal carboxylates; soaps
7. Esters; named as alkyl carboxylates
   a. Fischer esterification (alcohol + acid)
   b. mechanism: nucleophilic attack by the alcohol on the carbonyl carbon of the protonated acid. Esters derived from *acyl* group of acid and *alkoxyl* group of alcohol.
   c. saponification (hydrolysis with alkali); reaction with ammonia $\longrightarrow$ amides, with alcohols $\longrightarrow$ transesterification. Mechanistically, all involve nucleophilic attack at the carbonyl carbon.
   d. Reduction $\longrightarrow$ primary alcohols
   e. polyesters, Dacron from dimethyl terephthalate and ethylene glycol
8. Acid halides and anhydrides; nomenclature; more reactive toward nucleophiles than esters or acids
   a. halides from acids and PX_3, PX_5 or $SOCl_2$
   b. anhydrides from acyl halides and salts, or from diacids and heat (acetic anhydride from ketene and acetic acid)
   c. hydrolysis, alcoholysis, ammonolysis
   d. Friedel-Crafts aromatic ketone synthesis
9. Amides; from other acid derivatives and ammonia; from ammonium salts and heat; resonance affects geometry
10. Urea, carbamic acid, biuret
11. Formic acid and oxalic acid; both are reducing agents.

Exercises and Problems

1. Write structural formulas for each of the following *acids:*
   a. propionic
   b. 3-methylpentanoic
   c. 2-chlorobutanoic
   d. p-toluic
   e. m-hydroxybenzoic

   f. oxalic
   g. phthalic
   h. formic
   i. cyclobutanecarboxylic
   j. β-bromobutyric

2. Write structural formulas for each of the following *acid derivatives:*
   a. methyl propionate
   b. butanoyl bromide
   c. propionamide
   d. phenyl benzoate
   e. benzonitrile

   f. acetic anhydride
   g. ammonium formate
   h. p-bromobenzamide
   i. oxalyl chloride
   j. calcium acetate

3. Name each of the following *acids:*

   a. $(CH_3)_2CHCH_2CH_2CO_2H$

   f. ─CO_2H

   b. $CH_3CHBrCH(CH_3)CO_2H$

   g. $CH_3CF_2CO_2H$

   c. [structure with CO_2H and NO_2 on benzene ring]

   h. [naphthalene structure with CO_2H]

   d. $CH_3CH(C_6H_5)CO_2H$

   i. $HO_2CCH_2CH(CH_3)CH_2CO_2H$

   e. $CH_2{=}CHCO_2H$

   j. HO_2C─[benzene ring with Cl, Cl, Cl, Cl substituents]─CO_2H

4. Name each of the following *acid derivatives:*
   a. $[CH_3(CH_2)_2CO_2]_2{}^-Ca^{2+}$
   b. $(CH_3)_2CHCO_2C_6H_5$
   c. CH_3CH_2COCl

   d. $ClCH_2CONH_2$

   e. CH_3─[benzene ring]─$CO_2{}^-Na^+$

   f. $(C_6H_5CO)_2O$
   g. $CF_3CO_2CH_3$
   h. $HCO_2CH_2CH_3$

   i. [bicyclic anhydride structure with Cl, Cl, Cl, Cl and O, O, O]

   j. $CH_3O_2CCO_2CH_3$

5. Write equations to show the reaction, if any, of benzoic acid with
   a. Na^+OH^-
   b. K
   c. $Ca(OH)_2$
   d. NH_3, followed by heat

   e. $SOCl_2$
   f. CH_3CH_2OH, H^+
   g. HCl
   h. H_2O

6. In each of the following pairs of acids, which would be expected to be the stronger, and why?
   a. CH_2ClCO_2H and CH_2BrCO_2H
   b. $o\text{-}BrC_6H_4CO_2H$ and $m\text{-}BrC_6H_4CO_2H$
   c. CCl_3CO_2H and CF_3CO_2H
   d. $C_6H_5CO_2H$ and $p\text{-}CH_3OC_6H_4CO_2H$
   e. $ClCH_2CH_2CO_2H$ and $CH_3CHClCO_2H$

7. Give equations which illustrate a good method to synthesize each of the following acids:
   a. butanoic acid from 1-butanol
   b. butanoic acid from n-propyl alcohol (2 ways)
   c. p-chlorobenzoic acid from p-chlorotoluene
   d. succinic acid (Table 10-2) from ethylene
   e. cyclopentanecarboxylic acid from cyclopentane
   f. 2-methoxyacetic acid from ethylene oxide
   g. pentanedioic acid from cyclopentene
   h. 2-methylbutanoic acid from 2-butene

8. Show how each of the following compounds can be prepared from the appropriate acid:
   a. propionyl bromide
   b. ethyl pentanoate
   c. n-butyramide
   d. phthalic anhydride
   e. calcium oxalate
   f. phenylacetamide
   g. pentanoic anhydride
   h. isopropyl benzoate
   i. m-nitrobenzoyl chloride
   j. formamide

9. Show all the steps in the mechanism for
   a. the reaction of ammonia with an ester (eq 10-24).
   b. transesterification (eq 10-25).
   Use equations 10-22 and 10-23 as a model, but certain proton transfers analogous to step 3 of equation 10-20 will also be necessary.

10. Write equations which clearly show all steps in the mechanism of
    a. equation 10-33
    b. equation 10-34
    c. the last step in equation 10-35

11. Write out all the steps in the mechanisms for equations 10-36 through 10-41. Each reaction begins by nucleophilic attack on a carbonyl carbon atom.

12. Esters react with an excess (2 moles) of a Grignard reagent to give tertiary alcohols. Recalling that Grignard reagents behave as nucleophiles toward carbonyl groups (sec 9.6f), write out the steps for the reaction of methyl benzoate with excess methylmagnesium bromide to give, after hydrolysis, 2-phenyl-2-propanol.

13. Explain the following differences in reactivity toward nucleophiles:
    a. Esters are less reactive than ketones.
    b. A given acid chloride is more reactive than the anhydride of the same acid.
    c. Benzoyl chloride is less reactive than cyclohexanecarbonyl chloride.

14. Explain what might happen to the shape of a Dacron molecule 100 units long if the starting dimethyl terepthalate were only 98% pure, and contained 2% of the *meta* isomer (dimethyl isophthalate).

15. Draw a portion of the structural formula for a polyester which might be obtained from the reaction of phthalic anhydride with glycerol (a glyptal resin).

16. The rearrangement of ammonium cyanate to urea (eq 10-50) is one of the oldest isomerization reactions known. Suggest a plausible mechanism for the reaction (HINT: begin by transferring a proton from the ammonium ion to the cyanate ion).

17. Show how each of the following conversions could be accomplished:
    a. *n*-butyryl chloride to methyl *n*-butyrate
    b. propionic anhydride to propionamide
    c. butanoic acid to 1-butanol
    d. 1-butanol to butanoic acid
    e. propionyl bromide to propionamide
    f. acetyl chloride to acetic anhydride
    g. acetic anhydride to acetophenone
    h. oxalic acid to diethyl oxalate
    i. urea to biuret
    j. benzoyl chloride to benzophenone

18. Complete each of the following equations, giving the structures and names of the main organic products:
    a. benzoic acid + ethylene glycol + H^+
    b. $C_6H_5CH_2MgBr$ + CO_2, followed by H_3O^+
    c. *p*-nitrobenzoyl chloride + sodium acetate
    d. $CH_3CH = CHCO_2H$ + excess $KMnO_4$ + H^+
    e. *n*-propylbenzene + $K_2Cr_2O_7$ + H^+
    f. phthalic anhydride + methanol + H^+
    g. pentanedioic acid + thionyl chloride
    h. cyclopropanecarboxylic acid + NH_4OH, then heat
    i. methyl 3-butenoate + $LiAlH_4$
    j. *n*-butyryl chloride + toluene + $AlCl_3$

19. Hydrolysis of an ester, $C_5H_{10}O_2$, gave an acid **A** and an alcohol **B**. Reaction of **B** with PBr_3 gave an alkyl bromide **C**; reaction of **C** with KCN, followed by acid hydrolysis gave **A**. Give the structure and name of the original ester, identify **A**, **B** and **C**, and write equations for all reactions mentioned.

20. A liquid, $C_6H_{12}O_2$, undergoes hydrolysis to an acid **A** and an alcohol **B**. Oxidation of **B** converts it to **A**. Suggest a structure for the original compound and write equations for all reactions mentioned.

CHAPTER ELEVEN

FATS, OILS, WAXES, AND DETERGENTS

The conversion of animal fats, by heating with water and wood ashes (alkali), into soap is one of the oldest known organic chemical reactions. The development of this reaction from the time of the ancient Greeks into the important chemical industry it is today has been climaxed in recent years by the production of many new synthetic detergents. The German chemist Liebig once stated that "of two countries with an equal amount of population, we may declare, with positive certainty, that the wealthiest and most highly civilized is that which consumes the greatest weight of soap." In addition to their use in the manufacture of detergents, drying oils, and waxes, lipids constitute one of the three major classes of foods in the human diet.

Fats, oils, and waxes belong to the group of naturally occurring organic materials called **lipids** (from the Greek *lipos,* fat). Lipids are those constituents of plants or animals which are insoluble in water but soluble in ether or other relatively non-polar organic solvents. They are distinguished in this way from the other two major classes of foodstuffs, the proteins and carbohydrates. Lipids may be sub-divided into two groups, depending on whether or not they can be hydrolyzed with aqueous base (saponified). Fats, oils, and waxes are saponifiable, whereas some lipids, such as the steroids (Chapter 18), are non-saponifiable.

We are all familiar with the many uses of fats and oils for frying, for making pastries, and for concocting salad dressings. The main sources of fats and oils in our diet are milk products (cream, butter), animal fats (lard, bacon fat), solid vegetable fats (oleomargarine), and liquid vegetable oils (corn oil, cottonseed oil, soybean oil, etc.). In addition to their use as foods, fats and oils are important raw materials for the manufacture of soaps, synthetic detergents, glycerol, drying oils, oilcloth, linoleum, paints, and varnishes. Waxes, often used as protective coatings, may be produced naturally by plants (for example the wax which permits us to polish apples) or insects (for example, beeswax).

11.1 THE STRUCTURE OF FATS AND OILS

Typical animal fats or vegetable oils are insoluble in water, but dissolve slowly in boiling aqueous alkali. Glycerol and salts of a mixture of carboxylic acids are isolated from these alkaline solutions. Some of the acids are saturated, some are unsaturated, and most have an even number of carbon atoms (usually twelve to twenty carbon atoms). Thus *fats and oils are esters of glycerol with carboxylic acids* **(glycerides).**

$$
\begin{array}{c}
\underset{\text{a glyceride}}{
\begin{array}{l}
CH_2-O-\overset{\displaystyle O}{\overset{\displaystyle \|}{C}}-R \\[4pt]
CH-O-\overset{\displaystyle O}{\overset{\displaystyle \|}{C}}-R \\[4pt]
CH_2-O-\overset{\displaystyle O}{\overset{\displaystyle \|}{C}}-R
\end{array}}
\quad\xrightarrow[\text{heat}]{3\,Na^+OH^-}\quad
\underset{\text{glycerol}}{
\begin{array}{l}
CH_2-OH \\[4pt]
CH-OH \\[4pt]
CH_2-OH
\end{array}}
\;+\;3\,R\overset{\displaystyle O}{\overset{\displaystyle \|}{C}}-O^-Na^+
\end{array}
\tag{11-1}
$$

salts of
carboxylic
acids

The distinction between fats and oils is based primarily upon the differences in melting points. Fats are *solid* esters of glycerol, whereas oils are *liquid* esters of glycerol (at room temperature). This distinction is not sharp since it depends upon climate, weather, and other physical variables. The melting point of a fat or oil depends upon the nature of the R groups of the fatty acid portion of the molecule. When these groups are saturated, the glycerol esters are generally solids. Double bonds in the R groups tend to lower the melting point. Thus, the R groups in an oil are usually highly un-saturated, whereas in fats fewer double bonds are present. The double bonds of the unsaturated fatty acids usually have the *cis* configuration. The more common saturated and unsaturated acids obtained from fats are listed in Table 11-1.

Triesters of glycerol are known as **glycerides.** *Simple* glycerides are esters in which all three hydroxyl groups of glycerol are esterified with the same acid.

$$
\begin{array}{ccc}
& \overset{\displaystyle O}{\underset{\displaystyle \|}{}} & \\
\text{CH}_2\text{OC(CH}_2)_{14}\text{CH}_3 & & \text{CH}_2\text{OC(CH}_2)_{16}\text{CH}_3 \\
\overset{\displaystyle O}{\underset{\displaystyle \|}{}} & & \overset{\displaystyle O}{\underset{\displaystyle \|}{}} \\
\text{CHOC(CH}_2)_{14}\text{CH}_3 & \text{simple} & \text{CHOC(CH}_2)_{16}\text{CH}_3 \\
\overset{\displaystyle O}{\underset{\displaystyle \|}{}} & & \overset{\displaystyle O}{\underset{\displaystyle \|}{}} \\
\text{CH}_2\text{OC(CH}_2)_{14}\text{CH}_3 & & \text{CH}_2\text{OC(CH}_2)_{16}\text{CH}_3 \\
\text{glyceryl tripalmitate} & & \text{glyceryl tristearate} \\
\text{(palmitin)} & & \text{(stearin)}
\end{array}
$$

Table 11-1
Common Acids Derived from Fats

NAME OF ACID	MOLECULAR FORMULA	STRUCTURAL FORMULA	MP, °C
Lauric	$C_{11}H_{23}COOH$	$CH_3(CH_2)_{10}COOH$	44
Myristic	$C_{13}H_{27}COOH$	$CH_3(CH_2)_{12}COOH$	58
Palmitic	$C_{15}H_{31}COOH$	$CH_3(CH_2)_{14}COOH$	63
Stearic	$C_{17}H_{35}COOH$	$CH_3(CH_2)_{16}COOH$	70
cis-Oleic	$C_{17}H_{33}COOH$	$\begin{array}{c}\text{H}\qquad\text{H}\\ \diagdown\text{C}=\text{C}\diagup\\ CH_3(CH_2)_7\qquad (CH_2)_7COOH\end{array}$	13
cis,cis-Linoleic	$C_{17}H_{31}COOH$	$\begin{array}{c}\text{H}\qquad\text{H H}\qquad\text{H}\\ \diagdown\text{C}=\text{C}\qquad\text{C}=\text{C}\diagup\\ CH_3(CH_2)_4\quad CH_2\quad (CH_2)_7COOH\end{array}$	−5
cis,cis,cis-Linolenic	$C_{17}H_{29}COOH$	$\begin{array}{c}\text{H}\qquad\text{H H}\qquad\text{H H}\qquad\text{H}\\ \diagdown\text{C}=\text{C}\quad\text{C}=\text{C}\quad\text{C}=\text{C}\diagup\\ CH_3CH_2\quad CH_2\quad CH_2\quad (CH_2)_7COOH\end{array}$	−11

Figure 11-1
Scale model of a long-chain fatty acid, stearic acid.

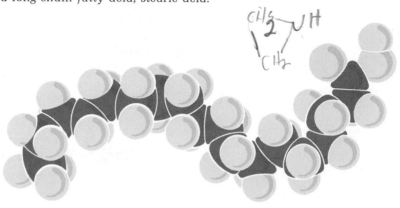

In mixed glycerides, glycerol may be esterified with two or three different acids. Both simple and mixed glycerides are present in natural fats and oils.

$$^{\alpha}CH_2-O\overset{O}{\overset{\|}{C}}(CH_2)_{14}CH_3$$
$$^{\beta}CH-O\overset{O}{\overset{\|}{C}}(CH_2)_{16}CH_3 \quad \text{mixed}$$
$$^{\alpha'}CH_2-O\overset{O}{\overset{\|}{C}}(CH_2)_{14}CH_3$$

β-stearo-α, α'-dipalmitin

$$CH_2-O\overset{O}{\overset{\|}{C}}(CH_2)_{14}CH_3$$
$$CH-O\overset{O}{\overset{\|}{C}}(CH_2)_{16}CH_3$$
$$CH_2-O\overset{O}{\overset{\|}{C}}(CH_2)_{7}CH=CH(CH_2)_{7}CH_3$$

glyceryl palmitostearoöleate

In general, a natural fat or oil does not consist of a pure glyceride, but rather of complex mixtures of glycerides. The composition of a fat is usually expressed in terms of the acids which may be obtained from it by hydrolysis. Certain fats or oils give mainly one or two acids, as with olive oil (83% oleic acid, 6% palmitic acid, 4% stearic acid, and 7% linoleic acid) or palm oil (43% palmitic acid, 43% oleic acid, 10% linoleic acid, and 4% stearic acid). Others are very complex: butter fat, for example, contains esters of at least fourteen different acids.* Butter is different from most other fats in

---

*3% butyric, 1.4% caproic, 1.5% caprylic, 2.7% capric, 0.7% lauric, 12.1% myristic, 25.3% palmitic, 9.2% stearic, 1.3% arachidic, 0.4% lauroleic, 1.6% myristoleic, 4% palmitoleic, 29.6% oleic, and 3.6% linoleic.

that it contains appreciable amounts of glycerides of the lower molecular weight fatty acids, including butyric acid.

Some natural products contain glycerides which, on hydrolysis, yield fatty acids with other functional groups or unusual structures. Examples include **ricinoleic acid** from castor oil, **lactobacillic acid** present in certain bacteria, and **sterculic acid,** isolated from the seeds of a tropical tree.

$$CH_3(CH_2)_5CHCH_2 \overset{\displaystyle \underset{H}{\overset{H}{C}}=\underset{}{\overset{H}{C}}}{} (CH_2)_7CO_2H$$
$$\underset{OH}{}$$

ricinoleic acid

$$CH_3(CH_2)_5 \overset{\displaystyle \underset{H}{\overset{CH_2}{C}}\text{——}\underset{}{\overset{H}{C}}}{} (CH_2)_9CO_2H$$

lactobacillic acid

$$CH_3(CH_2)_7 \overset{\displaystyle \overset{CH_2}{C}\text{===}C}{} (CH_2)_7CO_2H$$

sterculic acid

11.2 REACTIONS OF FATS AND OILS

We shall consider three fundamental reactions of fats and oils, each of which has been discussed previously for simpler esters (review sec 3.4b and 10.6c).

11.2a HYDROLYSIS OF FATS AND OILS

Boiling with aqueous alkali hydrolyzes fats and oils to glycerol and salts of fatty acids (eq 11-1). The salts of long chain fatty acids are soaps (sec 11.3a). The reaction, called **saponification** (from the Latin *sapon,* soap) is used industrially to manufacture soap. Equation 11-2 illustrates the process with a specific example.

$$\begin{array}{l} CH_2O\overset{O}{\overset{\|}{C}}(CH_2)_{14}CH_3 \\[2mm] CHO\overset{O}{\overset{\|}{C}}(CH_2)_{14}CH_3 \\[2mm] CH_2O\overset{O}{\overset{\|}{C}}(CH_2)_{14}CH_3 \end{array} + 3\,Na^+OH^- \overset{heat}{\longrightarrow} \begin{array}{l} CH_2OH \\[2mm] CHOH \\[2mm] CH_2OH \end{array} + 3\,CH_3(CH_2)_{14}CO_2^-Na^+$$

$$(11\text{-}2)$$

| glyceryl tripalmitate (from palm oil) | glycerol | sodium palmitate (a soap) |

11.2b HYDROGENATION OF FATS AND OILS

Glycerides that have unsaturated R groups in the acid portion of the fat molecule can be converted to saturated glycerides by catalytic hydrogenation. This is illustrated by the conversion of olein to stearin, wherein hydrogen adds to the carbon-carbon double bonds (compare eq 11-3 with 3-10).

$$
\begin{array}{l}
CH_2OC(CH_2)_7CH{=}CH(CH_2)_7CH_3 \\[4pt]
CHOC(CH_2)_7CH{=}CH(CH_2)_7CH_3 \\[4pt]
CH_2OC(CH_2)_7CH{=}CH(CH_2)_7CH_3
\end{array}
\xrightarrow[\text{Ni catalyst}]{\substack{3\ H_2 \\ \text{heat}}}
\begin{array}{l}
CH_2OC(CH_2)_{16}CH_3 \\[4pt]
CHOC(CH_2)_{16}CH_3 \\[4pt]
CH_2OC(CH_2)_{16}CH_3
\end{array}
$$

olein   stearin

$$(11\text{-}3)$$

Since most people who live in the United States or northern Europe prefer to cook with fats rather than oils, the hydrogenation of vegetable oils is carried out on a large commercial scale. By decreasing the amount of unsaturation in the oil, the melting point is gradually raised until the material becomes a solid fat. These vegetable fats are sold for kitchen use. The process of hydrogenation of oils to fats is sometimes called **hardening.** There is some evidence that it is important to maintain a balance between the amount of saturated and unsaturated glycerides in the diet to lessen the chance that lipid material will be deposited in vascular systems, e.g., arteries, causing arteriosclerosis (hardening of the arteries). The association of unsaturated fats and low cholesterol levels has revitalized the use of oils in cooking.

Oleomargarine and various other butter substitutes are generally mixtures of vegetable oils or animal fats which have been partially hydrogenated to a consistency resembling that of butter. The most common starting materials are cottonseed, soybean, or peanut oil. The product is frequently churned with milk and artificially colored to simulate the flavor and appearance of butter.

11.2c HYDROGENOLYSIS OF FATS AND OILS

The ester links in glycerides may be hydrogenolyzed (compare with eq 10-26). With a saturated glyceride, this leads to long-chain saturated alcohols which can be used to manufacture certain synthetic detergents. Two moles of hydrogen are used per ester linkage hydrogenolyzed.

$$
\begin{array}{c}
\text{CH}_3(\text{CH}_2)_{14}-\overset{\overset{\displaystyle O}{\|}}{\text{C}}-\text{OCH}_2 \\[4pt]
\text{CH}_3(\text{CH}_2)_{14}-\overset{\overset{\displaystyle O}{\|}}{\text{C}}-\text{OCH} \quad + 6\ \text{H}_2 \xrightarrow[\substack{\text{copper chromite}\\ \text{catalyst}}]{\text{heat, pressure}} \\[4pt]
\text{CH}_3(\text{CH}_2)_{14}-\overset{\overset{\displaystyle O}{\|}}{\text{C}}-\text{OCH}_2
\end{array}
$$

palmitin

$$
3\ \text{CH}_3(\text{CH}_2)_{14}\text{CH}_2\text{OH} \ + \ \begin{array}{c} \text{CH}_2\text{OH} \\ | \\ \text{CHOH} \\ | \\ \text{CH}_2\text{OH} \end{array} \qquad (11\text{-}4)
$$

cetyl alcohol   glycerol

Long-chain *unsaturated* alcohols may be obtained by the analogous hydrogenolysis of unsaturated glycerides, employing a catalyst such as zinc chromite which causes hydrogenolysis of the ester linkage without hydrogenation of the carbon-carbon double bonds. For example, olein yields oleyl alcohol:

$$
\text{CH}_3(\text{CH}_2)_7\text{CH}{=}\text{CH}(\text{CH}_2)_7\text{CH}_2\text{OH}
$$

Figure 11-2
Fats can be obtained by hydrogenation of the olefinic double bonds in vegetable oils. The beaker on the left contains clear oil before hydrogenation; on the right, the same oil hardened by hydrogenation. (Courtesy of The Proctor & Gamble Company.)

11.3 DETERGENTS

For centuries, soap was the major useful cleansing agent. However, in recent years there has been a phenomenal upsurge in the manufacture of synthetic detergents. Sales of synthetic detergents have overtaken and rapidly passed the annual sales of ordinary soap. Yet much is to be learned from the chemistry of soap itself, and we can more easily understand the structural requirements and advantages of synthetics if we have a knowledge of the chemistry of ordinary soaps.

11.3a PREPARATION OF SOAPS

The most widely used ordinary soaps are *sodium salts of long-chain carboxylic acids.* Potassium salts form soaps which are softer and more soluble, but being more expensive, they are used only to a limited extent in liquid soaps and shaving creams. Aluminum salts of fatty acids form gels which can be used as thickeners for lubricating greases; mixed with gasoline, they form gels used in incendiary bombs and flame throwers (napalm). Calcium, magnesium, and zinc salts of fatty acids are water insoluble, but see use in face and dusting powders.

Soaps are prepared by the saponification of natural or hardened fats with a slight excess of alkali (caustic soda) in a heated open kettle. When the reaction is complete, salt is added to precipitate the soap as thick curds. The water layer, containing salt, glycerol, and excess alkali, is drawn off and the glycerol is recovered by vacuum distillation. The crude soap curds contain some salt, alkali, and glycerol as impurities. These are removed by boiling with water and reprecipitating with salt several times. Finally, the curds are boiled with sufficient water to form a smooth mixture which, on standing, gives a homogeneous upper layer of soap. This may be sold without further processing as an inexpensive industrial soap. Its quality will depend in part on the chain length and degree of unsaturation of the fatty acids. Various fillers such as sand, pumice, or sodium carbonate may be added to make scouring soaps. Other treatments transform the crude soaps into toilet soaps, powdered or flaked soaps, medicated soaps, laundry soaps, liquid soaps, and by blowing air in, even floating soaps.

11.3b MECHANISM OF SOAP ACTION

Most dirt sticks to clothes by a thin oily film. If the oil can be removed, the solid dirt particles can be washed away. The cleansing ability of soaps is due to their chemical structure, which permits them to emulsify oils, greases, fats, and other organic molecules.

The most important structural feature of a soap is that one end of the molecule is highly polar or ionic, whereas the remainder of the molecule is nonpolar or hydrocarbon-like (Figure 11-3).

Figure 11-3
Scale model of a typical soap molecule.

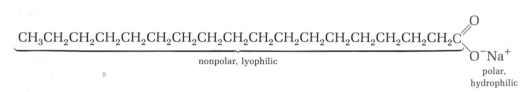

$$CH_3CH_2CH_2CH_2CH_2CH_2CH_2CH_2CH_2CH_2CH_2CH_2CH_2CH_2CH_2CH_2CH_2C \overset{\displaystyle O}{\underset{\displaystyle O^-Na^+}{}}$$

nonpolar, lyophilic

polar, hydrophilic

sodium stearate, an ordinary soap

The polar end of the molecule tends to make it water-soluble (hydrophilic or attracted to water), whereas the nonpolar end of the molecule tends to make it oil-soluble (lyophilic or attracted to fats). It has been shown that, if a droplet of a fatty acid is placed upon the surface of water, it spreads out to form a thin film one molecule thick. The molecules in this film are arranged so that the carboxyl (polar) end dips into the water, and the molecules stand on end (Figure 11-4). When soap molecules (salts, which have a considerably more polar end than do fatty acids) dissolve in water, they form opalescent or colloidal rather than true solutions. These soap solutions contain aggregates of soap molecules or **micelles** (Figure 11-5). In order to form such solutions, the hydrocarbon part of the molecule should not be too short or too long (generally twelve to eighteen carbon atoms).

Oil and water do not mix. But when a small amount of oil and a soap solution are shaken, an emulsion of the oil in the soap solution is formed. The soap molecules surround the fine oil droplets, their hydrocarbon "tails" being soluble in the oil (Figure 11-6). The ionic or hydrophilic ends then stabilize the droplets in the water solution,

Figure 11-4

*Carboxylic acid film on water showing orientation of the polar end of
each molecule toward the surface of the water.*

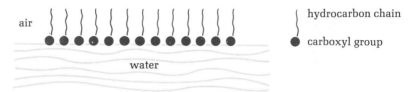

air

hydrocarbon chain

carboxyl group

water

Figure 11-5

A soap micelle.

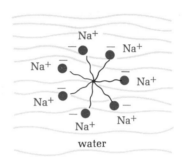

water

Figure 11-6

*Oil droplets become stably emulsified when surrounded by soap mole-
cules.*

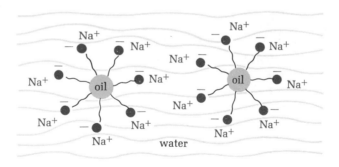

water

the negative surface charge of the droplets preventing their coales-
cence. Another striking property of soap solutions is their unusually
low surface tension. This increases the "wetting" power of the soap
solution over that of water. It is a combination of the emulsifying
power and surface action of soap solutions which enables them to
remove dirt, grease, and oil particles.

11.3c THE CHEMISTRY OF SOAPS

Since soaps are the sodium or potassium salts of weak acids, they are converted by stronger mineral acids into the free fatty acids.

$$C_{17}H_{35}C\underset{\substack{\\ O^-Na^+}}{\overset{\displaystyle O}{\diagdown}} + H^+Cl^- \longrightarrow C_{17}H_{35}C\underset{\substack{\\ OH}}{\overset{\displaystyle O}{\diagdown}} \downarrow + Na^+Cl^- \quad (11\text{-}5)$$

sodium stearate $\qquad\qquad\qquad\qquad$ stearic acid

Soaps, therefore, cannot be used in acidic solutions, because a scum or precipitate of the water-insoluble fatty acid will form.

A scum is also formed when ordinary soaps are used in hard water. Here, the calcium, magnesium, or ferric ions present form insoluble salts which precipitate and adhere to the fabric being washed (or to the rim of the tub).

$$2\ C_{17}H_{35}C\underset{\substack{\\ O^-Na^+}}{\overset{\displaystyle O}{\diagdown}} + Ca^{++} \longrightarrow (C_{17}H_{35}COO)_2{}^-Ca^{++}\downarrow + 2\ Na^+ \quad (11\text{-}6)$$

calcium stearate

This precipitation is undesirable not only because of the difficulty in cleansing, but also because it wastes soap.

11.3d SYNTHETIC DETERGENTS (SYNDETS) OR SURFACE-ACTIVE AGENTS (SURFACTANTS)

In recent years, detergents have been produced with cleansing power equal to that of ordinary soaps which avoid some of their disadvantages. Synthetic detergent molecules are designed according to the principles discussed in section 11.3b; that is, they contain a nonpolar organic portion with a highly polar group at one end of the molecule. The polar group may be negatively charged, as in ordinary soaps, or it may be neutral or positively charged.

Several hundred varieties of synthetic detergents are now manufactured commercially. Alkyl hydrogen sulfates constitute one of the more important types. Long-chain aliphatic alcohols, available from the hydrogenolysis of fats and oils (eq 11-4), can be sulfated with sulfuric acid. The resulting alkyl hydrogen sulfate (eq 6-33 and 11-7), when neutralized (eq 11-8), gives a sodium salt useful as a detergent.

$$CH_3(CH_2)_{10}CH_2OH + HOSO_2OH \longrightarrow$$

lauryl alcohol

$$CH_3(CH_2)_{10}CH_2OSO_2OH + H_2O \quad (11\text{-}7)$$

lauryl hydrogen sulfate

$$CH_3(CH_2)_{10}CH_2OSO_2OH + Na^+OH^- \longrightarrow$$
$$CH_3(CH_2)_{10}CH_2OSO_2O^-Na^+ + HOH \quad (11\text{-}8)$$

sodium lauryl sulfate

The sodium salts of alkylarylsulfonic acids constitute another important type of syndet. They are manufactured by sulfonating an alkylbenzene (eq 5-8) followed by neutralization of the resulting

$$R\text{---}\langle\text{benzene ring}\rangle\text{---}SO_2O^-Na^+$$

sodium alkylarylsulfonate

Figure 11-7
Excessive suds formation such as that shown here called attention to the need for chemists to develop biodegradable detergents which can be converted into nonpollutants by microorganisms. (Grant Heilman)

arylsulfonic acid. For years, the most widely used synthetic detergents were of this type, where R was a C-12 highly branched carbon chain derived from propylene tetramer (eq 3-37). Benzene was alkylated with this tetramer (Friedel-Crafts reaction, eq 5-25), and the resulting alkylbenzene was sulfonated and neutralized. Unfortunately, the detergent thus produced, with its highly branched R-group, was not readily destroyed by microorganisms. It accumulated in ground waters and rivers, and caused undesirable foaming in sewage disposal plants (Figure 11-7). Microorganisms attack long carbon chains by first converting the terminal methyl group to a carboxyl group. Then they consume the chain, two carbons at a time, by further oxidation. Branching prevents or retards this process. Thus detergents with straight chains are *biodegradable,* and do not cause water pollution, whereas those with highly branched chains are *nonbiodegradable* and cause difficulties. Currently, C-10 to C-12 1-alkenes are used in place of propylene tetramer; these give 2-phenylalkanes in the Friedel-Crafts reaction with benzene, and lead to biodegradable detergents.

The sodium salts of alkylsulfuric or sulfonic acids are especially useful detergents because they are effective in hard as well as soft water. This is because they do not form insoluble salts with calcium, magnesium, or ferric ions as do ordinary soaps.

Nonionic syndets usually have effective hydrogen bonding groups at one end of a hydrocarbon chain. Examples include monoesters of polyhydric alcohols, such as pentaerythrityl stearate

$$CH_3(CH_2)_{16}\overset{\displaystyle O}{\overset{\|}{C}}-OCH_2\overset{\displaystyle CH_2OH}{\underset{\displaystyle CH_2OH}{\overset{|}{\underset{|}{C}}}}-CH_2OH$$

pentaerythrityl stearate

or polyethers derived from ethylene oxide.

$$RCH_2(OCH_2CH_2)_nOH \qquad R-\overset{\displaystyle O}{\overset{\|}{C}}-(OCH_2CH_2)_nOH$$

(n = 8–12; R = 11–17)

The latter are particularly useful as low-sudsing detergents for automatic washers.

Cationic or invert detergents are typified by the alkylammonium salts in which one of the four alkyl groups attached to the nitrogen has a long carbon chain.

$$CH_3(CH_2)_{14}CH_2-\overset{\displaystyle CH_3}{\underset{\displaystyle CH_3}{\overset{|}{\underset{|}{N^+}}}}-CH_3 \ \ Cl^-$$

hexadecyltrimethylammonium chloride

Many invert soaps have a high germicidal action.

The cleansing power of detergents can be enhanced by various additives. Phosphates are used to break up suspended dirt particles and to form complexes with undesirable metal ions present in hard water. While effective in enhancing detergent action, phosphates can accumulate and pose serious pollution problems.

11.4 DRYING OILS

Certain vegetable oils, particularly those which contain a high proportion of unsaturated glycerides, possess the property of forming a hard tough film when exposed in thin layers to the air. The process is referred to as **drying,** although it is not drying in the sense of evaporation, but rather a complicated series of reactions of the oil with the oxygen of the air which leads to the formation of a dry film. The phenomenon involves both oxidation and polymerization.

The most common drying oil is linseed oil (obtained from flax seed) although other oils (tung, soybean, etc.) are also used. Linseed oil contains large percentages of the glycerides of linoleic and linolenic acids, both of which are unsaturated. Raw linseed oil "drys" rather slowly but the process may be hastened by boiling the raw oil with soluble cobalt, manganese, or lead salts.

Oil paints are suspensions of inorganic pigments in a drying oil which also contains a thinner and drier. The thinner is usually turpentine or a mixture of volatile petroleum hydrocarbons, which permit even spreading of a thin layer of the pigment and drying oil. The drier is a solution of metal salts (Co, Mn, Pb) which catalyze the oxidation and polymerization of the drying oil on the painted surface.

Oilcloth is made by applying several coats of a linseed oil-paint to woven fabric. *Linoleum* is made from a mixture of ground cork, boiled linseed oil, and rosin which has been pressed together and "dried." Over a billion pounds of drying oils are used annually in the United States.

11.5 ANALYSIS OF FATS AND OILS

Many properties of fats and oils are determined by the carbon chain length and by the degree of unsaturation of the chains. Since fats and oils are obtained from natural sources, their composition is variable. Therefore, to have some control over the raw materials used in the manufacture of soaps, synthetic detergents, drying oils, and other products derived from fats and oils, many analytical methods have been devised. Two of the more important of these are described here.

Fats and oils may be saponified quantitatively. The **saponification number,** which is defined as the number of milligrams of potassium hydroxide required to hydrolyze one gram of a fat or oil, gives an indication of the average molecular weight of the fat, or of the average length of the carbon chain of the fatty acid portions. The higher the saponification number of a fat, the greater the percentage of short-chain, low-molecular weight glycerides it contains.

The degree of unsaturation of a fat or oil may be expressed in terms of its **iodine number;** that is, the number of grams of iodine which will combine with one hundred grams of the fat or oil. Iodine, in an alcoholic solution of mercuric chloride adds to the carbon-carbon double bonds of the fatty acid portion of the molecule. The more double bonds (greater unsaturation) in the fat, the greater will be its iodine number.

The saponification and iodine numbers of several fats and oils are given in Table 11-2. Note the high saponification number of butter fat (which contains many glycerides of the lower fatty acids). Note also its low iodine number, which indicates that it contains few double bonds and is a solid at room temperature. Linseed oil, which is used as a drying oil, is highly unsaturated (has a high iodine number).

11.6 WAXES

Waxes differ from fats and oils in that they are not esters of glycerol, but rather are principally esters of long-chain even-numbered fatty acids with long-chain, even-numbered *monohydric* alcohols. There is only one ester linkage in each wax molecule.

$$R-\overset{\overset{\textstyle O}{\|}}{C}-OR'$$

a wax (R and R′ range from 16 to 36 carbons in length)

Waxes may be distinguished from fats in that they *cannot* be transformed into water-soluble products when boiled with alkali, since unlike glycerol, the long-chain alcohols are insoluble in water. To-

Table 11-2
Analysis of Some Fats and Oils

FAT OR OIL	SAPONIFICATION NO.	IODINE NO.
Butter fat	220–240	22–35
Tallow	190–195	35–40
Lard	195–200	48–64
Olive oil	190–195	80–85
Cottonseed oil	190–195	103–111
Linseed oil	190–195	170–185

gether with the esters, waxes frequently contain small quantities of saturated hydrocarbons, free fatty acids or alcohols, and steroids.

Beeswax, the material from which the bee builds honeycomb cells, melts at 60–80°. Saponification yields mainly C_{26} and C_{28} acids and C_{30} and C_{32} straight chain alcohols. Some hydroxyacids, such as 14-hydroxypalmitic acid, are also obtained. In addition, about 10–15% of the wax consists of hydrocarbons, with $C_{31}H_{64}$ predominating. *Spermaceti* is a wax which can be obtained by chilling the oil from the head of the sperm whale. It consists mainly of the ester cetyl palmitate, $C_{15}H_{31}CO_2C_{16}H_{33}$. The most valuable natural wax is *carnauba wax*, the coating on the leaves of Brazilian palm trees. Its high melting point (80–87°), hardness, and imperviousness to water make it particularly useful in automobile and floor polishes. The wax is a mixture of esters, chiefly of C_{24} and C_{28} acids with C_{32} and C_{34} l-alkanols. Bifunctional acids [$HO(CH_2)_nCO_2H$; n = 17–29] and alcohols [$HO(CH_2)_nOH$; n = 22–28] are also obtained from the saponification of carnauba wax, suggesting that in the original wax, polyesters may also be present and may contribute to its unique properties.

Waxes are generally more brittle, harder, and less greasy than fats. They are used to make polishes, cosmetics, ointments and other pharmaceutical preparations, candles, and phonograph records. In nature, waxes coat the leaves and stems of plants which grow in arid regions, thus reducing evaporation. Similarly, insects with a high surface-to-volume ratio are often coated with a natural protective wax.

11.7 METABOLISM OF FATS

Amongst the various foodstuffs, fats provide the body with maximum energy (9,500 calories per gram) per quantity of material oxidized, the value being approximately twice that for an equal weight of proteins or carbohydrates. The body not only oxidizes fats but also synthesizes them, especially from carbohydrates, as a steady diet rich in the latter will soon visibly demonstrate. Indeed, it has been established that body fats, proteins, and carbohydrates are in a dynamic state of degradation and synthesis. Feeding experiments with animals show, for example, that radioactive acetate (labelled with carbon-14 isotope) is rapidly incorporated into body fats, even when the animals maintain constant body weight. Depot fat is therefore constantly being used up and remanufactured in the body.

If a fat molecule were simply allowed to burn with oxygen, a large amount of energy would be liberated all at once as heat, a form of energy that cannot be used in metabolism. There is an important reason why oxidation in the cell, as contrasted with combustion, must proceed under rigorously controlled conditions. Living cells cannot use heat to perform cell functions because heat energy can do work only if it flows from a warm region to a cooler one, as in a heat engine. But for all practical purposes, there is no temperature differential in the living cell. The cell recovers the liberated energy from the oxidation of foodstuffs, not primarily as heat, but as chemical energy which can do work in a constant-temperature system. To obtain energy in a useful form, the cell oxidizes its fuels in a stepwise manner. The agents of this controlled combustion are enzymes, frequently assisted by smaller organic molecules called coenzymes. In this section the organic chemical reactions by which these catalysts degrade (catabolize) and synthesize (anabolize) fats will be briefly described.

Lipids enter the body through the mouth and pass to the stomach, but are little affected by its acidic environment. They are absorbed primarily in the small intestines, where they are emulsified by salts of the bile acids (which behave like soaps) and are hydrolyzed to fatty acids and glycerol by various water-soluble enzymes (*lipases*). From the intestines, the hydrolyzed lipids enter the bloodstream and are transported to other organs, mainly the liver, for further metabolism. Recombination may occur, or the glycerol, which structurally resembles a carbohydrate, may enter the carbohydrate metabolic cycle. The fatty acids may be degraded ultimately to carbon dioxide and water to furnish energy.

A large number of chemical reactions are involved in these conversions. Two facts suggest that the breakdown and synthesis of fats involves two-carbon units: (a) the common fatty acids have an even number of carbon atoms, and (b) if acetate, with two carbon atoms per molecule, is fed to an animal, it is rapidly converted to the larger fatty acids. The acetyl donor and acceptor essential to these processes was shown by Fritz Lipmann* to be a complex molecule called **coenzyme A** (A for acetylation). Parts of this molecule, the adenine,

coenzyme-A (CoA-SH)

* Nobel prize in medicine and physiology, 1953.

D-ribose, and pyrophosphate, are also present in nucleic acids (Chapter 18). The major structural feature of coenzyme A that is most important in fat metabolism is the mercaptan or —SH group (sec 6.9), and for this reason we shall represent the structure by the symbol CoA—SH.

The biological cell oxidation of a fatty acid is achieved by a process called "*beta* oxidation," which refers to the observation that initial oxidation of a fatty acid occurs at the carbon *beta* to the carboxyl group of the acid. The acid is first converted to a *thioester* of coenzyme A, as illustrated in equation 11-9 for stearic acid.

$$CH_3(CH_2)_{14}CH_2CH_2\overset{\overset{\displaystyle O}{\|}}{C}\text{—OH} + \text{CoA—SH} \overset{enzyme}{\rightleftharpoons}$$

stearic acid

$$CH_3(CH_2)_{14}CH_2CH_2\overset{\overset{\displaystyle O}{\|}}{C}\text{—S—CoA} + H_2O \quad (11\text{-}9)$$

stearylcoenzyme A
(a thioester)

The thioester is then oxidized to an α,β-unsaturated ester which is hydrated, and the resulting β-hydroxyester is oxidized to a β-ketoester. Each of these steps is reversible, and is catalyzed by a specific enzyme.

$$CH_3(CH_2)_{14}CH_2CH_2\overset{\overset{\displaystyle O}{\|}}{C}\text{—S—CoA} \underset{enzyme}{\overset{-2H}{\rightleftharpoons}} CH_3(CH_2)_{14}CH\text{=}CH\overset{\overset{\displaystyle O}{\|}}{C}\text{—S—CoA}$$

$$\Bigg\updownarrow \begin{matrix} + \text{ HOH} \\ enzyme \end{matrix} \qquad (11\text{-}10)$$

$$CH_3(CH_2)_{14}\overset{\overset{\displaystyle O}{\|}}{C}CH_2\overset{\overset{\displaystyle O}{\|}}{C}\text{—S—CoA} \underset{enzyme}{\overset{-2H}{\rightleftharpoons}} CH_3(CH_2)_{14}\underset{\underset{\displaystyle OH}{|}}{CH}CH_2\overset{\overset{\displaystyle O}{\|}}{C}\text{—S—CoA}$$

a β-keto thioester

In the final step in the sequence (eq 11-11), the β-keto thioester reacts with a second molecule of coenzyme A to produce acetylcoenzyme A and the thioester of palmitic acid, which has two carbon atoms less than the original stearic acid.

$$CH_3(CH_2)_{14}\overset{\overset{\displaystyle O}{\|}}{C}\text{—}CH_2\text{—}\overset{\overset{\displaystyle O}{\|}}{C}\text{—S—CoA} + \text{CoA—SH} \overset{enzyme}{\rightleftharpoons}$$

$$CH_3(CH_2)_{14}\overset{\overset{\displaystyle O}{\|}}{C}\text{—S—CoA} + CH_3\overset{\overset{\displaystyle O}{\|}}{C}\text{—S—CoA} \quad (11\text{-}11)$$

palmitylcoenzyme A   acetylcoenzyme A

In this way, the 18-carbon stearic acid is degraded to the 16-carbon palmitic acid and the 2-carbon fragment acetic acid. Palmitylcoenzyme A now goes through the processes outlined in equations 11-10 and 11-11 to produce a 14-carbon acid and another 2-carbon fragment. The sequence is repeated until the entire chain is degraded. From stearic acid, nine units of acetylcoenzyme A can be formed in this way. Acetylcoenzyme A may be used by the body for carbohydrate metabolism; for the formation of acetylcholine (sec 12.5b), a regulator of nerve impulses; and for many other biosyntheses in plants and animals. From this scheme it is clear why detergents whose nonpolar portions are linear are more biodegradable than those with branched chains.

The build-up or anabolism of fats in the body begins with acetylcoenzyme A. Under the influence of a suitable enzyme, it reacts with carbon dioxide to form malonylcoenzyme A. Reaction with a second molecule of acetylcoenzyme A, followed by decarboxylation gives acetoacetylcoenzyme A, a β-keto thioester (eq 11-12). The latter now feeds into the end of the sequence of reversible reactions in equation

$$CH_3\overset{O}{\underset{}{C}}-S-CoA \xrightleftharpoons[\text{enzyme}]{CO_2} H-O-\overset{O}{\underset{}{C}}-CH_2\overset{O}{\underset{}{C}}-S-CoA$$

malonylcoenzyme A

$$\text{enzyme} \; \Big\Updownarrow \; CH_3\overset{O}{\underset{}{C}}-S-CoA$$

$$CH_3-\overset{O}{\underset{}{C}}-CH_2\overset{O}{\underset{}{C}}-S-CoA \xrightleftharpoons[\text{enzyme}]{-CO_2} H-O-\overset{O}{\underset{}{C}}-\underset{\underset{CH_3}{\overset{|}{C=O}}}{\overset{|}{CH}}-\overset{O}{\underset{}{C}}-S-CoA$$

acetoacetylcoenzyme A

$$+ \; CoA-SH$$

(11-12)

11-10 to be reduced eventually to butyrylcoenzyme A, an ester with two more carbon atoms than acetylcoenzyme A. This ester replaces *the second* acetylcoenzyme A in equation 11-12 and transfers a butyryl group to malonylcoenzyme A. Repetition of this sequence can lead to the build-up of long-chain fatty acids from two-carbon fragments.

Thus, the biosynthesis and the oxidation of fatty acids takes place with the incorporation or the loss of two carbon atoms at each stage. Each step either requires or liberates energy in small increments and in a well-defined manner, allowing for careful regulation at an essentially constant temperature.

New Concepts, Facts, and Terms

1. Fats and oils are triesters of glycerol; the acid portions of oils are less saturated than those of fats.
2. The most common acids derived from fats and oils are listed in Table 11-1.
3. Glycerides—may be simple (all acid portions identical) or mixed.
4. Reactions of fats and oils:
    a. Hydrolysis with alkali gives glycerol + soaps.
    b. Hydrogenation converts oils to fats.
    c. Hydrogenolysis gives long-chain alcohols useful in making synthetic detergents.
5. Detergents:
    a. Preparation of soaps.
    b. Soap molecules must have a polar end and a nonpolar portion. The latter dissolves in oil particles, and the former keeps emulsified particles from coalescing.
    c. Soap reacts with mineral acids and with ions such as calcium and magnesium to precipitate from solution.
    d. Synthetic detergents (syndets) are constructed on the same principles as soaps, but the polar portion may be negative, neutral, or positive. Straight-chain nonpolar portions enhance biodegradability.
6. Drying oils contain a high proportion of unsaturated glycerides. These oxidize and polymerize in air and are used in paints, oilcloth, and linoleum.
7. Analysis of fats and oils: saponification number gives a measure of chain length, whereas iodine number gives a measure of unsaturation.
8. Waxes are esters of long-chain alcohols with long-chain acids.
9. Metabolism of fats: degradation and synthesis are accomplished via two-carbon fragments. Acetylcoenzyme A, a complex mercaptan, is the carrier of the two-carbon fragment.

Exercises and Problems

1. Write structural formulas for each of the following:
    a. potassium palmitate
    b. magnesium oleate
    c. sodium myristyl sulfate
    d. trilaurin
    e. glyceryl butyropalmitoöleate
    f. β-palmito-α,α'-distearin
    g. myristyl linoleate
    h. sodium p-octylbenzenesulfonate

2. Write equations for the saponification, hydrogenation, and hydrogenolysis of glyceryl linolenate.

3. Name the following esters (assume no chain branching):

a. $CH_2OCOC_{15}H_{31}$
$CHOCOC_{15}H_{31}$
$CH_2OCOC_{15}H_{31}$

b. $CH_2OCOC_{13}H_{27}$
$CHOCOC_{17}H_{35}$
$CH_2OCOC_{13}H_{27}$

c. $C_{17}H_{31}CO_2C_{13}H_{27}$

d. $CH_2OCOC_{17}H_{33}$
$CHOCOC_{17}H_{35}$
$CH_2OCOC_{11}H_{23}$

4. Describe, with an example of your own design, the structural features essential to a good detergent.

5. Explain, with the help of an equation, why ordinary soaps would be ineffective in dish water which contains some fruit juices.

6. Explain, with equations, why ordinary soaps do not work satisfactorily in hard water and how synthetic detergents overcome this defect.

7. How could the chain lengths of oleic, linoleic, and linolenic acids be readily proven?

8. How could the positions of the double bonds in oleic, linoleic, and linolenic acids be determined? Give equations. (HINT: review section 3.4e).

9. Write equations for the preparation of an alkylbenzenesulfonate detergent from propylene tetramer. Explain why this detergent might be a pollutant.

10. Write equations for the preparation of an alkylbenzenesulfonate detergent from 1-decene. Explain why this detergent might be more readily biodegraded than the detergent from propylene tetramer (Problem 9).

11. Write equations for the synthesis of a polyether detergent (sec 11.3d) starting with 1-hexadecanol and ethylene oxide.

12. What important structural feature is necessary to drying oils? How are drying oils structurally related to fats? Give three practical uses of drying oils.

13. How can the drying properties of a drying oil be destroyed and how can they be increased?

14. Using the definition in section 11.5, calculate the saponification number of
a. glyceryl tributyrate   b. glyceryl tripalmitate

15. What is the average molecular weight of a fat with a saponification number of 190? (Remember, 3 moles of KOH are needed to saponify 1 mole of fat.)

16. Using the definition in section 11.5, calculate the iodine number of
a. glyceryl trioleate   b. glyceryl trilinolenate

17. If a fat has a molecular weight of 800 and an iodine number of 95.3, what is the average number of double bonds per fat molecule?

18. Illustrate with structural formulas the primary difference between a fat or oil, and a wax. Write an equation for the saponification of cetyl palmitate, the principal component of spermaceti.

19. Write out (using eq 11-10 through 11-13 as guides) the steps in the biological conversion of
    a. butanoic acid to hexanoic acid
    b. hexanoic acid to butanoic acid

20. If an organism were fed acetic acid labelled with radioactive carbon in the methyl group (i.e., $C^{14}H_3CO_2H$), what would be the expected distribution of the label in the palmitic acid isolated from saponification of the fat synthesized by that organism? Explain.

CHAPTER TWELVE

AMINES AND DIAZONIUM COMPOUNDS

The most typical organic bases are related structurally to the inorganic base ammonia. They contain alkyl or aryl groups attached to a nitrogen atom and are known as amines. Amines are widely distributed in nature, both in plants and animals. The odor of decaying fish is typical of low molecular weight amines. Amino groups are present in many biologically important molecules, notably amino acids and proteins. One of the two raw materials used to produce the synthetic fiber nylon is an aliphatic diamine. Amines are used to manufacture local anesthetics, sulfa drugs, and many other medicinals.

The element nitrogen occurs in many classes of organic compounds, some of which have already been discussed. These include nitro compounds, nitriles, amides, and certain nitrogen derivatives of aldehydes and ketones. In this chapter we will consider several additional groups of nitrogen-containing compounds, focusing particular attention on *amines,* organic derivatives of ammonia.

12.1 NOMENCLATURE OF AMINES

Amines are classified as *primary, secondary,* or *tertiary* depending upon whether one, two, or three of the hydrogen atoms of ammonia have been replaced by organic groups (R). The R groups may be

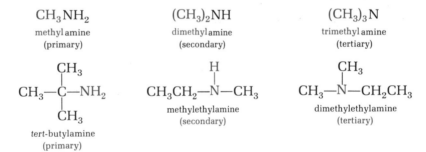

| a primary | a secondary | a tertiary |
| amine | amine | amine |

aliphatic or aromatic, and in secondary or tertiary amines, they may be identical or different from one another. The R groups may also be joined to form rings.

Simple amines are named by prefixing the names of the alkyl groups to the word **amine**, as in

$$CH_3NH_2 \qquad (CH_3)_2NH \qquad (CH_3)_3N$$

methylamine   dimethylamine   trimethylamine
(primary)   (secondary)   (tertiary)

$$CH_3-\overset{\displaystyle CH_3}{\underset{\displaystyle CH_3}{\overset{|}{\underset{|}{C}}}}-NH_2 \qquad CH_3CH_2-\overset{\displaystyle H}{\overset{|}{N}}-CH_3 \qquad CH_3-\overset{\displaystyle CH_3}{\overset{|}{N}}-CH_2CH_3$$

tert-butylamine   methylethylamine   dimethylethylamine
(primary)   (secondary)   (tertiary)

In more complex cases, an amino group may be named as a substituent on a chain, as in

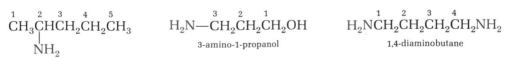

$$\overset{1}{C}H_3\overset{2}{C}H\overset{3}{C}H_2\overset{4}{C}H_2\overset{5}{C}H_3 \qquad \overset{3}{H_2N}-\overset{3}{C}H_2\overset{2}{C}H_2\overset{1}{C}H_2OH \qquad \overset{1}{H_2N}CH_2\overset{2}{C}H_2\overset{3}{C}H_2\overset{4}{C}H_2NH_2$$

$$\underset{\displaystyle NH_2}{|}$$

2-aminopentane   3-amino-1-propanol   1,4-diaminobutane

Aromatic amines are frequently named as derivatives of **aniline,** the most important commercial amine. Occasionally, when there is

aniline
(primary)

N-methylaniline
(secondary)

N,N-dimethylaniline
(tertiary)

p-bromoaniline
(primary)

N-methyl-2,4-dimethylaniline
(secondary)

some question, a capital N- may be used to indicate that a group is attached to nitrogen, and not located elsewhere. Other special names are sometimes used—for example, aminotoluenes are known as toluidines, and the prefixes *phenyl* or *phenylene* are common.

p-toluidine

diphenylamine

p-phenylenediamine

12.2 GENERAL CHARACTERISTICS OF AMINES

The simple aliphatic amines resemble ammonia. The lower members of the series are gases which are readily soluble in water to give basic solutions. The more volatile amines have odors similar to ammonia, but less pungent and more fishlike. Some properties of amines are given in Table 12-1.

Amines can form intramolecular hydrogen bonds, although the N—H$\cdots$N bond is not as strong as the O—H$\cdots$O bond because nitrogen is less electronegative than oxygen [compare, for example, the boiling points of CH_3NH_2 ($-6.5°$) and CH_3OH ($65°$)]. All three classes of amines form hydrogen bonds with water (N$\cdots$H—O); for this reason most of the amines listed in Table 12-1 are quite soluble in water. Hydrogen bonds involving nitrogen are particularly important in determining the structures of enzymes and other proteins, and of nucleic acids (Chapters 17 and 18).

Although only three groups are attached to nitrogen, spectroscopic studies have shown that the bond angles in amines and ammonia are closer to the tetrahedral angle expected for sp^3 bonding than to the 90° which would be expected for pure p bonds or the 120° expected if amines were planar. But at ordinary temperatures, most

Figure 12-1
Models of methylamine, dimethylamine, and trimethylamine.

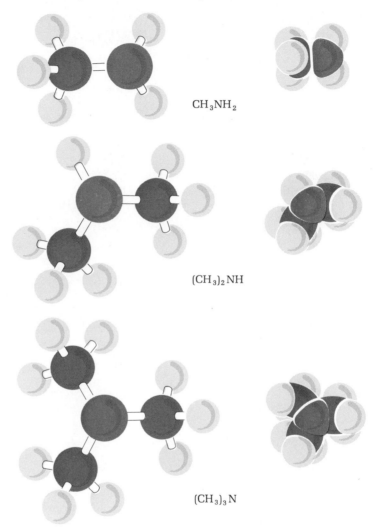

CH_3NH_2

$(CH_3)_2NH$

$(CH_3)_3N$

Figure 12-2
The pyramidal structure of an amine, trimethylamine.

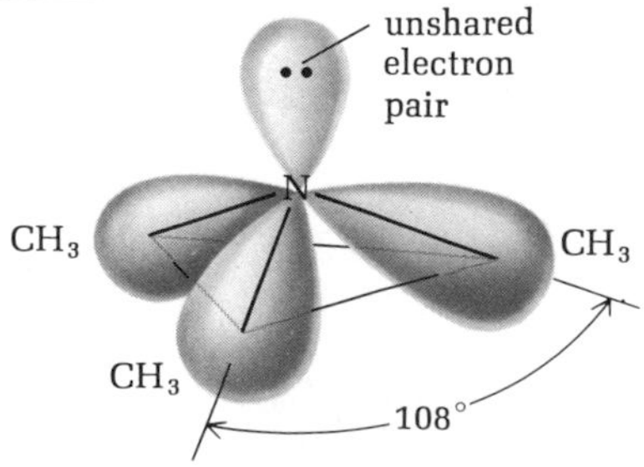

unshared
electron
pair

N

CH_3

CH_3

CH_3

108°

amines interconvert from one pyramidal form to its mirror image
many times per second as shown.

$$\ddot{N} \underset{R_1 \quad \overset{|}{R_2} \diagdown R_3}{\overset{}{}} \rightleftharpoons \left[R_1 - \overset{\cdot\cdot}{N} \diagup\diagdown \overset{R_3}{\underset{R_2}{}} \right] \rightleftharpoons \overset{R_1}{\underset{\ddot{}}{\diagdown}} N \overset{R_3}{\underset{\diagup R_2}{}}$$

planar transition
state

Table 12-1

Some Properties of Amines

NAME	FORMULA	BP, °C	DISSOCIATION CONSTANT, K_b
Ammonia	NH_3	−33.4	2.0×10^{-5}
Methylamine	CH_3NH_2	−6.5	44×10^{-5}
Dimethylamine	$(CH_3)_2NH$	7.4	51×10^{-5}
Trimethylamine	$(CH_3)_3N$	3.5	5.9×10^{-5}
Ethylamine	$CH_3CH_2NH_2$	16.6	47×10^{-5}
n-Propylamine	$CH_3CH_2CH_2NH_2$	48.7	38×10^{-5}
n-Butylamine	$CH_3CH_2CH_2CH_2NH_2$	77.8	40×10^{-5}
Aniline	$C_6H_5NH_2$	184	4.2×10^{-10}
Methylaniline	$C_6H_5NHCH_3$	195.7	7.1×10^{-10}
Dimethylaniline	$C_6H_5N(CH_3)_2$	193.5	11×10^{-10}
Ethylenediamine	$H_2NCH_2CH_2NH_2$	116.5	8.5×10^{-5}
Hexamethylenediamine	$H_2NCH_2CH_2CH_2CH_2CH_2CH_2NH_2$	204.5	85×10^{-5}
Pyridine	C_5H_5N	115.3	23×10^{-10}

12.3 THE BASE STRENGTHS OF AMINES

The unshared electron pair on the nitrogen atom in ammonia and
amines is responsible for their basic properties. A neutral ammonia
molecule can share these electrons with a proton to form a positive
ammonium ion.

$$H:\overset{\cdot\cdot}{\underset{H}{N}}:H + H:\overset{\cdot\cdot}{\underset{\cdot\cdot}{O}}: \rightleftharpoons \left[H:\overset{H}{\underset{H}{\overset{\cdot\cdot}{N}}}:H \right]^{+} + \left[:\overset{\cdot\cdot}{\underset{\cdot\cdot}{O}}:H \right]^{-} \qquad (12\text{-}1)$$

ammonium
ion

Likewise, all classes of amines react reversibly with water to produce
substituted ammonium ions.

$$R-\overset{\cdot\cdot}{N}H_2 + H:\overset{\cdot\cdot}{\underset{H}{O}}: \rightleftharpoons \left[R-\overset{+}{\overset{H}{\underset{H}{N}}}:H\right]^+ + \left[:\overset{\cdot\cdot}{\underset{\cdot\cdot}{O}}:H\right]^- \qquad (12\text{-}2)$$

an alkylammonium
ion

Aqueous solutions of ammonia and amines are basic (alkaline), since hydroxide ion is produced as a result of the equilibria shown in equations 12-1 and 12-2. The value of the equilibrium constant for these equations, called the *dissociation constant* K_b of the base, gives a measure of the base strength of the amine.

The basic dissociation constant K_b for ammonia is written as the equilibrium constant in equation 12-1.

$$K_b = \frac{[NH_4^+][OH^-]}{[NH_3]} = 2 \times 10^{-5}$$

Similar expressions may be written for the amines. For example, the dissociation constant of methylamine is

$$CH_3\overset{\cdot\cdot}{N}H_2 + HOH \rightleftharpoons [CH_3NH_3]^+ + [OH]^- \qquad (12\text{-}3)$$

$$K_b = \frac{[CH_3NH_3]^+[OH^-]}{[CH_3NH_2]} = 4.4 \times 10^{-4}$$

The larger K_b (that is, the smaller the integer in the negative exponent) the greater the concentration of OH^- and the stronger the base.

The dissociation constants for a number of simple amines are listed in Table 12-1. In general, aliphatic amines are slightly stronger bases than ammonia, because alkyl groups are electron-releasing (sec 3.4c, 5.4b, 10.2c) and can stabilize the positive charge on nitrogen in the corresponding ammonium ions. Aromatic amines, on the other hand, are much weaker bases, primarily because of resonance which is possible in the free base, but not in the arylammonium ion. This stabilizes the base relative to the arylammonium ion and the equilibrium concentration of the ion is therefore lower.

resonance structures of aniline

the positive charge in the anilinium ion resides primarily on the nitrogen atom and cannot be delocalized by resonance

12.4 THE PREPARATION OF AMINES

Amines may be prepared by a variety of methods. The choice of method may depend on the class of amine (primary, secondary, or tertiary) and whether it is aliphatic or aromatic. Several of the more general methods are described here.

12.4a ALKYLATION OF AMMONIA AND AMINES

Ammonia and amines react with alkyl halides to form alkyl-ammonium halides. The reaction (eq 12-4) is a bimolecular nucleo-philic displacement (S_N2 reaction; sec 6.5c). The unshared electron pair on nitrogen displaces the halide ion from the alkyl halide. Treatment of the product, an alkylammonium halide, with a base

$$H_3N: + R-\overset{..}{\underset{..}{X}}: \longrightarrow RNH_3^+ + :\overset{..}{\underset{..}{X}}:^- \qquad (12\text{-}4)$$

an alkylammonium halide

liberates the free amine. This step is essentially the reverse of equation 12-2.

$$RNH_3^+X^- + Na^+OH^- \longrightarrow RNH_2 + HOH + Na^+X^- \quad (12\text{-}5)$$

As with other S_N2 displacements, the reaction works best when R is a primary alkyl group.

Unfortunately, the reaction does not always stop cleanly with the replacement of only one hydrogen in the ammonia or amine by an alkyl group. This is because the alkylammonium ion (produced in eq 12-4) is in equilibrium with the ammonia present as a reactant, since ammonia and alkylamines have comparable base strengths. The amine (as produced in eq 12-6) can then serve

$$\overset{..}{N}H_3 + RNH_3^+ \rightleftharpoons NH_4^+ + RNH_2 \qquad (12\text{-}6)$$

as a nucleophile, leading to a dialkylammonium ion, and ultimately the secondary amine R_2NH. This can react in like fashion, leading

$$R\overset{..}{N}H_2 + R-\overset{..}{\underset{..}{X}}: \longrightarrow R_2NH_2^+ :\overset{..}{\underset{..}{X}}:^- \qquad (12\text{-}7)$$

a dialkylammonium halide

to tertiary amine, R_3N. Thus, when ammonia and ethyl chloride are allowed to react, one obtains a mixture of ethylamine, diethylamine, and triethylamine. This mixture can be separated by careful fractional distillation, and all three of these amines are produced commercially in this way.

One, of course, need not begin with ammonia. Methylaniline and dimethylaniline, which are two important raw materials in the manufacture of certain dyes, are obtained by the alkylation (methylation) of aniline.

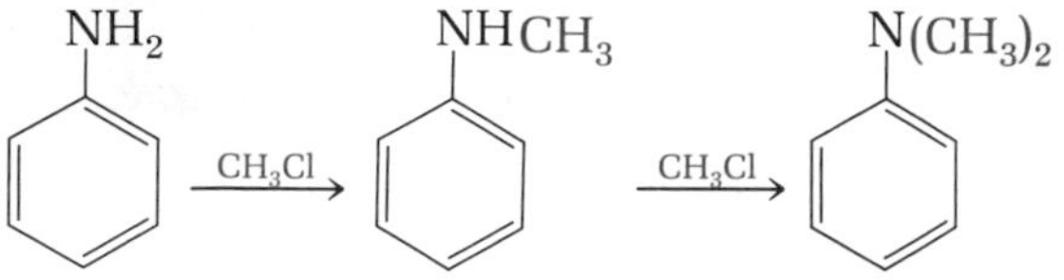

aniline methylaniline dimethylaniline (12-8)

12.4b REDUCTION OF OTHER NITROGEN-CONTAINING COMPOUNDS

Several classes of unsaturated nitrogen-containing compounds give amines on reduction. Either catalytic (H_2 and a metal catalyst) or chemical reducing agents are used.

Nitriles (sec 8.2) and *oximes* (sec 9.6d) are reduced to primary amines.

$$\underset{\text{nitrile}}{R{-}C{\equiv}N} + 4\,[H] \xrightarrow[\text{heat, pressure}]{H_2,\ \text{Ni catalyst}} \underset{\text{a primary amine}}{RCH_2NH_2} \qquad (12\text{-}9)$$

$$\underset{\text{an oxime}}{(R_1)(R_2)C{=}N{-}OH} \xrightarrow{2\,[H]} \underset{\text{a hydroxylamine}}{(R_1)(R_2)CH{-}NHOH} \xrightarrow[-H_2O]{2\,[H]} \underset{\text{a primary amine}}{(R_1)(R_2)CH{-}NH_2} \qquad (12\text{-}10)$$

Nitro compounds (sec 2.7b, 5.4) also give primary amines. This method is particularly useful for aromatic amines. Aniline is manufactured commercially from nitrobenzene by reduction with iron and steam. In the laboratory, hydrochloric acid may replace steam as the

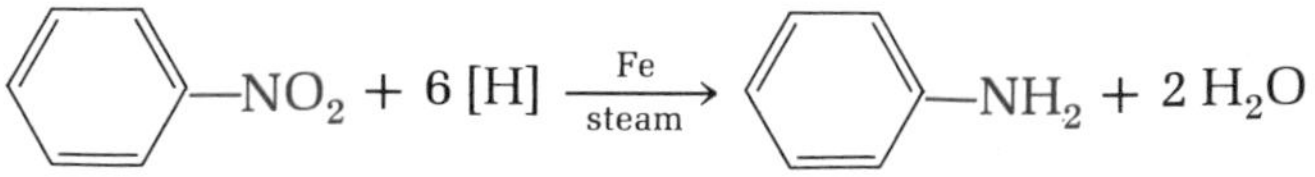

nitrobenzene aniline (12-11)

source of hydrogen, and more reactive (but also more expensive) metals such as tin or zinc may be used.

Figure 12-3
Scale model of aniline.

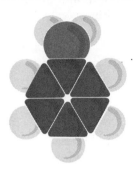

12.4c AMINES FROM AMIDES: THE HOFMANN REARRANGEMENT

The German chemist A. W. Hofmann, who did much to unravel the early chemistry of amines, discovered in 1881 that amides ($RCONH_2$), when treated in alkaline solution with bromine, gave good yields of primary amines (RNH_2). The over-all reaction is

$$\underset{\text{an amide}}{R-\overset{\overset{\text{O}}{\|}}{C}-NH_2} + Br_2 + 4\,Na^+OH^- \longrightarrow \underset{\text{an amine}}{RNH_2} + 2\,Na^+Br^- +$$

$$Na_2^+CO_3^{2-} + 2\,H_2O \qquad (12\text{-}12)$$

At first glance it seems as if CO were somehow magically plucked from the amide. However, the reaction is known to proceed in several rational stages. The first (eq 12-13) involves a base-catalyzed bromination of the amide, on nitrogen.

$$\underset{\text{an amide}}{R-\overset{\overset{\text{O}}{\|}}{C}-\overset{\cdot\cdot}{N}H_2} \xrightarrow{Na^+OBr^-} \underset{\text{N-bromoamide}}{R-\overset{\overset{\text{O}}{\|}}{C}-\overset{\cdot\cdot}{N}HBr} + Na^+OH^- \qquad (12\text{-}13)$$

Because of the electron-withdrawing effect of the bromine, N-bromoamides are stronger acids than unsubstituted amides, and in base they ionize to salts.

$$R-\overset{\overset{\text{O}}{\|}}{C}-\underset{\underset{\text{H}}{|}}{\overset{\cdot\cdot}{N}}-Br + OH^- \longrightarrow R-\overset{\overset{\text{O}}{\|}}{C}-\overset{\cdot\cdot}{\underset{\cdot\cdot}{N}}-Br + H_2O \qquad (12\text{-}14)$$

The step which follows is the key to the entire sequence. The bromoanion loses bromide ion; as it does so, a rearrangement occurs, the net result of which is to move the R group from carbon to nitrogen. The product is an **isocyanate:**

$$R-C(=O)-N(Br) \longrightarrow R-N=C=O + Br^- \qquad (12\text{-}15)$$

an alkyl isocyanate

It will be noted from equation 12-12 that the R group is attached to a carbon atom in the reactant, but to a nitrogen atom in the product. Equation 12-15 explains how this transfer occurs. The final step involves hydrolysis of the isocyanate to an amine.

$$R-\ddot{N}=C=O + HOH \longrightarrow \left[R-\underset{H}{\overset{\ddot{}}{N}}-\overset{O}{\overset{\|}{C}}-OH \right] \xrightarrow{-CO_2} R-\ddot{N}H_2 \quad (12\text{-}16)$$

an isocyanate   a carbamic acid   an amine

The intermediate carbamic acids are unstable, and readily lose carbon dioxide.

 Steps in this mechanism have been verified experimentally. N-bromoamides can be isolated, and they do form salts which on heating rearrange to isocyanates. In some cases, the Hofmann rearrangement can be stopped at the isocyanate, and independent experiments show that isocyanates are readily hydrolyzed to amines.

The Hofmann rearrangement can be used to decrease the length of a carbon chain by one carbon atom, since the amine produced has one less carbon atom than the starting amide.

12.5   THE REACTIONS OF AMINES

The unshared pair of electrons on the nitrogen atom of amines is the key to their reactivity. It accounts for their basicity and for their behavior as nucleophiles.

12.5a   SALT FORMATION

Amines, like ammonia, react with acids to form salts (review sec 12.3). Equation 12-18 illustrates the reaction for a tertiary amine.

$$\ddot{N}H_3 + \textcircled{H}:\ddot{C}l: \longrightarrow NH_4^+ + :\ddot{C}l:^- \qquad (12\text{-}17)$$

$$R_3\ddot{N} + \textcircled{H}:Cl: \longrightarrow R_3NH^+ + :\ddot{C}l:^- \qquad (12\text{-}18)$$

a trialkyl-
ammonium ion

The reaction can be used to separate amines from neutral or acidic compounds. A solution of the compounds in an organic solvent (say, ether) is extracted with dilute acid. The amines react with the acid to form salts which dissolve in the aqueous layer. After the layers are separated, the aqueous layer is made alkaline, whereupon the free amine is liberated. The sequence is illustrated for p-toluidine:

$$CH_3-\underset{\substack{\text{p-toluidine}\\\text{(water-insoluble)}}}{\boxed{}}-\overset{\cdot\cdot}{N}H_2 + HCl \xrightarrow{H_2O}$$

$$CH_3-\underset{\substack{\text{p-toluidinium chloride}\\\text{(water-soluble)}}}{\boxed{}}-NH_3^+ \; Cl^- \qquad (12\text{-}19)$$

$$CH_3-\boxed{}-NH_3^+Cl^- + Na^+OH^- \longrightarrow$$

$$CH_3-\boxed{}-NH_2 + H_2O + Na^+Cl^- \qquad (12\text{-}20)$$

Amines which occur naturally in plants (for example, nicotine or strychnine) can be extracted in this way.

12.5b ALKYLATION: QUATERNARY AMMONIUM COMPOUNDS

The replacement of one, two, or all three of the hydrogens in ammonia by alkyl groups was discussed in section 12.4a. The reaction, an S_N2 displacement of X^- from an alkyl halide, R—X, is initiated by the unshared electron pair on nitrogen (eq 12-4 or 12-7). But tertiary amines, even though they have no remaining hydrogens on nitrogen which might be replaced by alkyl groups, *do* have an unshared electron pair on nitrogen, and therefore might also be expected to displace X^- from R—X. In fact, tertiary amines react with alkyl halides to form products in which all four hydrogens of the ammonium ion have been replaced by organic groups. Such compounds are called **quaternary ammonium compounds.**

$$\underset{\substack{\text{a tertiary}\\\text{amine}}}{R-\overset{\displaystyle R}{\underset{\displaystyle R}{N:}}\curvearrowright R\overset{\cdot\cdot}{\underset{\cdot\cdot}{X}}:} \longrightarrow \underset{\text{a tetraalkylammonium halide}}{R-\overset{\displaystyle R}{\underset{\displaystyle R}{N^+}}-R \;\; :\overset{\cdot\cdot}{\underset{\cdot\cdot}{X}}:^-} \qquad (12\text{-}21)$$

As a specific example, trimethylamine reacts quantitatively with methyl iodide to give the salt, tetramethylammonium iodide.

$$(CH_3)_3N: + CH_3I \longrightarrow (CH_3)_4N^+I^- \qquad (12\text{-}22)$$

trimethylamine tetramethylammonium iodide

Such salts, in which one of the four alkyl groups contains a long carbon chain, are used as detergents and germicides (sec 11.3d).

Quaternary salts can be converted to quaternary *bases* by treatment in aqueous solution with silver hydroxide.

$$\begin{bmatrix} & CH_3 & \\ CH_3-N-CH_3 \\ & CH_3 & \end{bmatrix}^+ I^- + AgOH \longrightarrow$$

$$\begin{bmatrix} & CH_3 & \\ CH_3-N-CH_3 \\ & CH_3 & \end{bmatrix}^+ OH^- + AgI\downarrow \qquad (12\text{-}23)$$

tetramethylammonium hydroxide

The precipitate of silver halide is removed by filtration. Upon evaporation of the filtrate, white deliquescent crystals of the quaternary base remain. These substances are strong bases, similar to sodium or potassium hydroxide.

Several quaternary ammonium compounds have physiological activity. **Choline,** or trimethyl-β-hydroxyethylammonium hydroxide, exists in combination with phospholipids which make up part of brain and spinal cord tissue. These compounds are called **lecithins.**

$$\begin{bmatrix} & CH_3 & \\ CH_3-N-CH_2CH_2OH \\ & CH_3 & \end{bmatrix}^+ OH^-$$

choline

$$CH_2-O-\overset{\overset{\displaystyle O}{\|}}{C}-R$$

$$CH-O-\overset{\overset{\displaystyle O}{\|}}{C}-R$$

$$CH_2-O-\overset{\overset{\displaystyle O^-}{\diagup}}{\underset{\diagdown O}{P}}-OCH_2CH_2-\overset{\overset{\displaystyle CH_3}{|}}{\underset{|}{\overset{+}{N}}}-CH_3$$
$$CH_3$$

an α-lecithin
(a phospholipid)

Choline is essential to growth and is involved in fat transport and in carbohydrate and protein metabolism. It is also the precursor of **acetylcholine,** which is involved in the transmission of nerve impulses to ganglion cells, and also of motor nerve impulses to the fibers of voluntary muscles, resulting in muscle contraction. The "nerve gases" developed during World War II get their lethal punch

by damaging cholinesterase, an enzyme that hydrolyzes acetylcholine to choline.

$$\left[CH_3-\underset{\underset{CH_3}{|}}{\overset{\overset{CH_3}{|}}{N}}-CH_2CH_2-O-\overset{\overset{O}{||}}{C}-CH_3 \right]^{+} OH^{-}$$

acetylcholine

12.5c ACYLATION: FORMATION OF AMIDES

Primary and secondary amines react with carboxylic acids and their
derivatives in a manner exactly analogous to ammonia (review
sec 10.8). The products are amides with one or two alkyl groups,
respectively, on the nitrogen atom. Equation 12-24 compares the
reactions of acetic anhydride with ammonia, a primary amine, and
a secondary amine. The net result in each case is the replacement

$$CH_3\overset{\overset{O}{||}}{C}-O-\overset{\overset{O}{||}}{C}CH_3 \quad \begin{cases} \xrightarrow{NH_3} & CH_3\overset{\overset{O}{||}}{C}-NH_2 \\ & \text{acetamide} \\ \xrightarrow{RNH_2} & CH_3\overset{\overset{O}{||}}{C}-NHR + CH_3\overset{\overset{O}{||}}{C}-OH \quad (12\text{-}24) \\ & \text{an N-alkylacetamide} \\ \xrightarrow{R_2NH} & CH_3\overset{\overset{O}{||}}{C}-NR_2 \\ & \text{an N,N-dialkylacetamide} \end{cases}$$

acetic anhydride

of one of the hydrogens on the amine nitrogen by an acyl (in this
case, acetyl) group. Since tertiary amines have no such hydrogen
atom, they cannot be converted to amides.

Acetanilide, the amide derived from acetylation of aniline, is used
in medicine as an antipyretic (fever-reducing substance).

$$CH_3\overset{\overset{O}{||}}{C}O\overset{\overset{O}{||}}{C}CH_3 + H_2N-\bigcirc \longrightarrow$$

$$CH_3\overset{\overset{O}{||}}{C}-NH-\bigcirc + CH_3CO_2H \quad (12\text{-}25)$$

acetanilide

N,N-**Diethyl-*m*-toluamide,** the active ingredient in insect repellants
such as "Off" (Figure 12-4) is an amide of a secondary amine.

$$CH_3 - C_6H_4 - \overset{O}{\underset{\|}{C}} - \overset{..}{N}(CH_2CH_3)_2$$

N,N-diethyl-m-toluamide
(meta-Delphene)

Since amides can be reduced to amines with lithium aluminum
hydride, one can use them to convert primary to secondary amines:

$$R - \overset{O}{\underset{\|}{C}} - Cl + H_2N - R' \longrightarrow R - \overset{O}{\underset{\|}{C}} - NHR' \xrightarrow{\ \text{LiAlH}_4\ }$$

a primary   an amide
amine

$$RCH_2 - NHR' \qquad (12\text{-}26)$$

a secondary
amine

Figure 12-4
A swarm of mosquitoes shows no interest in the man's arm above the
black line, where meta-Delphene repellent has been applied, although
they feast on the untreated part of his arm. It is remarkable that if the
methyl group is in either of the other isomeric positions (ortho or para)
the compound is ineffective as an insect repellent. (USDA photo. Cour-
tesy of the Hercules Powder Company.)

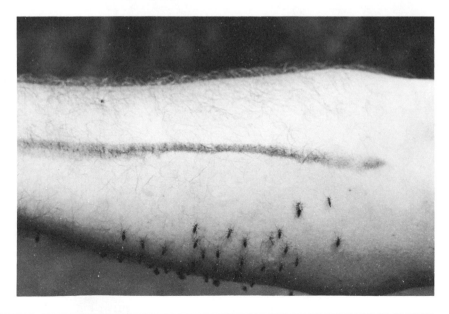

The sequence can be repeated to convert a secondary amine to a tertiary amine. This method is particularly useful for synthesizing amines which have two or three different R groups attached to the nitrogen.

Primary and secondary amines form not only **carboxamides** (from carboxylic acid derivatives), but also form amides with derivatives of other types of acids. For example, sulfonyl or phosphonic chlorides react to form sulfonamides or phosphonamides respectively.

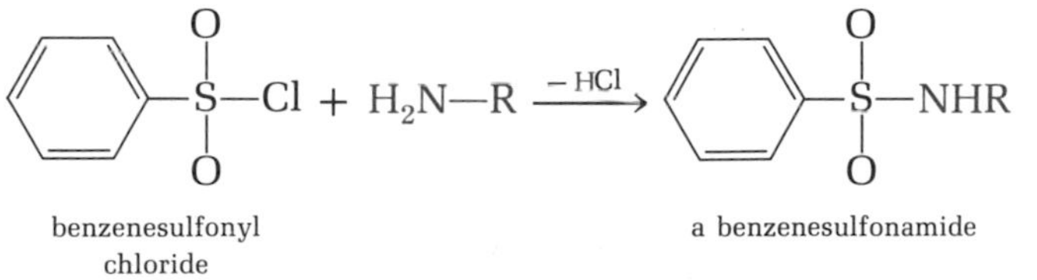

$$\text{benzenesulfonyl chloride} \quad\quad \text{a benzenesulfonamide} \tag{12-27}$$

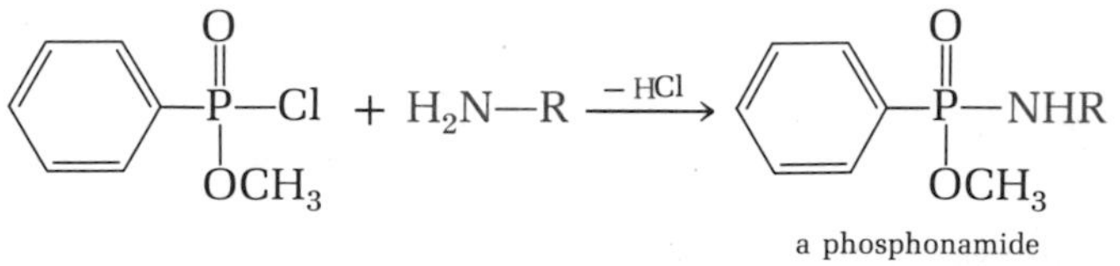

$$\text{a phosphonamide} \tag{12-28}$$

Sulfanilamide, the first of the sulfa drugs, can be prepared from acetanilide by a series of steps which involve sulfonation, ammonolysis of the sulfonic acid chloride, and selective hydrolysis of the carboxamide function.

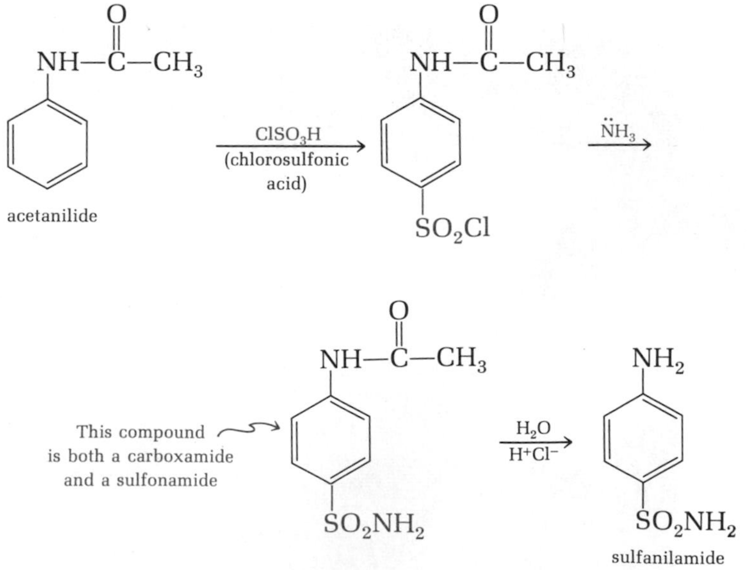

acetanilide

This compound is both a carboxamide and a sulfonamide

sulfanilamide

$$\tag{12-29}$$

The **Hinsberg test,** used to distinguish between the three classes of amines, is based on their reaction with benzenesulfonyl chloride.

Primary amines give a sulfonamide which, because of the remaining hydrogen on the nitrogen, is acidic and soluble in alkali.

benzenesulfonyl
chloride

an alkylsulfonamide
(soluble in base)

$$\left[\text{⬡-SO}_2\ddot{\text{N}}\text{R}\right]^- \text{Na}^+ \xrightarrow{\text{H}^+} \text{⬡-SO}_2\text{NHR} \qquad (12\text{-}30)$$

(insoluble in acid)

Secondary amines also give a sulfonamide which, however, has no acidic hydrogen and is insoluble in base.

$$\text{R}_2\text{NH} + \text{⬡-SO}_2\text{Cl} \longrightarrow \text{⬡-SO}_2\text{NR}_2 \qquad (12\text{-}31)$$

a dialkylsulfonamide
(insoluble in base or acid)

Tertiary amines, having no hydrogen on the nitrogen, do not react with the reagent.

$$\text{R}_3\text{N} + \text{⬡-SO}_2\text{Cl} \xrightarrow{\text{Na}^+\text{OH}^-} \text{no reaction} \qquad (12\text{-}32)$$

In practice, the amine to be tested is shaken with benzenesulfonyl chloride and alkali. Primary amines yield a clear solution which, on acidification, precipitates the alkylsulfonamide. Secondary amines yield an insoluble dialkylsulfonamide which is unaffected by acid. Tertiary amines also may give an insoluble compound (the unreacted amine) which, however, dissolves on acidification (forming a soluble amine salt).

12.5d   REACTION WITH NITROUS ACID

Reaction with nitrous acid provides a test which distinguishes between primary, secondary, and tertiary aliphatic amines. Nitrous acid, HONO, is stable only in solution at low temperatures, and is prepared as needed by the reaction of sodium nitrite with a strong acid.

$$\text{Na}^+\text{NO}_2^- + \text{H}^+\text{Cl}^- \xrightarrow{\text{0}°} \text{H}-\text{O}-\ddot{\text{N}}=\underset{..}{\overset{..}{\text{O}}}: + \text{Na}^+\text{Cl}^- \qquad (12\text{-}33)$$

sodium nitrite   nitrous acid

Primary aliphatic amines react to liberate nitrogen gas. The nitrogen-nitrogen bond, present in the nitrogen gas which is evolved, is formed in the first step by a nucleophilic addition of the amine to the nitrogen-oxygen double bond (much like the addition of nucleophiles to carbon-oxygen double bonds, sec 9.6d).

$$\underset{\substack{\text{a primary}\\\text{amine}}}{\overset{\overset{\displaystyle H}{\vert}}{\underset{\underset{\displaystyle H}{\vert}}{R-\overset{\displaystyle..}{N}:}}} \quad \underset{\substack{\text{nitrous}\\\text{acid}}}{\overset{\overset{\displaystyle }{}}{\underset{\underset{\displaystyle OH}{\vert}}{\overset{..}{N}=\overset{..}{O}:}}} \longrightarrow \underset{\underset{\displaystyle H}{\vert}}{R-\overset{\displaystyle\overset{H}{\vert}}{N^+}} - \underset{\underset{\displaystyle OH}{\vert}}{\overset{..}{N}} - \overset{..}{\underset{..}{O}} : ^- \qquad (12\text{-}34)$$

Subsequent protonation followed by elimination of two molecules of water leads to the formation of an alkyldiazonium ion.

$$R - \overset{+}{N} \equiv N :$$

an alkyldiazonium ion

Such ions are quite unstable,* readily lose nitrogen, and generate a carbonium ion.

$$R \overset{+}{\frown} N \equiv N : \longrightarrow \quad \underset{\text{a carbonium ion}}{R^+} \quad + \underset{\text{nitrogen}}{:N \equiv N:} \qquad (12\text{-}35)$$

The organic product is derived from this carbonium ion. If the solvent for the reaction is water, alcohols will be formed.

$$R^+ + H_2O \longrightarrow ROH + H^+ \qquad (12\text{-}36)$$

The carbonium ion may also rearrange, or lose a proton to form an alkene; its ultimate fate depends upon its structure and the reaction conditions. For example, if the amine is *n*-propylamine, the organic products will include 1-propanol, 2-propanol, and propene.

$$CH_3CH_2CH_2NH_2 \xrightarrow{\text{HONO}} \begin{Bmatrix} CH_3CH_2CH_2OH \\ CH_3\underset{\underset{\displaystyle OH}{\vert}}{CHCH_3} \\ CH_3CH{=}CH_2 \end{Bmatrix} + N_2\uparrow + H_2O \qquad (12\text{-}37)$$

Secondary amines form nitrosoamines which separate from the aqueous solution as a yellow oily layer.

$$\underset{\substack{\text{a secondary}\\\text{amine}}}{\overset{\displaystyle R}{\underset{\displaystyle R'}{>}}N{-}H} + HONO \longrightarrow \underset{\text{a nitrosoamine}}{\overset{\displaystyle R}{\underset{\displaystyle R'}{>}}N{-}N{=}O} + H_2O \qquad (12\text{-}38)$$

---

*Solutions of aromatic diazonium ions (R = an aryl group) are moderately stable. Their chemistry is discussed in section 12.8.

Tertiary amines simply dissolve in the aqueous acid without evolution of nitrogen, usually giving complex products. The evolution of a gas, formation of a yellow oily layer, or the solution itself may be taken as criteria for the class of amine tested.

12.5e AROMATIC SUBSTITUTION

Like phenols, aromatic amines react strongly with electrophiles. The amino group is ring-activating and *ortho-para*-directing as a consequence of resonance structures which increase the electron density at these positions (sec 12.3). Thus 2,4,6-tribromoaniline can be prepared by simply shaking aniline with bromine water (compare with eq 6-38).

$$\text{aniline} + 3\ Br_2 \xrightarrow{H_2O} \text{2,4,6-tribromoaniline} + 3\ HBr \qquad (12\text{-}39)$$

One can moderate the reactivity of the amino group by acylating it (eq 12-25). The acetamido group is also ring-activating and *ortho-para*-directing, but less so than the amino group because of the competing amide resonance (sec 10.8) which decreases the availability of the unshared electron pair on nitrogen to the aromatic ring. For example, p-bromoaniline can be prepared according to the sequence in equation 12-40.

aniline → acetanilide

p-bromoacetanilide → p-bromoaniline (12-40)

Electrophilic substitutions on anilines are sometimes complicated by the fact that the reagents are strong acids which form salts with the amino group. For example, sulfonation of aniline first gives the salt, anilinium hydrogen sulfate. If the salt is heated to 200°, sulfonation occurs to give sulfanilic acid (*p*-aminobenzenesulfonic acid).

$$
:NH_2 \xrightarrow{\text{H--OSO}_3\text{H}} NH_3^+OSO_3H^- \xrightarrow{200°} NH_2 \quad + H_2O \quad (12\text{-}41)
$$

anilinium hydrogen
sulfate

sulfanilic acid
SO_3H

Actually sulfanilic acid contains both acidic (SO_3H) and basic (NH_2) functional groups; the proton is therefore attached to the amino nitrogen rather than to the oxygen of the sulfonic acid group, because the former is more basic. This gives the molecule a *dipolar* or *inner salt* structure.

$$
H_3\overset{+}{N}\text{---}\underset{}{\bigcirc}\text{---}SO_3^-
$$

dipolar structure of sulfanilic acid

Dipolar structures can be important in all molecules which contain acidic and basic functions, as for example in amino acids (Chapter 17), peptides, and proteins.

12.6 DIAMINES

The simplest aliphatic diamine, **ethylenediamine,** may be prepared by the reaction of ammonia with ethylene dichloride.

$$
\begin{array}{l} CH_2\text{---}Cl \\ | \\ CH_2\text{---}Cl \end{array} + 4\,NH_3 \longrightarrow \begin{array}{l} CH_2\text{---}NH_2 \\ | \\ CH_2\text{---}NH_2 \end{array} + 2\,NH_4^+Cl^- \quad (12\text{-}42)
$$

ethylenediamine

It is a water-soluble basic liquid which, with long-chain acids, forms salts which are useful emulsifying agents.

Two rather foul-smelling diamines, **putrescine** (1,4-diaminobutane) and **cadaverine** (1,5-diaminopentane), are formed when animal flesh decays. Although not highly poisonous themselves, putrescine and cadaverine belong to a group of compounds known as **ptomaines,** nitrogen compounds formed from the decomposition of proteins by bacteria.

$$H_2NCH_2CH_2CH_2CH_2NH_2 \qquad H_2NCH_2CH_2CH_2CH_2CH_2NH_2$$
putrescine   cadaverine

Hexamethylenediamine (1,6-diaminohexane) is one of the two major starting materials in the manufacture of *nylon,* and is prepared in tonnage quantities for this purpose. The other raw material is adipic acid, the straight-chain dicarboxylic acid with six carbon atoms. When hexamethylenediamine and adipic acid react they form a polysalt, called Nylon salt. On heating, this salt is converted to a **polyamide** (compare with eq 10-46).

$$HO_2C(CH_2)_4CO_2H + H_2N(CH_2)_6NH_2 \longrightarrow$$
adipic acid   hexamethylenediamine

$$^-O_2C(CH_2)_4CO_2^- \ H_3\overset{+}{N}(CH_2)_6\overset{+}{N}H_3 \xrightarrow[-H_2O]{200-300^\circ}$$

$$-NH \left[\overset{O}{\overset{\|}{C}}(CH_2)_4 \overset{O}{\overset{\|}{C}} NH(CH_2)_6 NH \right]_n \overset{O}{\overset{\|}{C}} - \qquad (12\text{-}43)$$
Nylon 6-6
(a polyamide)

The resulting polyamide, with molecular weights ranging from 10,000 to 25,000, becomes molten at 260–270°. The molten material can be drawn into threads which, upon cooling to room temperature, can be further stretched to about four times their original length. This "cold drawing" process orients the polymer molecules so that their long axis is parallel to the fiber axis. Hydrogen bonds of the N—H····O type at regular intervals cross-link nitrogen and oxygen atoms on adjacent chains and give the fiber strength. This variety of nylon is known as **Nylon 6-6,** because the diamine and dicarboxylic acid each contain six carbon atoms.

The raw materials for Nylon 6-6 can be prepared in several ways. For example, hexamethylenediamine can be obtained in three steps from butadiene.

$$CH_2{=}CH{-}CH{=}CH_2 \xrightarrow{Cl_2} \underset{\underset{Cl}{|}}{CH_2}{-}CH{=}CH{-}\underset{\underset{Cl}{|}}{CH_2} \xrightarrow[-2\,NaCl]{2\,NaCN}$$

$$NCCH_2CH{=}CHCH_2CN \xrightarrow[\substack{metal \\ catalyst}]{5\,H_2} H_2N(CH_2)_6NH_2 \qquad (12\text{-}44)$$
hexamethylenediamine

The last step involves not only reduction of the carbon-carbon double bond, but also the two C≡N bonds (compare eq 12-9).

Adipic acid can be prepared from benzene.

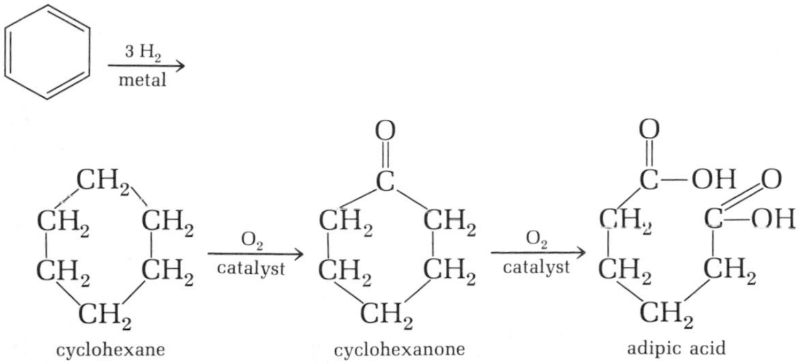

$$\text{benzene} \xrightarrow[\text{metal}]{3\ H_2}$$

$$\text{cyclohexane} \xrightarrow[\text{catalyst}]{O_2} \text{cyclohexanone} \xrightarrow[\text{catalyst}]{O_2} \text{adipic acid} \tag{12-45}$$

Hexamethylenediamine can also be obtained from adipic acid, via the amide and nitrile:

$$HO_2C(CH_2)_4CO_2H \xrightarrow[\text{heat}]{NH_3} H_2N\overset{O}{\overset{\|}{C}}(CH_2)_4\overset{O}{\overset{\|}{C}}NH_2 \xrightarrow[\text{catalyst}]{350°}$$

$$\underset{\text{adipic acid}}{} \qquad \underset{\text{adipamide}}{}$$

$$N{\equiv}C(CH_2)_4C{\equiv}N \xrightarrow[\text{catalyst}]{4\ H_2} H_2NCH_2(CH_2)_4CH_2NH_2 \tag{12-46}$$

$$\underset{\text{adiponitrile}}{} \qquad \underset{\text{hexamethylenediamine}}{}$$

It is not necessary to have a diamine and diacid to produce a polyamide; a molecule with an amino group at one end and a carboxyl at the other can, by reacting with itself in proper fashion, produce a polyamide. **Caprolactam,** the cyclic amide of 6-aminocaproic acid, produces a polyamide known as **Nylon-6.**

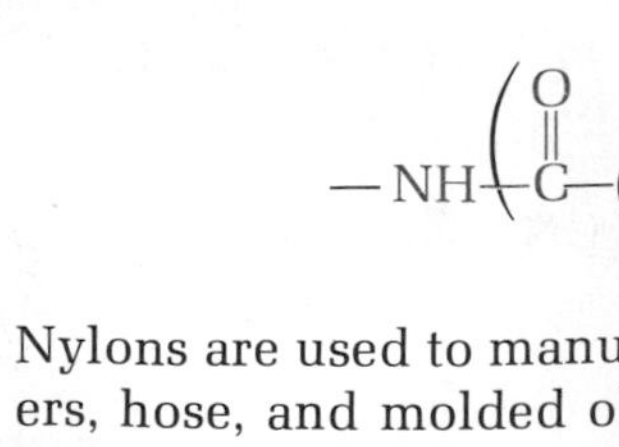

caprolactam

$$-NH{\left(\overset{O}{\overset{\|}{C}}{-}(CH_2)_5{-}NH{-}\overset{O}{\overset{\|}{C}}{-}(CH_2)_5{-}NH\right)}_{/n}\overset{O}{\overset{\|}{C}}- \tag{12-47}$$

Nylon-6

Nylons are used to manufacture textile fabrics, carpets, rope, sweaters, hose, and molded objects.

Figure 12-5
The spinnaker of this sailboat is made of nylon, while the mainsail is of
Dacron polyester fiber. Among the advantages of these fibers is the
fact that mildew, which is the action of fungi and bacteria, will not
grow on sailcloth of either fiber. (Sue Cummings, courtesy of Hood Sail-
makers.)

Aromatic diamines are usually prepared from nitro compounds
by reduction.

$$
\text{m-dinitrobenzene} \xrightarrow[\text{HCl}]{\text{Sn}} \text{m-phenylenediamine}
$$

m-dinitrobenzene   m-phenylenediamine

(12-48)

Benzidine, an aromatic diamine used in the manufacture of direct dyes for cotton, is produced from hydrazobenzene by treatment with acid. A possible mechanism for this unusual rearrangement is shown.

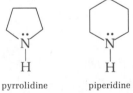

(12-49)

The precursor, hydrazobenzene, is made from nitrobenzene by reduction with zinc and alkali.

12.7 HETEROCYCLIC AMINES

Heterocyclic amines contain at least one nitrogen atom incorporated in a ring. They are usually known by common names. If the ring is saturated, as in **pyrrolidine** or **piperidine,** the compounds behave chemically like typical secondary amines.

pyrrolidine   piperidine

More interesting are the unsaturated analogs, for they are quite common in natural products, and have special aromatic properties.

In **pyrrole** the unshared pair of electrons on nitrogen together with the 4π electrons from the two double bonds combine to form a planar aromatic 6π electron system similar to that in benzene (Figure 5-3).

pyrrole

Since the unshared pair on nitrogen is part of the aromatic system, it is not as available to a proton as is the electron pair of most amines. Consequently, pyrrole is an exceedingly weak base ($K_b = 2.5 \times 10^{-14}$) compared with, say, the closely related secondary amine pyrrolidine ($K_b = 2 \times 10^{-4}$). Indeed pyrrole is actually a weak acid ($K_a = 10^{-15}$), stronger than alcohols but weaker than phenols.

The pyrrole ring system is present in two important natural coloring materials, the **chlorophyll** of green plants, and **hemin** of red blood cells (sec 18.5b). The pyrrolidine ring is present in two common amino acids, proline and hydroxyproline (Table 17-1), as well as in certain alkaloids such as nicotine, cocaine, and atropine (sec 18.5a). A benzo derivative of pyrrole, called **indole,** together with 3-methylindole **(skatole)** is formed during protein putrefaction.

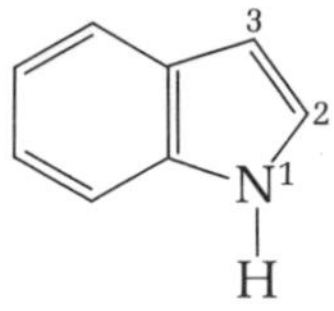

indole

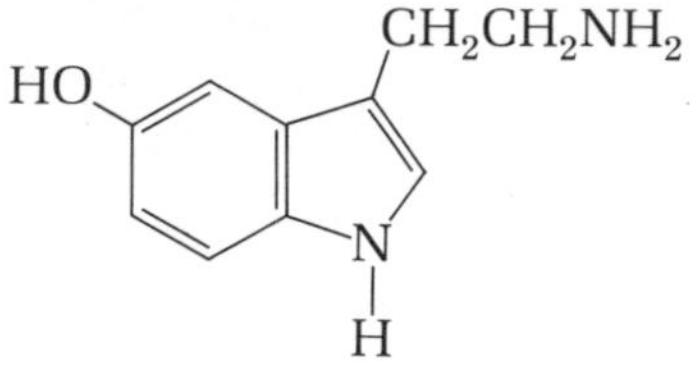

serotonin

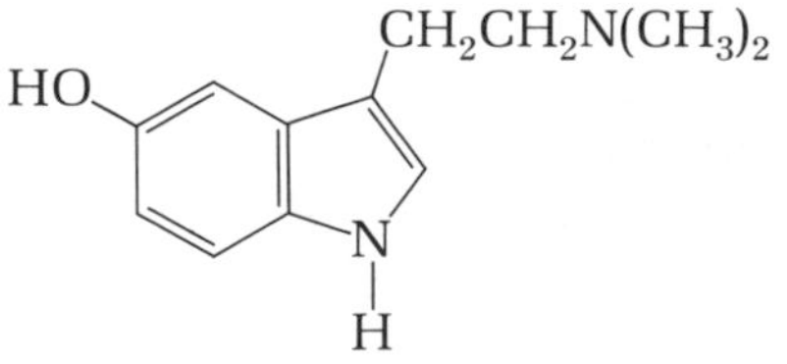

bufotenine

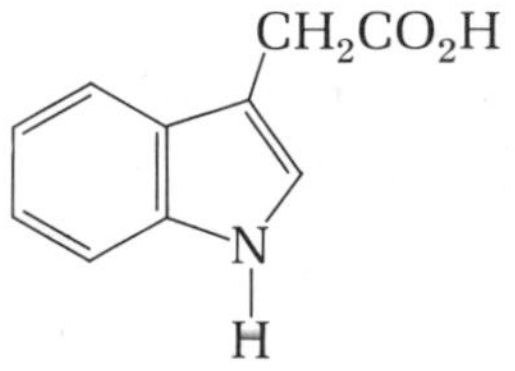

3-indoleacetic acid

Both contribute to the odor of feces. The indole ring system is present in one of the common amino acids, tryptophan (Table 17-1), and in the vasoconstrictor **serotonin,** the hallucinogen **bufotenine,** and the plant growth hormone **3-indoleacetic acid,** amongst other natural products.

Pyridine is the heterocyclic analog of benzene, with one CH group

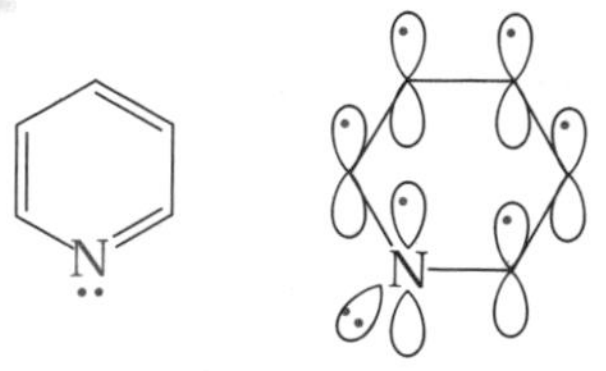

pyridine

replaced by a nitrogen atom. The aromatic 6π electron system is derived from six p orbitals, one on each ring atom. Unlike pyrrole, the unshared electron pair on nitrogen in pyridine is *not* part of the aromatic system and is therefore available for accepting a proton.

$$+ \ H^+ \ \rightleftharpoons \qquad \qquad (12\text{-}50)$$

pyridine   pyridinium ion

Pyridine, although basic ($K_b = 2.3 \times 10^{-9}$), is much less so than most aliphatic tertiary amines ($K_b \simeq 10^{-5}$). The orbital in which the unshared electron pair resides has greater s character and the pair of electrons is held closer to the nitrogen (which is sp^2, rather than the usual sp^3 hybridized).

Like benzene (eq 5-20), pyridine resists oxidation with permanganate, whereas alkyl side-chains can be converted to carboxyl groups.

$$\xrightarrow{\text{KMnO}_4} \qquad \qquad (12\text{-}51)$$

β-picoline   nicotinic acid
(3-methylpyridine)   (pyridine-3-carboxylic acid)

Pyrrole and pyridine behave very differently from one another in electrophilic aromatic substitution; pyrrole is much more reactive than benzene, whereas pyridine is much less reactive. The reason becomes clear if one considers the possible resonance contributors in each case:

resonance contributors to pyrrole   resonance contributors to pyridine

In pyrrole, the nitrogen atom feeds electrons into the ring, increasing its reactivity toward electrophiles. In pyridine, the nitrogen withdraws electrons from the ring, decreasing its reactivity toward electrophiles. Pyrrole undergoes electrophilic substitution readily, usually at the 2-position. Pyridine undergoes electrophilic substitution only under severe conditions, usually at the 3-position.

$$+ \ HONO_2 \xrightarrow[\substack{-10°}]{\text{acetic anhydride}} \qquad -NO_2 + H_2O \qquad (12\text{-}52)$$

2-nitropyrrole

$$\text{(pyridine)} + HONO_2 \xrightarrow[300°]{\text{sulfuric acid}} \text{(3-nitropyridine)} + H_2O \qquad (12\text{-}53)$$

3-nitropyridine

In contrast with its behavior toward electrophiles, pyridine is readily attacked by nucleophiles. For example with sodium amide in liquid ammonia, 2-aminopyridine is formed (eq 12-54). The first step is like nucleophilic addition to a carbon-oxygen double bond (sec 9.6b), but the reaction is completed by rearomatization with loss of hydride ion. This reacts with the ammonia to liberate hydrogen and re-form sodium amide.

$$\xrightarrow[\text{liquid } NH_3]{Na^+NH_2^-} \xrightarrow{NH_3} \text{(2-aminopyridine)} + H_2 + NaNH_2 \qquad (12\text{-}54)$$

2-aminopyridine

The pyridine ring occurs in many important natural products such as nicotine, vitamin B_6, and the coenzyme NAD (nicotinamide adenine dinucleotide), an important biological oxidizing agent (Chapter 18). The benzo derivatives **quinoline, isoquinoline,** and **acridine** have been isolated from coal tar.

quinoline   isoquinoline   acridine

Their ring systems occur in several alkaloids such as quinine and papaverine (sec 18.5a).

Heterocyclic amines with more than one nitrogen atom in a ring are well known; many such rings occur in nature. Some examples are the **imidazole, pyrimidine,** and **purine** rings. The imidazole ring

imidazole   pyrimidine   purine

occurs in the amino acid histidine (Table 17-1) and seems frequently to be involved in the active sites of enzymes. The pyrimidine ring is present in thiamin (vitamin B_1), and the purine ring forms part of certain alkaloid molecules such as caffeine (sec 18.5a). The pyrimidine and purine rings form the main framework of the bases found in nucleic acids (sec 18.6).

12.8   AROMATIC DIAZONIUM COMPOUNDS

In this section we consider a class of compounds which can be readily prepared from primary aromatic amines and are extremely useful and versatile in the synthesis of other aromatic compounds.

Primary aromatic amines react with cold nitrous acid to form **diazonium salts.** The process is known as **diazotization.** For example aniline reacts with nitrous acid and hydrochloric acid at $0°$ to yield a solution of benzenediazonium chloride.

$$\text{C}_6\text{H}_5\text{—NH}_2 + \text{HONO} + \text{H}^+\text{Cl}^- \xrightarrow[\substack{\text{aqueous} \\ \text{solution}}]{0-5°} \text{C}_6\text{H}_5\text{—N}_2{}^+\text{Cl}^- + 2\,\text{H}_2\text{O}$$

benzenediazonium
chloride

$$(12\text{-}55)$$

The mechanism by which the diazonium ion is formed is as described in section 12.5d, especially equation 12-34. But unlike alkyldiazonium ions which usually decompose spontaneously even at very low temperatures (eq 12-35), aromatic diazonium ions are moderately stable at low temperatures (say, $<10°$) and can be used in synthesis, by reaction with other reagents.

The main contributors to the resonance hybid in diazonium ions are:

C₆H₅—N⁺≡N: ⟷ C₆H₅—N=N⁺:

The positive charge is distributed over both nitrogens. The charge may also be delocalized into the aromatic ring, through structures such as:

—N⁺≡N: ⟷ —N=N⁻: ⟷ —N=N⁻:

These structures help explain why aromatic diazonium ions are more stable than alkyldiazonium ions (where such charge delocalization is not possible). They are also useful in explaining the reactions of diazonium compounds.

12.8a   NOMENCLATURE AND PROPERTIES

In naming diazonium compounds, the name of the aromatic compound from which they are derived is prefixed to the word *diazonium,* and this is followed by the name of the anion, as in the following examples:

$$CH_3-\underset{}{\bigcirc}-\overset{+}{N}\equiv N\colon Cl^- \qquad\qquad Cl-\underset{}{\bigcirc}-\overset{+}{N}\equiv N\colon OSO_3H^-$$

p-toluenediazonium chloride   p-chlorobenzenediazonium hydrogen sulfate

The salts are ionic, water-soluble, and stable in acid solution. The solid salts explode when heated or subjected to mechanical shock. Some anions, such as cyanide ion, form covalent compounds with diazonium ions. Benzenediazocyanide, for example, is not an electrolyte and is soluble in organic solvents.

$$\bigcirc-N_2{}^+Cl^- + Na^+CN^- \longrightarrow \bigcirc-\overset{\cdot\cdot}{N}=\overset{\cdot\cdot}{N}-CN + Na^+Cl^-$$

benzenediazocyanide

$$(12\text{-}56)$$

12.8b REPLACEMENT OF NITROGEN

The great stability of the nitrogen molecule contributes to the ease with which it can be displaced from diazonium compounds. If an aqueous solution of a diazonium ion is heated, nitrogen is evolved and a phenol is formed.

$$\bigcirc-N_2{}^+Cl^- + H-OH \xrightarrow{\text{heat}}$$

$$\bigcirc-OH + N_2\uparrow + H^+Cl^- \quad (12\text{-}57)$$

phenol

The reaction is analogous to the reaction of primary amines with nitrous acid (eq 12-37) and may proceed by an S_N1 mechanism.

$$\bigcirc-\overset{+}{N}\equiv N\colon \xrightarrow{-N_2} \bigcirc^+ \xrightarrow{H-OH} \bigcirc-OH + H^+ \quad (12\text{-}58)$$

The diazonium group can be replaced by any of the halogens. Heating with cuprous chloride or bromide produces aryl halides (the **Sandmeyer** reaction).

$$\underset{}{\bigcirc}\overset{CH_3}{-N_2{}^+Cl^-} \xrightarrow{Cu_2Cl_2} \underset{}{\bigcirc}\overset{CH_3}{-Cl} + N_2 \qquad (12\text{-}59)$$

o-toluenediazonium chloride   o-chlorotoluene

If an aqueous solution of a diazonium sulfate is heated with potassium iodide, the aryl iodide is formed. Fluorides are prepared by adding the diazonium salt solution to fluoboric acid. The tetra-fluoborate salt which precipitates is dried and heated to yield the aryl fluoride (**Schiemann** reaction).

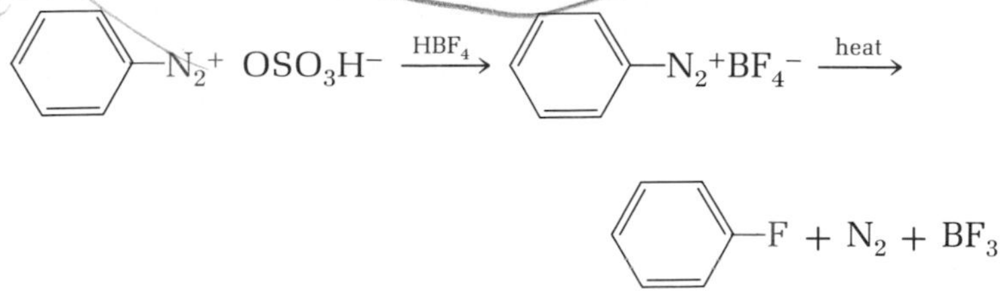

$$\text{Ph}-N_2^+\ OSO_3H^- \xrightarrow{HBF_4} \text{Ph}-N_2^+BF_4^- \xrightarrow{heat}$$

$$\text{Ph}-F + N_2 + BF_3 \quad (12\text{-}60)$$

fluorobenzene

If cuprous cyanide is used in place of a halide in the Sandmeyer reaction, the product is an aryl cyanide.

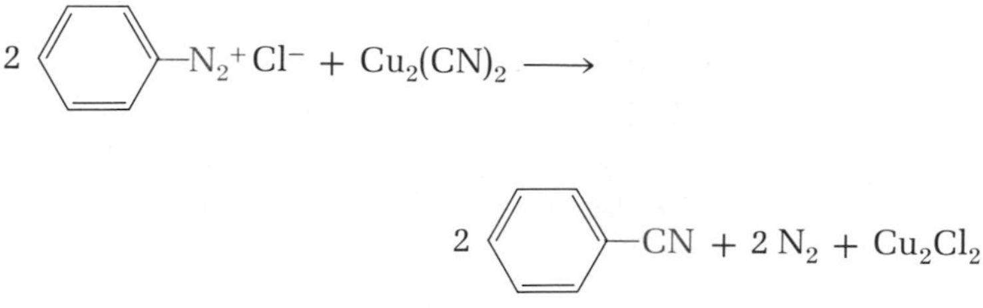

$$2\ \text{Ph}-N_2^+Cl^- + Cu_2(CN)_2 \longrightarrow$$

$$2\ \text{Ph}-CN + 2\,N_2 + Cu_2Cl_2 \quad (12\text{-}61)$$

benzonitrile

Since nitriles are readily hydrolyzed to acids (sec 10.4b), this reaction enables one to convert an aromatic primary amine to an aromatic acid.

Finally, certain reducing agents enable one to replace the diazonium group with a hydrogen. One of the more effective is hypophosphorous acid, H_3PO_2. The reaction is useful for obtaining aromatic compounds with substituents in orientations which are difficultly accessible by more direct routes. For example:

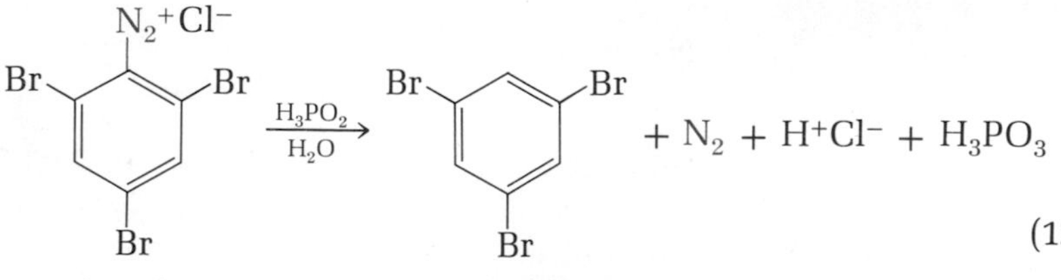

$$\xrightarrow[H_2O]{H_3PO_2} \quad + N_2 + H^+Cl^- + H_3PO_3$$

$$(12\text{-}62)$$

2,4,6-tribromobenzene-diazonium chloride   1,3,5-tribromobenzene

Diazotization of the readily available 2,4,6-tribromoaniline (eq 12-39) followed by removal of the amino group (eq 12-62) gives a product with *meta*-substituted bromines, whereas direct bromination of benzene would give *ortho*- and *para*-oriented bromines (sec 5.4b).

In summary, the amino group of primary aromatic amines can be replaced, via diazonium compounds, with hydroxyl, cyano, halogen, carboxyl, or hydrogen.

12.8c REDUCTION, RETAINING NITROGEN

Certain reducing agents (for example, stannous chloride in hydrochloric acid) convert aryldiazonium ions to arylhydrazines.

$$\text{C}_6\text{H}_5{-}\overset{+}{\text{N}}{\equiv}\text{N}\colon \text{Cl}^- \xrightarrow{4\,[\text{H}]} \text{C}_6\text{H}_5{-}\overset{\cdot\cdot}{\text{N}}\text{H}{-}\overset{\cdot\cdot}{\text{N}}\text{H}_2 + \text{H}^+\text{Cl}^- \qquad (12\text{-}63)$$

phenylhydrazine

Phenylhydrazine, discovered by Emil Fischer in 1875, is a useful reagent in elucidating the chemistry of carbonyl compounds (sec 9.6d) and sugars (sec 16.5b).

12.8d COUPLING REACTIONS

The diazonium ion is a weak electrophile and can attack *strongly activated* aromatic rings. If phenols or aromatic amines are added to *alkaline* or *neutral* solutions of diazonium salts, a coupling reaction occurs, the nitrogen being retained. The overall equation for coupling with phenol is

$$\text{C}_6\text{H}_5{-}\overset{+}{\text{N}}{\equiv}\text{N}\colon \text{Cl}^- + \text{H}{-}\text{C}_6\text{H}_4{-}\text{OH} \xrightarrow{\text{alkali}}$$

benzenediazonium
chloride   phenol

$$\text{C}_6\text{H}_5{-}\text{N}{=}\text{N}{-}\text{C}_6\text{H}_4{-}\text{OH} + \text{H}^+\text{Cl}^- \qquad (12\text{-}64)$$

p-hydroxyazobenzene

The products are **azo compounds;** they contain the —N=N— or azo group. Azo compounds are colored; many of them can be used for dyestuffs, particularly if they are provided with other functional groups which make it possible to fix them to fabrics. **Congo red,** for example, is an azo dye which can be prepared by coupling diazotized benzidine (eq 12-49) with 1-aminonaphthalene-4-sulfonic acid. It is used as a direct dye for cotton.

Congo red

The mechanism of diazo coupling is essentially the same as that of other electrophilic aromatic substitutions. Equation 12-64, for example, may be subdivided into the following steps, beginning with the resonance contributor in which the positive charge is on the nitrogen remote from the aryl ring, and recalling that the phenol may be converted to phenoxide ion in alkaline solution:

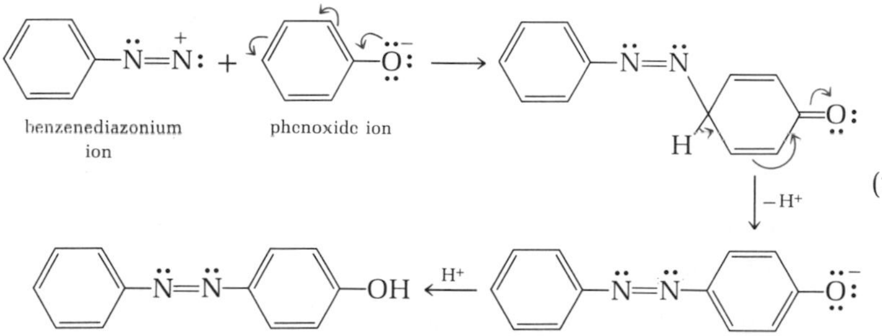

benzenediazonium ion

phenoxide ion

(12-65)

$-H^+$

Since the aromatic compound which undergoes electrophilic attack must be strongly activated, it usually contains an *o,p*-directing group (often —OH or —NH$_2$). Coupling usually occurs *para* to the —OH or —NH$_2$ unless that position is blocked by another substituent, in which case *ortho* coupling takes place.

12.8e   ARYNES

Certain groups *ortho* to a diazonium group may be lost coincidentally with nitrogen loss. For example, the diazonium salt of *o*-aminobenzoic acid, when heated, gives (amongst other products) biphenylene.

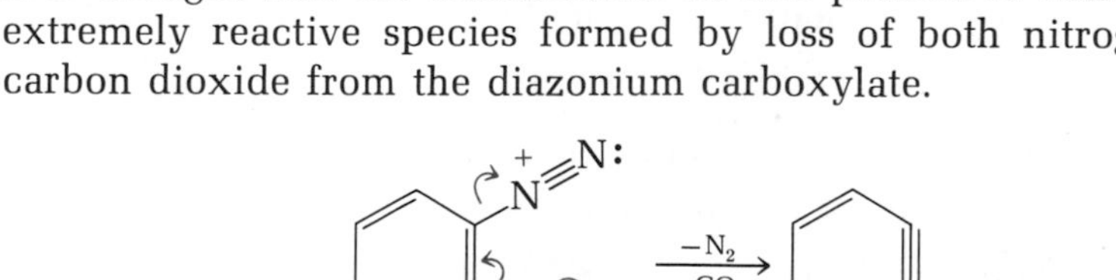

anthranilic acid

a diazonium carboxylate

biphenylene

(12-66)

It is thought that the intermediate in this process is **benzyne,** an extremely reactive species formed by loss of both nitrogen and carbon dioxide from the diazonium carboxylate.

$-N_2$
$-CO_2$

benzyne

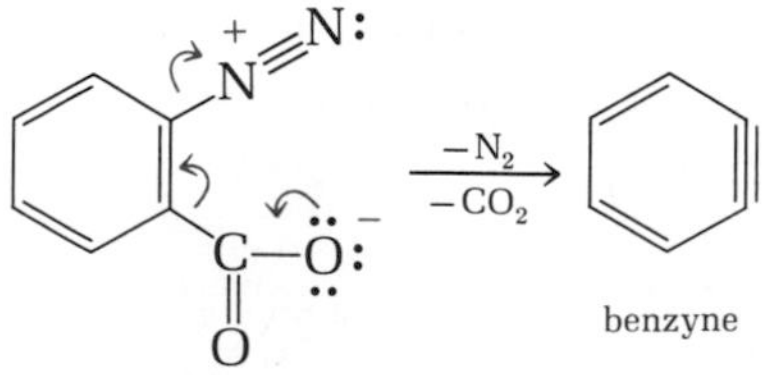

(12-67)

A possible molecular orbital structure for benzyne is shown in Figure 12-6. In addition to the usual benzene orbitals which mainly occupy the space above and below the ring plane, a new π bond

Figure 12-6

Bonding in benzyne. The additional π bond lies in the plane of the carbon atoms.

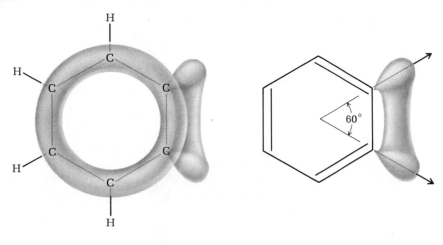

is formed between the two carbon atoms which do not have hydrogen atoms attached to them. The bond is weak because the two sp^2 orbitals which must overlap to form it are tilted away from one another by 60°.

Benzynes are extremely reactive. In the absence of other reagents they may dimerize to biphenylenes. If generated in the presence of a diene they readily undergo Diels-Alder addition (sec 4.9).

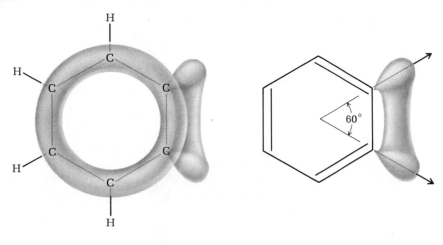

(12-68)

cyclopentadiene

benzonorbornene

Other ways of generating arynes have been devised, and they now contribute a potent weapon in the arsenal of the synthetic organic chemist.

New Concepts, Facts, and Terms

1. Three classes of amines: primary, secondary, and tertiary
2. Nomenclature (sec 12.1)
3. Amines form hydrogen bonds: N—H$\cdots$N and O—H$\cdots$N.

4. Base strengths (K_b) are approximately 10^{-5} for aliphatic amines, 10^{-10} for aromatic amines.
5. Preparation of amines:
   a. S_N2 displacements on alkyl halides by ammonia or amines
   b. reduction of nitriles, oximes and nitro compounds
   c. Hofmann rearrangement of amides (with Br_2 and base) gives primary amines with one less carbon atom.
6. Reactions of amines:
   a. Salt formation, by reaction, with acids.
   b. Alkylation of tertiary amines gives quaternary ammonium salts; choline.
   c. Acylation of primary or secondary amines gives amides; reduction of amides gives a route to secondary and tertiary amines.
   d. Sulfonyl halides, with primary or secondary amines, give sulfonamides; Hinsberg test.
   e. Nitrous acid distinguishes between the three classes of amines.
   f. Amino and amide groups are *o,p*-directing in electrophilic aromatic substitution.
7. Diamines; nylons are polyamides; benzidine rearrangement.
8. Heterocyclic amines; pyrroles (5-membered ring) are aromatic, exceedingly weak bases, reactive in electrophilic substitution; pyridines (6-membered ring) are aromatic, fairly basic, and unreactive in electrophilic substitution. Both ring systems are prominent in nature. Quinoline, isoquinoline, imidazole, pyrimidine, purine
9. Aromatic diazonium compounds:
   a. Prepared from primary aromatic amines and nitrous acid.
   b. Structure is $ArN_2^+X^-$.
   c. The nitrogen is readily replaced by OH, halogens, CN, H.
   d. Reduction gives arylhydrazines.
   e. Electrophilic substitution (coupling) with phenols and amines gives azo compounds, useful as dyes.
   f. Arynes; benzyne—a reactive intermediate which reacts mainly by cycloaddition.

Exercises and Problems

1. Give an example of each of the following:
   a. a primary amine
   b. a cyclic secondary amine
   c. a tertiary aromatic amine
   d. a quaternary ammonium salt
   e. an aryldiazonium salt
   f. a heterocyclic amine
   g. an azo compound
   h. a nitrosoamine

2. Write structural formulas for each of the following compounds:
   a. *m*-chloroaniline
   b. *sec*-butylamine
   c. 2-aminohexane
   d. dimethyl-*n*-propylamine
   e. benzylamine
   f. 1,2-diaminopropane
   g. *N,N*-dimethylaminocyclohexane
   h. tetraethylammonium bromide
   i. triphenylamine
   j. *o*-phenylenediamine

3. Write a correct name for each of the following compounds:

a. Br—⟨ ⟩—NH$_2$

b. (CH$_3$)$_2$CHNHCH$_3$

c. (CH$_3$CH$_2$)$_2$NCH$_3$

d. (CH$_3$)$_4$N$^+$Cl$^-$

e. CH$_3$CH(OH)CH$_2$CH$_2$NH$_2$

f. CH$_3$NHCH$_2$CH$_2$OH

g. C$_6$H$_5$N$_2$$^+OSO_3H^-$

h. (naphthalene with —NH$_2$)

i. (cyclopentane with —NH$_2$)

j. H$_2$NCH$_2$CH$_2$CH$_2$NH$_2$

4. Draw the structures for, name, and classify as primary, secondary, or tertiary
   a. the eight isomeric amines, C$_4$H$_{11}$N.
   b. the five amines, C$_7$H$_9$N, each of which contains a benzene ring.

5. Explain why
   a. cyclohexylamine is a much stronger base than aniline.
   b. p-nitroaniline is a much weaker base than aniline.

6. Suggest an explanation for the fact that dimethylamine has a higher boiling point (Table 12-1) than trimethylamine, though the latter compound has the higher formula weight.

7. Write out the mechanism for the formation of triethylamine from ammonia and ethyl chloride (sec 12.4a), showing clearly how the necessary nucleophile is obtained for each successive alkylation step. How could monoalkylation (i.e., formation of ethylamine) be favored?

8. Starting with benzene, toluene, or any alcohol with four carbon atoms or fewer, and any essential inorganic reagents, outline steps for the synthesis of
   a. n-butylamine
   b. p-toluidine
   c. 1-aminopentane
   d. N-ethylaniline
   e. m-aminobenzoic acid
   f. tri-n-butylamine
   g. 1,4-diaminobutane
   h. ethyl n-propylamine

9. Show how each of the following conversions could be accomplished:
   a. n-butyramide to n-propylamine
   b. n-butyraldehyde to 1-aminobutane
   c. benzyl bromide to 2-phenylethylamine
   d. nitrobenzene to m-chloroaniline
   e. 3-pentanone to 3-aminopentane

10. Write out each step in the mechanism for the hydrolysis of an alkyl isocyanate to an amine (eq 12-16). (HINT: Begin with nucleophilic attack by water on the carbonyl carbon.)

11. Write equations for the reaction of aniline with each of the following reagents:
    a. hydrochloric acid
    b. acetyl chloride
    c. acetic anhydride
    d. ethylene oxide
    e. benzenesulfonyl chloride and base
    f. sodium nitrite and aqueous sulfuric acid, 0°
    g. excess methyl iodide
    h. aqueous bromine

12. Write out in detail the steps in the mechanism for the reaction of ethylamine with acetic anhydride (eq 12-24). Suggest an explanation for the fact that even if excess acetic anhydride is used, no diacylation occurs (that is, no $(CH_3CO)_2NCH_2CH_3$ is formed).

13. Devise a synthesis of
    a. choline from trimethylamine
    b. *N,N*-diethyl-*m*-toluamide from *m*-toluic acid

14. Isopropylamine, methylethylamine, and trimethylamine are isomers. Show by means of equations how they can be distinguished by their reactions with (a) nitrous acid and (b) benzenesulfonyl chloride.

15. a. Write equations which explain clearly how an alkyldiazonium ion can be obtained from the product of equation 12-34.
    b. Write a possible mechanism for the formation of a nitrosoamine from a secondary amine (eq 12-38). (HINT: Use eq 12-34 as a guide.)

16. Write equations which describe the reaction of sulfanilic acid (dipolar structure, sec 12.5e) with acid; with base. What term is used to describe this behavior?

17. a. Show how both raw materials for Nylon 6-6 might be synthesized from 1,4-dichlorobutane.
    b. Starting with phenol as the only organic material and any necessary inorganic reagents, write equations showing how nylon might be prepared.

18. Write structural formulas for each of the following:
    a. 2-methylpyrrolidine   d. 5-hydroxyindole
    b. 4-chloropiperidine   e. quinoline
    c. 2,4,6-trimethylpyridine   f. 2-aminoimidazole

19. Pyrrole reacts strongly with electrophiles, especially in alkaline solution. For example, with iodine and base, 2,3,4,5-tetraiodopyrrole is produced directly (much like the reaction of aniline or phenol with bromine water). Using the concept of resonance, suggest an explanation for these observations.

20. Write equations for the reaction of *p*-toluenediazonium hydrogen sulfate with each of the following:

a. cuprous cyanide	e. cuprous chloride
b. water, heat	f. fluoboric acid, heat
c. N,N-dimethylaniline, base	g. o-cresol, alkali
d. potassium iodide	h. hypophosphorous acid in D_2O

21. The dye Congo red (sec 12.8d) can be synthesized from benzidine and 1-aminonaphthalene-4-sulfonic acid. Write a sequence of equations for this synthesis.

22. Suggest a practical, simple method for separating a mixture of aniline (bp 184°), n-butylbenzene (bp 183°), and valeric acid (bp 189°) into its pure components. All three compounds are liquids at room temperature and are not very soluble in water.

23. Describe a simple chemical test which would distinguish between the following pairs of compounds. Tell exactly what you would do and see.
    a. N-methylaniline and p-toluidine
    b. aniline and acetanilide
    c. methyl-n-propylamine and dimethylethylamine
    d. n-propylamine and methylethylamine

24. Write equations to describe how each of the following conversions might be accomplished:
    a. aniline to p-chloroaniline
    b. o-toluidine to o-toluic acid
    c. o-toluidine to o-bromobenzoic acid
    d. 1-butene to sec-butylamine
    e. benzene to fluorobenzene
    f. isobutyric acid to isopropylamine

25. A neutral substance **A,** $C_{14}H_{12}NOCl$, after reflux with hydrochloric acid, gave two products, **B** $(C_7H_5O_2Cl)$ and **C** $(C_7H_{10}NCl)$. Compound **B,** on treatment with PCl_3 followed by ammonia, gave **D,** C_7H_6NOCl which, with sodium hypobromite, gave **E,** C_6H_6NCl. Diazotization of **E,** followed by warming, gave p-chlorophenol.

    Compound **C** was soluble in water; when this solution was made alkaline, compound **F** (C_7H_9N) was isolated. Compound **F** gave a yellow oil with nitrous acid, and with benzenesulfonyl chloride gave an alkali-insoluble benzenesulfonamide.

    Deduce structures for **A–F,** and write equations for all reactions mentioned.

CHAPTER THIRTEEN

SPECTROSCOPY AND ORGANIC STRUCTURE

One of the most useful developments of recent years in organic chemistry has been the application of spectroscopy to the solution of structural problems. Infrared, ultraviolet, nuclear magnetic resonance, and mass spectra can be obtained routinely with instruments which often require little more than button-pushing. These spectra enable one to deduce the structural formula of a compound in a fraction of the time required by classical methods. They also permit one to solve more subtle structural problems which cannot be investigated in any other way.

In previous chapters the most important functional groups found in organic molecules were discussed. Before proceeding to compounds which contain two or more different groups, we take a diversion here to describe how various spectroscopic methods are used to determine organic structures. It is possible to use these techniques to determine in just a few minutes whether a substance has, for example, a hydroxyl or a carbonyl group; if it has a carbonyl group, whether it is an aldehyde, ketone or ester; if it is a cyclic ketone, whether the ring is 4-, 5- or 6-membered; whether the carbon skeleton contains an ethyl, or an isopropyl, or a *t*-butyl group, etc. Furthermore, most spectroscopic methods require only small amounts of material (a few milligrams are usually adequate) and are nondestructive; that is, the sample can be recovered. Spectroscopic methods now provide the most powerful tool the organic chemist has for determining structures.

13.1 GENERAL PRINCIPLES

13.1a ENERGY, RADIATION, AND ABSORPTION SPECTRA

Electromagnetic radiation (for example, cosmic rays, ultraviolet or visible light, infrared radiation, radar, radio waves, etc.) can be described in terms of a wave motion. The waves have associated with them a certain frequency (that is, so many waves per unit of distance) and a wavelength. Wavelength and frequency are inversely proportional; the longer a wave, the fewer waves one can have in a given distance. This inverse relationship is expressed in the equation

$$\nu = \frac{c}{\lambda} \tag{13-1}$$

where ν is the frequency of the radiation and λ is its wavelength. The proportionality constant is c, the velocity of light.

The energy associated with any electromagnetic radiation is directly proportional to its frequency (and consequently, inversely proportional to its wavelength). This relationship is expressed mathematically in the equation

$$E = h\nu = \frac{hc}{\lambda} \tag{13-2}$$

where E is energy. The proportionality constant h, known as *Planck's constant* is, like the velocity of light, a universal constant.

Absorption spectroscopy, which is most useful to organic chemists, is performed in the following way. The substance being studied is placed between a radiation (i.e., energy) source and a detector. The spectrometer measures the difference between the amount of energy which impinges on the sample (E_1) and the amount which is transmitted by it (E_2) (Figure 13-1). Often, the per cent of energy transmitted by the sample [that is, $(E_1 - E_2)/E_1 \times 100$] is plotted—usually automatically, by the instrument—as the energy (E_1) is varied over some range of the electromagnetic spectrum (Figure 13-2). The curve obtained is called an *absorption spectrum*. The *position* of an absorption band along the energy axis is a function of the molecular structure of the substance; the *intensity* of the band (i.e., per cent transmittance) is a function not only of the structure, but also of the number of molecules in the sample (length of sample tube and density or concentration of the sample). The intensity is usually expressed in terms of the *molar absorption coefficient* ε, given by the equation

$$\log (I_0/I) = \varepsilon c l \tag{13-3}$$

where I_0 is the intensity of the incident radiation, I is the intensity of the radiation transmitted by the sample, c is the concentration

Figure 13-1
Schematic drawing of an absorption spectrometer.

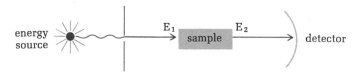

Figure 13-2
Schematic drawing of an absorption spectrum.

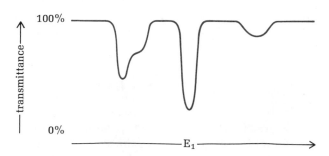

Figure 13-3
Energy levels E_0, E_1, E_2, and
possible transitions between
them.

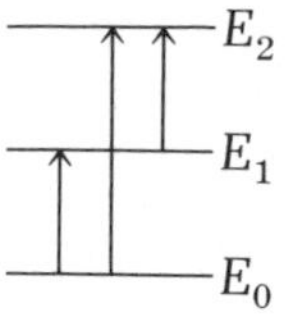

of the absorbing species in moles/liter and l is the length of the sample, usually in centimeters.

13.1b   THE QUANTUM THEORY AND ENERGY LEVELS

One fundamental tenet of the quantum theory is that an atom or molecule can only have discrete quanta of energy. That is, a molecule may have an energy E_1, or E_2, or E_3, but not a continuum of energy values between them (Figure 13-3). Thus, using equation 13-2, we see that each possible energy corresponds to a particular frequency.

$$E_0 = h\nu_0, \; E_1 = h\nu_1, \; E_2 = h\nu_2, \text{ etc.}$$

If a molecule with energy E_0 is subjected to electromagnetic radiation, it can only absorb energy from that radiation in quanta sufficient to raise its energy to E_1 or to E_2, etc. Therefore the frequencies of radiation which the molecule could absorb are limited to discrete values, given by the equations

$$\frac{E_1 - E_0}{h} = \nu_1 - \nu_0 \quad \text{or} \quad \frac{E_2 - E_0}{h} = \nu_2 - \nu_0, \text{ etc.} \quad (13\text{-}4)$$

That is, the molecule could absorb energy of a frequency corresponding to the difference between frequencies ν_1 and ν_0 or between ν_2 and ν_0, etc., but could not absorb energies corresponding to other frequency differences. These molecules would have an absorption spectrum (Figure 13-4) which would consist of a series of very sharp lines corresponding to absorption of energy for each possible transition. In practice, absorption spectra usually consist of broad bands (Figure 13-2) rather than sharp lines. This is because several transitions with rather close energy differences are usually possible, and most instruments are not capable of resolving these into separate lines; instead, several absorption lines merge to form an absorption "band."

Figure 13-4

Absorption spectrum of a molecule at energy E_0, with transitions to E_1, E_2, and E_3.

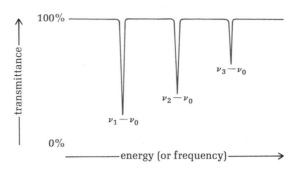

13.1c TYPES OF ENERGY TRANSITIONS

Most molecules have many possible energy levels. Transitions between them require greater or lesser amounts of energy, depending on the type of transition. We consider here only those transitions which result in spectra that give the organic chemist useful information regarding a molecule's structure.

1. Electronic Transitions. An electron in an atom or molecule may, on being irradiated, acquire sufficient energy to "jump" from the orbital it normally occupies to a higher energy, normally vacant orbital. These electronic transitions, as they are called, usually require the absorption of energy in the range of 30–150 kilocalories per mole of substance. If translated into wavelength using equation 13-2, this energy corresponds to light in the region of 8000 Å to 2000 Å, encompassing the visible (8000 Å to 4000 Å) and near ultraviolet (4000 Å to 2000 Å) regions of the spectrum. Consequently, visible and ultraviolet absorption spectroscopy give us information about electronic transitions in a molecule.

2. Vibrational and Rotational Transitions. Consider first a simple, diatomic molecule such as hydrogen chloride. The two atomic nuclei are not at a constant, fixed distance from one another, but rather they vibrate with a certain frequency along the internuclear axis. If the molecule absorbs energy, it may vibrate with a higher frequency which would, on the average, increase the mean internuclear distance slightly. Once again, the frequencies (i.e., energies) are quantized, and transitions between them are possible. Larger molecules also have vibrational energy levels, although each level may not necessarily be associated with a particular bond.* The energy required for vibrational transitions, usually in the range of 1–10 kilocalories per mole, is less than that for electronic transitions. This energy corresponds to the infrared region of the spectrum [wavelengths of 20,000 Å to 150,000 Å, or 2–15 microns (one micron, μ, = 10,000 Å = .0001 cm)].

In addition to vibrations, rotational motions are also possible, and are quantized. Transitions between these levels require even less energy than vibrational transitions, usually from only one kilocalorie to perhaps one calorie per mole. They occur when a molecule is irradiated in the far infrared or microwave region of the spectrum (wavelengths of about .005 cm to 10 cm).

The relationship between electronic (e), vibrational (v), and rotational (r) energy levels is illustrated in Figure 13-5. If a molecule is irradiated with light of sufficient energy to cause an electronic transition (say, from e_0 to e_1), clearly enough energy will also be

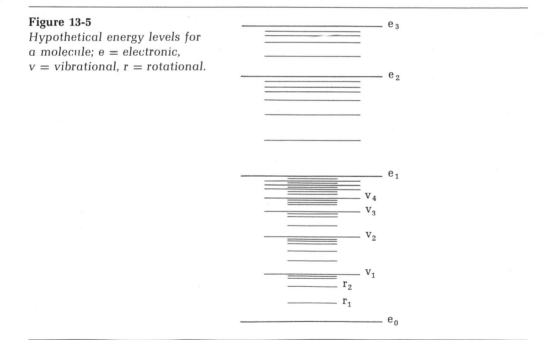

Figure 13-5
Hypothetical energy levels for a molecule; e = electronic, v = vibrational, r = rotational.

* Imagine, for example, a suspended network of six balls (or atoms) connected in some way by springs. If one of the balls is plucked, not only will it and the balls directly attached to it be set in motion, but all will vibrate. The extent to which this happens will depend on the mass of the balls and the tightness of the springs.

present to cause many vibrational and rotational transitions. Consequently, visible and ultraviolet spectra will have very broad absorption bands. But if, as in infrared spectroscopy, the energy available is sufficient only to allow vibrational transitions within one electronic state (say e_0), the bands will be sharper. And in rotational spectra, the absorption lines will become very sharp, since only one type of transition is possible.

3. *Nuclear Spin Transitions.* Certain atomic nuclei have a spin associated with them which, since the nuclei are charged, gives rise to a magnetic moment. The nuclei can be thought of as being like tiny bar magnets. When placed in a homogeneous magnetic field, the magnetic moments tend to align *with* the field; alignment *against* the field represents a higher energy state, and transitions between these states are possible if energy is supplied to the nuclei. Once again, the energy is quantized, and only discrete quantities can be absorbed. These amounts are very small and will of course depend on the strength of the homogeneous magnetic field. For hydrogen nuclei, in instruments which are now in most common use, the energy is only 0.0057 calories per mole, which corresponds to frequencies in the radio range (60 megacycles or 6×10^7 sec^{-1}) or waves which are about 5 meters long. For a given kind of nucleus, the precise amount of energy depends not only on the strength of the homogeneous magnetic field, but also upon the local magnetic fields caused by nearby electrons and nuclei. Nuclear magnetic resonance (nmr) spectroscopy, which involves the study of such nuclear spin transitions, can give, for each nucleus with a magnetic moment in a molecule, information about the number and kinds of other nuclei which surround it. This is enormously helpful in determining organic structures, as we shall see.

13.2 ULTRAVIOLET AND VISIBLE SPECTROSCOPY

Ultraviolet and visible spectra arise from electronic transitions (sec 13.1c). The less firmly an electron is held by a molecular framework the easier it is to "excite" it (or cause a transition) to a higher energy level. Consequently, π-electrons are more easily excited than σ-electrons, and one generally associates uv and visible spectra with unsaturated molecules. Saturated hydrocarbons, alcohols, and ethers absorb at much shorter wavelengths (i.e., higher energies) than unsaturated compounds. They are transparent in the usual ultraviolet and visible regions of the spectrum (200–800 nm*) and consequently are useful as solvents for spectral determinations throughout this range.

* 10 Å = 1 nm (nanometer). The unit of wavelength currently used in uv and visible spectroscopy is the nanometer (1 nm = 10^{-9} meters = 10^{-7} centimeters = 1 millimicron).

Table 13-1

Absorption Maxima of Some Unsaturated Hydrocarbons

COMPOUND	$\lambda_{max}(nm)$	ε
Ethylene	171	15,500
1-Butene	185	20,000
1,3-Butadiene	217	20,900
1,3-Pentadiene	224	25,500
1,4-Pentadiene	187	40,000
1,3,5-Hexatriene	258	35,000
1,3,5,7-Octatetraene	287	52,000
trans-β-Carotene (sec 4.10)	478	122,000

Simple alkenes absorb at very short wavelengths, with maxima below 200 nm (Table 13-1). However, when two or more double bonds are conjugated, the π-electrons are delocalized over a larger number of atoms, therefore they are more easily excited. Consequently, the absorption maximum, λ_{max}, shifts toward longer wavelengths (compare conjugated butadiene, hexatriene, and octatetraene, Table 13-1). Conjugation is essential for this shift, as is strikingly seen by comparing the spectra of 1,3- and 1,4-pentadiene. If a sufficient number of double bonds are conjugated, the maximum may be shifted into the visible region in which case the compound acquires a visible color. Thus β-carotene, with 11 conjugated double bonds, is yellow (λ_{max} = 478 nm).

Another example of a similar effect is seen in the spectra of the linear fused aromatic hydrocarbons. Whereas benzene, naphthalene, and anthracene are colorless, naphthacene is orange and pentacene is purple.

benzene
λ_{max} 260 nm

naphthalene
280 nm

anthracene
375 nm

naphthacene
450 nm

pentacene
575 nm

Conjugation effects are not limited to hydrocarbons. α,β-Unsaturated carbonyl compounds (aldehydes, ketones, esters, etc.) absorb at longer wavelengths than the corresponding saturated analogs. The exact position of the absorption maximum depends on the structure; alkyl groups attached to the double bond shift the maximum to longer wavelengths, as the following examples illustrate:

190 nm
saturated

213 nm
unsaturated

224 nm

237 nm

220 nm
1-methyl added

236 nm
2-methyls added

13.2a DYES AND INDICATORS

Organic dyes usually have rather extended conjugated systems, since they must absorb visible light. The visual color seen by an observer is the complement to the color absorbed by the dye; that is, it is the sensation produced by all wavelengths of light to which the eye is sensitive *minus* those wavelengths absorbed by the dye.

Dyes must retain their color on prolonged exposure to light and should also resist removal from the substance they are used to color (i.e., should remain "fast"). Many dyes, therefore, contain not only a **chromophore**—a group or arrangement of atoms which is responsible for the color—but also a group which permits the dye to be bound to the fabric it is to dye.

A dye may be applied to a textile fiber in many ways. *Direct* or *substantive* dyes are applied from a hot aqueous dye solution. Simple acidic or basic dyes, for example, are substantive to wool and silk since their fibers are proteins which contain free basic or acidic groups that can form salts with the dye molecules. Congo red (sec 12.8d) is a direct dye for cotton. Certain dyes may be adsorbed by a fabric in a colorless form, then acquire the desired color by oxidation when the fabric is dried in air. These are known as *vat* dyes. In *mordant* dyeing, the cloth is first impregnated with a mordant—a substance that can be fixed to the fiber and later dyed. Frequently the mordant is a heavy metal salt (Cu, Co, Cr) which can form a complex with the dye. *Ingrain* dyeing involves formation of the dye by chemical reaction directly on the fiber. For example, a fabric may be impregnated with an azo dye by carrying out the diazo coupling reaction (sec 12.8d) on the fiber. In this process, the cloth is first passed through an alkaline solution of the phenol or amine, then through a solution of the diazonium salt. *Fluorescent brighteners* are colorless substances which have an affinity for fibers and have a blue fluorescence when exposed to ultraviolet light. Finally, *food and cosmetic* dyes must, in addition to the usual properties, be non-toxic.

Literally thousands of dyes are known, varying widely in chemical constitution as well as color. Only a few examples may be cited here. *Azo dyes* form the largest single class; they account for one-third of all dyes produced. They are made by coupling a diazotized aromatic amine with a phenol or amine (sec 12.8d). Examples are

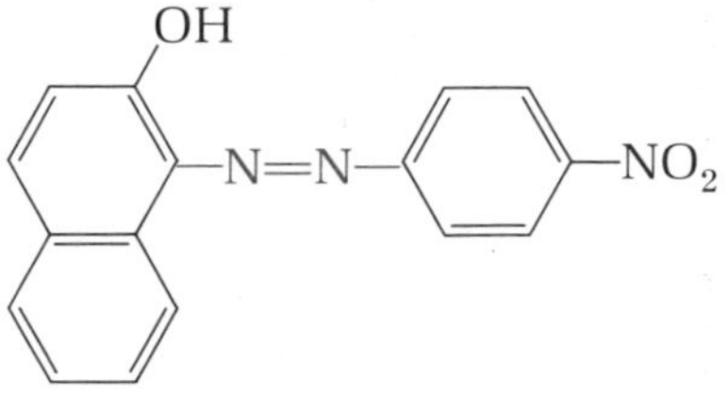

para red
(an ingrain dye)

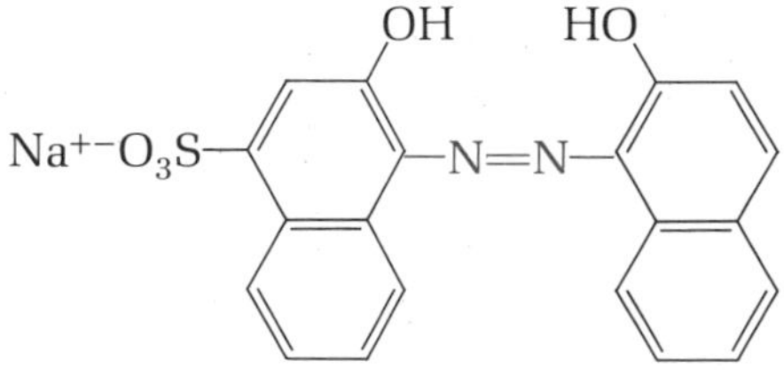

mordant black 11
(a mordant dye)

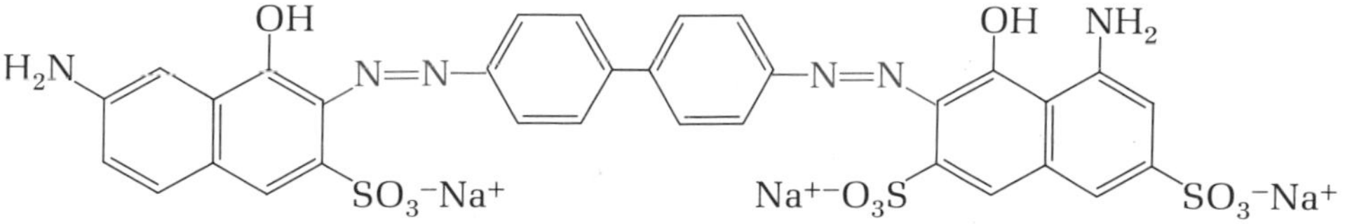

direct blue 2
(a direct dye)

Note the very extended conjugated system, including the nitrogen-nitrogen double bond, in all of these structures.

Probably the oldest known organic dye is the vat dye **indigo,** which was used over four thousand years ago to dye Egyptian mummy cloth. Though originally obtained from plants, the dye is now produced synthetically. Here the fabric is first dipped in the yellow,

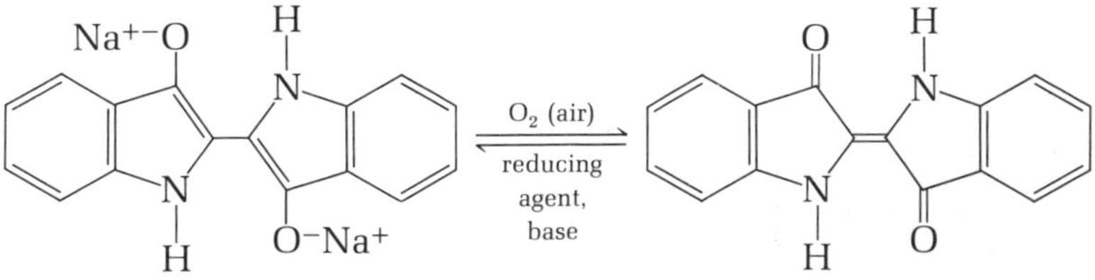

$$(13\text{-}5)$$

indigo white
(water-soluble)

indigo
(water-insoluble)

water-soluble, reduced form of the dye (indigo white), then exposed to air whereupon oxidation to the insoluble blue dye occurs. Many variants of indigo (for example, with halogens or other substituents in the aromatic rings) are in use.

Many dyes have, as their basic structural unit, three aromatic groups attached to a single carbon atom (*triphenylmethane dyes*). **Malachite green,** for example, is an intense bluish green dye made from dimethylaniline and benzaldehyde.

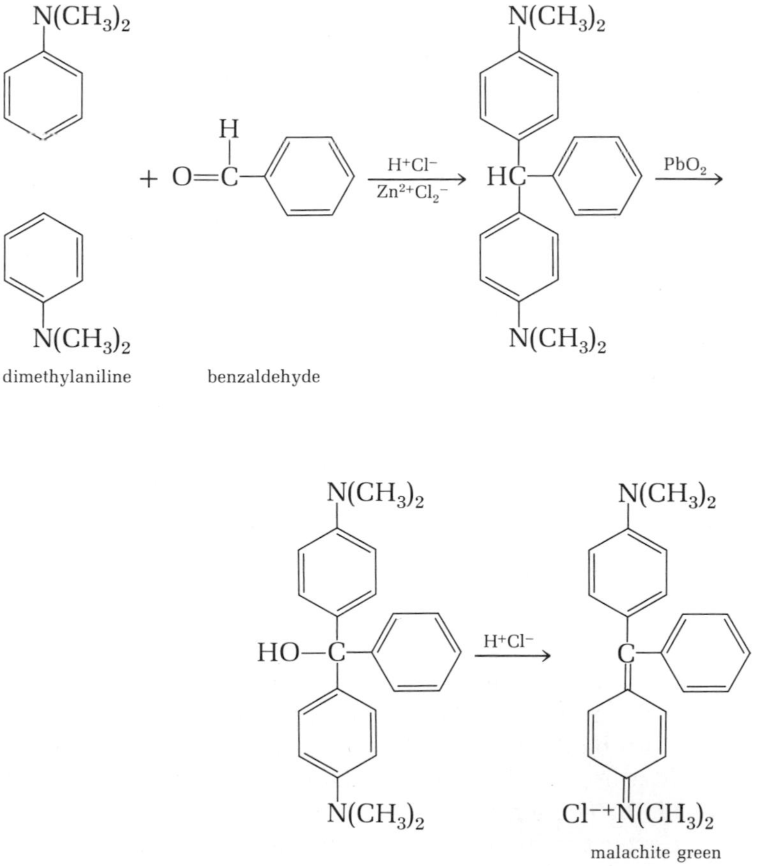

dimethylaniline   benzaldehyde

(13-6)

malachite green

Oxidation of the triarylmethane produces the colorless tertiary alcohol which is converted to the dye by acid. Malachite green is used to dye acrylic fibers and leather. Considerable color variation can be obtained by changing the substituents in the aromatic rings.

If a dye contains acidic or basic groups, its color can usually be altered by changing the acidity (pH) of the medium. If the color change is sharp, the dye can be used as an acidity *indicator*. One of the best known examples is **phenolphthalein,** a dye of the triphenylmethane type. It can be synthesized from phenol and phthalic anhydride.

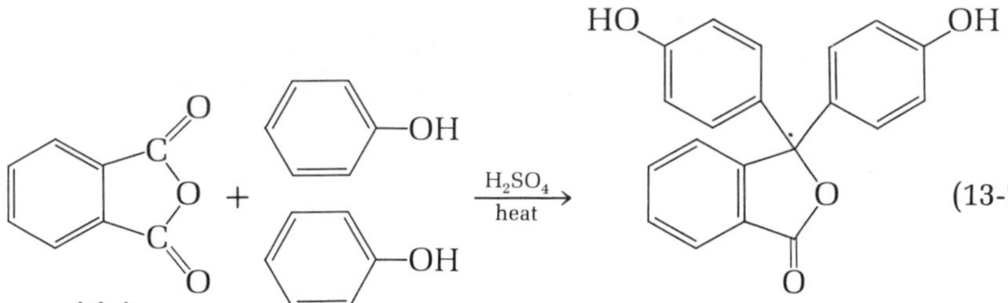

$$\text{phthalic anhydride} + \text{phenol} \xrightarrow[\text{heat}]{\text{H}_2\text{SO}_4} \text{phenolphthalein} \tag{13-7}$$

Although phenolphthalein itself is colorless (pH <8.3), its solutions become red when made alkaline (pH >10) because of the extended conjugation in the dianion that is produced when the phenolic protons are lost and the cyclic ester (lactone) ring opens.

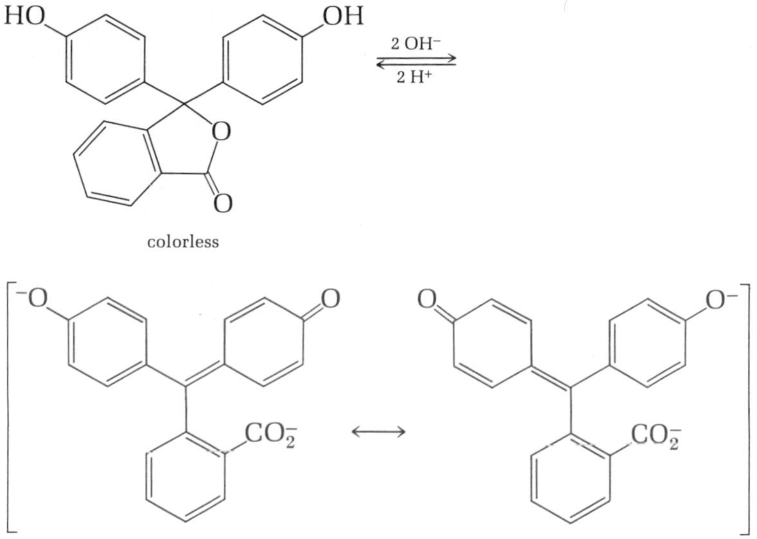

colorless

$$\xrightleftharpoons[\text{2 H}^+]{\text{2 OH}^-}$$

red $\tag{13-8}$

13.2b PHOTOCHEMISTRY

When a molecule absorbs sufficient energy to undergo an electronic transition, a new chemical species is produced. This new species, an "excited state" molecule, will have quite different chemical properties from those of the "ground state" molecule. Normally, the light source in visible and ultraviolet spectrometers is of such low intensity that the number of "excited" molecules produced during the measurement of a spectrum is very small. However, high intensity ultraviolet (and visible) light sources are available which can "excite" a sufficient number of molecules to bring about useful chemical reactions.

The organic chemist can use such reactions to synthesize compounds which may be difficult to prepare by other methods. For example, irradiation of alkenes with UV light leads to cyclobutanes.

$$\text{(13-9)}$$

"Ground state" alkenes do not undergo such reactions. Instead, the reaction involves production of an "excited state" alkene molecule which then reacts with a "ground state" alkene molecule to form a cyclobutane. Similarly, carbonyl compounds may react with alkenes to form cyclic ethers.

$$\text{(13-10)}$$

It can be shown that this reaction involves "excited" carbonyl molecules and "ground state" alkenes, since the reaction occurs well with light that has sufficient energy to excite the carbonyl compounds, but not the alkenes. The reaction in equation 13-10 will not occur in the absence of UV light. Photochemistry, the study of chemical reactions initiated by light, is an important branch of chemistry. Various biologically important processes, such as photosynthesis, phototropy, vision, and photomutation, involve photochemical reactions.

13.3 INFRARED SPECTROSCOPY

The position of an infrared band is usually given in reciprocal centimeters (cm^{-1}; that is, the number of waves per centimeter, sometimes called the wave *number*) and is designated by the Greek letter ν (nu). Alternatively, wave*length* may be used, in which case the common unit is *microns,* μ. The inverse of the wavelength in microns, when multiplied by 10,000, gives the wave number in cm^{-1}.

One reason infrared spectroscopy has been so useful in determining organic structures is that certain functional groups vibrate within certain frequency ranges, regardless of the gross structure of the molecule in which they are present. Table 13-2 lists some of the more common **group frequencies.** These values were deduced experimentally by determining the infrared spectra of a large number of compounds whose structures were already known. All carbonyl compounds examined, for example, had a band in the region 1660–1750 cm^{-1}; the absence of a band in that region of the spectrum of

Table 13-2

Infrared Stretching Frequencies of Some Functional Groups

GROUP	FREQUENCY RANGE (cm^{-1})
O—H	3500–3700
N—H	3300–3500
C—H	2840–2980
S—H	2550–2600
C=O	1630–1750
C=N	1470–1690
C=C	1620–1680
C≡N	2210–2260
C≡C	2100–2260

an unknown compound would clearly rule out the presence of a carbonyl group in the molecule, whereas the presence of a strong band in that region would indicate that such a group (or possibly C=N or C=C) was present.

Figure 13-6 shows the infrared spectra of two ketones, cyclopentanone and cyclohexanone. The spectra are nearly identical in the region from 4000 to 1500 cm^{-1} and show only two major group frequencies, C—H stretch at about 3000 cm^{-1} and C=O stretch at about 1700 cm^{-1}. Most ketones, aldehydes, and esters will have similar spectra in this region, because they all contain C—H and C=O bonds. However, the two spectra in Figure 13-6 differ markedly in the region below 1500 cm^{-1}. Bands in this region are due to various **bending** vibrations, or combinations of bending and stretching vibrations. They reflect various scissoring, rocking, twisting, and wagging motions of the molecule, and form a pattern which is *unique* for every molecule. This region is therefore sometimes called the **fingerprint** region of the spectrum, and can be used to identify a specific substance.

The relative positions of the group frequencies in Table 13-2 are subject to fairly simple interpretation. Consider two atoms joined by a bond. The **stretching frequency** of that bond will depend on the masses of the atoms (when the molecule absorbs radiation energy, light atoms will move more rapidly than heavy atoms) and on the "tightness" of the bond (it will be more difficult to stretch two atoms held together by several pairs of electrons than those which are bound by only one). Thus, bonds which involve the light element H (O—H, C—H etc.) all vibrate with higher frequencies than bonds between heavier elements. For bonds between like atoms, the vibrational frequency is higher the greater the number of bonds. (Compare C≡C with C=C, or C≡N with C=N; single bonds between atoms of nearly equal weight, such as C—C, C—N, C—O are not shown in Table 13-2 because

Figure 13-6

The infrared spectra of two ketones.

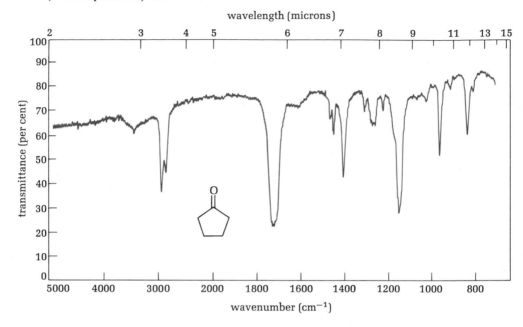

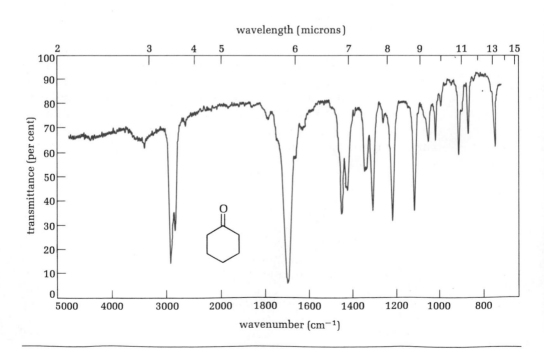

these motions are usually coupled with other molecular vibrations. They generally come in the 1000–1300 cm⁻¹ region of the spectrum.)

Careful study of group frequencies permits one to make finer distinctions as to molecular structure. For example, carbonyl group frequencies can be further subdivided into the following regions (amongst others):

		cm⁻¹
![structure] —C—H (O)	aldehyde	1720–1740
—C=C—C—H (O)	conjugated or aromatic aldehyde	1680–1715
—C— (O)	ketone	1705–1725
—C=C—C— (O)	conjugated or aromatic ketone	1660–1700
—C—OR (O)	ester	1735–1750
—C—N (O)	amide	1630–1690

Ring size may affect the exact position of a band (cyclohexanone, 1715; cyclopentanone, 1750; cyclobutanone, 1775).

13.3a INFRARED SPECTROSCOPY AND HYDROGEN BONDS

Some of the most direct evidence for hydrogen bonds comes from infrared spectroscopy (review sec 6.3). Very dilute solutions of alcohols or phenols in an inert solvent (for example, CCl_4) show a sharp O—H band in the region 3580–3650 cm⁻¹. As the solutions are made more concentrated, a broad band begins to appear at lower frequencies in the region 3200–3550 cm⁻¹. Eventually, the sharp band disappears and is replaced by the broad band. This concentration effect is interpreted in terms of the association of alcohol (or phenol) molecules as the solutions become more concentrated.

$$\text{>O—H} \xrightarrow[\text{concentration}]{\text{increasing}} \cdots\text{O—H}\cdots\text{O—H}\cdots$$

"free" hydroxyl   hydrogen-bonded hydroxyl
3580–3650 cm⁻¹   3200–3550 cm⁻¹

Hydrogen-bonded hydroxyl, containing a less "tight" bond with variable length, absorbs at lower frequencies and with a broader range of frequencies than the "free" hydroxyl group.

Hydrogen bonds to other elements (N, S, halogen) show similar effects in their infrared spectra.

13.4 NUCLEAR MAGNETIC RESONANCE SPECTRA

The most generally useful type of nuclear magnetic resonance **(nmr)** spectroscopy for assigning organic structures involves hydrogen nuclei, i.e., protons. Here, a sample of the compound under study is placed (usually as a solution in an inert solvent which does not contain protons, such as CCl_4, $CDCl_3$, or $CD_3\overset{\overset{\displaystyle O}{\|}}{C}CD_3$) in the center of a radio-frequency (rf) coil between the poles of a powerful magnet with an extremely homogeneous magnetic field. The spinning protons may align with or against the field. The difference between the two energy levels produced in this manner is given by the equation

$$\Delta E = \gamma hH/2\pi \qquad (13\text{-}11)$$

where γ is a constant characteristic of the kind of nucleus (protons, in the present case), h is Planck's constant and H is the strength of the magnetic field. The energy supplied to the nuclei by the rf coil can be varied gradually; when the energy corresponds to ΔE, it will be absorbed by the nuclei in the lower energy state (aligned with the field) as they are promoted to the higher energy state (against the field). The absorption plotted against the frequency gives an nmr spectrum. In practice, it is usually easier to apply a constant amount of energy to the nuclei (via the rf coil) and vary the strength of the magnetic field. One then determines the value of H which corresponds to absorption of the radio frequency. Currently most instruments use a radio frequency of 60 MegaHertz (MHz).

13.4a CHEMICAL SHIFT

One might assume from the discussion thus far that all protons in a given molecule, when placed in the same external magnetic field, would absorb at precisely the same frequency. Fortunately, this is not the case. The electrons which surround the various nuclei shield them, in part, from the external magnetic field. The extent of this shielding depends on the molecular environment of the proton, so that distinguishable hydrogens in a molecule will be shielded to different extents. The more strongly shielded a particular proton, the larger the external field required to produce a magnetic field right at the nucleus such that resonance (absorption of the rf frequency) will occur.

Chemical shifts are measured relative to the resonance position of a standard. Tetramethylsilane [$(CH_3)_4Si$, usually referred to as **TMS**] is the most common reference compound, because it has a single, strong resonance line (all 12 protons in TMS are equivalent) and is chemically inert; that is, it may be added to a solution of the compound under study without the likelihood of chemical reaction occurring. Chemical shifts are usually measured in either δ (delta) or τ (tau) units, where $\tau = 10 - \delta$ and

$$\delta = \frac{\Delta\nu \times 10^6}{\text{radio frequency}} \qquad (13\text{-}12)$$

$\Delta\nu$ is the difference in absorption frequency of the sample and the reference compound (TMS) in Hertz (Hz). As stated above, the radiofrequency is usually $60\ \text{MHz} = 60 \times 10^6$ Hz. Therefore, if an absorption were to occur 60 Hz downfield from TMS, the δ and τ values for that absorption would be

$$\delta = \frac{60 \times 10^6}{60 \times 10^6} = 1.00;\ \tau = 10 - 1 = 9.00$$

Put still another way, a shift ($\Delta\nu$) of 60 Hz in a total radio frequency of 60 MHz is a shift of one part in a million. δ can therefore be considered as the chemical shift from TMS, expressed in ppm.

Figure 13-7 shows the nmr spectrum of p-xylene. The spectrum is very simple, consisting of only two lines, at δ 2.20 and 6.95 ppm

---

Figure 13-7
Nmr spectrum of para-xylene.

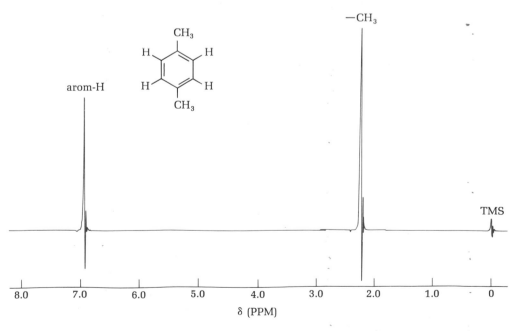

from TMS. These presumably correspond to the two types of hydrogens in the molecule, the methyl protons and the aromatic protons. Furthermore, *the intensity of an nmr signal is directly proportional to the number of protons responsible for it.* The ratio of the peak areas at δ 2.20 and 6.95 is 3:2 or 6:4; thus the peak at δ 2.20 is due to the methyl protons, and that at δ 6.95 is due to the aromatic protons.

These assignments can be verified by studying other compounds with similar structures. Most methyl benzenes have a peak around δ 2.2, and most compounds with aromatic protons have a peak around δ 6.5–8.0. Tables of chemical shifts have been compiled from studies of numerous compounds with known structures. Table 13-3 gives some examples.

It is noted from Table 13-3 that the O—H and N—H resonance positions are variable. Protons bound to carbon almost always give well-defined signals. However, protons bound to oxygen (in alcohols, phenols, acids) or nitrogen (in amines) sometimes give rather broad signals. This is because such protons may move from one oxygen (or nitrogen) atom to another, through hydrogen-bonding mechanisms. The position of the band will vary with concentration, since the extent of hydrogen bonding depends on concentration. By studying the nmr spectra of alcohols (and similar compounds) as a function of temperature, one can obtain information about the rates of such exchange processes.

To summarize, nmr spectra tell us (a) the "kinds" of protons present in a molecule, from their chemical shifts, and (b) the numbers of protons of each kind, from the relative areas of the peaks.

Table 13-3
Typical Chemical Shifts of Protons

TYPE OF PROTON	δ (ppm)	TYPE OF PROTON	δ (ppm)
CH_3—R	.85– .95	CH_3—O	3.5– 3.8
R—CH_2—R	1.20–1.35	CH_2=	4.6– 5.0
R_3—CH	1.40–1.65	—CH=C	5.2– 5.7
CH_3—C=C	1.6 –1.9	Ar H	6.6– 8.0
CH_3—C=O	2.1 –2.6	H—C(=O)—N	7.9– 8.1
CH_3—Ar	2.25–2.5	R—C(=O)—H	9.5–9.65
CH_3—S—	2.1 –2.8	Ar—C(=O)—H	9.7–10.0
CH_3—N	2.1 –3.0	O—H	variable
H—C≡C	2.45–2.65	N—H	variable

Figure 13-8
Nmr spectrum of diethyl ether.

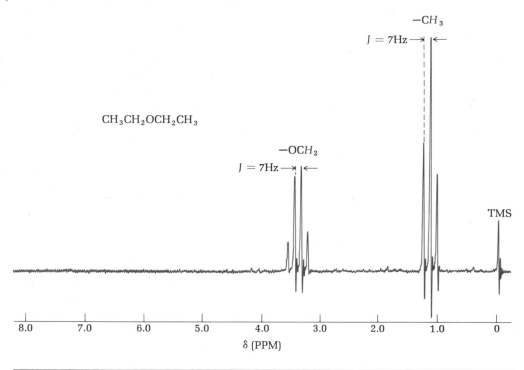

13.4b SPIN-SPIN SPLITTING

From what has been said thus far, one might expect the nmr spectrum of diethyl ether ($CH_3CH_2OCH_2CH_3$) to consist of two lines, one at about δ 0.9, and one at about δ 3.8, with relative areas of 6:4 (Table 13-3). In fact, one does see absorptions in these regions with the expected total area ratios, *but they are not single lines* (Figure 13-8). The signal for the methyl groups appears as a triplet, with relative intensities 1:2:1; and the signal for the methylene (CH_2) groups is a quartet, with relative intensities 1:3:3:1. These **spin-spin splittings,** as they are called, give us a great deal of information about the structure of a molecule. They arise in the following way.

Consider two protons with different δ-values attached to adjacent

carbon atoms: $-\overset{|}{\underset{\underset{H_a}{|}}{C}}-\overset{|}{\underset{\underset{H_b}{|}}{C}}-$. H_a would normally resonate at a particular δ-value. However, its close neighbor H_b may be aligned in either of two ways, with or against the external magnetic field. Since H_b is like a tiny magnet, it affects the field "seen" by H_a. If aligned *with* the external field, H_b slightly increases the field "seen" by H_a, whereas if aligned *against* the external field, H_b slightly decreases the field "seen" by H_a. Therefore H_a will experience two slightly different fields as a consequence of being adjacent to H_b and will appear as a doublet rather than as a singlet, with resonances slightly above and slightly below its usual δ-value. The same is true for H_b as it experiences two slightly different magnetic fields as a consequence of being adjacent to H_a. Two protons which affect each other in this way are said to be **coupled;** the extent of coupling, called the **coupling constant** *J*, is the number of Hertz (Hz) that the signals are split.

One can now readily rationalize the splittings seen in the nmr spectrum of diethyl ether (Figure 13-8). Each CH_3 proton is adjacent to two CH_2 protons. These may both be aligned with the field, both against the field, or one may be aligned with and one against the field, the latter arrangement being possible in two ways:

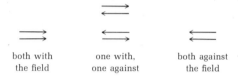

| both with | one with, | both against |
| the field | one against | the field |

Consequently, the methyl protons experience three slightly different magnetic fields and appear as a triplet; the middle field has twice the probability of the other two, causing the triplet to appear with relative intensities 1:2:1.

In completely analogous fashion, each methylene proton is adjacent to three methyl protons which as a group may be aligned in one of four ways:

Consequently, the methylene protons appear as a 1:3:3:1 quartet. The two sets of protons have identical coupling constants ($J = 7$ Hz, Figure 13-8), showing that they are mutually coupled.

The rule for the splitting pattern is very simple: there will be $n + 1$ lines, where n is the number of adjacent protons with identical coupling constants. Two other rules are important. Protons with identical chemical shifts (for example, the three protons of a methyl group) do not normally split one another's absorption lines. Also, the coupling constant usually diminishes sharply with distance. As a first approximation, protons which are not on adjacent atoms do not "split" one another. A few typical coupling constants are illustrated by the following formulas:

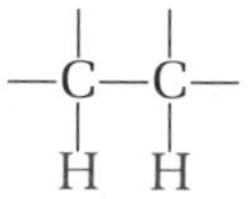

Hydroxyl protons usually do not "split" the signals of protons on an adjacent carbon atom. Thus the methyl signal in methanol is usually a singlet, rather than a doublet. The reason is that the O—H proton exchanges so quickly through hydrogen-bonding mechanisms that it does not remain on the oxygen long enough for the methyl protons to "see" the two different spin states. Similarly, the nmr spectrum of ethanol is as shown.

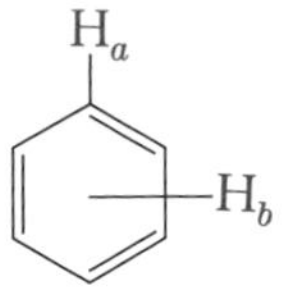

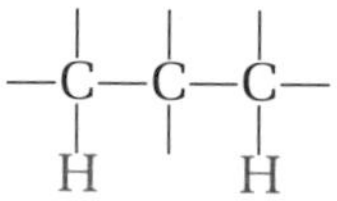

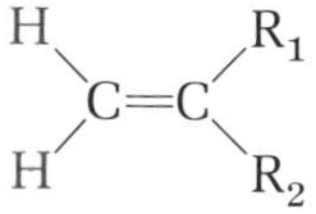

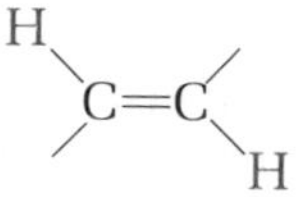

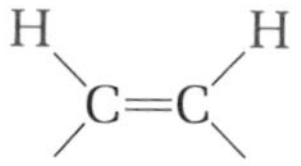

13.4c SOLUTION OF A STRUCTURAL PROBLEM

It is clear that nmr spectroscopy affords the organic chemist with a great deal of information about the structure of a molecule. Consider, for example, the three ethers with the formula $C_4H_{10}O$.

$$CH_3CH_2OCH_2CH_3 \qquad CH_3OCH_2CH_2CH_3 \qquad CH_3OCH(CH_3)_2$$

An nmr spectrum would quickly distinguish between these structures. If the substance were diethyl ether, the spectrum would consist of a quartet and triplet, area ratio 4:6, as already described (Figure 13-8), and no other signals. The two remaining ethers are also readily distinguished, since the methyl signal (around δ 0.9, for the colored hydrogens) in one would be a triplet, and the other a doublet. Furthermore, their areas would correspond to 3 and 6 protons respectively. The reader may wish, for practice, to sketch out the expected spectra of these compounds.

13.5 MASS SPECTROMETRY

In mass spectrometry, the molecules of a substance (in the vapor state, under conditions of high vacuum) are bombarded with high-energy electrons. The result is the ejection of an outer electron from the molecule, to give a positive ion.

$$M \; + \; e \; \longrightarrow \; M^+ \; + \; 2\,e \qquad (13\text{-}13)$$

molecule	bombarding electron	molecule-ion

The symbol M^+ stands for the molecule-ion (that is, the original molecule minus one electron). The molecule-ion then passes between the poles of a powerful magnet which can deflect the beam to an extent which depends on the mass (and charge) of the ion. A **mass spectrometer** is an instrument which converts molecules to ions, separates them into a spectrum according to their mass-to-charge ratio (m/e), and determines the relative amounts of each ion present.

Mass spectrometers clearly can be used to obtain precise molecular weights. However, they are of much greater use than this because of a process called *fragmentation*. If the bombarding electrons have sufficient energy, they not only produce molecule-ions but frag-ment-ions as well; that is, the original molecule breaks into smaller pieces, some of which are also ions and get sorted by the spectrom-eter on the basis of their mass-to-charge ratio. Consequently, a **mass spectrum** consists of a series of signals of varying intensity at differ-ent mass/charge ratios. Figure 13-9 shows a typical mass spectrum.

The fragmentation patterns frequently give useful structural in-formation. For example the $C_5H_{11}OH$ alcohols shown, all with the same molecular weight (88), exhibit large peaks at m/e 31, 45, and 59 respectively. These are due to cleavage of the molecule as shown

$$CH_3CH_2CH_2CH_2 \!\!\mid\!\! CH_2OH \qquad CH_3CH_2CH_2 \!\!\mid\!\! \underset{OH}{CH} \!\!-\!\! CH_3 \qquad CH_3CH_2 \!\!\mid\!\! \overset{CH_3}{\underset{OH}{C}} \!\!-\!\! CH_3$$

$$\text{31} \qquad\qquad\qquad\qquad \text{45} \qquad\qquad\qquad\qquad \text{59}$$

by the dashed lines. There is essentially no peak at m/e 59 in the mass spectra of the first two alcohols.

Through the study of the mass spectra of a large number of known compounds, many generalizations regarding fragmentation patterns have now been developed which can be used to gain much structural information from the mass spectrum of a compound whose structure is unknown.

Figure 13-9
The mass spectrum of 4-octanone.

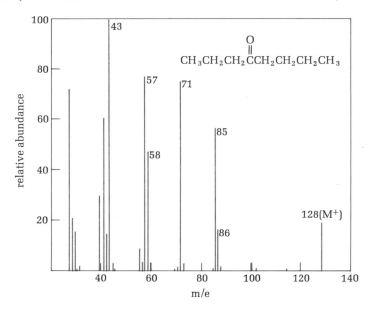

New Concepts, Facts, and Terms

1. Energy (E) and radiation frequency (ν) are related by the equation $E = h\nu$, where $h =$ Planck's constant; an *absorption spectrum* is a plot of the amount of radiation absorbed (or transmitted) by a substance as the radiation energy (that is, the frequency or wavelength) is varied.
2. Energy transitions are quantized; they are not continuous, but occur in discrete jumps.
   a. Electronic transitions—"jump" of an electron from one orbital to another—give rise to ultraviolet (uv) and visible spectra.
   b. Vibrational and rotational transitions give rise to infrared (ir) spectra.
   c. Nuclear spin transitions give rise to nuclear magnetic resonance (nmr) spectra.
3. Ultraviolet and visible spectroscopy are usually important for unsaturated molecules. Conjugation of double bonds shifts the absorption to longer wavelengths (lower energy).
   a. Dyes have a chromophore (arrangement of atoms which impart color to the molecule) with an extended conjugated system.
   b. Dyes may be classified by method of application (direct or substantive, mordant, and ingrain dyes are examples).
   c. Dyes may be classified according to their chemical structure (azo, indigo, and triphenylmethane dyes are examples).

d. Indicators are dyes whose color changes with the acidity (pH) of the medium; phenolphthalein.

e. Photochemistry is the study of reactions of electronic excited states initiated by light.

4. Infrared spectra give information through *group frequencies* about the kinds of bonds present in a molecule. The *fingerprint region* is used to identify compounds.

   a. Band positions are usually designated by *wave number* (cm^{-1}) or *wavelength* (μ = microns).

   b. Hydrogen bonding shifts the ir absorption bands of O—H and N—H groups.

5. Nuclear magnetic resonance spectra, through the use of *chemical shift* and *spin-spin splitting* patterns, can be used to tell how many hydrogens of each kind are present in a molecule and what their nearest neighbors are. *TMS* (tetramethylsilane) is a reference compound; chemical shifts are measured in δ (delta) or τ (tau) units; coupling constants (J) are measured in Hertz (Hz).

6. Mass spectra give molecular weights directly and, through *fragmentation patterns,* give structural information as well.

Exercises and Problems

1. Use equation 13-2 and the data in Table 13-2 to calculate (approximately) the ratio of the energies required to stretch a carbon-carbon triple and double bond. Which requires the greater energy?

2. Explain why the bands in ultraviolet spectra are broader than those in infrared spectra.

3. The two possible cyclohexadienes (1,3- and 1,4-) have ultraviolet absorption maxima at 224 nm and 256 nm. Which is which?

4. Two naturally occurring terpenes, myrcene and ocimene, both have the formula $C_{10}H_{16}$ and, on complete hydrogenation, give the same hydrocarbon, 2,6-dimethyloctane, $C_{10}H_{22}$. Their structures appear to be double bond isomers, **A** and **B**.

A

B

Myrcene has a UV maximum at 224 nm and ocimene a maximum at 237 nm. Which is which?

5. Dehydration of the hydroxyketone **A** gives a mixture of two unsaturated

A

ketones. The one which predominates has a UV maximum at 236 nm. What is its structure? Would the minor product absorb at a longer or shorter wavelength than the major product?

6. Stilbene (1,2-diphenylethylene) can exist in either a *cis* or a *trans* form. The *cis* isomer absorbs at a shorter wavelength and less intensely (λ_{max} 278 nm, ε 9350) than the *trans* isomer (λ_{max} 294 nm, ε 24,000). Suggest a possible explanation for the difference.

7. Draw the various resonance contributors to the structure of malachite green cation (eq. 13-6). This dye is green in aqueous solutions at pH $>$ 2, but in very strong acids (pH $<$ 0.5), the solutions become yellow (that is, the λ_{max} occurs at a shorter wavelength). Suggest a possible explanation.

8. Irradiation of 1,5-hexadiene with ultraviolet light gives two products, neither of which has a double bond. Both are isomers of the starting diene. Suggest possible structures for the photoproducts (see equation 13-9).

9. Irradiation of 1,5-cyclooctadiene gives a mixture of three isomeric products, all of which are isomers of the starting diene and none of which has a double bond. Suggest possible structures of the photoproducts.

10. Note that a carbon-carbon double bond conjugated with a carbonyl group lowers the frequency of the C=O stretching vibration (sec 13.3). Suggest an explanation.

11. The O—H stretching frequency of p-hydroxybenzaldehyde becomes broad and shifts toward longer wavelength as the solution on which the ir spectrum is determined is made more concentrated (review sec 13.3a). In sharp contrast, the O—H stretching frequency of o-hydroxybenzaldehyde is hardly affected by a change in concentration. Suggest an explanation for these observations.

12. Do you expect the group stretching frequency for C=S to be higher or lower than the C=O frequency (Table 13-2)? Explain.

13. The nmr spectrum of methyl p-toluate consists of a singlet at δ 2.35 (3H), a singlet at δ 3.82 (3H), and two doublets at δ 7.15 and 7.87, $J = 8$ Hz. Use Table 13-3 to assign these peaks to the various protons in the structure.

14. An unknown substance, $C_5H_{10}O$, has a strong band in the ir spectrum at 1725 cm^{-1} and an nmr spectrum which consists of a quartet at δ 2.7

and a triplet at δ 0.9, $J = 7.5$ Hz, with relative areas 2:3. What is the structure of the compound?

15. Sketch, using the information in Table 13-3, the anticipated nmr spectra of
    a. acetaldehyde
    b. diisopropyl ether
    c. 1,1-dichloropropene

16. The nmr spectrum of a compound C_4H_9Br consists of a single sharp line. What is the structure of the compound? An isomer of this compound has an nmr spectrum which consists of a doublet at δ 2.7, a complex pattern centered at δ 1.9 and a doublet at δ 1.0 with relative areas 2:1:6. What is its structure?

17. A compound $C_5H_{10}O_3$ with a strong ir band at 1745 cm^{-1} has an nmr spectrum with a quartet at δ 4.15 and a triplet at δ 1.20, $J = 7$ Hz, relative areas 2:3. Suggest a structure for the compound.

18. The mass spectrum in Figure 13-9 shows, in addition to the molecular ion peak (M$^+$) at m/e 128, several strong fragmentation peaks. Account for the strong peaks at m/e 85, 71, 57, and 43.

19. A compound C_3H_6O has no bands in the ir around 3500 or 1720 cm^{-1}. What structures can be eliminated by these data? Suggest a possible structure and tell how you could determine whether it is correct.

20. A hydrocarbon has a molecular ion peak in its mass spectrum at m/e 102 and shows only two peaks in its nmr spectrum, area ratio 1:5 at δ 3.08 and 7.4 respectively. What is its structure?

CHAPTER FOURTEEN

BIFUNCTIONAL COMPOUNDS

Thus far we have concerned ourselves primarily with the chemistry of compounds containing a single functional group. However, many organic substances have two or more different functional groups within one molecule. In general, the chemistry of these substances is the sum of the chemistry of the individual functional groups, plus certain new properties due to the interaction of these groups when present in the same molecule. In this chapter we shall consider the chemistry of selected bifunctional molecules.

The number of possible compounds with more than one functional group is enormous. In this chapter discussion will be limited primarily to three types of functional groups—alcohols, aldehydes or ketones, and acids or their derivatives. We will first consider bifunctional molecules with identical groups; following this, the chemistry of several classes of compounds with two different functional groups will be described.

If there is a central theme to the study of bifunctional molecules, it is that their chemistry depends upon the distance between the functional groups. In general, when two functional groups in the same molecule are remote from one another they react independently, whereas if they are close together they usually influence one another's chemical behavior.

14.1 MOLECULES WITH TWO IDENTICAL FUNCTIONAL GROUPS

Several classes of compounds of this type have already been discussed (dienes, sec 4.7 to 4.9; diols, sec 6.8; di-, tri- and polyesters, sec 10.6d and Ch. 11; and diamines, sec 12.6). Attention here will be given to diacids and di-carbonyl compounds.

14.1a DICARBOXYLIC ACIDS

The names of some of these acids and a few of their reactions were described in Chapter 10, and these should be reviewed (Table 10-2, sec 10.10 and eq 10-4, 10-6, and 10-34). We consider here the acidity of dicarboxylic acids and their behavior on being heated.

Since a carboxyl group is electron-withdrawing, the presence of one carboxyl group close to another increases the ease of ionization of the first proton. Thus, the first ionization constant of oxalic acid

$$\text{HO}-\overset{\overset{\text{O}}{\|}}{\text{C}}-\overset{\overset{\text{O}}{\|}}{\text{C}}-\text{OH} \underset{}{\overset{K_{a_1} = 5.9 \times 10^{-2}}{\rightleftharpoons}} \text{HO}-\overset{\overset{\text{O}}{\|}}{\text{C}}-\overset{\overset{\text{O}}{\|}}{\text{C}}-\text{O}^- + \text{H}^+ \quad (14\text{-}1)$$

is about 3,000 times greater than that of acetic acid (compare, Table 10-3). Ionization of the second proton is more difficult, however, since the resulting anion has *two like* (negative) *charges* in close proximity.

$$\text{HO}-\overset{\overset{\text{O}}{\|}}{\text{C}}-\overset{\overset{\text{O}}{\|}}{\text{C}}-\text{O}^- \underset{}{\overset{K_{a_2} = 6.4 \times 10^{-5}}{\rightleftharpoons}} {}^-\text{O}-\overset{\overset{\text{O}}{\|}}{\text{C}}-\overset{\overset{\text{O}}{\|}}{\text{C}}-\text{O}^- + \text{H}^+ \quad (14\text{-}2)$$

This difference between the first and second ionization constants of dicarboxylic acids depends on the distance between the two carboxyl groups, and diminishes with increasing distance as the following examples illustrate:

		K_{a_1}	K_{a_2}
$HO_2C-CH_2-CO_2H$	malonic acid	1.5×10^{-3}	2.0×10^{-6}
$HO_2C-(CH_2)_4-CO_2H$	adipic acid	3.8×10^{-5}	3.9×10^{-6}

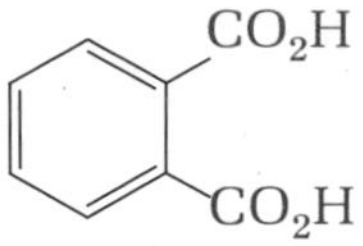

phthalic acid 1.0×10^{-3} 5.3×10^{-6}

terephthalic acid 3.1×10^{-4} 1.5×10^{-5}

The case of phthalic and terephthalic acids (which are identical except for the distance between the carboxyl groups) is particularly striking. The first and second ionization constants differ by factors of 190 and 20 respectively.

Intramolecular dehydration to form a cyclic anhydride occurs readily when two carboxyl groups are so situated that the anhydride formed contains a five- or six-membered ring. Thus succinic and glutaric acids easily form cyclic anhydrides when heated.

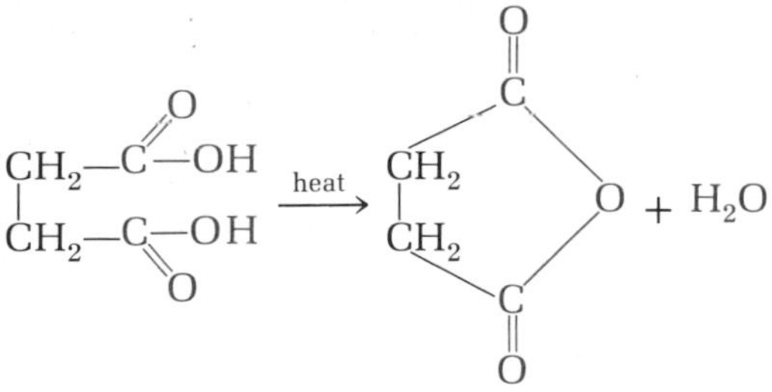

$$\text{(14-3)}$$

succinic anhydride

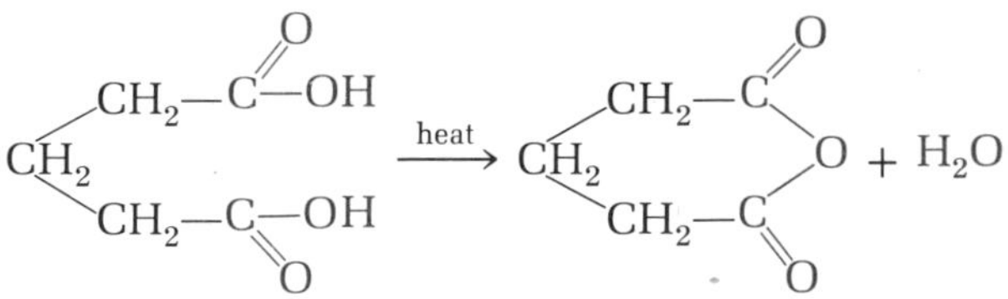

$$\text{(14-4)}$$

glutaric anhydride

Phthalic acid reacts analogously (eq 10-34), as does **maleic acid,** an unsaturated acid with *cis* carboxyl groups.

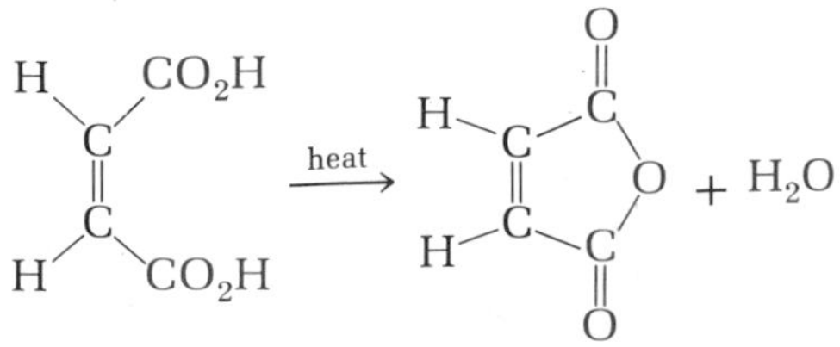

$$\text{(14-5)}$$

maleic acid   maleic anhydride

In contrast, **fumaric acid,** with *trans* carboxyl groups, cannot form an analogous cyclic anhydride, because it is not possible to have a stable molecule with a *trans* double bond in a five-membered ring.

$$(14\text{-}6)$$

fumaric acid   impossible structure

Acids that have widely separated carboxyl groups usually do not form cyclic anhydrides. When heated with a dehydrating agent, they give linear polymeric anhydrides by *inter*molecular dehydration.

$$HO_2C-(CH_2)_4-CO_2H \xrightarrow[\text{acetic anhydride}]{\text{heat}}$$

adipic acid

$$HO_2C-(CH_2)_4\!-\!\overset{O}{\underset{\|}{C}}\!-\!O\!-\!\overset{O}{\underset{\|}{C}}\!\!\left(CH_2\right)_4\!\!\right]_n\!CO_2H \quad (14\text{-}7)$$

polyadipic anhydride

The short-chain acids also behave differently on being heated. We have already seen (eq 10-57) that **oxalic acid** decomposes thermally to carbon dioxide, carbon monoxide, and water. The monomeric or dimeric cyclic anhydrides which might have been formed are as yet unknown compounds.

unknown   unknown
"oxalic anhydride"   "oxalic anhydride"
monomer   dimer

Malonic acid and substituted malonic acids, when heated above their melting points, lose carbon dioxide to give a monocarboxylic acid. The mechanism may involve a cyclic transition state leading

$$HO_2C-CH_2-CO_2H \xrightarrow{\text{heat}} CH_3-CO_2H + CO_2 \qquad (14\text{-}8)$$

malonic acid   acetic acid

$$HO_2C-\overset{R}{\underset{R'}{\overset{|}{\underset{|}{C}}}}-CO_2H \xrightarrow{\text{heat}} \overset{R}{\underset{R'}{>}}CHCO_2H + CO_2 \qquad (14\text{-}9)$$

a disubstituted   a disubstituted
malonic acid   acetic acid

to the "enediol" which subsequently ketonizes to give the mono-carboxylic acid.

$$
\begin{array}{c}
\underset{\substack{\\ \text{O} \quad \text{O}}}{\overset{\substack{R \quad R^1 \\ \diagdown C \diagup}}{O=C \overset{\frown}{} C-OH}} \\
\underset{H}{}
\end{array}
\xrightarrow{-CO_2}
\left[
\begin{array}{c}
\overset{R^1}{} \quad \overset{OH}{} \\
C=C \\
R \quad \quad OH
\end{array}
\right]
\longrightarrow
\begin{array}{c}
R^1 \qquad O \\
CH-C \diagup \\
R \qquad \diagdown OH
\end{array}
\tag{14-10}
$$

The decarboxylation of malonic acids is one important step in the metabolism of fats (eq 11-13). It is also the final step in an important synthesis of monocarboxylic acids known as the malonic ester synthesis.

14.1b   THE MALONIC ESTER SYNTHESIS

The diethyl ester of malonic acid (properly named diethyl malonate, but frequently known simply as malonic ester) is a useful starting point for certain organic syntheses. The methylene protons, because

$$
\underset{\substack{\text{diethyl malonate} \\ \text{(malonic ester)}}}{CH_3CH_2O-\overset{\overset{\displaystyle O}{\|}}{C}-CH_2-\overset{\overset{\displaystyle O}{\|}}{C}-OCH_2CH_3}
$$

of their location between two carbonyl groups, are considerably more acidic than one might expect for protons attached to a carbon atom. Their acidity is enhanced because the **enolate anion** produced when malonic ester is treated with a strong base (say, sodium ethoxide in ethanol) is resonance stabilized.

$$
CH_2(CO_2C_2H_5)_2 + {}^-OC_2H_5 \xrightarrow{-C_2H_5OH}
$$

$$
\left[
\begin{array}{ccc}
\underset{\underset{O}{\overset{\|}{C}-OC_2H_5}}{\overset{\overset{\displaystyle O}{\|}}{C}-OC_2H_5} &
\underset{\underset{O}{\overset{\|}{C}-OC_2H_5}}{\overset{\overset{\displaystyle O^-}{|}}{C}-OC_2H_5} &
\underset{\underset{O^-}{\overset{|}{C}-OC_2H_5}}{\overset{\overset{\displaystyle O}{\|}}{C}-OC_2H_5} \\
{}^-CH & CH & CH
\end{array}
\right]
\tag{14-11}
$$

enolate anion of
malonic ester

The negative charge can be distributed over *both* carbonyl oxygens, as well as on the central carbon atom, making the malonic ester enolate anion considerably more stable than the enolate anions of simple aldehydes or ketones (review sec 9.7). Indeed, diethyl malonate is nearly as strong an acid ($K_a = 10^{-13}$) as phenol ($K_a = 10^{-10}$).

The malonic ester enolate anion is a useful nucleophile in S_N2 displacements. It reacts with primary or secondary alkyl halides to form alkylmalonic esters.

$$\overset{\curvearrowleft}{X}\text{—}R\overset{\curvearrowright}{+}\text{—}CH(CO_2C_2H_5)_2 \longrightarrow RCH(CO_2C_2H_5)_2 + X^- \qquad (14\text{-}12)$$

an alkylmalonic ester

This ester can be hydrolyzed to the corresponding alkylmalonic acid which, on being heated, decarboxylates to a monocarboxylic acid (eq 14-8 or 14-9).

$$RCH(CO_2C_2H_5)_2 \xrightarrow[-2\,C_2H_5OH]{\text{NaOH}} RCH(CO_2^-)_2 \xrightarrow{2\,H^+}$$

$$RCH \overset{CO_2H}{\underset{CO_2H}{\diagdown}} \xrightarrow[-CO_2]{\text{heat}} R\overset{H}{\underset{H}{\overset{|}{\underset{|}{C}}}}\text{—}CO_2H \quad (14\text{-}13)$$

Thus, for example, the following reaction sequence could be used to synthesize hexanoic acid via malonic ester:

$$CH_2(CO_2C_2H_5)_2 \xrightarrow{^-OC_2H_5} {}^-CH(CO_2C_2H_5)_2$$

$$\Big\downarrow CH_3CH_2CH_2CH_2Br \qquad (14\text{-}14)$$

$$CH_3CH_2CH_2CH_2CH_2CO_2H \xleftarrow[\substack{\text{1. NaOH, then} \\ \text{acidify} \\ \text{2. heat}}]{} CH_3CH_2CH_2CH_2CH(CO_2C_2H_5)_2$$

hexanoic acid

Since the alkylmalonic ester (eq 14-12) still has one acidic hydrogen on the carbon between the two ester groups, the steps represented by equations 14-11 and 14-12 can be repeated to give a dialkylmalonic ester.

$$RCH(CO_2C_2H_2) \xrightarrow{^-OC_2H_5} R\bar{C}(CO_2C_2H_5)_2 \xrightarrow{R'\text{—}X} RR'C(CO_2C_2H_5)_2$$

$$(14\text{-}15)$$

This, on hydrolysis and decarboxylation, leads to a monocarboxylic acid with the general formula $RR'CHCO_2H$. In its most common form then, the malonic ester synthesis is used to prepare mono- or disubstituted acetic acids (acetic acid in which one or two of the methyl hydrogens are replaced by organic groups).

Malonic esters may also be acylated (to give ketoesters), a reaction which is important in certain metabolic processes (for example, see eq 11-12).

Finally, as we shall see later in this chapter, the malonic ester synthesis is just one example of a general synthetic method which makes use of compounds of the type $X\text{—}CH_2\text{—}Y$, where X and Y are electron-withdrawing groups capable of stabilizing the anion $X\text{—}CH^-\text{—}Y$.

14.1c   DICARBONYL COMPOUNDS

1,2-Dicarbonyl compounds may be prepared by the oxidation of aldehydes or ketones with selenium dioxide.

$$CH_3-\overset{\overset{\displaystyle O}{\|}}{C}-CH_3 + SeO_2 \longrightarrow CH_3-\overset{\overset{\displaystyle O}{\|}}{C}-\overset{\overset{\displaystyle O}{\|}}{C}H + H_2O + Se \qquad (14\text{-}16)$$

$$\text{acetone} \qquad\qquad\qquad \text{pyruvic aldehyde}$$

Glyoxal, the simplest member of the series, is manufactured commercially by the vapor phase oxidation of ethylene glycol (compare with eq 9-1). It is used to shrink-proof rayon fabrics. **Biacetyl**,

$$\underset{\underset{\displaystyle OH \quad\ OH}{|\qquad\ |}}{CH_2-CH_2} \xrightarrow[\substack{\text{Ag or Cu} \\ 300°}]{O_2} \underset{\underset{\displaystyle O \quad\ O}{\|\quad\ \|}}{HC-CH} + 2\,H_2O \qquad (14\text{-}17)$$

$$\text{glyoxal}$$

$CH_3COCOCH_3$, is present in small amounts in butter and is in part responsible for its characteristic color and flavor. 1,2-Dicarbonyl compounds are usually yellow (orange, or red), since they possess a conjugated system of double bonds. In acyclic systems, the preferred conformation has the oxygens furthest apart, to minimize the repulsion between the two adjacent dipoles.

preferred   less stable
conformation   conformation

Acetylacetone (2,4-pentanedione) is the best known 1,3-dicarbonyl compound. Having a methylene group between two carbonyl groups, it forms a resonance-stabilized enolate anion (analogous to that of malonic ester, sec 14.1b) when treated with base. Indeed, acetyl-

$$CH_3\overset{\overset{\displaystyle O}{\|}}{C}-CH_2-\overset{\overset{\displaystyle O}{\|}}{C}CH_3 \xrightarrow[-H^+]{\text{Base}}$$

acetylacetone
(2,4-pentanedione)

$$\left[CH_3\overset{\overset{\displaystyle O}{\|}}{C}-\overset{-}{C}H-\overset{\overset{\displaystyle O}{\|}}{C}CH_3 \longleftrightarrow CH_3-\overset{\overset{\displaystyle O^-}{|}}{C}=CH-\overset{\overset{\displaystyle O}{\|}}{C}CH_3 \right] \quad (14\text{-}18)$$

enolate anion of acetylacetone

acetone is a stronger acid ($K_a = 10^{-9}$) than phenol ($K_a = 10^{-10}$). It exists in part in the enol form, which is stabilized by hydrogen bonding and by the formation of a conjugated system of double bonds (C=C—C=O). Acetylacetone forms stable complexes with

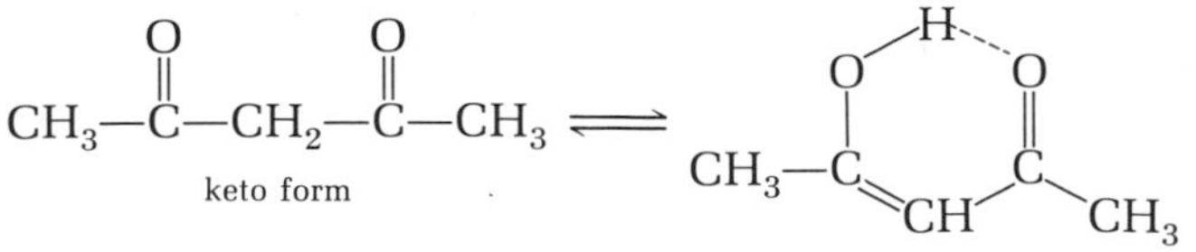

$$CH_3-\overset{\overset{\displaystyle O}{\|}}{C}-CH_2-\overset{\overset{\displaystyle O}{\|}}{C}-CH_3 \rightleftharpoons \qquad (14\text{-}19)$$

keto form

enol form

many metallic ions. Many of these complexes are soluble in organic solvents and can even be distilled without decomposition. Such properties are unusual in ordinary metallic salts.

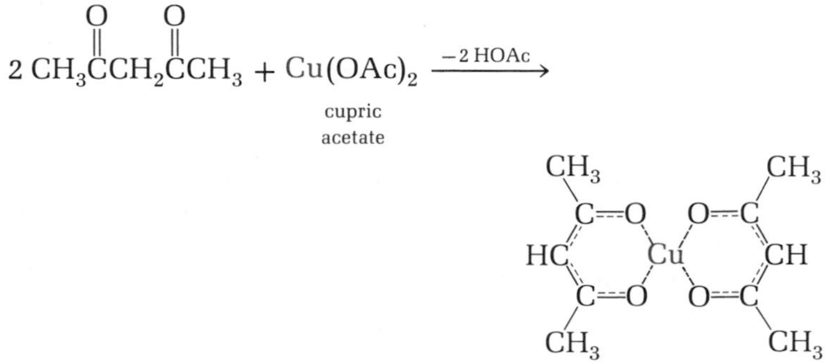

$$2\ CH_3\overset{\overset{\displaystyle O}{\|}}{C}CH_2\overset{\overset{\displaystyle O}{\|}}{C}CH_3 + Cu(OAc)_2 \xrightarrow{\ -2\ HOAc\ }$$

cupric
acetate

$$(14\text{-}20)$$

copper acetylacetonate

1,4-Dicarbonyl compounds readily cyclize to heterocyclic compounds with five-membered rings. Thus 2,5-hexanedione is dehydrated by acid to 2,5-dimethylfuran, and with ammonia, it gives 2,5-dimethylpyrrole.

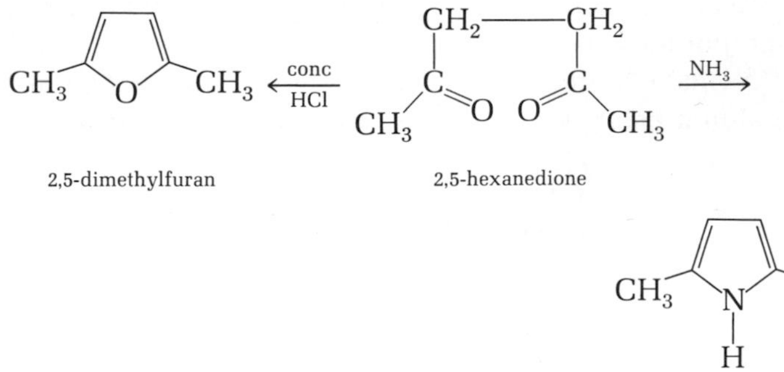

2,5-dimethylfuran   2,5-hexanedione

$$(14\text{-}21)$$

2,5-dimethylpyrrole

14.2 MOLECULES WITH TWO DIFFERENT FUNCTIONAL GROUPS

In this section we will discuss difunctional compounds with the three possible combinations of the three functional groups — OH, $\overset{\overset{\displaystyle O}{\|}}{C}$, and $\underset{\text{OH}}{\overset{\overset{\displaystyle O}{\|}}{C}}$ —and their closely related derivatives. Many compounds of this type occur in nature and are important in biological processes. For example, in the discussion of carbohydrate chemistry which is to come (Chapter 16) it will be important to understand how carbonyl and hydroxyl groups interact intramolecularly.

14.2a ALIPHATIC HYDROXY ACIDS: OCCURRENCE AND SYNTHESIS

The most important hydroxy acids, biologically speaking, have the hydroxyl group in the alpha (α), or C-2 position. Examples are **glycolic** and **lactic acids.**

$\underset{\underset{\text{OH}}{|}}{CH_2}-CO_2H \qquad CH_3-\underset{\underset{\text{OH}}{|}}{\overset{\alpha}{CH}}-CO_2H$

glycolic acid   lactic acid
(hydroxyacetic acid)   (α-hydroxypropionic acid)

α-Hydroxy acids can be made by hydrolysis of the corresponding chloro (or bromo) acids which, in turn, are obtained directly from the acids by halogenation in the presence of a small amount of phosphorus trihalide.

$$R-CH_2-CO_2H \xrightarrow[\text{PCl}_3]{\text{Cl}_2} R-\underset{\underset{\text{Cl}}{|}}{CH}-COCl \xrightarrow[\text{dilute base}]{-\text{OH}} R-\underset{\underset{\text{OH}}{|}}{CH}-CO_2H \quad (14\text{-}22)$$

an α-chloro   an α-hydroxy acid
acid chloride

The first step in this sequence is known as the **Hell-Volhard-Zelinsky** reaction and proceeds via the acyl halide. The combined electron-withdrawing effect of the chloro-carbonyl group and the halogen enhances the acidity of the α-hydrogens, so that halogenation occurs only at the α-carbon, possibly through the enol form.

$$R-\underset{\underset{\text{H}}{|}}{CH}-\overset{\overset{\displaystyle O}{\|}}{C}-Cl \rightleftharpoons \left[R-CH=C\overset{\diagup OH}{\diagdown_{Cl}}\right] \xrightarrow{Cl_2}$$

enol form

$$\left[R-\underset{\underset{\text{Cl}}{|}}{CH}-\underset{\underset{\text{Cl}}{|}}{C}\overset{\overset{\displaystyle O-H}{\diagdown}}{-Cl}\right] \xrightarrow{-\text{HCl}} R-\underset{\underset{\text{Cl}}{|}}{CH}-\overset{\overset{\displaystyle O}{\|}}{C}-Cl \quad (14\text{-}23)$$

The second step in equation 14-22 is a simple nucleophilic displacement reaction.

α-Hydroxy acids can also be prepared from cyanohydrins (sec 9.6e) by hydrolysis (sec 10.4b). Thus, lactic acid is synthesized from acetaldehyde through the sequence

$$CH_3-C\underset{H}{\overset{O}{\diagdown}} \xrightarrow{H^+CN^-} CH_3-\underset{H}{\overset{OH}{\underset{|}{\overset{|}{C}}}}-CN \xrightarrow[H^+]{HOH} CH_3-\underset{H}{\overset{OH}{\underset{|}{\overset{|}{C}}}}-C\underset{OH}{\overset{O}{\diagup}} \qquad (14\text{-}24)$$

lactonitrile   lactic acid

Lactic acid is formed when milk turns sour, because of the action of lactobacilli on the milk sugar, lactose. It is also produced in muscle tissue by enzymatic reduction of pyruvic acid (sec 14.2d). Lactic acid is used commercially (about 10 million pounds/year in the U.S.) as a food additive, to act as an acidulant and to improve the baking properties of bread dough.

Figure 14-1
Models of lactic acid.

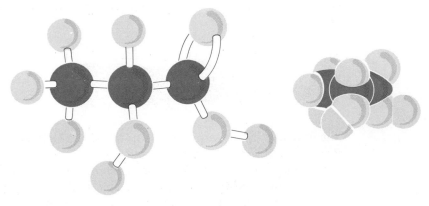

Although polyfunctional rather than bifunctional, several other biologically important α-hydroxy acids deserve brief mention here. **Malic acid** (hydroxysuccinic acid) is present in many fruit juices and was first isolated from unripe apples (Latin *malum*, apple). It is an important intermediate in carbohydrate metabolism. **Tartaric acid** is a dihydroxysuccinic acid. It is prepared from grape juice ferments, first being obtained in the form of its monopotassium salt, cream of tartar. Tartar is a sludgy precipitate which forms during the fermentation of grapes. Potassium hydrogen tartrate is used in medicine as a laxative; it is also the acid component of some baking powders. The sodium potassium salt is known as *Rochelle salts* and is used as a purgative. The acid itself is used in carbonated beverages and effervescent tablets. **Citric acid,** which is formed in the juices of fruits, particularly lemon, lime, and other citrus fruits, is an α-hydroxy tricarboxylic acid. It is added to many soft drinks and candies. Citric acid is an important intermediate in carbohydrate metabolism and is a normal constituent of blood serum and human urine.

CO_2H CO_2H

$$
\begin{array}{ccc}
CO_2H & CO_2H & \\
HC-OH & HC-OH & CH_2-CO_2H \\
CH_2 & HC-OH & HO-C-CO_2H \\
CO_2H & CO_2H & CH_2-CO_2H \\
\text{malic acid} & \text{tartaric acid} & \text{citric acid}
\end{array}
$$

Hydroxy acids with the hydroxyl group further removed from the carboxyl group than the α position may be prepared by reduction of the corresponding keto acids or by hydrolysis of the corresponding halogenated acids.

14.2b   ALIPHATIC HYDROXY ACIDS: REACTIONS

Since alcohols react with acids to form esters (eq 10-18), one might expect the same type of reaction from hydroxy acids, provided that the hydroxyl and carboxyl groups were located suitably in the molecule. Indeed, γ (gamma) and δ (delta) hydroxy acids cyclize spontaneously to 5- or 6-membered ring **cyclic esters,** called **lactones.**

$$
\underset{OH}{\overset{\gamma}{C}H_2\overset{\beta}{C}H_2\overset{\alpha}{C}H_2CO_2H} \xrightarrow{-H_2O}
\begin{array}{c}
CH_2-C\overset{O}{\diagup} \\
| \qquad\quad\;\, O \\
CH_2-\overset{\gamma}{C}H_2
\end{array}
\qquad (14\text{-}25)
$$

γ-hydroxybutyric acid   γ-butyrolactone

$$
\underset{OH}{\overset{\delta}{C}H_2\overset{\gamma}{C}H_2\overset{\beta}{C}H_2\overset{\alpha}{C}H_2CO_2H} \xrightarrow{-H_2O}
\begin{array}{c}
CH_2-C\overset{O}{\diagup} \\
CH_2 \qquad O \\
CH_2-\overset{\delta}{C}H_2
\end{array}
\qquad (14\text{-}26)
$$

δ-hydroxyvaleric acid   δ-valerolactone

The analogous reaction with an α- or β-hydroxy acid would lead to rather strained α- or β-lactones.

$$
R-\overset{\alpha}{C}H-C\overset{O}{\diagup}_{\diagdown O}
\qquad\qquad
R-CH-C\overset{O}{\diagup} \atop \underset{\beta}{C}H_2-O
$$

an α-lactone   a β-lactone

Although such structures can be prepared by special methods, they are highly strained and reactive, and are not produced directly from the corresponding hydroxy acids. α-Hydroxy acids undergo bimolecular esterification to form **lactides,** compounds which are cyclic diesters.

$$\text{(14-27)}$$

a lactide

The cyclization of α-, γ-, and δ-hydroxy acids occurs so readily that the compounds are usually kept in the form of their sodium salts, since acidification usually leads to the formation of lactones or lactides.

β-Hydroxy acids, which do not cyclize by either of the reaction paths just discussed, lose water on heating or treatment with strong acid to produce unsaturated acids. Usually the α,β-unsaturated acid predominates, since it contains a conjugated system of double bonds.

$$\underset{\text{β-hydroxybutyric acid}}{\overset{\gamma}{CH_3}-\overset{\beta}{\underset{\underset{OH}{|}}{CH}}-\overset{\alpha}{\underset{\underset{H}{|}}{CH}}-CO_2H} \xrightarrow[-H_2O]{H^+ \text{ or heat}} \underset{\substack{\text{crotonic acid} \\ \text{(an α,β-unsaturated acid)}}}{\overset{\gamma}{CH_3}\overset{\beta}{CH}=\overset{\alpha}{CH}-\overset{O}{\overset{||}{C}}-OH} \qquad \text{(14-28)}$$

The reaction which equation 14-28 typifies is simply the dehydration of an alcohol to an alkene (sec 3.5a).

14.2c PHENOLIC ACIDS

The most important phenolic acid is **salicylic acid** (*o*-hydroxybenzoic acid). It has been used in medicine for many years, either as the sodium salt or in several esterified forms. Salicylates are *anti-*

salicylic acid

acetylsalicylic acid
(as the sodium salt,
aspirin)

methyl salicylate
(oil of wintergreen)

phenyl salicylate
(salol)

pyretics—that is, they lower the body temperature in subjects having a fever, but have little effect if the temperature is normal. They are also mild analgesics which relieve certain types of pain, such as headaches, neuralgia, and rheumatism. Sodium salicylate has an irritating effect on the stomach lining; for this reason, various esters are now used. They pass unchanged through the stomach but are hydrolyzed in the alkaline medium of the intestines, liberating the salicylic acid. *Aspirin* is the most common salicylate used in medicine today. It is the sodium salt of acetylsalicylic acid, and can be prepared from salicylic acid and acetic anhydride (compare with eq 10-39).

salicylic acid   acetylsalicylic acid

$$+ \; CH_3CO_2H \qquad (14\text{-}29)$$

Annual production in the United States, in the neighborhood of 30 million pounds, is sufficient to produce over 30 billion standard tablets, or more than 150 tablets per person per year.

Salol, the phenyl ester of salicylic acid, is used as a coating for pills, to permit their contents to pass through the stomach into the intestines. This is desirable for certain drugs which may have an upsetting effect on the stomach or may be destroyed by its acidic contents.

Methyl salicylate is the chief component of oil of wintergreen. It is used as a flavoring agent and in rubbing liniments, where its mild irritating action on the skin provides a counterirritant for sore muscles.

Salicylic acid is manufactured industrially by heating the sodium salt of phenol with carbon dioxide under pressure.

sodium phenoxide

$$\qquad (14\text{-}30)$$

salicylic acid
(o-hydroxybenzoic acid)

14.2d KETO ACIDS AND ESTERS

Keto acids, especially those with the carbonyl group either α or β to the carboxyl group, are important intermediates in biological oxidations and reductions.

Pyruvic acid plays an important role in the metabolism of carbohydrates, where it serves as a source of acetyl groups for the production of acetylcoenzyme A (sec 11.7). In the muscle, pyruvic acid is reduced enzymatically to lactic acid. And in the fermentation of glucose, pyruvic acid is decarboxylated to acetaldehyde, which is then enzymatically reduced to ethanol.

Figure 14-2
Role of pyruvic acid in several biological processes.

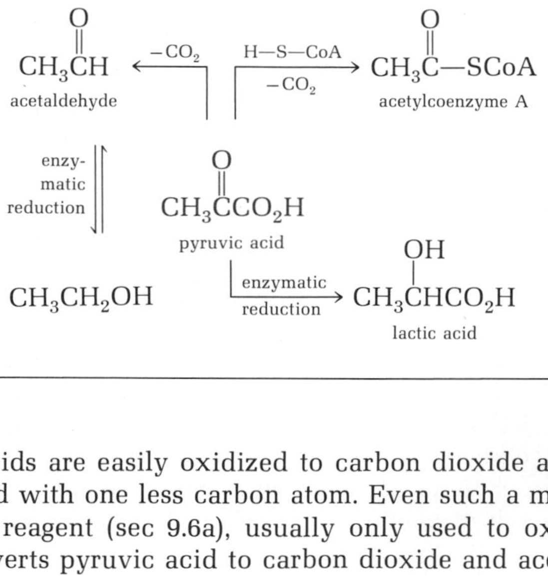

α-Keto acids are easily oxidized to carbon dioxide and the carboxylic acid with one less carbon atom. Even such a mild oxidant as Tollens' reagent (sec 9.6a), usually only used to oxidize aldehydes, converts pyruvic acid to carbon dioxide and acetic acid.

$$CH_3COCO_2^- + 2\ Ag(NH_3)_2^+ + 2\ OH^- \longrightarrow$$

pyruvate

$$CH_3CO_2^- + CO_2 + 2\ Ag^0 + 4\ NH_3 + H_2O \quad (14\text{-}31)$$

acetate

β-Keto acids are important intermediates in several metabolic processes. Their behavior is typified by **acetoacetic acid** which, like malonic acid (eq 14-8), loses carbon dioxide on being heated.

$$\underset{\text{acetoacetic acid}}{CH_3\overset{O}{\overset{\|}{C}}CH_2-\overset{O}{\overset{\|}{C}}-O-H} \xrightarrow{\text{heat}} \underset{\text{acetone}}{CH_3\overset{O}{\overset{\|}{C}}CH_3} + CO_2 \quad (14\text{-}32)$$

In this case, the other product is a ketone rather than an acid. The decarboxylation mechanism is analogous to that for malonic acid (eq 14-10). In patients suffering from diabetes, acetoacetic acid and acetone may accumulate in the blood and urine because of the improper completion of fat metabolism, which would normally convert the acetoacetic acid to two moles of acetate (sec 11.7).

Equation 14-32 is the final step in an important ketone synthesis which resembles in many respects the malonic ester synthesis of acids (sec 14.1b). **Ethyl acetoacetate** (also called **acetoacetic ester**) has a methylene group between two carbonyl groups and, like malonic ester (eq 14-11) or acetylacetone (eq 14-19), readily forms a resonance-stabilized enolate anion.

$$CH_3\overset{O}{\overset{\|}{C}}-CH_2-\overset{O}{\overset{\|}{C}}OC_2H_5 \xrightarrow[-C_2H_5OH]{^-OC_2H_5}$$

ethyl acetoacetate

$$\left[CH_3\overset{O}{\overset{\|}{C}}-\overset{-}{\overset{}{C}}H-\overset{O}{\overset{\|}{C}}OC_2H_5 \longleftrightarrow CH_3\overset{O^-}{\overset{|}{C}}=CH-\overset{O}{\overset{\|}{C}}OC_2H_5 \longleftrightarrow CH_3\overset{O}{\overset{\|}{C}}-CH=\overset{O^-}{\overset{|}{C}}OC_2H_5\right]$$

ethyl acetoacetate enolate anion

$$(14\text{-}33)$$

This anion can react as a nucleophile with alkyl halides, making possible the replacement of one or both (stepwise) methylene hydrogens by alkyl groups.

$$CH_3COCH_2CO_2C_2H_5 \xrightarrow[2.\ RX]{1.\ ^-OC_2H_5} CH_3COCHCO_2C_2H_5 \xrightarrow[2.\ R'X]{1.\ ^-OC_2H_5} CH_3CO\overset{R'}{\underset{R}{C}}CO_2C_2H_5$$

$$(14\text{-}34)$$

These esters may be hydrolyzed by base to the β-keto acids which, on heating (eq 14-32), decarboxylate to give mono- or disubstituted acetones.

$$CH_3\overset{O}{\overset{\|}{C}}\underset{R}{CHCO_2C_2H_5} \xrightarrow[2.\ heat]{1.\ ^-OH,\ then\ H^+} CH_3\overset{O}{\overset{\|}{C}}C\overset{H}{\underset{H}{\diagdown}}{-}R \qquad (14\text{-}35)$$

monosubstituted acetone

$$CH_3\overset{O}{\overset{\|}{C}}C\overset{R}{\underset{R'}{\diagdown}}CO_2C_2H_5 \xrightarrow[2.\ heat]{1.\ ^-OH,\ then\ H^+} CH_3\overset{O}{\overset{\|}{C}}C\overset{R}{\underset{R'}{\diagdown}}{-}H \qquad (14\text{-}36)$$

disubstituted acetone

β-Keto acids are usually synthesized by careful hydrolysis of the corresponding esters. These, in turn, may be synthesized by a base-catalyzed condensation of ordinary esters, quite analogous to the aldol condensation (sec 9.7a). For example, treatment of ethyl acetate with a strong base (such as sodium ethoxide in ethanol) gives ethyl acetoacetate.

$$CH_3\overset{O}{\overset{\|}{C}}-OC_2H_5 + H-CH_2\overset{O}{\overset{\|}{C}}OC_2H_5 \xrightarrow[C_2H_5OH]{^-OC_2H_5\ in}$$
ethyl acetate   ethyl acetate

$$CH_3\overset{O}{\overset{\|}{C}}CH_2\overset{O}{\overset{\|}{C}}OC_2H_5 + C_2H_5OH \quad (14\text{-}37)$$
ethyl acetoacetate

The mechanism involves formation of the enolate anion of the ethyl acetate and nucleophilic attack of that anion on the carbonyl group of a second ester molecule.

$$H-CH_2CO_2C_2H_5 + {}^-OC_2H_5 \rightleftharpoons {}^-CH_2CO_2C_2H_5 + C_2H_5OH \qquad (14\text{-}37a)$$

$$CH_3\overset{O}{\overset{\|}{C}}-OC_2H_5 + {}^-CH_2CO_2C_2H_5 \rightleftharpoons CH_3\overset{O^-}{\overset{|}{C}}CH_2CO_2C_2H_5 \qquad (14\text{-}37b)$$
$$\overset{}{\underset{OC_2H_5}{}}$$

$$CH_3\overset{O^-}{\overset{|}{C}}CH_2CO_2C_2H_5 \rightleftharpoons CH_3\overset{O}{\overset{\|}{C}}CH_2CO_2C_2H_5 + {}^-OC_2H_5 \quad (14\text{-}37c)$$
$$\underset{OC_2H_5}{}$$

All three steps in the mechanism are reversible. The overall reaction is driven to completion by the fact the final product, a β-ketoester, forms a more stable enolate anion than the starting, simple ester; that is, the ketoester formed in equation 14-37c is converted *irreversibly* to its anion, equation 14-33.

This base-catalyzed condensation of esters, known as the **Claisen condensation,** is a general synthetic method for β-ketoesters.

14.2e   HYDROXY CARBONYL COMPOUNDS

The simplest member of this series, **glycolic aldehyde,** has not been isolated in pure form, since it readily dimerizes. The product, which is cyclic, arises from the addition of the hydroxyl function of one molecule to the aldehyde function of another, producing a hemiacetal (sec 9.6c).

$$2 \text{ HO}-\text{CH}_2-\overset{\overset{\displaystyle O}{\|}}{\text{CH}} \rightleftharpoons \begin{array}{c} O \\ \text{CH}_2 \quad \text{CHOH} \\ | \qquad | \\ \text{HOCH} \quad \text{CH}_2 \\ O \end{array} \qquad (14\text{-}38)$$

glycolic aldehyde   dimer

(a *bis*-hemiacetal)

α-Hydroxyketones, having a secondary alcohol function, are known as **acyloins** (aliphatic) or **benzoins** (aromatic). **Acetoin** (together with biacetyl) contributes to the characteristic flavor of butter. Solutions of **benzoin** are used in medicine as an antiseptic. α-Hydroxyketones are easily oxidized to α-diketones (sec 14.1c).

$$\text{CH}_3-\overset{\overset{\displaystyle}{|}}{\underset{\underset{\displaystyle \text{OH}}{|}}{\text{CH}}}-\overset{\overset{\displaystyle O}{\|}}{\text{C}}-\text{CH}_3 \qquad\qquad \text{C}_6\text{H}_5-\overset{\overset{\displaystyle}{|}}{\underset{\underset{\displaystyle \text{OH}}{|}}{\text{CH}}}-\overset{\overset{\displaystyle O}{\|}}{\text{C}}-\text{C}_6\text{H}_5$$

acetoin   benzoin

β-Hydroxy aldehydes and ketones have already been discussed (sec 9.7a), since they are the products of aldol condensations.

Additional chemistry of hydroxycarbonyl compounds will be discussed at some length in Chapter 16, since carbohydrates are polyhydroxyaldehydes and ketones.

New Concepts, Facts, and Terms

1. Dicarboxylic acids; a carboxyl group enhances the acidity of a nearby second carboxyl group in the same molecule.
   a. Dicarboxylic acids form cyclic or polymeric anhydrides, depending on the distance between the two carboxyl groups. Exceptions are malonic acids and oxalic acid, which decarboxylate.
   b. The malonic ester synthesis; used to prepare acids of the type $\text{RCH}_2\text{CO}_2\text{H}$ or $\text{RR'CHCO}_2\text{H}$.
2. 1,2-Dicarbonyl compounds; glyoxal, pyruvic aldehyde, biacetyl. 1,3-Dicarbonyl compounds form more stable enols than monocarbonyl compounds; acetylacetone.
3. Hydroxyacids; glycolic, lactic, malic, tartaric, and citric acids all have a hydroxyl group α to the carboxyl group.
   a. α-Haloacids are prepared by the Hell-Volhard-Zelinsky reaction (RCO_2H, X_2, PX_3). With base, they give α-hydroxyacids.
   b. Hydrolysis of cyanohydrins gives α-hydroxyacids.
   c. On heating, α-hydroxyacids give lactides, β-hydroxyacids dehydrate to unsaturated acids, and γ- and δ-hydroxy acids give cyclic esters called lactones.
   d. Phenolic acids: salicylic acid, sodium acetylsalicylate (aspirin), methyl salicylate (oil of wintergreen)

4. Keto acids and esters; pyruvic acid, acetoacetic acid. The acetoacetic ester synthesis is used to prepare ketones of the type CH_3COCH_2R and $CH_3COCHRR'$. The Claisen condensation of esters is used to prepare β-ketoesters.
5. Hydroxy carbonyl compounds; acetoin, benzoin

Exercises and Problems

1. Write structural formulas for each of the following compounds:
   a. α-bromobutyric acid
   b. 2,3-butanedione
   c. salicylic acid
   d. ethyl 3-ketobutyrate
   e. diethyl ethylmalonate
   f. succinic acid
   g. methyl lactate
   h. glutaric anhydride
   i. γ-butyrolactone
   j. diethyl malate

2. Name each of the following compounds:
   a. $CH_2CH_2CH_2CO_2H$
      |
      OH
   b. $CH_3CH(CO_2H)_2$
   c. $HO_2CCH_2CH(CH_3)CO_2H$
   d. $CH_3COCH_2CO_2H$
   e. $CHO—CHO$
   f. CH_3COCO_2H
   g. $CH_2(CO_2C_2H_5)_2$
   h. $CH_3COCH_2COCH_3$
   i.

   j. $HOCH_2CO_2CH_3$

3. Using general formulas, write a structural formula for each of the following types of compounds:
   a. γ-hydroxyacid
   b. lactide
   c. γ-lactone
   d. cyclic anhydride
   e. β-bromoacid
   f. dialkylmalonic acid
   g. α-diketone
   h. β-hydroxyaldehyde (aldol)

4. Give an explanation for the fact that K_{a_1} for phthalic acid is larger (3X) than K_{a_1} for terephthalic acid, but that K_{a_2} for terephthalic acid is larger (3X) than K_{a_2} for phthalic acid.

5. *cis-* and *trans*-1,2-Cyclopropanedicarboxylic acids behave differently when heated. Write equations which illustrate this difference. Do you expect *cis-* and *trans*-1,2-cyclooctanedicarboxylic acids to show the same difference? Explain.

6. Write equations for the action of heat on
   a. *cis*-1,3-cyclohexanedicarboxylic acid
   b. 4-hydroxypentanoic acid
   c. 1,1-cyclopentanedicarboxylic acid
   d. 2-ketocyclopentanecarboxylic acid
   e. α-hydroxyphenylacetic acid
   f. 2-hydroxycyclohexanecarboxylic acid

7. Give equations which show how malonic acid could be synthesized in three steps, from acetic acid.

8. Show how each of the following could be synthesized from malonic ester and any other necessary reagents:
   a. 3-phenylpropanoic acid
   b. 2-methylbutanoic acid
   c. dicyclopentylacetic acid
   d. 4-pentenoic acid

9. Malonic ester can be used, with the dihalide 1,2-dichloroethane, to synthesize either adipic acid or cyclopropanecarboxylic acid. Write equations which illustrate these syntheses.

10. Explain why acetylacetone ($K_a = 10^{-9}$) is a stronger acid than malonic ester ($K_a = 10^{-13}$).

11. The nmr spectrum of acetylacetone, $CH_3COCH_2COCH_3$, might be expected to consist of two singlets, with relative areas 6:2. Indeed, two peaks with this ratio appear at δ 2.0 and 3.6. But in addition, the spectrum also contains sharp singlets at δ 1.8 and 5.6, and a broad band at δ 15.1, relative areas 6:1:1. Suggest an explanation for this observation.

12. The UV visible absorption spectra of *cyclic* 1,2-diketones show an interesting dependence on ring size, the wavelength of maximum absorption being longest (466 nm) when the ring is 5-membered. The maximum shifts to shorter wavelength as the ring size increases, reaching a minimum at the 7- and 8-membered rings (340 nm), then moves again toward longer wavelengths, reaching 384 nm for the 18-membered ring. Suggest a possible explanation for these observations.

13. Give a plausible mechanism for the acid-catalyzed dehydration of 2,5-hexanedione to 2,5-dimethylfuran (eq 14-21).

14. Write equations which show how lactic acid could be synthesized from each of the following:
   a. propionic acid   c. pyruvic acid
   b. acetaldehyde   d. 1,2-propanediol

15. Mandelic acid [($C_6H_5CH(OH)CO_2H$] can be isolated from bitter almonds (German, *mandel*) and is used in medicine for the treatment of urinary infections. Suggest a two-step method for its synthesis from benzaldehyde.

16. Barbiturates are used in medicine as sedatives, soporifics, and hypnotics. Typical of these is *veronal*, which has the formula

veronal

Barbiturates are synthesized by reaction of urea (sec 10.9) with substituted malonic esters. With this information, write equations which describe the synthesis of veronal from urea and malonic ester.

17. Write equations which illustrate the steps in the mechanism for
    a. equation 14-25   c. equation 14-28
    b. equation 14-27   d. equation 14-29

18. Write a mechanism for the reaction given in equation 14-32. What is the actual reaction product, and how is it converted to acetone?

19. Give equations for the synthesis of
    a. 2-pentanone from ethyl acetoacetate
    b. 3-methyl-4-phenyl-2-butanone from ethyl acetoacetate
    c. 2-methylcyclopentanone from ethyl 2-ketocyclopentanecarboxylate

20. Show all the steps in the mechanism of the Claisen condensation of ethyl propionate to give ethyl 2-methyl-3-ketopentanoate.

21. Several variations of the Claisen condensation are possible.
    a. Give the structure and equations for the formation of the "cross"-Claisen product from the condensation of ethyl acetate with ethyl formate.
    b. Give the structure and equations for the formation of the *intra-molecular* Claisen product (called the Dieckmann product) from diethyl adipate.

22. The Claisen condensation carried out using sodium ethoxide in ethanol as the base works well with ethyl butyrate, but fails with ethyl isobutyrate. Suggest an explanation.

23. Write out the steps in the mechanism for equation 14-38.

24. An acid with the molecular formula $C_4H_8O_3$ reacts with acetyl chloride, gives a positive iodoform test (sec 9.7b), and on heating gives a mixture of *three* isomeric acids, all with the formula $C_4H_6O_2$. Give correct structures and equations for all compounds and reactions described.

25. A phenolic acid $C_9H_8O_3$ **(A)** exists in two isomeric forms. Both rapidly decolorize permanganate and on moderate oxidation yield salicylic and oxalic acids as the only organic products. One isomer of **A** easily loses water, when heated, to yield $C_9H_6O_2$; the other fails to dehydrate under the same conditions. Suggest structural formulas for the isomers.

CHAPTER FIFTEEN

OPTICAL ISOMERISM

Molecules that are identical with one exception—in the way the atoms are arranged in space—are called **stereo-isomers.** Geometric **(cis-trans)** isomers provide an example (sec 3.2). In this chapter we discuss another type of stereoisomerism which occurs when a molecule and its mirror image are not identical. Such isomerism is commonly detected by the behavior of these molecules toward plane-polarized light. Many naturally occurring substances show optical isomerism, and this may markedly affect their biochemical behavior.

15.1 BASIC CONCEPTS

The symmetry of molecules can affect their chemical behavior. The kind of behavior we have in mind occurs when a molecule has the property of handedness, called **chirality** (Greek: *cheir* = hand). We all know that it is quite easy for two right-handed (or left-handed) persons to shake hands, but it is not possible to "shake" a right hand with a left hand in the usual way. Similarly, molecules may react differently with one another, depending on their chirality. Whether a molecule is **chiral** or **achiral** depends upon the presence or lack of certain elements of symmetry.

15.1a ELEMENTS OF SYMMETRY

The most important elements of symmetry are simple axes, centers, and planes of symmetry. An *n*-fold **axis of symmetry** is one which passes through a molecule in such a way that rotation about the axis by an angle of 360°/n brings the molecule into a position indistinguishable from its original one. A **center of symmetry** is a point within a molecule such that a straight line drawn from any part of the molecule to the center and extended an equal distance on the other side encounters an equivalent part of the molecule. A **plane of symmetry** is one which passes through a molecule in such a way that the part on one side of the plane is the exact reflection of the part on the other side (the plane acting as a mirror). These symmetry elements are illustrated in Figure 15-1.*

A **symmetry operation** is any operation performed on an object such that some new orientation of the object results which is indistinguishable from and superimposable on the original. One can view an object, close the eyes while a symmetry operation is performed, then open the eyes and notice no change. For example, if a molecule possesses a symmetry or mirror plane, then reflection in that plane will produce an object indistinguishable from the original; the process of reflection is called a symmetry operation, whereas the mirror plane is called the symmetry element.

Molecules with a center or plane of symmetry are superimposable (or "identical") with their mirror images. Such molecules are sometimes called symmetric, though the term **achiral** or without handedness is preferred (since, as we shall see, some molecules may have certain symmetry elements, such as a simple axis, yet be chiral or handed). Examples of simple achiral objects are a sphere, a cube

---

*A fourth symmetry element, the *n*-fold alternating axis, may be significant in some molecules. Rotation by 360°/n followed by reflection in a plane at right angles to the rotation axis gives a molecule identical with the original. Since molecules which possess only this symmetry element are relatively rare, they will be neglected in our discussion.

or other rectangular box, and an isosceles or equilateral triangle.

On the other hand, molecules which lack a center or plane of symmetry have mirror images which are *not* superimposable on the molecule. Such substances are **chiral.** They may be **asymmetric** (that is, without any symmetry element whatever) or **dissymmetric** (that is, without a plane or center of symmetry, but not completely devoid of all symmetry—a simple axis may be present). Well-known chiral objects are right- or left-handed gloves or shoes, screws or other "threaded" objects, and a helical or "spiral" staircase.

The mirror image relationship of the right and left hands, objects which lack a plane or center of symmetry, is shown in Figure 15-2.

Figure 15-1
Three symmetry elements of molecules. (Each molecule shown has more symmetry elements than just the one which is illustrated.)

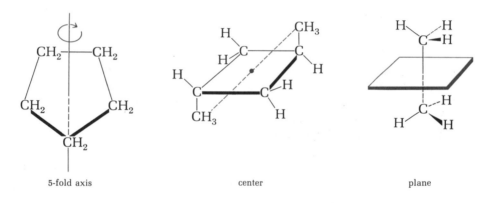

5-fold axis   center   plane

Figure 15-2
The mirror image relationship of the right and left hands. The objects are chiral and asymmetric—without any symmetry.

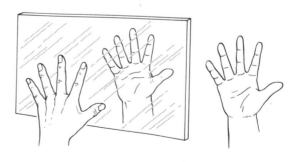

Figure 15-3

The mirror image of a clockwise helix is a counterclockwise helix. Although chiral, the cylindrical helix is not entirely without symmetry; it has a two-fold simple axis, shown in color at the right. Hence, the object is dissymmetric but not asymmetric. If rotated 180° about the axis, the coil at the right still remains a counterclockwise helix.

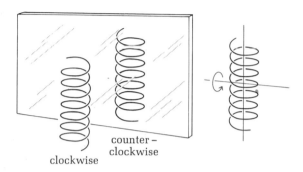

clockwise

counter–
clockwise

These objects are asymmetric—devoid of symmetry. Figure 15-3 shows the mirror image relationship of a clockwise and counterclockwise cylindrical helix. These objects are dissymmetric (not superimposable on their mirror images) but not asymmetric, since they possess a simple two-fold symmetry axis at right angles to the long axis through the center of the helix. The objects in both figures are chiral.

The most common criterion in molecules for determining whether or not a substance is chiral is the presence or absence of a plane of symmetry.

15.1b ENANTIOMERS

Two substances whose molecular structures are related as an object and its nonsuperimposable mirror image are said to be **enantiomers.** The left and right hands in Figure 15-2 and the clockwise and counterclockwise helices in Figure 15-3 have this relationship and are said to constitute enantiomeric pairs. Similar relationships are possible in molecules, as we shall see (sec 15.5).

The chirality (left- or right-handedness) of an object cannot be determined by its interaction with an achiral (symmetric) object. Thus a left- and right-handed baseball player can use the same bat or ball equally well (the bat is an achiral object, having an infinite number of symmetry planes through the length of the bat, and the ball is achiral, having a center of symmetry). But they cannot use the same gloves (the glove is a chiral object, being either left- or right-handed).

The same is true of mirror image chiral molecules. They will react identically with achiral molecules and will have similar achiral properties (melting and boiling points, spectra, etc.). However, they will behave differently toward other chiral molecules and will have different chiral physical properties. One of these physical properties involves the interaction with polarized light.

15.2   PLANE-POLARIZED LIGHT

Since optical isomerism is usually detected by determining whether a substance can alter the plane of vibration of plane-polarized light, it will be necessary to consider first some of the properties of light.

Ordinary light may be thought of as rays of different wave length vibrating in all possible planes perpendicular to the direction of their propagation. When white light passes through a suitably colored filter, **monochromatic light** (light of a uniform wavelength) is obtained. Such light may still vibrate in all possible planes. This is indicated in Figure 15-4, in which a beam of monochromatic light is directed out of the page toward the reader. A cross section of the beam would show waves with any axis from 0° to 360°. However, each plane of vibration may be considered to be the result of two component waves vibrating in perpendicular planes (Figure 15-5).

Figure 15-4
Ordinary light beam vibrating in all possible planes, coming toward the reader.

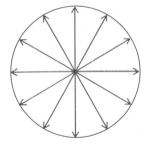

Figure 15-5
Resolution of light beam AB into its vertical (CD) and horizontal (EF) components.

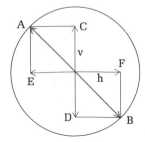

Figure 15-6

A monochromatic beam of light AB, initially vibrating in all directions, passes through a polarizing substance which "strains" the light so that only the vertical component emerges.

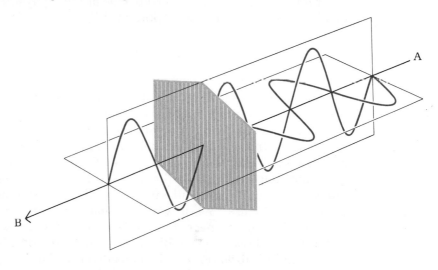

If the light beam were to pass through a substance which would permit only one component (say, the vertical) to pass through it, the resulting light beam would have all waves vibrating in a single plane; it would be described as **plane-polarized light** (Figure 15-6).

Several rather special substances such as tourmaline crystals, polaroid (crystals properly oriented and embedded in a transparent plastic), or a Nicol prism* especially constructed from Iceland spar can be used to polarize light. The phenomenon can readily be demonstrated experimentally. A beam of light will pass through two tourmaline crystals (or sheets of polaroid) only if the axes of the crystals are parallel. If, however, one crystal is turned at right angles to the first, light will be prevented from passing through the second crystal, and the latter will appear opaque (Figures 15-7 and 15-8).

15.3 OPTICAL ACTIVITY: ITS MEASUREMENT

A substance is **optically active** if it is capable of rotating the plane of plane-polarized light. It has been known since the early nineteenth century that many natural substances (carbohydrates, proteins,

*Named after William Nicol, the British physicist who invented it.

Figure 15-7
A beam of plane-polarized light as it passes through two tourmaline crystals with axes parallel (left) or perpendicular (right).

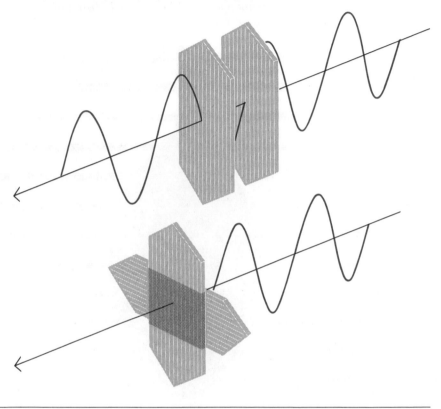

petroleum, and various cell constituents) alter the plane of polarization when a beam of polarized light passes through them. The extent of rotation is measured with a **polarimeter** (Figure 15-9).

Monochromatic light from some source (usually a sodium vapor lamp) is polarized by passing through a polarizing prism. The beam of polarized light then passes through a sample tube containing a solution of the substance to be tested. Upon emerging from the sample tube, the light passes through an analyzer, which is similar to the polarizing prism. If the test substance is optically inactive, the polarizer and analyzer will be parallel for maximum transmission of light (that is, angle α will be 0°). If the substance being tested is optically active (that is, if it rotates the plane of the polarized light α degrees), it will be necessary to rotate the analyzer an equal

Figure 15-8

A beam of light cannot be seen through two sheets of linear polarizers unless their axes are parallel. The region of overlap of these unaligned Polaroid polarizing discs appears opaque although each disc alone is transparent. (Courtesy of Polaroid Corporation.)

Figure 15-9

Diagrammatic sketch of a polarimeter.

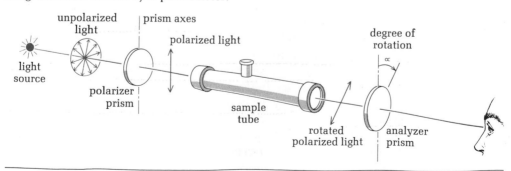

number of degrees in order to obtain the maximum intensity of transmitted light. The angle α through which the analyzer must be rotated is equivalent to the rotation of the light by the test substance. This angle is measured by means of a calibrated circular scale.*

The extent to which a given sample rotates polarized light depends not only upon its structure, but also upon the length of the sample tube and the concentration of the optically active substance in the inactive solvent. It also depends upon temperature and on the wavelength of the light used. These factors are usually standardized when it is necessary to compare the activity of different substances. The **specific rotation (α)** of an optically active substance in solution is expressed as

$$[\alpha]_{\lambda}^{t} = 100 \ \alpha/l\cdot c \ \text{(solvent)} \tag{15-1}$$

where t is the temperature, λ the wavelength of light in nanometers, α the observed rotation in degrees, l the length of the tube in decimeters, c the concentration in grams of solute per 100 ml of solvent, and the solvent is indicated in brackets. *Specific rotation is as discrete a property of an optically active compound as is its melting or boiling point.* Some optically active substances rotate plane-polarized light to the left and are called **levorotatory** or **negative;** others rotate plane-polarized light to the right and are said to be **dextrorotatory** or **positive.** This is usually indicated by a (−) or (+) sign placed before the name of the compound. Sometimes a small l or small d is used, but the algebraic signs are preferred. Being mirror images, enantiomers have specific rotations which are equal in magnitude, but opposite in sign.

15.4   OPTICAL ACTIVITY: AN EXPLANATION

The question naturally arises as to why certain substances are optically active and others are not. A precise explanation is complex, but briefly it can be stated that, since light is a form of electromagnetic radiation, some interaction is to be expected between a light beam and the electrons in a molecule. The effect produced by any one molecule will be small, but that of a collection of molecules can be appreciable.

A beam of plane-polarized light can be considered to be the result of two component beams of circularly polarized light moving in

---

* In practice, the analyzer is usually balanced at 90° to the polarizer at the start, and to a position which again blocks all the light when the substance being tested is in place. It is easier to detect when the beam is entirely blocked than when it is at maximum intensity. Currently, automatic instruments are available, in which a photoelectric attachment replaces the eye as the detector.

phase but with opposite chirality (Figure 15-10). If achiral molecules were placed in such a beam, they would interact equally with both components and would therefore have no net effect on the plane of polarization. *Achiral molecules are therefore optically inactive.* However, if identical chiral molecules were placed in the light beam, they would interact differently with the right- and left-handed beam components. Suppose, for example, they were to transmit the right-handed beam more readily than the left-handed beam. The left-handed beam would be slowed down relative to the right-handed beam, and the resulting beam would now be polarized in a different plane from the original (Figure 15-11). *A chiral substance is therefore*

Figure 15-10
A light beam moving in the vertical phase is the result of two circularly polarized beams moving in phase but in opposite senses (R = right-handed, L = left-handed).

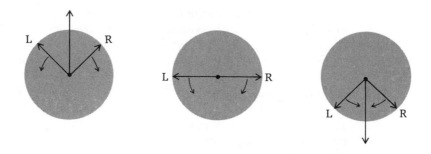

Figure 15-11
The beam of plane-polarized light at the right has interacted with the chiral molecules which retard the L component relative to the R component. The result is rotation of the plane of polarization to the right by α degrees.

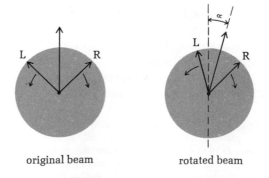

original beam   rotated beam

optically active. An equimolar mixture of two enantiomers (called a **racemic mixture**) would be optically inactive, however, since the rotation in one direction caused by one set of molecules would be precisely balanced by rotation in the opposite direction caused by its mirror image counterparts.

15.5 MOLECULAR DISSYMMETRY AND THE TETRAHEDRAL CARBON ATOM

The need for molecular dissymmetry in optically active molecules was clearly recognized in the mid-nineteenth century by Louis Pasteur, who did pioneering research in this field. Pasteur was aware of two constitutionally identical acids deposited in wine casks during fermentation. One was called **tartaric** acid (now called (+)-tartaric acid). It was optically active and dextrorotatory. The other, called **racemic** acid, was optically inactive. We now know that this was an equimolar mixture of two enantiomers, (+)- and (−)-tartaric acids. Pasteur noted that crystals of the sodium ammonium salt of (+)-tartaric acid were dissymmetric and all of the same chirality. Careful examination of crystals of the sodium ammonium salt of racemic acid showed that they too were dissymmetric, but that some were "right handed" and others were "left handed" (Figure 15-12). With a magnifying lens and pair of tweezers Pasteur carefully separated the mixture into two piles of crystals, each of like handedness. When he separately dissolved the two types of crystals in water, he found that each solution was optically active. Pasteur was the first to achieve, by this experiment, what we now call the **resolution** of a racemic mixture into its two enantiomers. He found that one of these was identical with (+)-tartaric acid; the other had an equal specific rotation but in the opposite direction and was therefore recognized as the mirror image of (+)-tartaric acid. Furthermore, Pasteur ascribed the optical activity to dissymmetry of the molecules

Figure 15-12

Enantiomorphous crystals of (+)- and (−)-sodium ammonium tartrate.

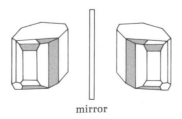

mirror

themselves, rather than to the crystals, since the dissymmetry of the crystals disappeared when they were dissolved, yet the solutions were optically active.

It was not until 1874, however, that the Dutch physical chemist, van't Hoff, and his former fellow student, the Frenchman Le Bel, simultaneously but independently made a bold hypothesis concerning the structure of organic molecules which would explain the optical activity of some and the inactivity of other organic compounds. It had already been established, by Kekulé and others, that carbon was usually tetravalent. Van't Hoff and Le Bel noted that it is a fact of solid geometry that, *if the four corners of a regular tetrahedron have four different things attached to them, two arrangements are possible.* These arrangements are nonsuperimposable mirror images (Figures 15-13 and 15-14). The chirality of the models in these figures is readily seen by glancing down one of the bonds (say, the **C—A** bond, as in Figure 15-15).

In contrast, if two or more of the corners of the tetrahedron are occupied by *identical* groups, the structure has a plane of symmetry, is achiral, and is identical with its mirror image (Figure 15-16).

Le Bel and van't Hoff were the scientists responsible for the suggestion that the valences of carbon were directed toward the corners of a regular tetrahedron. Any carbon, then, which has *four different groups* attached to it would be a source of dissymmetry in the molecule. Such a carbon, called an **asymmetric carbon atom,** would have two possible arrangements of the groups around it. The two structures would be non-superimposable mirror images. They would be expected to differ in some right-handed or left-handed manner, such as the *rotation of plane-polarized light to an equal extent but*

Figure 15-13
Four different groups may be arranged at the corners of a tetrahedron in two different ways.

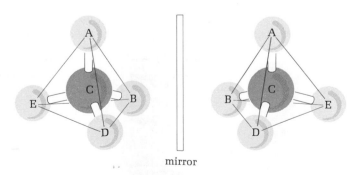

mirror

Figure 15-14
When the four different groups attached to an asymmetric atom are arranged as mirror images, the resulting molecules are not superimposable.
Each molecule is dissymmetric. The models may be twisted or turned in
any direction, but as long as no bonds are broken, only two of the four
attached groups will coincide.

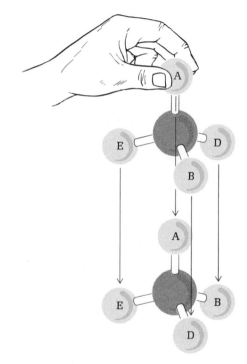

Figure 15-15
The chirality of enantiomers. Looking down the C—A bond, one must
proceed in a clockwise direction to spell BED for the model on the left,
but in a counterclockwise direction for its mirror image. The models at
the left of this figure and Figure 15-13 have identical structures.

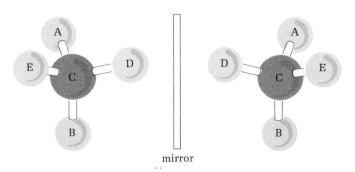

mirror

Figure 15-16

The model at the left with two identical groups A has a plane of symmetry which passes through atoms BCD and bisects angle ACA. Its mirror image is identical to itself as is seen by a 180° rotation of the mirror image about the C—B bond. Hence, the model is achiral.

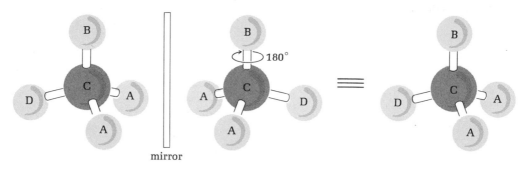

mirror

in opposite directions. In contrast, any carbon which had two identical groups attached to it would be incapable of imparting optical activity to the molecule.

At the time the van't Hoff-Le Bel theory was proposed, thirteen optically active compounds of established structure were known, and all contained at least one asymmetric carbon atom. Further experimentation confirmed the correctness of the theory (the brilliance of which is better appreciated when one recalls that electrons and the nature of atomic structure and chemical bonding were then unknown).

15.6 THE STEREOCHEMISTRY OF LACTIC ACID: CONVENTIONS FOR REPRESENTING THREE-DIMENSIONAL STRUCTURES

Lactic acid (sec 14.2a) is a naturally occurring optically active compound important in several biological processes. Its levorotatory form is found in souring milk whereas its enantiomer, (+)-lactic acid, is produced in living muscle when it performs work. The two lactic acids differ not only in the direction in which they rotate plane-polarized light, but in their biological properties as well. For example, the enzyme lactic acid dehydrogenase can distinguish between the two stereoisomers and will oxidize (+)-, but not (−)-lactic acid to pyruvic acid (sec 14.2d).

Lactic acid can be optically active because it has an asymmetric carbon atom, shown in color in Figure 15-17. The four different

Figure 15-17
The structures of the lactic acid enantiomers.

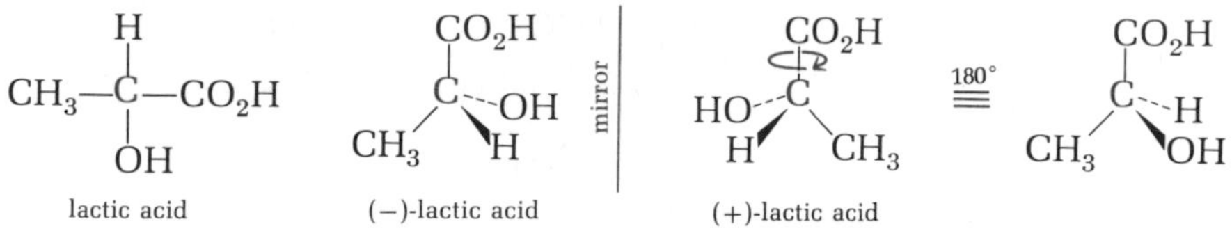

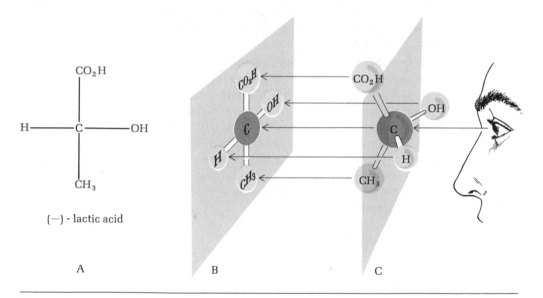

Figure 15-18
*The Fischer projection formula (**A**) stands for the structure in which horizontal groups project toward the viewer and vertical groups project away from the viewer (**C**).*

groups attached to that carbon can be arranged in two ways, giving two enantiomeric structures. The matter of how we know which is (+) and which is (−) will be dealt with later (sec 15.7).

Although three-dimensional formulas such as those in the right portion of Figure 15-17 are useful for designating stereoisomers, they are sometimes cumbersome to use, especially if a molecule contains several asymmetric carbon atoms. Conventions for representing such structures in two dimensions have been devised. One, proposed by the German chemist Emil Fischer, is illustrated in Figure 15-18.

Horizontal groups are considered to project toward the viewer, and vertical groups project away from the viewer; that is

$$
\begin{array}{ccc}
& CO_2H & & CO_2H \\
& | & & | \\
H & -C-OH & \equiv & H{\blacktriangleright}C{\blacktriangleleft}OH \\
& | & & | \\
& CH_3 & & CH_3
\end{array}
$$

If two groups in a projection formula are interchanged, the projection which results represents the enantiomer of the molecule symbolized by the original projection. Similarly, if a two-dimensional projection is rotated by 90° in the plane of the paper, the resulting projection will represent the enantiomer of the molecule symbolized by the original projection. If either of these operations is performed twice, a projection of the original structure is obtained. The validity of these statements can be verified by comparison with models (see also Problems 7-9).

The **Cahn-Ingold-Prelog** convention, named for the three chemists who proposed it, is used to describe, in the name of a compound, the arrangement of four different groups around an asymmetric carbon atom. The four groups are placed in a priority order (by a system which will be described shortly), $a \longrightarrow b \longrightarrow c \longrightarrow d$. The molecule is then observed from the side opposite the lowest priority group (d). If the remaining three groups ($a \longrightarrow b \longrightarrow c$) form a *clockwise* array, the arrangement is designated **R** (Latin: *rectus*-right); if counterclockwise, the arrangement is designated **S** (Latin: *sinister*-left).

$a \longrightarrow b \longrightarrow c$ clockwise $a \longrightarrow b \longrightarrow c$ counterclockwise

R   S

The priority order is assigned as follows: atoms attached to the asymmetric carbon are ordered according to atomic number—the higher the number, the higher the priority. Thus, I $\longrightarrow$ Br $\longrightarrow$ Cl $\longrightarrow$ F, or OH $\longrightarrow$ NH$_2$ $\longrightarrow$ CH$_3$ $\longrightarrow$ H. If two atoms are alike (for example, —CH$_3$ and —CH$_2$CH$_3$), then one goes out to the next atoms in the group and, if necessary, even further until a difference is observed. Thus, the priority order for an ethyl and methyl group would be —CH$_2$CH$_3$ $\longrightarrow$ —CH$_3$ because as one goes out from the attached carbon, one eventually encounters the difference between a carbon and a hydrogen: $-CH_2-CH_3 > -CH_2-H$. Additional rules are available for assigning priorities in more complex molecules.

The priority order for the four groups attached to the asymmetric carbon in lactic acid, using the above rules, is OH $\longrightarrow$ CO_2H $\longrightarrow$ CH_3 $\longrightarrow$ H. Consequently, (+)-lactic acid (Figure 15-17) is seen to have the **S configuration.**

(+)-**S**-lactic acid
(OH $\longrightarrow$ CO_2H $\longrightarrow$ CH_3 counterclockwise)

The symbols **R** and **S**, which refer to configuration, are not necessarily related to (+) and (−), which designate the experimentally observed sign of rotation of plane-polarized light.

15.7 RELATIVE AND ABSOLUTE CONFIGURATION

The term **configuration** refers to the arrangement of atoms in space which characterizes a particular stereoisomer. Enantiomers are said to have "opposite" configurations. Configurations (designated in the Cahn-Ingold-Prelog system by the symbols **R** and **S**) are distinct from conformations (sec 2.5), and a molecule with a particular configuration about a chiral center (asymmetric carbon atom) may assume many conformations. Whereas conformers may be interconverted by rotational motions (Figure 2-5), molecules whose configurations differ because of an asymmetric carbon atom cannot be interconverted without breaking and re-making one or more chemical bonds.

Two substances are said to be *configurationally related* if they can readily be interconverted without altering the relative positions of the groups attached to the chiral center. For example, **S**-lactic acid, if esterified with methanol, gives **S**-methyl lactate.

(15-2)

S-lactic acid   S-methyl lactate

Prior to 1949, no method had been devised to determine the *absolute* configuration of any dissymmetric molecule. However, the *relative* configurations of two different substances could still be established by interconversions such as that shown in equation 15-2. For example, when levorotatory or (−)-lactic acid was experimentally isolated from sour milk and esterified with methanol, the resulting methyl lactate had the same relative configuration as the (−)-lactic acid, since none of the bonds to the asymmetric carbon were broken during the reaction. It happened that the methyl lactate obtained in this experiment was also levorotatory. One could therefore say that (−)-lactic acid and (−)-methyl lactate have the same *relative* configurations, even though the *absolute* configuration of each had not yet been established.

Emil Fischer years ago suggested that *glyceraldehyde* be chosen as a standard reference compound, and that the configurations of other dissymmetric molecules be related to it through chemical transformations. He chose glyceraldehyde because it could readily be converted to carbohydrates and amino acids, two important classes of naturally occurring optically active substances. He arbitrarily assigned the dextrorotatory isomer, or (+)-glyceraldehyde, the configuration now known as **R**.

$$
\begin{array}{ccc}
& \text{CHO} & & \text{CHO} \\
& | & & | \\
\text{HOH}_2\text{C}\overset{\displaystyle\text{C}}{\diagdown}\overset{\text{OH}}{\underset{\text{H}}{}} & \equiv & \text{H}-\text{C}-\text{OH} \\
& & & | \\
& & & \text{CH}_2\text{OH}
\end{array}
$$

R-(+)-glyceraldehyde   Fischer projection formula for D*(+)-glyceraldehyde

In 1949, an ingenious application of X-ray diffraction to the crystals of certain optically active substances permitted the determination of absolute configurations. It turned out that Fischer had guessed correctly, and that (+)-glyceraldehyde does in fact have the **R**-configuration. This was fortunate, since the relative configurations of many molecules then also turned out to be their correct absolute configurations.

15.8 DIASTEREOMERS

When a molecule has two or more chiral centers, an important phenomenon emerges; it becomes possible to have two substances which are stereoisomers but which are not mirror images. *Stereoisomers which are not related as enantiomers are called* **diastereo-**

---

*He assigned it the symbol D and referred to its enantiomer as L. The Fischer symbols are still used in carbohydrate and amino acid chemistry (Chapter 16 and 17).

Figure 15-19

*The stereochemistry of molecules with two different asymmetric carbon atoms is illustrated with shoes (colored and black). Sets **AB** and **CD** are clearly mirror images (enantiomers) whereas sets **AC**, **AD**, **BC**, and **BD** are diastereoisomeric.*

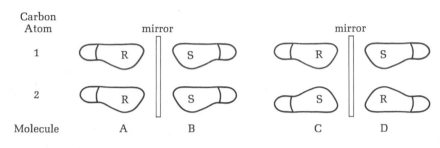

mers (or diastereoisomers).*

Consider, for example, a molecule with two asymmetric carbon atoms. The absolute configuration at each carbon may be either **R** or **S**. If we use colored and black letters to indicate the two different asymmetric carbon atoms, we see that four combinations are possible.

	Carbon Atom	
Molecule	(1)	(2)
A	R—	R } enantiomers
B	S—	S }
C	R	S } enantiomers
D	S—	R }

Since **R** is the mirror image of **S**, molecules **A** and **B** are enantiomers (the configuration at carbon 1 and at carbon 2 in **B** is the mirror image of the configurations at each of those carbon atoms in **A**). **C** and **D** constitute another pair of enantiomers. However, molecules **A** and **D**, though stereoisomers, are not mirror images. The configuration at carbon 2 is identical in both molecules; **A** and **D** are therefore diastereomers. Other pairs of diastereomers are **AC**, **BC**, and **BD**.

The relationship just described for molecules **A–D** is illustrated with colored and black shoes in Figure 15-19, where one can clearly see that **R—R** is the mirror image of **S—S**, etc.

---

* According to the definition of diastereomers given here, geometric (*cis-trans*) isomers are diastereomers. Some chemists see no reason for a special subcategory for such diastereomers, whereas others believe it is a useful distinction which should be retained.

As an example of a molecule with two different asymmetric carbon atoms, consider

α-hydroxy-β-methylsuccinic acid

The colored carbons are asymmetric; they are also different from one another in that the four groups attached to one of them are different from the four groups attached to the other. The four possible isomers are shown in three dimensions and as the corresponding Fischer projections in Figure 15-20. The projection formulas are obtained from the three-dimensional formulas by first rotating the rear asymmetric carbon 180° so that the carboxyl group is down, as it is on the front asymmetric carbon; the model is then viewed from above, or imagined to be pressed down on a flat surface. This is illustrated for formula **A** of Figure 15-20. In accord with the Fischer

convention, horizontal groups extend above the plane of the paper toward the viewer, and vertical groups extend below the paper plane, with the asymmetric carbon atoms lying in the plane of the paper. The formulas in Figure 15-20 correspond to those indicated in Figure 15-19. The pairs **AB** and **CD** are enantiomers, whereas the other possible pairs are diastereomers.

Note that α-hydroxy-β-methylsuccinic acid has *two* different asymmetric carbon atoms and exists in *four* stereoisomeric structures. It can be shown that *if a molecule has n different asymmetric carbon atoms it may exist in 2^n isomeric forms* **(the van't Hoff rule).** This rule will be particularly useful when we discuss carbohydrate chemistry (Chapter 16).

Because they are not mirror images, two compounds which are diastereomers will differ not only in chiral properties, but in achiral properties as well. They may differ in melting points, boiling points, and in their solubility in various solvents. If dissymmetric, they will interact with plane-polarized light but need not rotate the light in opposite directions nor have equal specific rotations. In short, though stereoisomers, they will behave as two different chemical substances.

Figure 15-20

The stereoisomers of α-hydroxy-β-methylsuccinic acid shown in "saw-horse" and Fischer projection formulas.

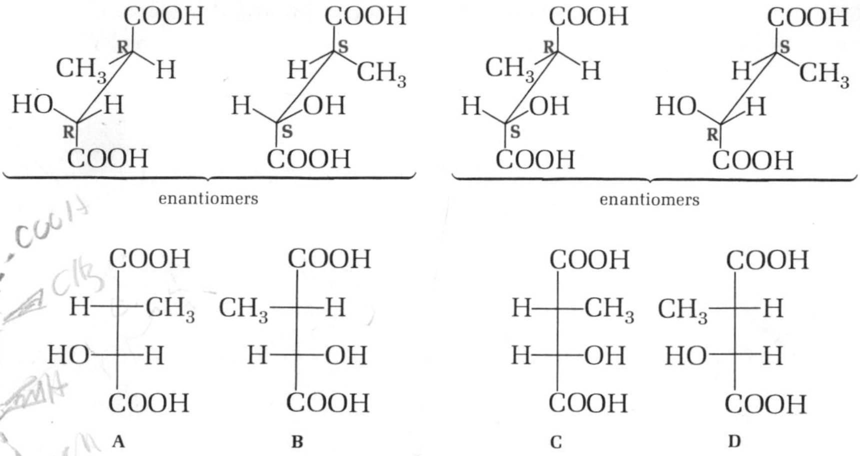

enantiomers   enantiomers

15.9 MESO STRUCTURES

It is possible for a molecule to possess a chiral center yet be achiral and optically inactive if it possesses in the same molecule another center of equal but opposite chirality. This situation can arise in a molecule with two *identical* asymmetric carbon atoms (that is, two carbon atoms each of which has the same four different groups attached to it). If carbon 1 and carbon 2 are indistinguishable (both shown in color), the chart in section 15.8 simplifies to

	Carbon Atom	
Molecule	(1) (2)≡(1)	
A	R—R	} enantiomers
B	S—S	
C	R—S ↙ meso form	

A and B are enantiomers, and each can be optically active. But C (which is a diastereomer of A and B) is optically inactive, since it has a plane of symmetry and is therefore identical with its mirror image.

plane of symmetry

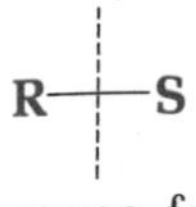

R—|—S

C (a meso form)

Figure 15-21
*The stereochemistry of molecules with two identical asymmetric carbon atoms is illustrated with shoes (both colored). Set **AB** forms a mirror image pair (enantiomers) whereas **C** and **D** are identical, possess a plane of symmetry, and constitute a meso form.*

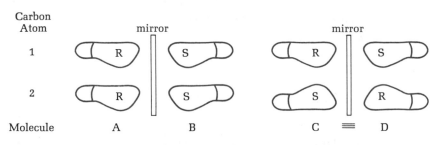

Carbon Atom		mirror				mirror	
1	R		S		R		S
2	R		S		S		R
Molecule	A		B		C	$\equiv$	D

The phenomenon can easily be seen with objects if we consider what happens to the sets of shoes in Figure 15-19 when we make both shoes (carbon atoms) identical; that is, both colored (or black). The result is seen in Figure 15-21. Whereas **A** and **B** still form a pair of enantiomers, **C** and **D** are identical (each consists of a right and left colored shoe). The plane of symmetry in **C** (or **D**) is shown below.

R

– – – – – – – – – – – plane of symmetry

S

C

Tartaric acid, the compound whose optical activity was first studied carefully by Louis Pasteur, has two *identical* asymmetric carbon atoms. Each asymmetric carbon (shown in color) has the same four

$$HO_2C-\underset{\underset{OH}{|}}{\overset{\overset{H}{|}}{C}}-\underset{\underset{OH}{|}}{\overset{\overset{H}{|}}{C}}-CO_2H$$

tartaric acid

groups attached, H, OH, COOH, and —CH(OH)COOH. Tartaric acid exists in three stereoisomeric forms, shown in Figure 15-22. Two of these (**A** and **B**) are optically active and form an enantiomeric pair. The third (**C**) has a center of symmetry in its staggered conformation, or a plane of symmetry in its eclipsed conformation, and is therefore optically inactive and a *meso* form. **C** is a diastereomer of **A** and **B.** The properties of the three forms of tartaric acid are shown in Table 15-1. Note that the enantiomers have identical properties except for the sign of rotation of plane polarized light, whereas the *meso* form, being a diastereomer of the enantiomers, differs from them in all its properties.

15.10 DISSYMMETRIC MOLECULES *WITHOUT* ASYMMETRIC CARBON ATOMS

It is possible for a molecule to be chiral (i.e., dissymmetric) without having an asymmetric carbon atom. Consider, for example, the

Figure 15-22
The stereoisomers of tartaric acid.

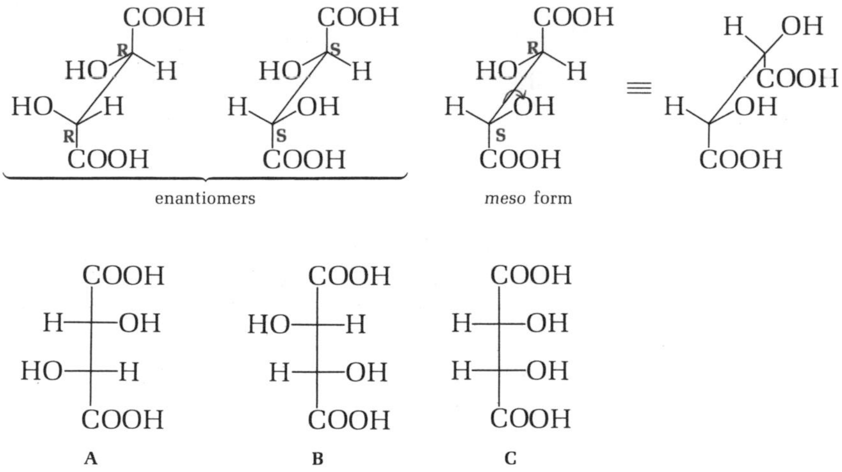

Table 15-1
Physical Properties of the Tartaric Acids

TARTARIC ACID	MELTING POINT, °C	SPECIFIC GRAVITY	SOLUBILITY, g/100g H_2O, 20° C	$[\alpha]_D^{20}$ in H_2O
RR	170	1.76	139	+12°
SS	170	1.76	139	−12°
RS	140	1.67	125	0°

staggered conformations of n-butane shown in Figure 15-23. Conformation **A** has no plane or center of symmetry and is dissymmetric. It is therefore not superimposable on its mirror image **B**. However, **B** can be made identical to **A** by a rotational motion about the central single bond. Since such rotations occur readily at room temperature, **A** and **B** cannot be separated and n-butane is optically inactive. However, if some means could be devised to slow down or prevent the interconversion of the two conformers, then optical activity would ensue.

Figure 15-23
*Conformations **A** and **B** of n-butane constitute a pair of enantiomers which, however, are readily interconvertible by a rotation of the C_2—C_3 single bond.*

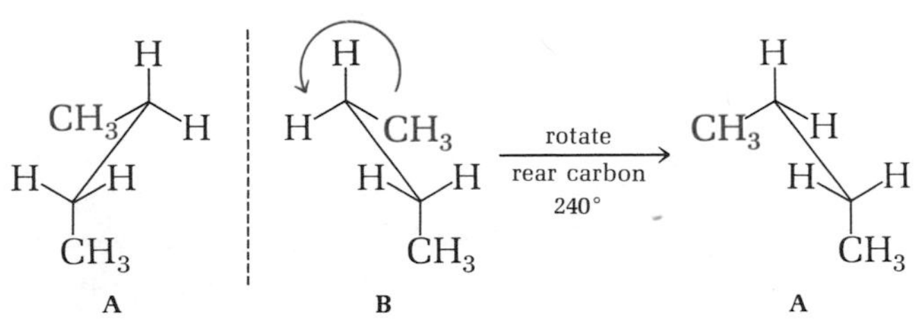

Several classes of molecules have been devised in which the interconversion of two dissymmetric conformers is prevented, either by making use of bonds which cannot freely rotate (that is, double bonds), or by placing large groups in positions which restrict what would otherwise be fairly free rotational motions.

The hydrocarbon 2,3-pentadiene (an **allene**, sec 4.7), although it has no asymmetric carbon atoms, exists in two enantiomeric forms,

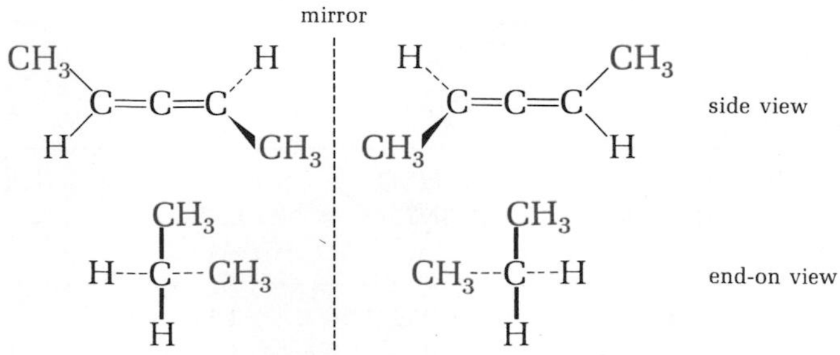

2,3-pentadiene

the chirality of which is perhaps best appreciated in the end-on view.*
Interconversion could be accomplished in theory by rotating one of
the methyl-bearing carbons 180° with respect to the other, but this
is not possible for double bonds at room temperature. Consequently,
the enantiomers are separable and each is optically active.

Certain ortho-substituted **biphenyls** behave similarly. The large

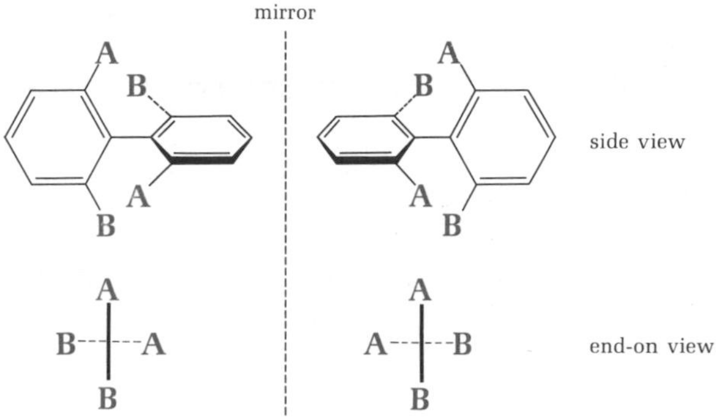

optically active biphenyls

groups in the ortho positions prevent the interconversion of the two
enantiomers which would otherwise occur by simple rotation of one
benzene ring 180° with respect to the other about the single bond
which joins the two rings. **A** and **B** are usually large groups, such
as NO_2 and CO_2H. The larger the groups, the more difficult it is to
interconvert the two enantiomers, although this can frequently be
accomplished by heat.

Hexahelicene exists in two forms which represent the first turns
of a right- and left-handed helix. The two benzene rings at the "top"

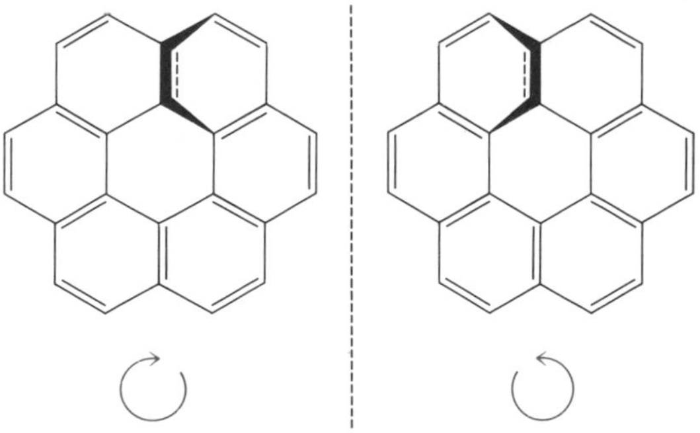

hexahelicene

* In the end-on view, the solid lines are bonds to the front carbon, which is shown;
the dashed lines are bonds to the rear carbon, which is directly behind the middle and
front carbons.

of the molecule are not chemically bonded to one another. They can only move past one another (a process which would interconvert the two enantiomers) by a motion which pulls them apart and enlarges the diameter of the helical loop. This does not occur readily at room temperature. Helical structures are important in several naturally occurring materials such as proteins (Chapter 17) and nucleic acids (Chapter 18).

15.11 SYNTHESIS OF DISSYMMETRIC MOLECULES

Most laboratory syntheses of compounds with one asymmetric carbon atom lead to racemic mixtures. The reason becomes clear if we consider a specific case. Equation 15-3 describes a synthesis of lactic acid from propionic acid (review eq 14-22).

$$CH_3CH_2CO_2H \xrightarrow[P]{Br_2} \underset{\underset{\alpha\text{-bromopropionic acid}}{Br}}{CH_3\overset{|}{C}HCO_2H} \xrightarrow{OH^-} \underset{\underset{\text{lactic acid}}{OH}}{CH_3\overset{|}{C}HCO_2H} \qquad (15\text{-}3)$$

Propionic acid has a plane of symmetry (the plane of this page, in eq 15-4) which makes both hydrogens on the α-carbon equivalent. The α-bromo acid which is obtained in the first step is therefore a racemic (50-50) mixture of the **R** and **S** enantiomers, and is optically inactive. It is not possible to obtain a dissymmetric product from symmetric reactants. Thus the laboratory synthesis of lactic acid according to equation 15-3 gives a racemic product and is said to be **symmetric.**

$$(15\text{-}4)$$

But, in sharp contrast, most biological syntheses are **asymmetric.** The reactions occur on enzyme surfaces which are themselves asymmetric. For example, when glycerol is phosphorylated by adenosine triphosphate (ATP) in the presence of the enzyme glycerokinase, only the **R**-(−)-α-glycerophosphate is produced.

$$(15\text{-}5)$$

How might one explain such a result? One possibility is shown in Figure 15-24. Suppose the enzyme surface has three active sites, **ABC,** arranged in a clockwise manner. Two of these (**A** and **B**) may bind the secondary and primary alcohol functions respectively, whereas **C** may bind the ATP which is to deliver phosphate to one of the primary alcohol groups. Clearly only the arrangement at the left allows the ATP to react. Consequently, only one of the two possible stereoisomeric glycerophosphates will be produced; the reaction will be 100% stereospecific and give only the product with the **R** configuration. Such stereospecificity is the rule, rather than the exception, in biological systems.

Figure 15-24
In the arrangement at the left, the primary alcohol is properly positioned to accept a phosphate group from ATP bond at C, whereas in the arrangement at the right, no phosphorylation can occur.

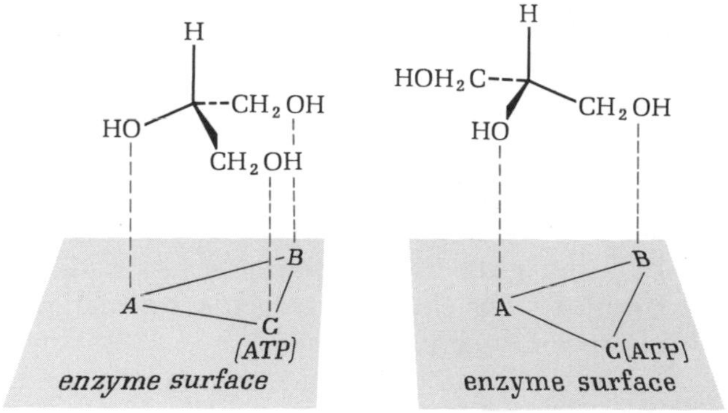

15.12 RESOLUTION OF RACEMIC MIXTURES

Since the products obtained from most laboratory syntheses of dissymmetric molecules are racemic mixtures, it becomes important to have methods for resolving such mixtures into their optically active components. The physical properties of enantiomers are identical, which means that they cannot be separated by ordinary physical methods, such as distillation, crystallization, or chromatography.

One method used by Pasteur, the mechanical separation of dissymmetric crystals, has already been described (sec 15.5). This procedure is obviously laborious, possible only in special cases, and is primarily of historic interest only.

The guiding principles behind other methods of resolution are (*a*) that enantiomers react differently with a dissymmetric reagent, and (*b*) that diastereomers (in contrast to enantiomers) differ in all physical properties (sec 15.8). Consider, for example, the reaction of a racemic acid with an optically active base (eq 15-6).

$$\text{racemic}\begin{cases} \textbf{R}\text{-acid} \\ \textbf{S}\text{-acid} \end{cases} + \textbf{R}\text{-base} \longrightarrow \begin{cases} \textbf{R}\text{—}\textbf{R} \text{ salt} \\ \textbf{S}\text{—}\textbf{R} \text{ salt} \end{cases} \qquad (15\text{-}6)$$

The product consists of two salts which, being diastereomers, will differ in many properties. They may, for example, have different solubilities in some solvent, enabling separation by recrystallization. The *separated* salts may then be reconverted to the original acids, now resolved. In practice, naturally occurring, optically active, weak bases, such as quinine, are used; the resolved acid can then be recovered from the salt by treating it with a strong acid (eq 15-7 and 15-8).

$$\textbf{R}\text{—}\textbf{R} \text{ salt} + \text{HCl} \longrightarrow \textbf{R}\text{-acid} + \textbf{R}\text{-base hydrochloride} \quad (15\text{-}7)$$

and

$$\textbf{S}\text{—}\textbf{R} \text{ salt} + \text{HCl} \longrightarrow \textbf{S}\text{-acid} + \textbf{R}\text{-base hydrochloride} \quad (15\text{-}8)$$

Similar resolutions of bases may be accomplished using an optically active acid as the resolving agent. Alcohols may be resolved by forming the diastereomeric esters with an optically active acid. By a judicious choice of reagents, the method is applicable to many classes of organic compounds.

The principle of this method can be used in many other ways. Resolutions have been accomplished by recrystallization from an optically active solvent, by chromatography using optically active (dissymmetric) adsorbents, and by feeding racemic mixtures to microorganisms (which contain optically active enzymes that react selectively with one enantiomer).

New Concepts, Facts, and Terms

1. Chirality (handedness) results when a molecule and its mirror image are nonsuperimposable; achiral = without handedness.
2. Symmetry elements: axis, center, plane. Dissymmetry, asymmetry.
3. Enantiomers—two molecules related as object and nonsuperimposable mirror image. Differ in chiral properties.
4. Plane-polarized light; optical activity; polarimeter; specific rotation; dextro- and levorotatory molecules; circularly polarized light

5. Tetrahedral carbon is asymmetric if four different groups are attached; two enantiomeric arrangements are possible. Racemic mixture (1:1) of two enantiomers is optically inactive.

6. Lactic acid has an asymmetric carbon; Fischer projection formulas; Cahn-Ingold-Prelog **R,S** nomenclature.

7. Relative and absolute configuration; **R**-(+)-glyceraldehyde

8. Diastereomers; stereoisomers not related as object and mirror image; differ in achiral as well as chiral properties. Van't Hoff rule—2^n isomeric forms if a molecule has n different asymmetric carbon atoms

9. Meso structure; optically inactive. Molecule with two identical asymmetric carbon atoms of opposite chirality; tartaric acid

10. Dissymmetric molecules without asymmetric carbon atoms; allenes, biphenyls, hexahelicene

11. Laboratory synthesis of molecules with one chiral center gives racemic mixtures; enzymatic syntheses are stereospecific.

12. Resolution of racemic mixtures

Exercises and Problems

1. Define the following terms:
   a. chiral molecule
   b. dissymmetric molecule
   c. asymmetric molecule
   d. enantiomers
   e. polarized light
   f. specific rotation
   g. asymmetric carbon atom
   h. chiral center
   i. meso form
   j. diastereomers
   k. racemic mixture
   l. plane of symmetry

2. Figure 15-1 shows the five-fold axis of symmetry in planar cyclopentane. This structure also has six planes of symmetry and five two-fold symmetry axes. Describe them.

3. Is a conical helix (shown) asymmetric or dissymmetric? Compare with Figure 15-3.

4. a. What is a necessary and sufficient condition for enantiomerism?
   b. What is a sufficient but not a necessary condition for enantiomerism?
   c. How can one determine from the formula of a compound whether or not it can exist in enantiomeric forms?

5. Circle the asymmetric carbon atoms in the formulas given on the following page:

a. $CH_3CH(Br)CH_2CH_3$

b. $CH_3CH(Cl)CH(Cl)CH_3$

c. $C_6H_5CH(OH)CO_2H$

d. $CH_2(OH)CH(OH)CH(OH)CH_2OH$

e.

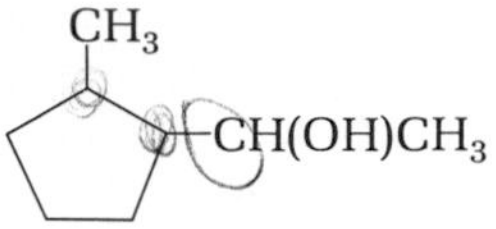

f.

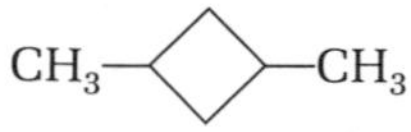

6. Which of the following substances are capable of existing in optically active forms?

   a. 1-chloro-2-propanol
   b. cyclopentanol
   c. 2-bromoethanol
   d. 1-chloro-1-phenylethane

   e. 2-aminopropionic acid
   f. *cis*-1,2-dimethylcyclobutane
   g. *trans*-1,2-dimethylcyclobutane
   h. 2,2-dibromopropane

7. The Fischer projection formulas shown represent the two enantiomers of lactic acid. Show that one can be obtained from the other *only* by

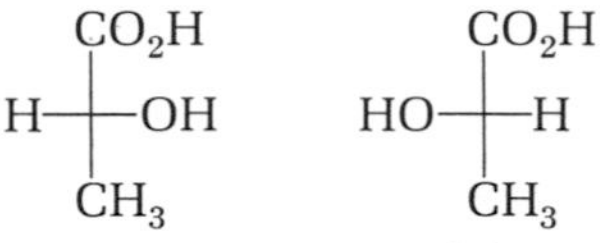

   an *odd* number of interchanges of groups (for example, try to obtain the (+) formula from the (−) formula by first interchanging the OH and CH_3 positions in the latter).

8. If the Fischer projection formula for (−)-lactic acid shown in Problem 7 is rotated 90° to the right, one obtains CH_3—CO$_2$H. Show that this changes the configuration to (+)-lactic acid (i.e., that it can be obtained from the (−) formula only by an *odd* number of group interchanges).

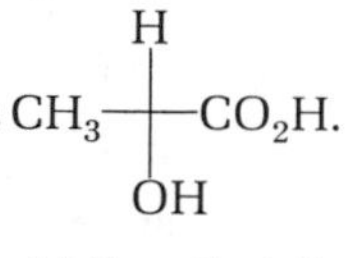

9. Which of the following Fischer projection formulas for lactic acid represents the (−)-enantiomer?

   a.

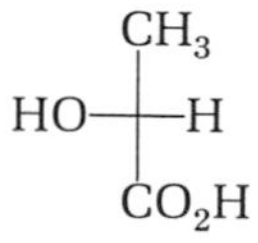

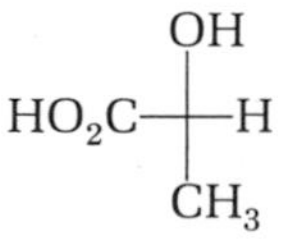

   b.

   c.

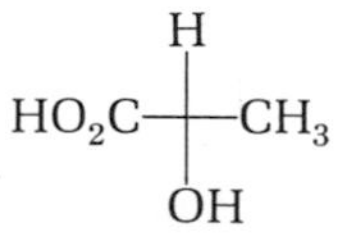

   d.

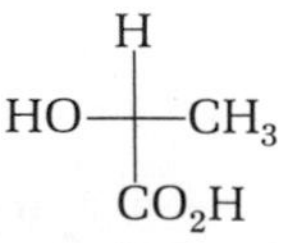

10. Place the following groups in proper priority order according to the Cahn-Ingold-Prelog **(R,S)** convention:

    a. CH_3, H, OH, CH_3CH_2—
    b. H, CH_3, C_6H_5, Cl
    c. CH_3, OH, $-CH_2Cl$, $-CH_2OH$
    d. $-C(CH_3)_3$, $-CH(CH_3)_2$, $-CH_2CH_3$, CH_3

11. Assume that the four groups in each part of Problem 10 are attached to a single asymmetric carbon atom. Draw a three-dimensional formula for the **R**-form of each molecule.

12. Draw Fischer projection formulas for the compounds in Problem 11, in each case placing the lowest priority group at the "bottom" of the formula. What do you notice about the clockwise or counterclockwise arrangement of the remaining three groups in their proper priority order?

13. Draw a structural formula for an optically active compound with the molecular formula
    a. $C_4H_{10}O$ c. $C_4H_8(OH)_2$
    b. $C_5H_{11}Br$ d. C_6H_{12}

14. Might the fat, glyceryl stearopalmitoöleate, be optically active? What effect would catalytic addition of a mole of hydrogen to this glyceride have on its optical behavior?

15. Show that the **R,S** designations in Figures 15-20 and 15-22 are correct.

16. Two possible configurations for a molecule with 3 asymmetric carbons are **R—R—R** and its mirror image, **S—S—S**. Verify the van't Hoff rule for compounds with three and four different asymmetric carbons by writing out all the remaining possibilities.

17. Draw a Fischer projection formula for **R,R**-2,3-dibromobutane. Now draw a Fischer projection formula for its enantiomer; its diastereomer. What type of structure does the latter compound have?

18. In each of the following reactions would you expect the product to be optically active or inactive? Explain.
    a. $CH_3CH{=}CHCH_3 + Cl_2 \longrightarrow$
    b. $(+)\text{-}C_6H_5CH(Br)CH_3 + CN^- \longrightarrow$
    c. $(+)\text{-}CH_2{=}CH{-}CH(OH)CH_2CH_3 + H_2 \xrightarrow{Pt}$
    d. $CH_3CH{=}O + HCN \longrightarrow$

19. When maleic acid is treated with bromine, the resulting addition product, 2,3-dibromosuccinic acid, can be resolved into enantiomers. However, addition of bromine to fumaric acid gives a 2,3-dibromosuccinic acid which cannot be resolved. What do these observations indicate regarding the mechanism of bromine addition?

20. A derivative of one of the tartaric acids is optically active but yields optically inactive products when esterified with methyl alcohol or when hydrolyzed. Deduce the stereochemical formula which fits these data, and explain, using equations.

21. Can this ortho-substituted biphenyl be optically active? Explain.

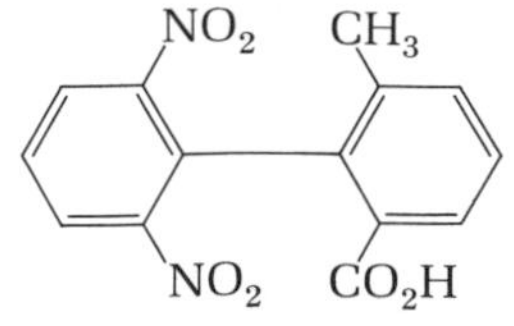

22. Which conformations of *n*-butane shown in Figure 2-6 are chiral? Which conformations have a symmetry element which makes them achiral?

23. When racemic (±) 2-chlorobutane is chlorinated, one obtains some 2,3-dichlorobutane which consists of 71% meso isomer and 29% racemic (±) isomers. Explain why the mixture need not be 50:50 meso and racemic 2,3-dichlorobutane (it will help if you draw three-dimensional structures in seeking an explanation).

24. When the (−)-isomer of **A** is converted to its ethyl ester **B**, the latter

$$CH_3-\overset{\overset{\displaystyle CO_2H}{\vert}}{\underset{\underset{\displaystyle CH_2OH}{\vert}}{C}}-CH_2CH_3$$

A

can be oxidized to an acid **C**. Catalytic hydrogenolysis (eq 10-26) of **C** yields **D** which is still optically active. Write equations, using Fischer projection formulas, for the conversion of **A** ⟶ **B** ⟶ **C** ⟶ **D**. How are **A** and **D** related?

25. Paper chromatography of a racemic mixture sometimes results in separation of the enantiomers. Explain.

CHAPTER SIXTEEN

CARBOHYDRATES

Carbohydrates constitute a major class of naturally occurring organic compounds. They include sugars, starches, and cellulose; they are essential to the maintenance of plant and animal life. Carbohydrates provide the raw materials for many industries, among them clothing, paper, fermentation, confections, and film. Although carbohydrate formulas may at first seem complex, their chemistry is primarily that of the carbonyl and hydroxyl groups.

The name **carbohydrate** originated from the empirical formulas of compounds in this class, many of which can be represented as $C_m(H_2O)_n$ (hydrates of carbon). For example, the molecular formula for glucose is $C_6H_{12}O_6$ or $C_6(H_2O)_6$ and for sucrose, or cane sugar, $C_{12}H_{22}O_{11}$ or $C_{12}(H_2O)_{11}$. Although these molecular formulas bear little relationship to the structures of the molecules, the general name persists.

Structurally, carbohydrates are polyhydroxy aldehydes or polyhydroxy ketones, or they are substances which, on hydrolysis, give these classes of compounds. The chemistry of carbohydrates involves, therefore, the chemistry of aldehydes or ketones, of alcohols, and of the hemiacetals and acetals which can be formed from them.

16.1 CLASSIFICATION

Carbohydrates are conveniently classified into three major groups: **mono-, oligo-, and poly**saccharides. Monosaccharides are the simplest carbohydrate units. Oligosaccharides contain two or more (but a small number; *oligo* = few) of such units linked through acetal or ketal functions, and polysaccharides contain many (hundreds or thousands) of such units. Partial hydrolysis of polysaccharides gives oligosaccharides, whereas complete hydrolysis yields monosaccharides. Common examples of each type are shown in equation 16-1.

$$\text{polysaccharides} \xrightarrow[\text{H}^+]{\text{H}_2\text{O}} \text{oligosaccharides} \xrightarrow[\text{H}^+]{\text{H}_2\text{O}} \text{monosaccharides}$$

$$\text{Examples} \begin{cases} \text{starch} & \text{sucrose} & \text{glucose} \\ \text{cellulose} & \text{lactose} & \text{fructose} \\ & \text{maltose} & \text{ribose} \end{cases}$$

$$(16\text{-}1)$$

The functional group which links two or more monosaccharide units together is called a **glycosidic** linkage.

Common names are used extensively in carbohydrate chemistry. The suffix *-ose* signifies the carbohydrate structure. The most common monosaccharides (also called simple sugars) have five or six carbon atoms, hence are called **pent**oses or **hex**oses. They may also be classified as **ald**oses or **ket**oses, depending on whether they are polyhydroxyaldehydes or ketones, respectively. Combination names, such as aldopentose or ketohexose, convey both types of information.

16.2 THE MONOSACCHARIDE GLUCOSE

The most important monosaccharide is **glucose** (also called *dextrose*, because the form which occurs naturally is optically active and dextrorotatory). It constitutes about 0.1% of the blood of mammals and is essential to life. Glucose, either free or combined with other molecules, is probably the most abundant organic compound. It is the ultimate hydrolysis product of starch and cellulose.

16.2a THE ACYCLIC STRUCTURE OF GLUCOSE

The experiments which led to the determination of the structure of glucose are instructive. The empirical formula CH_2O was deduced from the compound's percentage composition; the molecular weight showed that its molecular formula was $(CH_2O)_6$ or $C_6H_{12}O_6$. Complete reduction gave *n*-hexane, showing that the six carbons formed a consecutive, unbranched chain.

Mild oxidation of glucose (with bromine water) gave *gluconic acid,* a monocarboxylic acid with the formula $C_6H_{12}O_7$. This implied the presence of an aldehyde group, since only the aldehyde function can be oxidized to an acid by gaining one oxygen atom without losing any hydrogen atoms ($-CHO \longrightarrow -COOH$). The aldehyde function was confirmed by a positive silver mirror test (eq 9-14). The aldehyde group, therefore, had to occupy one end of the six-carbon chain.

Further oxidation of gluconic acid gave *glucaric acid,* a dicarboxylic acid with the formula $C_6H_{10}O_8$. This indicated the presence of a primary alcohol function, since oxidation occurred with the loss of two hydrogens and gain of one oxygen ($-CH_2OH \longrightarrow -CO_2H$; $C_6H_{12}O_7 \longrightarrow C_6H_{10}O_8$).

Reduction of glucose (with sodium amalgam) gave *sorbitol,* $C_6H_{14}O_6$. This reacted with acetic anhydride to form a hexa-acetate. Consequently, sorbitol must have six hydroxyl groups, one on each carbon atom. (Compounds with two hydroxyl groups on a single carbon atom are rare; those which are known usually lose water to produce a carbonyl group.) This suggested that glucose itself had five hydroxyl groups, and consistent with this deduction, glucose formed a penta-acetate with acetic anhydride.

These chemical properties of glucose can all be accommodated by a structure having a six-carbon chain with an aldehyde function at one end and a hydroxyl group on each of the remaining carbon atoms. The reactions and structural deductions are summarized in Table 16-1.

Though these experiments established the gross structure of glucose, they left several details to be accounted for. One was the optical activity. The structure has four different asymmetric carbon atoms, (carbons 2-5), which permit 2^4 or 16 possible stereoisomers (van't Hoff rule, sec 15.8). It was necessary, then, to determine which of

$$\underset{OH \quad\; OH \; OH \; OH \; OH}{\overset{6 \qquad 5 \quad\;\; 4 \quad\; 3 \quad\; 2 \quad\;\; 1}{CH_2-CH-CH-CH-CH-CH=O}}$$

Table 16-1
Reactions Leading to a Structure for Glucose

	FACT	CONCLUSION
	→ Percentage composition	CH_2O
	→ Molecular weight	$C_6H_{12}O_6$
	→ n-Hexane	C—C—C—C—C—C (continuous chain)

CH=O
CHOH
CHOH
CHOH
CHOH
CH₂OH

Glucose

reduction

$$\begin{array}{c}CO_2H\\|\\(CHOH)_4\\|\\CH_2OH\end{array}$$ Gluconic acid $\xrightarrow[\text{strong oxidation}]{HNO_3}$ $$\begin{array}{c}CO_2H\\|\\(CHOH)_4\\|\\CO_2H\end{array}$$ Glucaric acid

—CHO and —CH₂OH at the ends of the six-carbon chain

Br₂, H₂O mild oxidation

$\xrightarrow{Ag(NH_3)_2{}^+}$ Silver mirror — Confirms —CHO group

reduction (Na/Hg) $\rightarrow$ $$\begin{array}{c}CH_2OH\\|\\(CHOH)_4\\|\\CH_2OH\end{array}$$ Sorbitol $\xrightarrow{(CH_3CO)_2O}$ $$\begin{array}{c}CH_2OCOCH_3\\|\\(CHOCOCH_3)_4\\|\\CH_2OCOCH_3\end{array}$$ Sorbitol hexa-acetate

Six —OH groups in sorbitol, five —OH groups in glucose

$\xrightarrow{(CH_3CO)_2O}$ $$\begin{array}{c}CH=O\\|\\(CHOCOCH_3)_4\\|\\CH_2OCOCH_3\end{array}$$ Glucose penta-acetate

Confirms five —OH groups in glucose

these sixteen isomers was the naturally occurring (+)-glucose. This task was accomplished through a series of ingenious experiments carried out by Emil Fischer and his students.* The correct Fischer projection formula for (+)-glucose is shown. Sometimes the

$$\begin{array}{c}{}^1CHO\\|\\H—C—OH\\|\\HO—C—H\\|\\H—C—OH\\|\\H—{}^5C—OH\\|\\CH_2OH\end{array}$$

Fischer projection formula for D(+)-glucose

$$\begin{array}{c}CHO\\|\\\_\_\\\_\_\\\_\_\\|\\CH_2OH\end{array}$$

Abbreviated Fischer projection formula

$$\begin{array}{c}CHO\\|\\H—C—OH\\|\\CH_2OH\end{array}$$

Fischer projection formula for D(+)-glyceraldehyde

*For an especially clear account, see C. R. Noller, *Chemistry of Organic Compounds*, 3rd edition, W. B. Saunders Company, 1965, pp. 383–386.

Figure 16-1
Emil Fischer (1852–1919), the illustrious German organic chemist who is noted for his pioneering work in the field of carbohydrates, proteins, and dyestuffs. He received the Nobel prize in 1902. (Courtesy of The Bettmann Archive, Inc.)

formula is abbreviated to show only the disposition of the hydroxyl groups on the asymmetric carbon atoms as short lines to the right or left. The prefix D is used to denote that the configuration at the carbon atom adjacent to the primary alcohol (in glucose, C-5) has the same configuration as that of D(+)-glyceraldehyde (in the Cahn-Ingold-Prelog notation, **R**). The Fischer projection is most easily visualized if the molecule is arranged so that the carbon chain forms an arc and is viewed from "outside" that arc (Figure 16-2). In this case, whenever attention is focused on one of the asymmetric carbon atoms, the two carbon atoms of the chain attached vertically to it recede from the viewer, and horizontal groups come toward the viewer. Students are advised to verify these points with molecular models.

16.2b   THE CYCLIC STRUCTURE OF GLUCOSE

The structure just deduced for glucose, though it successfully rationalizes the data in Table 16-1, does not explain certain other types of chemical behavior, and must be modified to account for them.

Figure 16-2
*The "arc" structure of D(+)-glucose, which, if stretched and flattened,
gives the Fischer projection formula.*

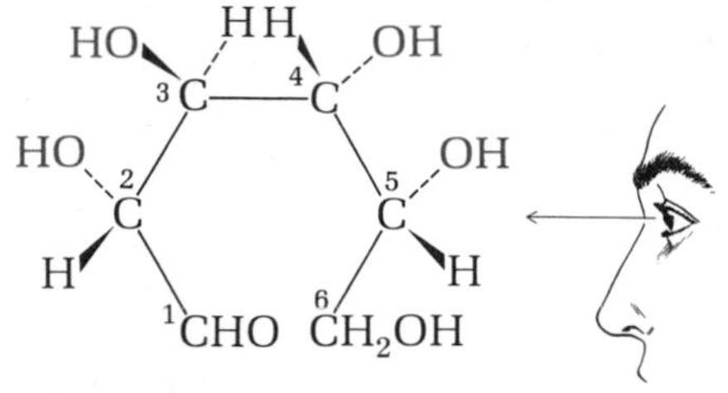

If, for example, glucose is an aldehyde, it ought to be possible to
convert it to an acetal by reaction with two moles of methanol
(eq 16-2; review sec 9.6c). When Emil Fischer tried the experiment

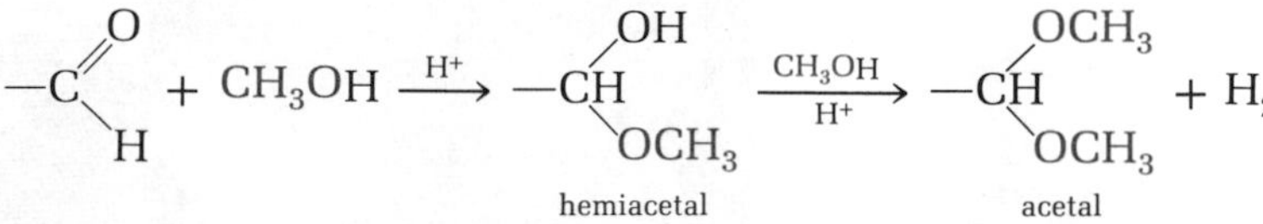

$$-\overset{O}{\underset{H}{C}} + CH_3OH \xrightarrow{H^+} -\overset{OH}{\underset{OCH_3}{CH}} \xrightarrow[H^+]{CH_3OH} -\overset{OCH_3}{\underset{OCH_3}{CH}} + H_2O \quad (16\text{-}2)$$

hemiacetal   acetal

with D(+)-glucose, he isolated two crystalline products. Surprisingly,
each contained only *one* methoxyl group. Both products had acetal-
like properties; for example, they were stable toward base but were
hydrolyzed by acid to glucose and methanol. These observations
could be explained if, as a result of the interaction of one of the
five hydroxyl groups with the aldehyde function, glucose itself were
a hemiacetal. Glucose would then react with only one mole of methanol to form an acetal.

Examination of models shows that the hydroxyl group on carbon
five of glucose can easily be brought within bonding distance of the
carbonyl carbon atom. Interaction between these groups leads to a
cyclic hemiacetal with a six-membered ring containing one oxygen
atom and five carbon atoms (eq 16-3). The hemiacetal reacts with

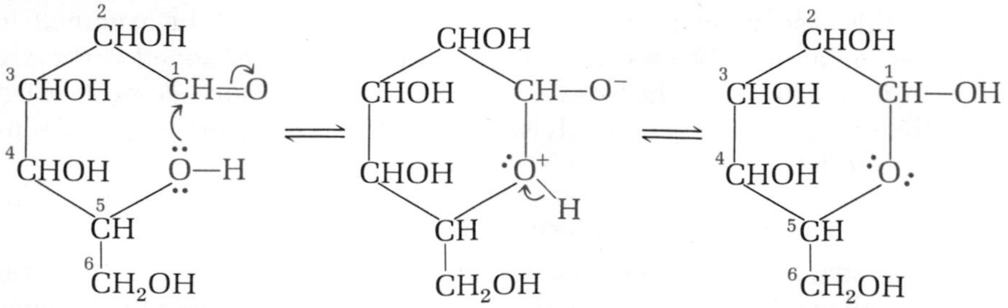

open-chain aldehyde formula
of a hexose

cyclic, hemiacetal formula
of a hexose

$$(16\text{-}3)$$

one mole of methanol to form an acetal (eq 16-4). Only the hydroxyl

$$
\begin{array}{c}
\overset{2}{C}HOH \\
\overset{3}{C}HOH \quad \overset{1}{C}H\!-\!OH \\
\overset{4}{C}HOH \quad O \\
\overset{5}{C}H \\
\overset{6}{C}H_2OH
\end{array}
\;+\; CH_3OH \underset{}{\overset{H^+}{\rightleftharpoons}}\;
\begin{array}{c}
\overset{2}{C}HOH \\
\overset{3}{C}HOH \quad \overset{1}{C}H\!-\!OCH_3 \\
\overset{4}{C}HOH \quad O \\
\overset{5}{C}H \\
\overset{6}{C}H_2OH
\end{array}
\;+\; H_2O \qquad (16\text{-}4)
$$

hemiacetal of a hexose   acetal of a hexose

group at carbon one is replaced by a methoxyl group, since it is a hemiacetal hydroxyl, whereas those at carbons two, three, four, and six are ordinary alcohol groups and require more vigorous reaction conditions for conversion to ethers.

The acetals of sugars are called **glycosides;** for glucose, such derivatives would be called *gluco***sides.** This type of structure, as we shall see, is used to link monosaccharide units into oligo- and polysaccharides.

Monosaccharides exist predominantly as cyclic hemiacetals. The oxygen-containing ring may be five- or six-membered, depending on which hydroxyl group in the chain reacts with the aldehyde function. For a hexose, the two possible structures are

$$
\begin{array}{c}
\overset{2}{C}HOH \\
\overset{3}{C}HOH \quad \overset{1}{C}HOH \\
\overset{4}{C}H\!-\!\!-\!\!-\!O \\
\overset{5}{C}HOH \\
\overset{6}{C}H_2OH
\end{array}
\qquad\qquad
\begin{array}{c}
\overset{2}{C}HOH \\
\overset{3}{C}HOH \quad \overset{1}{C}HOH \\
\overset{4}{C}HOH \quad O \\
\overset{5}{C}H \\
\overset{6}{C}H_2OH
\end{array}
$$

furanose   pyranose
structure   structure

They are referred to as the **furanose** and **pyranose** structures, respectively, these names being derived from two well-known, oxygen-containing heterocyclic structures with the same sized rings, **furan** and **pyran.** Though both occur, the pyranose structures of carbohydrates are by far the more common of the two.

$$
\begin{array}{c}
CH\!-\!\!-\!CH \\
\parallel \quad\; \parallel \\
CH \quad CH \\
\diagdown O \diagup
\end{array}
\qquad\qquad
\begin{array}{c}
CH_2 \\
CH \quad CH \\
\parallel \quad\; \parallel \\
CH \quad CH \\
\diagdown O \diagup
\end{array}
$$

furan   pyran

16.2c MUTAROTATION

Fischer isolated two methyl acetals from D(+)-glucose. Why two? It is a matter of stereochemistry, for one notes that in the hemiacetal structure of a hexose, carbon atom one has four different groups attached to it and therefore has become asymmetric. Two stereoisomers are possible. Equation 16-5 shows these relationships for D(+)-glucose.*

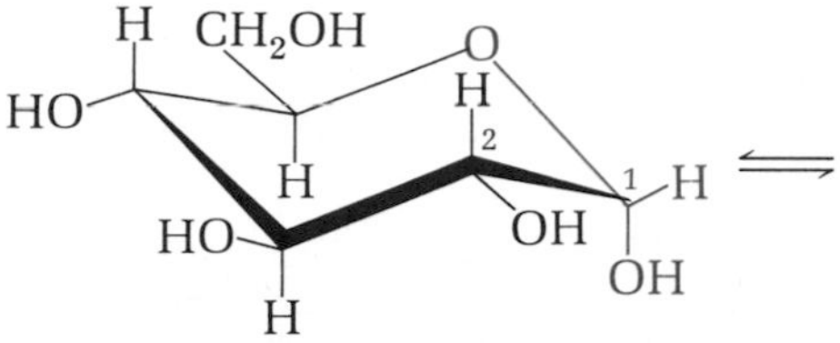

α -D(+)-glucose
(hemiacetal form)

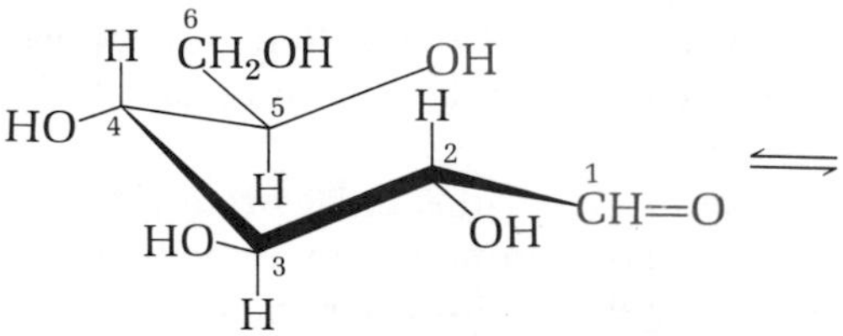

D(+)-glucose
(acyclic, aldehyde form)

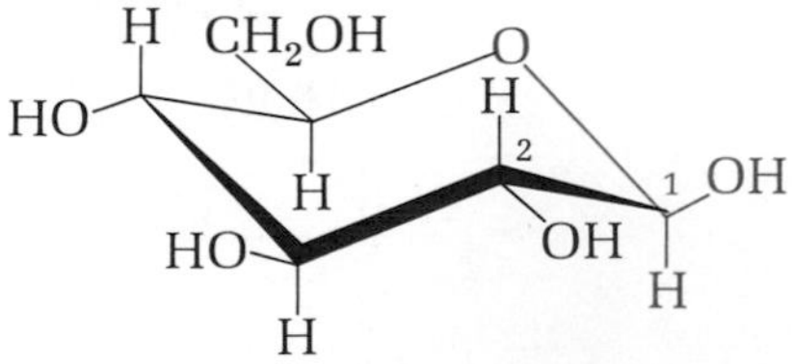

β -D(+)-glucose
(hemiacetal form)

(16-5)

The two cyclic structures are referred to as the α- and β-forms (or more rigorously, α- and β-D(+)-glucopyranose). The hemiacetal carbon atom is referred to as the **anomeric** carbon atom, and the two stereoisomers which differ only in their geometry at this carbon atom are called **anomers.** They are diastereomers, not enantiomers, since they have identical configurations at carbons two, three, four, and five but opposite configurations (**S** and **R**, respectively) at carbon one. Being diastereomers, they have different achiral (as well as chiral) properties. When D(+)-glucose is crystallized from methanol, the pure α-form is obtained, whereas if acetic acid is used as the

---

*The six-membered oxygen-containing rings (eq 16-3 and 16-4) are usually written in the cyclohexane chair conformation (cf. sec 2.10).

solvent, the β-form crystallizes. The α-form has a specific rotation of +113°, whereas the β-form has a specific rotation of only +19°. When either pure form is dissolved in water, the rotation gradually changes until it reaches an equilibrium value of +52°. This process of change in rotation is called **mutarotation**; it can be explained as follows: When either pure form (α or β) is dissolved in water, the equilibrium in equation 16-5 is set up. The concentration of open-chain aldehyde formed at equilibrium is extremely small (<0.1%), and the equilibrium mixture contains about 37% of the α and 63% of the β-form. The β-form is slightly preferred since in it, the hydroxyl group on carbon one occupies an equatorial, rather than an axial position (review sec 2.10).

It is now clear that the two products which Fischer obtained from D(+)-glucose and methanol were the acetals corresponding to the α- and β-structures of glucose.

methyl α-D-glucoside

methyl β-D-glucoside

Unlike the free sugars, the glucosides cannot mutarotate. That is, when either isomer is dissolved in water, it remains as such and is not transformed into an equilibrium mixture of the two isomers. This is because the anomeric carbon atom now has an alkoxyl group in place of the hydroxyl group; equilibration through the acyclic aldehyde form is no longer possible.

16.3   OTHER IMPORTANT ALDOSES

Formulas for the aldoses can be built up from the simplest monosaccharide, D(+)-glyceraldehyde, by adding one —CHOH group to the chain at a time. Table 16-2 shows the Fischer projection formulas of the sugars, through the hexoses, systematically constructed in this way. In each row the part of the structure which has the identical stereochemistry as the structure from which it was derived is shown

Table 16-2
Fischer Projection Formulas and Genealogy Chart for the D-aldoses through Six Carbon Atoms

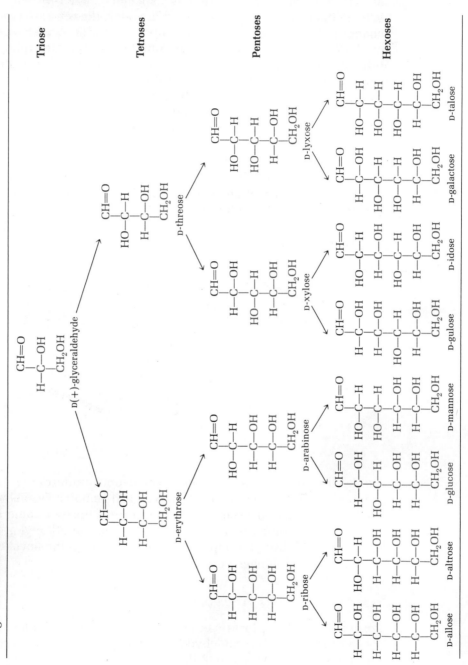

in color. Each of the 15 structures in the table has the D- (or **R**) configuration at the asymmetric carbon which is furthest from the aldehyde group. A similar set of 15-sugars, the enantiomers of those in the table, can be derived from L(−)-glyceraldehyde. All of the sugars in the table, except glyceraldehyde, can exist as cyclic hemi-acetals.

Though all of the sugars in Table 16-2 are known, only a few are very common. In addition to D(+)-glucose, the hexoses **D(+)-man-nose** and **D(+)-galactose** are common. Mannose differs from glucose only in the configuration at C-2. It is obtained most readily by the hydrolysis of the vegetable ivory nut, a polysaccharide material used to manufacture buttons. D-Galactose is obtained, along with glucose, by the hydrolysis of milk sugar (lactose). It differs from glucose only in the configuration at C-4.

Of the pentoses, **D(−)-ribose** is perhaps the most important. Although a D-sugar, it is levorotatory. It is one of the hydrolysis products of RNA (**r**ibose **n**ucleic **a**cid), compounds important in the transfer of genetic information and in cellular protein synthesis. **2-Deoxy-ribose,** in which C-2 of ribose is reduced to a —CH$_2$-group, is the sugar constituent of DNA (**d**eoxyribose **n**ucleic **a**cid), the other major genetic material. Both sugars occur in nucleic acids in the furanose structure, the α-configuration of which is shown.

α-D-ribofuranose   α-D-2-deoxyribofuranose

L(+)-Arabinose and **D(+)-xylose** are also common pentoses, and may be obtained by hydrolysis of a variety of plant polysaccharides, such as those found in corn cobs, oat hulls, and straw.

16.4   FRUCTOSE, A KETOHEXOSE

D(−)-Fructose is the most abundant ketose. It is sometimes called **levulose,** due to its large negative specific rotation (−92.4° after mutarotation to equilibrium). Also known as *fruit sugar,* it occurs free, along with glucose and sucrose, in many fruits. Ordinary cane sugar (sucrose) is a disaccharide composed of glucose and fructose monosaccharide units. The polysaccharide *inulin,* a starch-like substance present in many plants and most easily obtained from dahlia bulbs, primarily gives fructose on hydrolysis.

Fructose has a primary alcohol function at each end of the chain, and a keto group at carbon two. The configurations at carbons three, four, and five are identical with those of D(+)-glucose. Fructose is difficult to crystallize for it tends to form syrups, but from ethanol one can obtain pure crystals of the β-pyranose form. An aqueous solution of the crystals mutarotates and contains pyranose and furanose forms, two of which are shown.

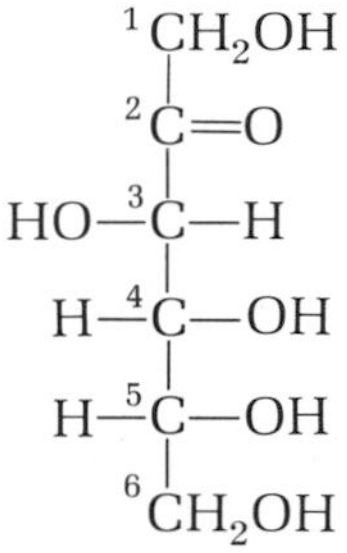

D-(−)-fructose
keto form

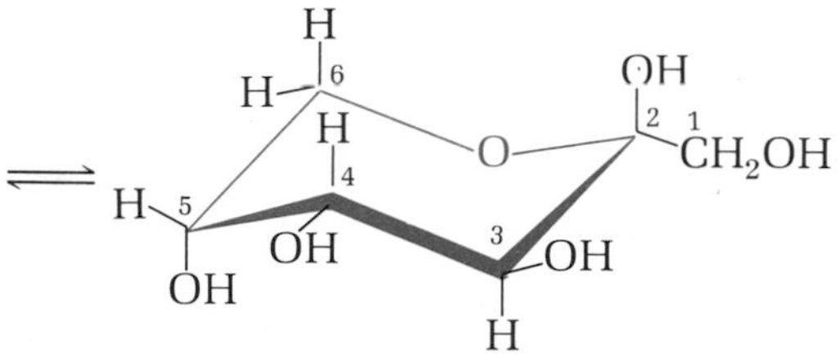

β-D-fructopyranose (a hemiketal)

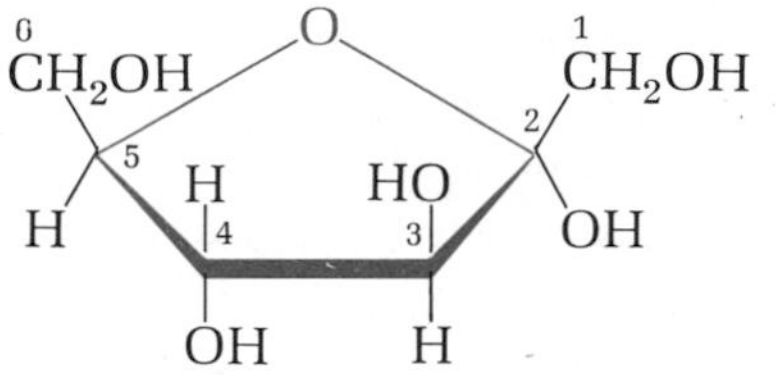

(16-6)

α-D-fructofuranose (a hemiketal)

Fructose is approximately three times as sweet as glucose.

16.5 SOME REACTIONS OF MONOSACCHARIDES

Monosaccharides show many of the reactions of their component functional groups. As alcohols, they react with acids or their derivatives to form esters; they may be converted to ethers; and they may be dehydrated. As aldehydes or ketones, they may be oxidized (to acids) or reduced (to alcohols), and they undergo many nucleophilic addition reactions (review sec 9.6). As already discussed, they form hemiacetals and acetals. With HCN they form cyanohydrins (sec 9.6e), and they react with many nitrogen bases to form oximes and related compounds (sec 9.6d). In this section we call attention to three of the more significant reactions of monosaccharides.

16.5a REDUCING ACTION

Those sugars which are α-hydroxyaldehydes or ketones, as for example glucose or fructose, readily reduce various metallic ions. The most common reagents for this purpose are **Fehling's** or **Benedict's** solutions, which consist of cupric ion, Cu^{2+}, complexed with tartrate or citrate ions, respectively. Such solutions are deep blue. When they react with a reducing sugar, a brick-red precipitate of cuprous oxide, Cu_2O, is produced (eq 16-7).

$$
\begin{array}{l}
CH{=}O \\
| \\
(CHOH)_4 \\
| \\
CH_2OH
\end{array}
+\ Cu^{II}\ +\ 4\ OH^-\ \longrightarrow\ Cu_2O\ +\
\begin{array}{l}
CO_2H \\
| \\
(CHOH)_4 \\
| \\
CH_2OH
\end{array}
+\ 2\ H_2O\quad(16\text{-}7)
$$

deep blue   brick red
in solution   precipitate

Even though sugars are predominantly in the cyclic hemiacetal form, they can reduce Cu^{2+} to Cu_2O because they are in equilibrium with the oxidizable α-hydroxycarbonyl form. But if the hemiacetal is converted to an acetal, as for example in the methyl glucosides, this equilibrium is no longer possible, and the glucoside cannot reduce Fehling's or Benedict's solution. This distinction is sometimes useful in determining structures of di- or polysaccharides. Tollens' reagent (sec 9.6a) can also be used for this purpose, as has already been described for glucose (Table 16-1).

Sugars which react positively with Fehling's, Benedict's, or Tollens' reagent are called *reducing sugars*. All of the sugars in Table 16-2 come in this category. The most common nonreducing sugar is sucrose (sec 16.6a).

16.5b   OSAZONES

The reaction of monosaccharides with phenylhydrazine, discovered by Emil Fischer, was of fundamental importance in determining their structure. It is possible to isolate the normal phenylhydrazone of an aldose or ketose. More commonly, however, additional phenyl-hydrazine is consumed in oxidizing the adjacent hydroxyl group to a carbonyl, which then forms a second phenylhydrazone link. Such bis-phenylhydrazones are called **osazones.**

$$
\begin{array}{l}
CH{=}O \\
| \\
CHOH \\
\{
\end{array}
\xrightarrow{C_6H_5NHNH_2}
\begin{array}{l}
CH{=}N{-}NHC_6H_5 \\
| \\
CHOH \\
\{
\end{array}
\xrightarrow{2\ C_6H_5NHNH_2}
\begin{array}{l}
CH{=}N{-}NHC_6H_5 \\
| \\
C{=}N{-}NHC_6H_5 \\
\{
\end{array}
$$

(16-8)

sugar   phenylhydrazone   osazone

The reaction does not proceed further down the sugar chain because, apparently, hydrogen bonding permits stabilization in the form of a cyclic structure.

$$
\begin{array}{c}
\overset{1}{CH}{=}N \\
\overset{2}{\sim\sim\sim C} \qquad N{-}C_6H_5 \\
N\cdots H \\
| \\
NHC_6H_5
\end{array}
\ \rightleftharpoons\
\begin{array}{c}
\overset{1}{CH}{-}N \\
\overset{2}{\sim\sim\sim C} \qquad N{-}C_6H_5 \\
N{-}H \\
| \\
NHC_6H_5
\end{array}
\qquad(16\text{-}9)
$$

cyclic osazone structure

Since osazone formation involves only the carbonyl and hydroxyl groups on carbon atoms one and two of the monosaccharide, sugars with the same configurations at the remaining carbon atoms will give the same osazone. It was in this way that the configurations of carbons three, four, and five of D-glucose, D-mannose, and D-fructose were shown to be identical. Each of these sugars, when treated with phenylhydrazine, gave the identical osazone, called D-glucosazone. This implied that the only difference between these sugars must lie in the structure or stereochemistry of the first two carbon atoms.

$$
\begin{array}{ccc}
\text{CHO} & \text{CHO} & \text{CH}_2\text{OH} \\
\text{H—C—OH} & \text{HO—C—H} & \text{C=O} \\
\text{HO—C—H} & \text{HO—C—H} & \text{HO—C—H} \\
\text{H—C—OH} & \text{H—C—OH} & \text{H—C—OH} \\
\text{H—C—OH} & \text{H—C—OH} & \text{H—C—OH} \\
\text{CH}_2\text{OH} & \text{CH}_2\text{OH} & \text{CH}_2\text{OH} \\
\text{D(+)-glucose} & \text{D(+)-mannose} & \text{D(−)-fructose}
\end{array}
\qquad (16\text{-}10)
$$

$$
\begin{array}{c}
\text{CH=NNHC}_6\text{H}_5 \\
\text{C=NNHC}_6\text{H}_5 \\
\text{HO—C—H} \\
\text{H—C—OH} \\
\text{H—C—OH} \\
\text{CH}_2\text{OH} \\
\text{D-glucosazone}
\end{array}
$$

Sugars are sometimes difficult to crystallize, since they tend to form syrups, especially when impure. However, the osazones are beautifully crystalline yellow compounds, which are sometimes of value in identifying particular monosaccharides.

16.5c GLYCOSIDES

Perhaps the most important key to understanding the structure and reactions of carbohydrates is a thorough comprehension of glycoside chemistry. The methyl glucosides have already been discussed (sec 16.2b and 16.2c). These are specific examples of a general class of compounds.

Sugars which have a hemiacetal (or hemiketal) structure can react with alcohols or with amines to form O- and N-glycosides, respectively. The glycosides may have the α- or β-geometry, as illustrated.

an α-sugar
(hemiacetal structure)

α-O-glycoside

β-O-glycoside

α-N-glycoside

β-N-glycoside

$$(16\text{-}11)$$

Particular glycosides are named according to the specific sugar from which they are derived (glucoside, mannoside, fructoside, riboside, etc.).

The alcohol portion (OR) of an **O-glycoside** may be derived from a simple alcohol such as methanol, as in the methyl glucosides already discussed, or it may be derived from a more complex alcohol. Indeed, the alcohol component may be derived from one of the hydroxyl groups of another monosaccharide. It is precisely in this way that oligo- and polysaccharides are constructed from monosaccharide units. Often naturally occurring alcohols are present in cells in combination with some sugar (usually glucose) as a glycoside. The pigments of flowers, the flavorings of vanilla beans or almonds, and many steroids are among the natural products which occur as glycosides. The many hydroxyl groups of the sugar portion of glycosides are undoubtedly important in solubilizing these pigments, etc., in the cellular protoplasm.

O-glycosides, being acetals, can be hydrolyzed by acids to their sugar and alcohol components. The same reaction is involved in the acid-catalyzed hydrolysis of oligo- and polysaccharides to monosaccharides.

Hydrolysis of O-glycosides may also be catalyzed by enzymes, which are usually extremely stereospecific in their reactions (Figure 15-24). The enzyme **emulsin,** for example, catalyzes the hydrolysis of methyl β-D-glucoside (but not the α-isomer) whereas the enzyme

maltase only catalyzes the hydrolysis of α-glucosides. One can use

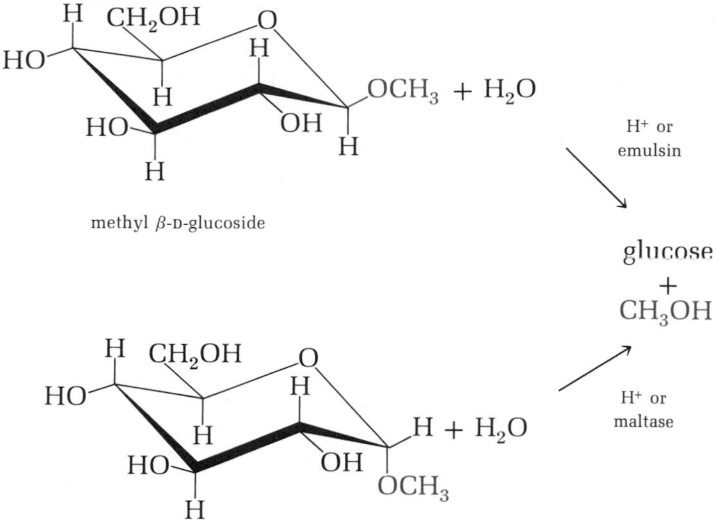

methyl β-D-glucoside

$OCH_3 + H_2O$

H⁺ or emulsin

glucose
+
CH_3OH

H⁺ or maltase

methyl α-D-glucoside

$H + H_2O$

(16-12)

these reactions to determine the stereochemistry at the anomeric carbon atom in naturally occurring glycosides.

The most important **N-glycosides** are the ribonucleosides and deoxyribonucleosides which make up the nucleic acids. It is through such linkages that the famous nitrogen bases of DNA and RNA are joined to the sugar moieties. The structure and chemistry of these substances will be discussed in greater detail in Chapter 18.

16.6 DISACCHARIDES

Disaccharides are glycosides in which one monosaccharide acts as the hemiacetal and a second monosaccharide furnishes the alcohol grouping. Disaccharides are therefore acetals, formed from two monosaccharides by the elimination of one mole of water. Conversely, hydrolysis of a disaccharide either by water in the presence of a mineral acid catalyst or by an enzymatic method yields two moles of monosaccharides. For two hexoses, the molecular equation is

$$C_6H_{12}O_6 + C_6H_{12}O_6 \rightleftharpoons C_{12}H_{22}O_{11} + H_2O \quad (16-13)$$

monosaccharide   monosaccharide   disaccharide

The structure determination of a particular disaccharide must include (*a*) which monosaccharide functions as the hemiacetal and which as the alcohol component of the glycoside, (*b*) which of the several hydroxyls in the monosaccharide functioning as the alcohol component is involved in the glycosidic linkage, (*c*) the configuration (α or β) of the glycosidic linkage, and (*d*) the ring size (furanose or pyranose) in each monosaccharide unit.

Among the most important disaccharides are **sucrose, lactose, maltose,** and **cellobiose.**

16.6a   SUCROSE

Sucrose is the most important disaccharide. It is the ordinary table sugar which we eat every day. It occurs in all photosynthetic plants where it appears to serve as an easily transported energy source. Its two main commercial sources are sugar cane and the sugar beet, each furnishing about half of the world's supply. The juices, which contain about 14 to 20% sucrose, are put through a rather extensive purification process to remove odoriferous and colored impurities. Pure sucrose is obtained as white crystals through concentration of the purified syrups by vacuum removal of the water. The principal use of sucrose is for food.

Hydrolysis of sucrose, either by acids or by the enzyme *sucrase,* gives *equal* quantities of D-glucose and D-fructose. Sucrose must therefore be made up of one unit of each of these monosaccharides. The following chemical tests enable us to decide how the two units must be linked.

Sucrose does not reduce Fehling's or Benedict's solution, does not form an osazone (except on prolonged boiling, when glucosazone is formed due to hydrolysis of the sucrose), and does not mutarotate. Since both glucose and fructose are reducing sugars, they must be joined together in sucrose in such a way as to destroy the possibility of forming a free aldehyde or ketone group. The glycosidic link therefore must involve splitting out a mole of water between the hydroxyl groups on each anomeric carbon atom of the cyclic forms of glucose and fructose, respectively. Thus, C-1 of glucose and C-2 of fructose are joined, through an oxygen atom, in the glycosidic link. In this way, each anomeric carbon is converted from a hemi-acetal to an acetal structure, thus ensuring that sucrose is a *non-reducing* sugar, incapable of mutarotation or osazone formation.

The size of the rings in each monosaccharide portion can be determined by first converting all the hydroxyls in sucrose to meth-oxyls (with methyl sulfate and base). Hydrolysis of the resulting octamethylsucrose gives a tetramethylglucose and a tetramethyl-fructose. The tetramethylglucose has methyls missing at carbons one and five; since carbon one was used to attach the fructose unit,

carbon five must be involved in ring formation. Similarly, the tetramethylfructose has methyls missing at carbons two and five, the
latter of which must be involved in ring formation.

Finally, the stereochemistry of the glycosidic link can be established by experiments with enzymes, certain of which hydrolyze α
or β glycosidic linkages preferentially. In sucrose, the link is α at
glucose carbon one and β at fructose carbon two.

The structure of sucrose which accounts for all these facts follows.

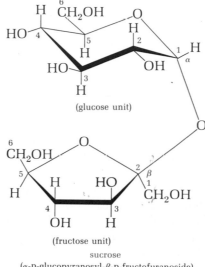

(glucose unit)

(fructose unit)

sucrose

(α-D-glucopyranosyl-β-D-fructofuranoside)

16.6b LACTOSE

Lactose is the sugar present in milk; human milk contains five to
eight percent; cow's milk, four to six percent. On hydrolysis, lactose
gives equal quantities of glucose and galactose. Lactose is a reducing
sugar, a fact which indicates that it still has a potential aldehyde
group. By methods analogous to those described above in more detail
for sucrose, it has been established that the linkage of the two units
is between carbon one of galactose and carbon four of glucose. This

β(can be α)

D-galactose unit

D-glucose unit

lactose

[4-O-(β-D-galactopyranosyl)-D-glucopyranose]

aldehyde form

$$(16\text{-}14)$$

leaves carbon one of the glucose unit free to be in either the α or β form, in equilibrium with the open-chain aldehyde structure. Thus, lactose reduces Fehling's solution, undergoes mutarotation, and forms an osazone.

16.6c MALTOSE AND CELLOBIOSE

Maltose is formed by the action of the enzyme *diastase* (from malt) on starch. Further hydrolysis of maltose, catalyzed by the enzyme *maltase* (from yeast) gives only glucose (eq 16-15). Consequently maltose must consist of two glucose units. Further experiments

$$\text{starch} \xrightarrow[\text{H}_2\text{O}]{\text{diastase}} \text{maltose} \xrightarrow[\text{H}_2\text{O}]{\text{maltase}} \text{glucose} \qquad (16\text{-}15)$$

show that the two glucose units are joined at carbons one and four respectively, and that the geometry at the anomeric carbon is α.

D-glucose unit   D-glucose unit   may be α or β

maltose
[4-O-(α-D-glucopyranosyl)-D-glucopyranose]

Cellobiose is a stereoisomer of maltose. It is obtained by the partial hydrolysis of cellulose. Enzymatic hydrolysis of cellobiose by the enzyme *emulsin* gives only glucose (eq 16-16). The latter hydrolysis

$$\text{cellulose} \xrightarrow{\text{H}_3\text{O}^+} \text{cellobiose} \xrightarrow[\text{H}_2\text{O}]{\text{emulsin}} \text{glucose} \qquad (16\text{-}16)$$

is not catalyzed by maltase, although the hydrolysis products of maltose and cellobiose are identical (D(+)-glucose). Thus, the only difference between maltose and cellobiose is the geometry at the anomeric carbon atom; cellobiose has the β geometry.

D-glucose unit   D-glucose unit

cellobiose *
[4-O-(β-D-glucopyranosyl)-D-glucopyranose]

The distinctions in structure between maltose and cellobiose give a clue to the structural differences between the two polysaccharides from which they are derived, i.e. starch and cellulose.

16.7 POLYSACCHARIDES

Polysaccharides are carbohydrates with a very high molecular weight, ranging from the tens of thousands into the millions. They are usually insoluble in most solvents and sometimes quite difficult to purify. Even when pure, they consist of molecules with a range of molecular weights. Some polysaccharides, such as starch or cellulose, give a single monosaccharide (in this case, glucose) on complete hydrolysis; others, for example gum arabic (a plant gum used in adhesives), may include several types of monosaccharides in the polymer chain.

The biological functions of polysaccharides vary considerably. Starch and glycogen are reserve foodstuffs for plants and animals; cellulose and chitin are structural materials for plants and crustacea, respectively.

16.7a STARCH

Starch is the reserve carbohydrate in many plants and comprises large percentages of cereals, potatoes, corn, and rice. Under the microscope, the appearance of the granules of starch from these different sources varies both in shape and size. Chemically, however, they are similar.

Complete hydrolysis of starch yields glucose in essentially the theoretical amount, but partial hydrolysis gives maltose as well. This

---

* In order to allow the β-glycosidic bond to be drawn more easily, the glucose unit on the right in this formula has been "flipped" over. Thus, the anomeric carbon of this unit (C1) has the β configuration as drawn, even though the hydroxyl group points downward.

Figure 16-3

Amylose (top) and amylopectin. Chain branching in the latter, through 1,6 as well as 1,4 linkages, decreases its water solubility.

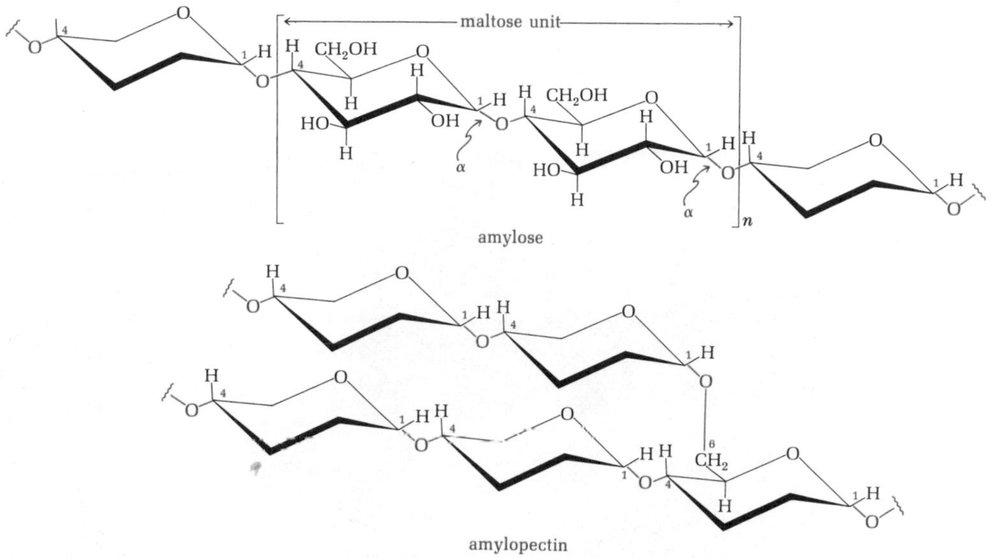

amylose

amylopectin

shows that starch is a polymer of glucose units, joined mainly through carbons one and four by an α-glycosidic linkage, as in maltose. In addition, the chains are branched through a modest number of glucose units joined at carbons one and six (rather than one and four).

Many of the differences in the properties of starch samples can be accounted for by variation in chain length and degree of branching. Starch can be separated into two main fractions by treatment with hot water. The soluble component (10–20%) is **amylose;** the insoluble component (80–90%) is **amylopectin** (Figure 16-3). The molecular weight of amylose ranges from 10,000 to 50,000 (60 to 300 glucose units). Amylopectin ranges from 50,000 to 1,000,000 in molecular weight (300 to 6000 glucose units), but the chain *lengths* vary only from 24 to 30 glucose units. Amylopectin is a highly branched polymer. Probably because of this rather loose, highly branched structure, starch granules swell in water, eventually forming a colloidal solution.

Partial hydrolysis of starch transforms it into **dextrins,** polysaccharides of smaller molecular weight than starch. They are more readily digested than starch and are used, mixed with maltose, in infant foods (e.g. Dextrimaltose). A dried mixture of dextrins, maltose, and milk is the preparation used for making malted milk.

Figure 16-4

Amylose starch from a new kind of corn is compared with amylopectin starch which is found in ordinary corn. Scientists of the United States Department of Agriculture found that amylose, when treated chemically, forms a fibrous material (right) that can be spun into fibers or made into film. Given the same treatment, amylopectin (left) forms a powdery substance with no film or fiber-forming properties. (United States Department of Agriculture.)

Dextrins are sticky when wet and are used in manufacturing mucilage for postage stamps and envelopes. In laundering, starched materials become stiff and shiny due to the transformation of the starch to dextrins by the heat of the iron.

16.7b GLYCOGEN

When starch is ingested, it is hydrolyzed enzymatically in a stepwise fashion. Initiated in the mouth by the enzyme *amylase*, present in saliva, hydrolysis is continued by additional amylase in the pancreatic juices. The maltose produced this way is, in turn, hydrolyzed to glucose with the aid of the enzyme *α-glucosidase* present in the intestines. The glucose is absorbed from the intestines into the blood and transported to the liver, muscles, and other sites where it is converted to another glucose polymer, **glycogen,** and stored.

Glycogen, the reserve carbohydrate of animals, is found mainly in the liver and muscles. It resembles starch in appearance but has a smaller molecular weight. Glycogen helps maintain the proper amount of glucose in the blood by removing and storing excess glucose derived from ingested food or by supplying it to the blood when it is needed by body cells for energy.

16.7c CELLULOSE

Cellulose is the main structural material of plant life, being the chief ingredient of cell walls, of cotton, wood pulp, linen, straw, corn cobs, and many other materials. Its chemical constitution has been elucidated by experiments involving hydrolysis. Complete hydrolysis gives glucose. More careful hydrolysis gives the disaccharide, cellobiose. The cellulose molecule is composed of long chains of cellobiose molecules joined together by β-linkages (Figure 16-5).

Figure 16-5
Partial structure of a cellulose molecule showing the β-linkage of glucose units.

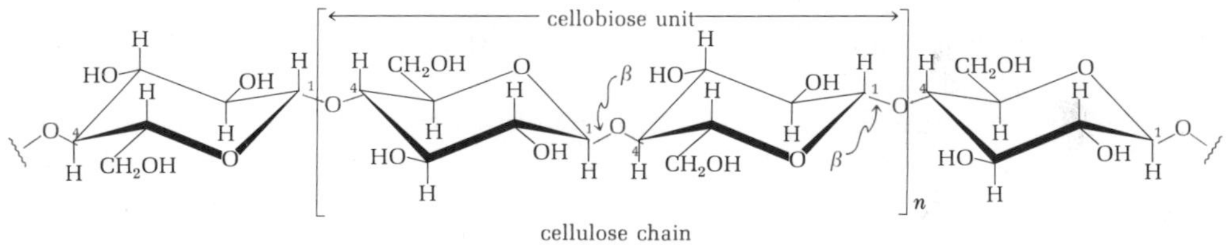

cellobiose unit

cellulose chain

The molecular weight of cellulose is probably between 300,000 and 500,000 (or 1,800 to 3,000 glucose units per molecule). X-ray examination of cellulose disclosed that it consists of linear chains made up of cellobiose units in which the ring oxygens alternate forward and back positions (Figure 16-5). Cellulose fibers consist of bundles of such chains, about 70 to 80 Å in diameter, held together by hydrogen bonds between hydroxyls on adjacent chains. Thus cellulose is unaffected by most solvents, in contrast to starch which swells under similar conditions.

The digestive systems of man and most other animals do not contain the necessary enzymes for hydrolyzing β-glucosidic linkages.

For this reason, they cannot digest cellulose. Certain bacteria and other microorganisms in the digestive tracts of ruminants and termites, however, can decompose cellulose and use it as food. The striking ability of the human body to digest starch (α-glucoside) but not cellulose (β-glucoside) once again emphasizes the stereospecificity of biochemical processes.

Cotton is about 95% cellulose, with impurities of fats and waxes. When the impurities are removed by washing the cotton with ether, the fibers will take up water. The product is then known as *absorbent cotton. Linen,* obtained from flax, is a somewhat less pure form of cellulose than cotton.

Cellulose for the manufacture of paper is obtained from wood pulp. Wood contains fibers of cellulose and hemi-cellulose (which has a lower molecular weight) bound by a high-molecular weight polymeric substance known as lignin. The latter may be removed from wood pulp with sodium hydroxide and sodium sulfite. The more completely the lignin is removed, the better is the grade of paper obtained. Better papers are treated with various sizing agents (rosin, dextrins, glue) to prevent the soaking up or blotting of ink.

16.7d CELLULOSE DERIVATIVES

The formula of cellulose (Figure 16-5) shows that each glucose unit contains three hydroxyl groups. Cellulose, therefore, has chemical properties similar to those of a trihydric alcohol; it forms esters with acids, ethers with other alcohols, and in general exhibits the chemistry of alcohols. Several important cellulose derivatives are prepared in this way.

Cellulose reacts with nitric acid to form **cellulose nitrates** much as does glycerol (eq 6-43). The number of hydroxyl groups nitrated per glucose unit determines the nature of the product obtained. When most of the hydroxyl groups are nitrated, the product is *guncotton.* Highly nitrated cotton is, like nitrated glycerol, an efficient explosive and is used in the manufacture of smokeless powders. The lower nitrates of cotton are called *pyroxylin. Celluloid,* made from pyroxylin, camphor, and alcohol, is highly combustible but not explosive. *Collodion* is a solution of pyroxylin in a mixture of ether and alcohol. When spread over a surface (such as a wound), the solvents evaporate, leaving a thin protective film.

segment of a cellulose nitrate molecule

Cellulose acetate is prepared from cellulose and acetic anhydride, which reacts with the hydroxyl groups to form ester linkages. Cellulose acetate rayon is made by forcing a viscous solution of cellulose acetate in acetone through fine openings, in a current of warm air which evaporates the acetone. The cellulose acetate is obtained in the form of long fibers which can be spun and woven. Cellulose acetate does not burn readily and is used in making motion-picture film.

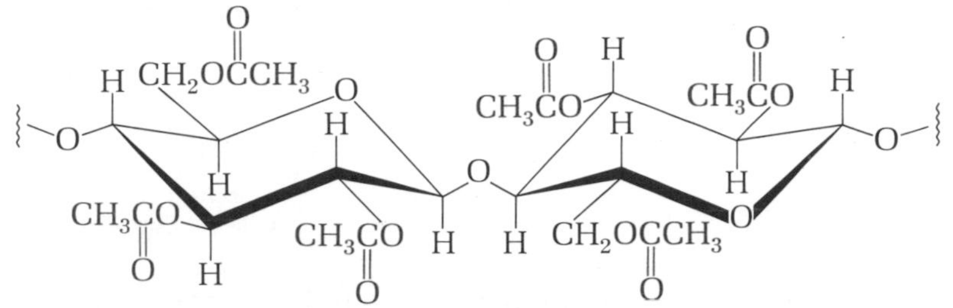

segment of a cellulose acetate molecule

Cellulose ethers are made by the action of ethyl chloride or ethyl sulfate on alkaline cellulose. The alkali converts the hydroxyl groups to alkoxide groups which then form ethers with the alkyl halide or sulfate (sec 7.2b).

$$RO^-Na^+ + C_2H_5Cl \xrightarrow{-Na^+Cl^-}$$

cellulose
alkoxide

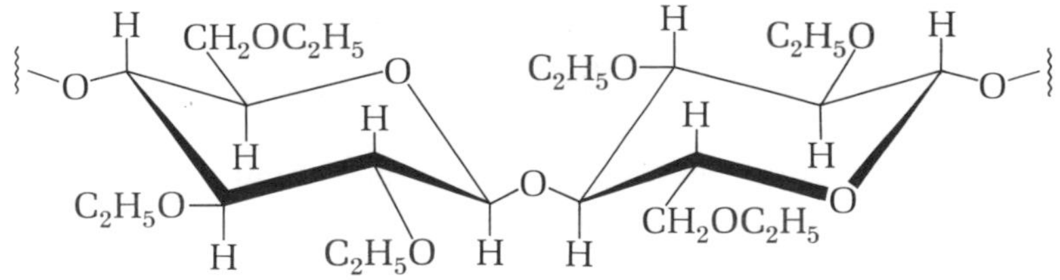

segment of an ethyl cellulose molecule

(16-17)

Ethyl cellulose is used to manufacture films, coatings, and plastics.

In the formulas given above for cellulose nitrate, esters, and ethers all three hydroxyl groups of each glucose unit are shown as having been modified. In practice, complete reaction is not always practical or desirable, and indeed the properties of the products can be varied to suit particular needs by controlling the extent to which hydroxyl groups are replaced by other functional groups.

Cellulose itself can be converted to fibers called **rayon.** The cellulose is converted to a soluble derivative, called a **xanthate,** by reaction with alkali and carbon disulfide (eq 16-18; RO⁻ in this equation stands for alkoxide groups, three on each glucose unit in cellulose). The cellulose xanthate is soluble in water and gives a viscous solu-

$$RO^-Na^+ + S{=}C{=}S \longrightarrow RO{-}\overset{\overset{\displaystyle S}{\|}}{C}{-}S^-Na^+ \qquad (16\text{-}18)$$

cellulose xanthate

tion (hence the name *viscose*) which, when forced through small openings into aqueous acid, regenerates the cellulose in fine threads (eq 16-19). These threads are spun into yarn.

$$RO{-}\overset{\overset{\displaystyle S}{\|}}{C}{-}S^-Na^+ + NaHSO_4 \longrightarrow ROH + CS_2 + Na_2SO_4 \quad (16\text{-}19)$$

regenerated
cellulose

Cellophane is produced in a similar manner, except that the xanthate is extruded into the acid bath in thin sheets. These are made water-resistant by a coating of a transparent cellulose nitrate lacquer.

16.8 CARBOHYDRATES IN THE CELL

Carbohydrates are vital to many cellular processes. In green plants, they are synthesized by a sequence of reactions referred to as **photo-synthesis.** The process is endothermic; energy from sunlight is absorbed, with the aid of the pigment chlorophyll, and used to transform carbon dioxide and water to carbohydrates and oxygen.

$$nCO_2 + nH_2O \xrightarrow[\text{chlorophyll}]{\text{sunlight}} (CH_2O)_n + nO_2 \qquad (16\text{-}20)$$

In the reverse process of **carbohydrate metabolism,** animals ingest carbohydrates and convert them (with the help of oxygen breathed from the air) to carbon dioxide (which is exhaled) and water. Since the net result is the reverse of equation 16-20, this process must be exothermic; it provides a major energy source for the body to perform work.

The detailed chemistry of these two processes involves an extraordinarily complex web of reactions, though each step has analogies in conventional organic chemistry. Many of the same chemical intermediates are involved in both processes, and virtually all steps require the catalytic action of enzymes and the frequent participation of other cell constituents, especially phosphates, present in minor but vital amounts in all cells.

The principle used to determine the sequence in which organic compounds are produced from carbon dioxide during photosynthesis is relatively simple, though the experiments were exacting and difficult. Plants were exposed to an atmosphere which contained labeled carbon dioxide ($C^{14}O_2$, where radioactive carbon, mass 14 instead of 12, is present) and were illuminated for varying periods of time. With short exposures (a few seconds) only the first compounds which incorporate carbon dioxide became labeled with radioactive carbon; longer exposures produced more compounds containing C^{14}, until eventually the carbohydrates produced by photosynthesis became uniformly labeled. Work of this type by M. Calvin and co-workers (Berkeley, 1963 Nobel Prize) showed that the key step which incorporates carbon dioxide involves the five-carbon ketopentose ribulose, as its diphosphate:

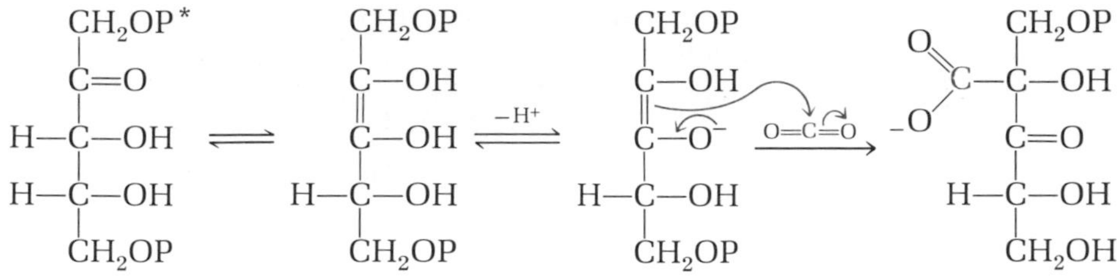

ribulose diphosphate

unstable
C_6 intermediate

(16-21)

The unstable C_6 intermediate rapidly undergoes a reverse aldol condensation (sec 9.7a) to produce two moles of 3-phosphoglyceric acid, which is one of the first carbohydrate fragments in which the radioactive carbon appears.

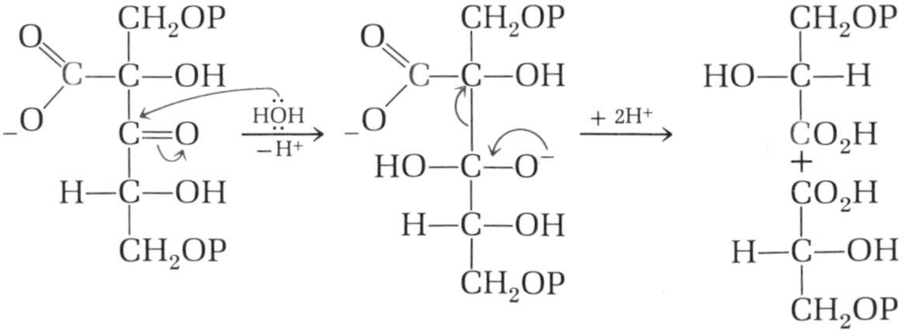

(16-22)

two moles of
3-phosphoglyceric acid

$*P = -\overset{O}{\overset{\|}{P}}(OH)_2$ or one of its ionized forms

It would seem from this sequence that one would obtain one labeled and one unlabeled 3-phosphoglyceric acid; since these cannot be separated or distinguished, all of the acid appears carboxyl-labeled, but with half the radioactivity.

The C_3 sugar is then converted, through a sequence of steps, to hexoses and to starch, as well as to other biosynthetic materials (sec 18.3). The role of light in photosynthesis is to provide an energy source for the synthesis of certain coenzymes required for these reactions.

In the process of carbohydrate metabolism, ingested di- and poly-saccharides are hydrolyzed, primarily by enzymes of the saliva and small intestines, to monosaccharides. Glucose is converted, in several phosphorylation and isomerization steps, to fructose-1,6-diphosphate. A key step in the metabolic process is again a reverse aldol condensation, in which fructose diphosphate is converted to two three-carbon sugar units:

$$(16-23)$$

fructose-1,6-diphosphate   3-phosphoglyceraldehyde   pyruvic acid

A proton is lost from the hydroxyl on carbon 4 and is gained by carbon 3. A sequence of reactions which involves isomerization, phosphorylation, and oxidation converts each of these three-carbon fragments to pyruvic acid, an important intermediate in several biological processes (sec 14.2d, and Figure 14-2). The most important of these is its conversion, through enzymatic decarboxylation, to acetylcoenzyme A. This, in turn, is essential for fat metabolism (sec 11.7), for terpene and steroid biosynthesis (sec 18.3), and for further carbohydrate metabolism through the citric acid cycle. The latter is a cyclical process, involving citric acid as one of several intermediates, which completes the oxidation of the ingested carbohydrate to carbon dioxide and water.

Space limitations do not permit us to do justice to the intricate beauty of these processes. We therefore urge the interested reader to refer to any of the many excellent texts in biochemistry for further details.*

New Concepts, Facts, and Terms

1. carbohydrates—polyhydroxyaldehydes or ketones, or closely related substances. Mono-, oligo-, and polysaccharides; pentose, hexose; aldose, ketose
2. glucose, gluconic and glucaric acids, sorbitol; the Fischer projection formula for D(+)-glucose
3. the cyclic, hemiacetal structure of D(+)-glucose; glycosides and glucosides; furanose and pyranose
4. mutarotation; α- and β-forms of D(+)-glucose; anomers; the chair conformation of sugars
5. the generic tree of D-aldoses (Table 16-2); mannose, galactose, ribose and deoxyribose
6. fructose, a ketohexose
7. reactions of monosaccharides; reducing sugars, Fehling's and Benedict's tests; osazones; O- and N-glycosides
8. disaccharides
   a. sucrose (glucose and fructose units); a nonreducing sugar
   b. lactose (glucose and galactose units)
   c. maltose (two glucose units, α-linkage) from starch
   d. cellobiose (two glucose units, β-linkage) from cellulose
9. polysaccharides
   a. starch, amylose, amylopectin, dextrins (α-glucose units)
   b. glycogen
   c. cellulose (cellobiose, β-glucose units)
10. cellulose derivatives; nitrate, acetate, ethers, xanthate; viscose rayon
11. carbohydrates in the cell; photosynthesis, metabolism

Exercises and Problems

1. Define, explain, or give the structural formula for an example of each of the following:
   a. aldohexose   f. furanose
   b. ketopentose   g. pyranose
   c. monosaccharide   h. glycoside
   d. disaccharide   i. osazone
   e. polysaccharide   j. reducing sugar

*The account given in Chapter II of T. A. Geissman and D. H. G. Crout, *Organic Chemistry of Secondary Plant Metabolism,* Freeman, Cooper, and Company, San Francisco, 1969 is especially lucid and simple, and is satisfying to the organic chemist since rational mechanisms for most steps are presented. This book will serve as an excellent source to teachers who wish to present this material in this course.

2. Write balanced equations for the reaction of D(+)-glucose (use either an acyclic or cyclic structure, whichever seems most appropriate) with each of the following:
   a. acetic anhydride   e. methanol, H+
   b. bromine water   f. hydrogen cyanide
   c. hydrogen, catalyst   g. phenylhydrazine
   d. hydroxylamine   h. Fehling's reagent

3. a. What is the Cahn-Ingold-Prelog (sec 15.6) priority order of groups at C-5 in D(+)-glucose? Draw a structure which shows that the configuration at this carbon atom is **R**.
   b. Determine the correct C-I-P designation (**R** or **S**) for carbons two, three, and four of D(+)-glucose.
   c. Show that the configuration at the anomeric carbon atom of α- and β-D-glucose is **S** and **R**, respectively.

4. Explain, using formulas, what is meant by the term "D-sugar."

5. Write out the steps in the mechanism for equation 16-4. Draw resonance structures for the intermediate which explain why only the hydroxyl group at carbon one is replaced by a methoxyl group.

6. Whether one starts with pure α- or pure β-D-glucose in the reaction with methanol and H+, one obtains a mixture of the methyl α- and β-D-glucosides. Write equations for the reaction mechanism which explain this observation.

7. Write a mechanism for the interconversion (mutarotation) of α- and β-D-glucose (eq 16-5). Explain why the corresponding methyl glucosides cannot mutarotate.

8. Explain the meaning of each part of the name methyl β-D(+)-glucopyranoside.

9. Draw the Fischer projection formula for
   a. L(−)-glucose   b. L(+)-ribose

10. D(+)-Glyceraldehyde can be converted, by a sequence of reactions, to a mixture of D-erythrose and D-threose. Being diastereomers, these tetroses can be separated by selective crystallization. One can readily tell which isomer is which by oxidizing each of them, separately, with nitric acid to a tartaric acid, and examining the optical rotation of each. Explain.

11. Using the formulas given in Table 16-2, draw structural formulas for
    a. α-D-gulopyranose   b. β-D-arabinofuranose

12. How many different osazones can be formed from the four D-pentoses? The eight D-hexoses? Which of these sugars give identical osazones? Explain.

13. Write out the steps in the mechanism for the hydrolysis of methyl α- and β-D-glucosides (eq 16-12). Explain why these substances are hydrolyzed by acids but not by bases.

14. Show how the specific rotations given in section 16.2c support the contention that at equilibrium in water solution, 63% of D(+)-glucose is in the β-form and 37% in the α-form.

15. Using the information given in section 16.6a, draw the structure of octamethylsucrose, and of the tetramethylglucose and tetramethylfructose which are obtained from it by hydrolysis.

16. Write equations which clearly show the mechanism for the acid-catalyzed hydrolysis of lactose to galactose and glucose.

17. When maltose is treated with dimethyl sulfate and base, it is converted to a compound which has eight methoxyl groups. Acid-catalyzed hydrolysis of the latter gives 2,3,4,6-tetra-O-methylglucose, 2,3,6-tri-O-methylglucose and a mole of methanol. Write equations for the reactions and show how they are consistent with the formula for maltose given in section 16.6c.

18. Given the structural formulas for four disaccharides, **A–D**, (all bonds designated by a line go to hydrogen atoms).

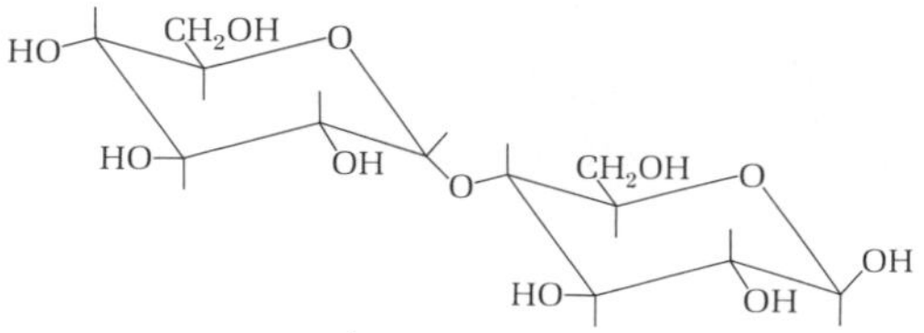

A

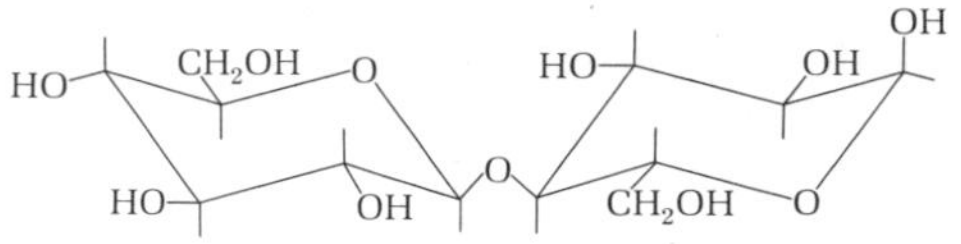

B

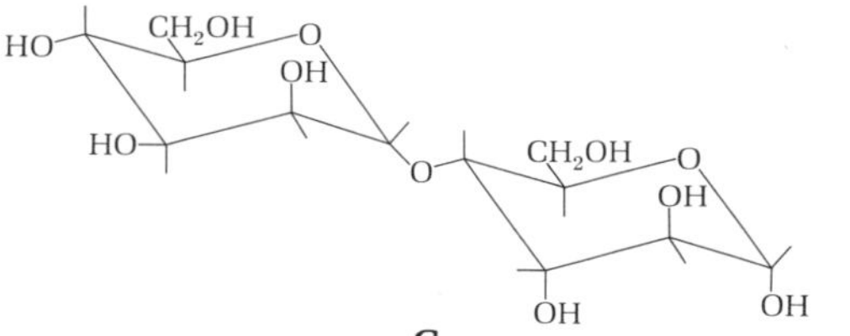

C

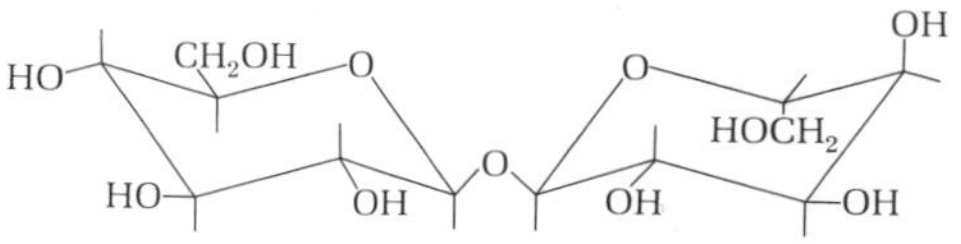

D

Which structure(s)
a. represent reducing sugars?
b. would be hydrolyzed by emulsin?
c. contain at least one D-glucose unit?
d. is **4**-(β-D-glucopyranosyl)-D-glucose?

19. Why is it not possible to synthesize cellulose ethers by the acid-catalyzed reaction of cellulose with an alcohol (i.e., according to the procedures given in sec 7.2a)?

20. a. Give satisfactory mechanisms for the reactions in equations 16-18 and 16-19.
    b. Rewrite equations 16-18 and 16-19 using for R at least two units of the cellulose chain.

CHAPTER SEVENTEEN

AMINO ACIDS AND PROTEINS

Proteins, as their name implies (Greek: *protos* = first), are of prime importance in biological systems. They are vital to the structure, function, and reproduction of living matter. They include such diverse materials as hemoglobin, enzymes, and hormones. Hair, skin, muscles, nerves, tendons, nails, feathers, horns, and hoofs are almost entirely protein, as are egg whites, antibodies, and silk. Because of their importance, proteins are the subject of extensive research, and our knowledge of their structure, properties, and functions is growing rapidly.

Proteins are polymers composed of amino acid units joined by amide linkages. When a protein is hydrolyzed by heating with aqueous acid it is converted to a mixture of α-amino acids. Considering the great variety of proteins found in nature, it is certainly remarkable that only about twenty different α-amino acids are common in protein hydrolyzates. These amino acids constitute the building blocks of proteins.

17.1   NATURALLY OCCURRING AMINO ACIDS

Only α-amino acids are common in protein hydrolyzates: that is, the amino group is on the carbon atom adjacent or α to the carboxyl group.

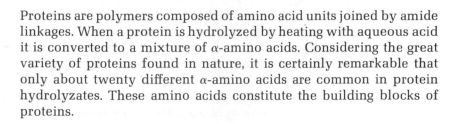

an α-amino acid

The simplest of these (R = H) is aminoacetic acid, whose common name is **glycine.** When R ≠ H, the α carbon will be asymmetric. With the exception of glycine, all the amino acids derived from proteins are optically active, and all have the L configuration (relative to glyceraldehyde); that is, in the Fischer projection formulas, the amino group appears on the left (Figure 17-1).

Figure 17-1
Naturally occurring α-amino acids have the L-configuration.

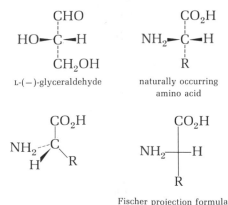

Fischer projection formula

The twenty most common amino acids are listed in Table 17-1. All have common names, and three- or four-letter abbreviations of the names are used to represent the amino acids in protein or peptide structures. Unfortunately the names are not systematic, but they are used instead of IUPAC names since the latter are quite cumbersome. To facilitate memorizing the names and formulas, they have been grouped in several subcategories in the table.

Table 17-1
Names and Formulas of the Common Amino Acids*

NAME	ABBREVIATION	FORMULA	R
		A. One Amino Group and One Carboxyl Group	
1. Glycine	Gly	$H-\underset{\underset{NH_2}{\vert}}{CH}-CO_2H$	
2. Alanine	Ala	$CH_3-\underset{\underset{NH_2}{\vert}}{CH}-CO_2H$	
3. Valine	Val	$CH_3\underset{\underset{CH_3}{\vert}}{CH}-\underset{\underset{NH_2}{\vert}}{CH}-CO_2H$	R=H or alkyl
4. Leucine	Leu	$CH_3\underset{\underset{CH_3}{\vert}}{CH}CH_2-\underset{\underset{NH_2}{\vert}}{CH}-CO_2H$	
5. Isoleucine	Ileu	$CH_3CH_2\underset{\underset{CH_3}{\vert}}{CH}-\underset{\underset{NH_2}{\vert}}{CH}-CO_2H$	
6. Serine	Ser	$\underset{\underset{OH}{\vert}}{CH_2}-\underset{\underset{NH_2}{\vert}}{CH}-CO_2H$	R contains an alcohol function
7. Threonine	Thr	$CH_3\underset{\underset{OH}{\vert}}{CH}-\underset{\underset{NH_2}{\vert}}{CH}-CO_2H$	
8. Cysteine	CySH	$\underset{\underset{SH}{\vert}}{CH_2}-\underset{\underset{NH_2}{\vert}}{CH}-CO_2H$	Three sulfur-containing amino acids
9. Cystine	CyS—SCy	$\left(S-CH_2-\underset{\underset{NH_2}{\vert}}{CH}-CO_2H\right)_2$	
10. Methionine	Met	$CH_3S-CH_2CH_2-\underset{\underset{NH_2}{\vert}}{CH}-CO_2H$	

- -

11. Proline   Pro

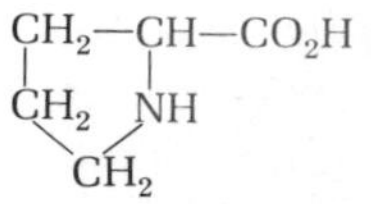

$$CH_2-CH-CO_2H$$
$$CH_2 \quad NH$$
$$CH_2$$

Two cyclic
amino acids

12. Hydroxyproline   HPro

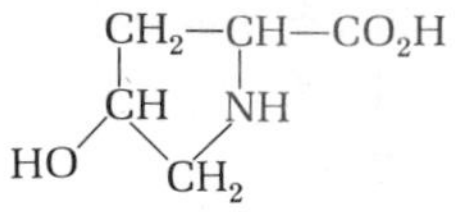

$$CH_2-CH-CO_2H$$
$$CH \quad NH$$
$$HO \quad CH_2$$

- -

13. Phenylalanine   Phe

$$\text{(phenyl)}-CH_2-CH-CO_2H$$
$$NH_2$$

14. Tyrosine   Tyr

$$HO-\text{(phenyl)}-CH_2-CH-CO_2H$$
$$NH_2$$

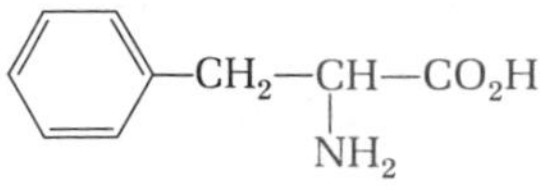

One hydrogen
in alanine is
replaced by
an aromatic
or heteroaromatic
(indole) ring

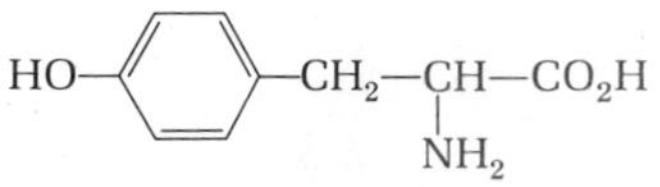

15. Tryptophan   Try

$$\text{(indole)}-CH_2-CH-CO_2H$$
$$NH_2$$

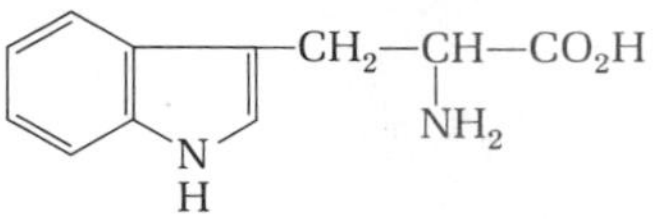

B. One Amino Group and Two Carboxyl Groups

16. Aspartic acid   Asp

$$HOOC-CH_2-CH-CO_2H$$
$$NH_2$$

17. Glutamic acid   Glu

$$HOOC-CH_2CH_2-CH-CO_2H$$
$$NH_2$$

C. One Carboxyl Group and Two Basic Groups

18. Lysine   Lys

$$CH_2CH_2CH_2CH_2-CH-CO_2H$$
$$NH_2 \qquad NH_2$$

19. Arginine   Arg

$$NH_2$$
$$C-NH-CH_2CH_2CH_2-CH-CO_2H$$
$$NH \qquad NH_2$$

For a brief
discussion of
the second
basic groups,
shown in color,
see the text

20. Histidine   His

$$CH=C-CH_2-CH-CO_2H$$
$$NH_2$$
$$N \qquad NH$$
$$CH$$

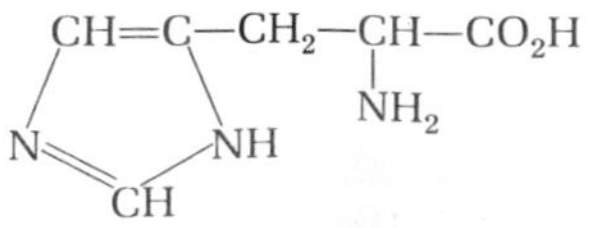

*Those amino acids which are necessary for humans but which must be ingested in
food (i.e., cannot be synthesized by the body), the so-called **essential** amino acids,
have abbreviations printed in color.

Most amino acids (entries 1–15) contain one amino and one carboxyl group. Among this type one finds simple R groups (1–5), and other functional groups such as alcohols (6, 7, and 12), sulfur compounds (8–10), a phenol (14), and aromatic rings (13–15).

Of the two amino acids with two carboxyl groups (16 and 17), glutamic acid, as its sodium salt (monosodium glutamate), is a flavor-enhancing agent sold under the trade name *Accent*.

The last three table entries have two basic groups, although in two of the compounds the second basic function is not a simple amino group. Arginine contains a **guanidine** group, and histidine contains an **imidazole** ring; each of these functions is basic and is readily protonated (eq 17-1 and 17-2).

$$\underset{\text{guanidine group}}{\overset{\displaystyle NH_2}{\underset{\displaystyle NH}{C-NH-R}}} + H^+ \rightleftharpoons \underset{\text{a guanidinium ion}}{\overset{\displaystyle NH_2}{\underset{\displaystyle +NH_2}{C-NH-R}}} \qquad (17\text{-}1)$$

$$\underset{\text{imidazole group}}{\overset{\displaystyle CH=C-R}{\underset{\displaystyle N \qquad NH}{\underset{\displaystyle CH}{}}}} + H^+ \rightleftharpoons \underset{\text{an imidazolium ion}}{\overset{\displaystyle CH=C-R}{\underset{\displaystyle HN^+ \qquad NH}{\underset{\displaystyle CH}{}}}} \qquad (17\text{-}2)$$

Proteins contain various numbers of the different amino acids listed in Table 17-1, though not all proteins contain all of the amino acids. The amino acid make-up of a particular protein determines its properties. Thus insoluble, structural proteins (such as those in hair, wool, tendons) contain high percentages of amino acids with non-polar side chains (entries 1–5, 11, 13) whereas more soluble proteins (such as enzymes, albumin, hemoglobin) contain large percentages of amino acids with polar, acidic, or basic side-chains (entries 6, 7, 16–20).

In addition to the twenty protein-derived amino acids listed in Table 17-1, many other amino acids have been isolated from natural sources (for example, from certain antibiotics and from hormones which are poly-amino acids). Some of these have the D configuration.

17.2 DIPOLAR PROPERTIES OF AMINO ACIDS

All of the amino acids in Table 17-1 are colorless, crystalline solids. They have high melting points (over 200°), at which they usually decompose. Most amino acids are relatively insoluble in organic solvents (alcohol, ether, benzene) but at least moderately soluble in water. These properties are characteristic of ionic rather than covalent compounds. The explanation is clear, if one recalls that the amino nitrogen is appreciably more basic than a carboxylate anion. Given a choice between the two, a proton should preferentially add to the former, so that an amino acid is best represented by the structure in which the proton is attached to nitrogen rather than to oxygen.

$$R-\underset{\underset{+NH_3}{|}}{CH}-C\underset{O}{\overset{O}{\diagup}}\Bigg\} -$$

dipolar structure of
an α-amino acid

This **dipolar structure** explains the salt-like or ionic properties of amino acids.*

Amino acids are **amphoteric;** they can either accept a proton from a strong acid or donate one to a strong base. The equilibria are expressed in equation 17-3 for an amino acid with no other acidic or basic groups.

$$\underset{\underset{NH_2}{|}}{RCHCOOH} \underset{}{\overset{\sim H^+}{\rightleftarrows}} \underset{\underset{NH_3^+}{|}}{RCHCOO^-} \underset{+OH^-}{\overset{+H^+}{\rightleftarrows}} \underset{\underset{NH_3^+}{|}}{RCHCOOH}$$

α-amino acid   dipolar ion   amino acid
            or zwitterion   in acid solution

$$+OH^- \updownarrow +H^+$$

$$\underset{\underset{NH_2}{|}}{RCHCOO^-} \tag{17-3}$$

amino acid
in basic solution

The net charge of the amino acid molecule is a function of the acidity of the solution. In an acid solution, the molecule bears a positive charge (on the nitrogen atom); in an alkaline solution it bears a negative charge (on the oxygen atoms). If an acidic solution of an amino acid is placed in an electric field, the organic ion will migrate toward the cathode (negative electrode); if an alkaline solution is electrolyzed, the organic ion will migrate toward the anode (positive electrode). Intermediate between these two extremes, it is possible to attain a solution of precise acidity (pH) so that the organic ion is dipolar and electrically neutral, and therefore will not migrate toward *either* electrode. This pH is known as the **isoelectric point** and is characteristic for each amino acid. This point may not come

---

* Such structures are sometimes called **zwitterions** (from the German, literally double ions).

at neutrality pH = 7.0) because the basic part of the molecule may not be as basic as the acid part is acidic, or vice versa. The isoelectric points for amino acids with one amino group and one carboxyl group (Table 17-1, entries 1–15) range between pH 5.5 and 6.5. In contrast, compounds with two acidic functions (16, 17) have isoelectric points near pH 3.0, whereas the basic amino acids (18–20) have isoelectric points at pH 7.5–10.5, arginine being the most basic of the three.

17.3 ANALYSIS OF AMINO ACID MIXTURES

The first stage in determining the structure of a protein involves hydrolysis, usually by heating with 6M hydrochloric acid, to the constituent amino acids. To determine the kinds and amounts of amino acids present in a protein, it is necessary to have methods for analyzing the mixtures of amino acids obtained when a protein is hydrolyzed. This requires (*a*) a procedure for separating the amino acids from one another, (*b*) a method for identifying each amino acid, and (*c*) a method for determining the amount of each amino acid present.

In recent years, machines have been developed which perform this task automatically. They operate in the following manner. The mixture of amino acids from the hydrolysis of a few milligrams of a protein is first placed on the top of a column packed with a material that selectively absorbs amino acids. This packing is composed of an insoluble resin which contains strongly acidic (sulfonic acid) groups. When the amino acids are protonated by the acidic groups on the resin, they become positively charged and the resin becomes negatively charged. A buffer solution of known pH is next pumped down the column. The amino acids pass through at different rates, depending on their structure and basicity, and are thus separated. The column effluent is met by a stream of **ninhydrin** (sec 17.3a), a reagent which reacts with amino acids to produce a blue dye. Therefore, the effluent is alternately colorless or blue, depending on whether or not an amino acid is being eluted from the column. The absorbance (review sec 13.1a) of the dye at 570 nm is automatically recorded as a function of the volume of effluent. The time of appearance of a peak is characteristic of the particular amino acid and is used to identify it (by calibration with known samples); the intensity of the absorption is a measure of amount of each amino acid present.

Other chromatographic techniques (thin layer, paper, or gas-liquid) can be used to analyze amino acid mixtures.*

---

*The appendix (pp. 117–130) of "Peptides and Amino Acids" by K. D. Kopple (W. A. Benjamin, Inc., 1966) contains an elementary discussion of these and other methods for separating and analyzing amino acids and proteins.

17.3a THE NINHYDRIN REACTION

Ninhydrin is a valuable reagent for detecting and for quantitatively determining the concentration of amino acids. It is the hydrate of a triketone, and the first step in its reaction with amino acids involves formation of an imine (Table 9-3).

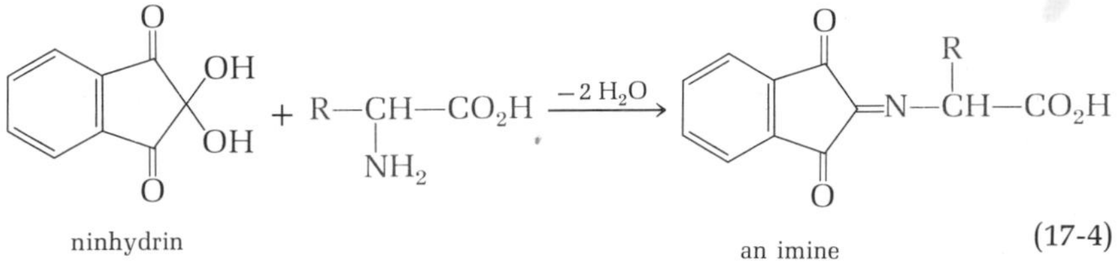

ninhydrin   an imine   (17-4)

Decarboxylation and hydrolysis leads to an amino derivative which condenses with a second mole of ninhydrin to produce a blue dye (eq 17-5). Resonance structures can be written for the dye which delocalize the

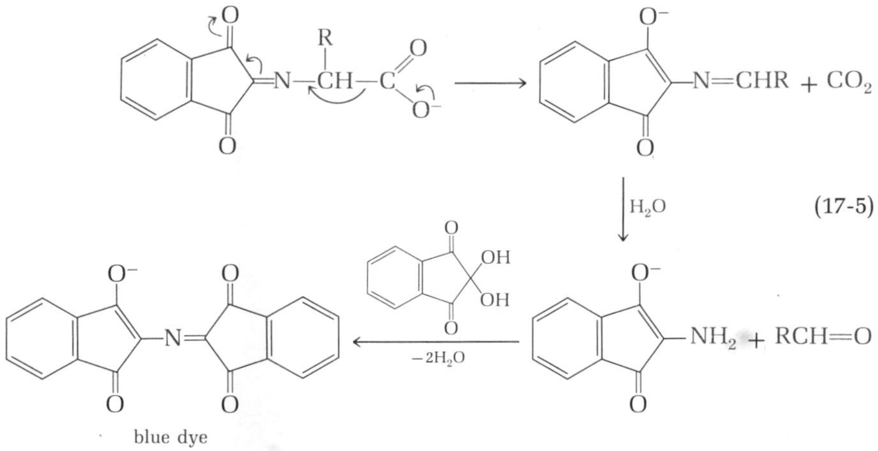

(17-5)

blue dye

negative charge over all four oxygen atoms. The color of the dye is extremely intense, so that 10^{-6} to 10^{-7} moles of amino acid can easily be determined quantitatively.

Since the only feature of the dye derived from the amino acid is the nitrogen atom, all amino acids with a primary amino group give the same dye. Proline and hydroxyproline behave somewhat differently and give a yellow dye. The absorbance can be determined with an amino acid analyzer by examining the effluent at 440 nm (rather than the usual 570 nm).

17.4 AMINO ACID SYNTHESIS

All of the common amino acids present in proteins are now available commercially. They are used as dietary supplements and in nutritional and clinical studies as well as in biochemical experiments.

Many synthetic methods for producing amino acids have been devised. We mention here only two, each of which is an extension of previously studied reactions.

α-Halogenated acids (sec 14.2a) react with excess ammonia to give α-amino acids. The reaction, which is illustrated for glycine (eq 17-6), is analogous to the preparation of amines from alkyl halides (sec 12.4a).

$$\underset{\substack{| \\ \text{Cl} \\ \text{chloroacetic acid}}}{\text{CH}_2\text{---CO}_2\text{H}} + 2\,\text{NH}_3 \longrightarrow \underset{\substack{| \\ \text{NH}_2 \\ \text{glycine}}}{\text{CH}_2\text{---CO}_2\text{H}} + \text{NH}_4{}^+\text{Cl}^- \quad (17\text{-}6)$$

A second method of synthesis is analogous to the synthesis of α-hydroxyacids from cyanohydrins (eq 14-24). When hydrogen cyanide reacts with an aldehyde in the presence of ammonia, the product is an aminonitrile (i.e., the hydroxyl group of the cyanohydrin is replaced by an amino group). Hydrolysis of the cyano group gives the amino acid. The reaction, known as the **Strecker synthesis,** is illustrated for alanine (eq 17-7).

$$\underset{\text{acetaldehyde}}{\overset{\overset{\text{O}}{\overset{\|}{}}}{\text{CH}_3\text{---C---H}}} \xrightarrow[\text{NH}_3]{\text{HCN}} \underset{\text{an aminonitrile}}{\overset{\overset{\text{NH}_2}{|}}{\text{CH}_3\text{---CH---CN}}} \xrightarrow[\text{H}^+]{\text{H}_2\text{O}} \underset{\text{alanine}}{\overset{\overset{\text{NH}_2}{|}}{\text{CH}_3\text{---CH---CO}_2\text{H}}} \quad (17\text{-}7)$$

Both these syntheses yield racemic amino acids which must be resolved if the optically active acids are desired.

17.5 REACTIONS OF AMINO ACIDS

Amino acids react as anticipated at either functional group. Their reactions with acids and bases have already been described (sec 17.2).

The primary amino group reacts with nitrous acid to liberate nitrogen gas (sec 12.5d). If the volume of nitrogen is measured, one

$$\underset{\substack{| \\ \text{NH}_2 \\ \text{glycine}}}{\text{CH}_2\text{---CO}_2\text{H}} + \underset{\substack{\text{nitrous} \\ \text{acid}}}{\text{HONO}} \longrightarrow \underset{\substack{| \\ \text{OH} \\ \text{glycolic acid}}}{\text{CH}_2\text{---CO}_2\text{H}} + \text{H}_2\text{O} + \text{N}_2\uparrow \quad (17\text{-}8)$$

can use the reaction to determine the number of free amino groups present in a peptide or protein (the Van Slyke method).

Amino acids react with alcohols to form esters. The ethyl esters are volatile, and many years ago Emil Fischer (Figure 16-1) took advantage of this fact to first isolate pure amino acids from protein hydrolyzates by fractional distillation of the esters.

$$R\!-\!\underset{\underset{NH_2}{|}}{CH}\!-\!CO_2H + HOCH_2CH_3 \xrightarrow[\text{heat}]{HCl} R\!-\!\underset{\underset{NH_2}{|}}{CH}\!-\!CO_2CH_2CH_3 + H_2O \qquad (17\text{-}9)$$

ethyl ester of an
amino acid

Amino acids can be converted to amides at either functional group, and it is this capability which is important in the formation of peptides and proteins. Recall that when ammonium (eq 10-45) or amine (eq 12-43) salts are heated, they eliminate water to form amides. A similar reaction can occur with amino acids which are *inner amine salts* (sec 17.2). If glycine, for example, is heated strongly, water may be eliminated to form the dipeptide glycylglycine (eq 17-10).

$$\underset{\text{glycine}}{\overset{+}{H_3N}\!-\!CH_2\!-\!CO_2^-} + \underset{\text{glycine}}{\overset{+}{H_3N}\!-\!CH_2CO_2^-} \xrightarrow[-H_2O]{\text{heat}}$$

$$\underset{\text{glycylglycine}}{\overset{+}{H_3N}\!-\!CH_2\!-\!\overset{\overset{\textstyle O}{\|}}{C}\!-\!NH\!-\!CH_2CO_2^-} \qquad (17\text{-}10)$$

Further reaction in the analogous fashion can lead to a linear polymer **(polyglycine)**. Alternatively, a cyclic diamide, called **diketopiperazine,** may be produced.

$$
\begin{array}{ccc}
\text{glycylglycine} & \xrightarrow[-H_2O]{\text{heat}} & \text{diketopiperazine}
\end{array} \qquad (17\text{-}11)
$$

glycylglycine   diketopiperazine

In practice, both reactions occur. Accordingly, equation 17-10 does not represent a practical method for making peptides.

17.6 PEPTIDES

The partial or incomplete hydrolysis of proteins leads not only to amino acids but to fragments which contain two or more amino acid residues linked together as they were in the original protein molecule. Such fragments are called **peptides;** they are called *di-, tri-, tetrapeptides,* etc., depending on the number of amino acid residues present.

Glycylglycine (eq 17-10) is a dipeptide in which both amino acid units are identical. If the two amino acids in a dipeptide differ, two structures are possible. This is illustrated for a dipeptide which consists of a glycine and an alanine unit.

$$
\overset{+}{H_3N}-CH_2-\overset{\overset{\displaystyle O}{\|}}{C}-NH-\underset{\underset{\displaystyle CH_3}{|}}{CH}-CO_2^-
\qquad\qquad
\overset{+}{H_3N}-\underset{\underset{\displaystyle CH_3}{|}}{CH}-\overset{\overset{\displaystyle O}{\|}}{C}-NHCH_2CO_2^-
$$

glycylalanine   alanylglycine
Gly—Ala   Ala—Gly

By convention, the formulas are written with the peptide bond in the direction shown, that is,

$$
\overset{\overset{\displaystyle O}{\|}}{\sim\!\!\sim\!C}-NH\!\sim\!\!\sim, \qquad not \qquad \sim\!\!\sim\!NH-\overset{\overset{\displaystyle O}{\|}}{C}\!\sim\!\!\sim
$$

Written this way, the amino acid at the left end of the chain has a free amino (or NH_3^+) group and is called the **N-terminal** amino acid, whereas the amino acid at the other end of the chain has a free carboxyl (or CO_2^-) group and is called the **C-terminal** amino acid.

The formulas of peptides are frequently abbreviated by linking together the abbreviations for their amino acid constituents, starting with the N-terminal amino acid. Thus the tripeptide Gly—Ala—Ser would contain two peptide bonds and have the formula

$$
\overset{+}{H_3N}-CH_2-\overset{\overset{\displaystyle O}{\|}}{C}\!\!\mid\!\!NH-\underset{\underset{\displaystyle CH_3}{|}}{CH}-\overset{\overset{\displaystyle O}{\|}}{C}\!\!\mid\!\!NH-\underset{\underset{\displaystyle CH_2OH}{|}}{CH}-CO_2^-
$$

Gly————————Ala————————Ser

N-terminal   C-terminal

One begins to appreciate the complexities of peptide and protein structures when one realizes that there are five additional tripeptides which can be constructed from these three amino acids. All are isomers and differ only in the sequence in which the amino acids are linked together. The number of possible isomers leaps dramatically with increasing chain length, being 24 for a tetrapeptide and

40,320 for an octapeptide in which all the amino acids are different.

Peptides are sometimes classified as *oligo*peptides or *poly*peptides, depending on whether there are a few or many amino acids per molecule. Many oligopeptides appear to be important biological materials in their own right. For example, a simple tripeptide (called TRH or thyrotropin-releasing hormone) has been identified as an important hormone secreted by the hypothalamus—a small section of tissue at the base of the brain. It appears to regulate the flow of another hormone from the nearby pituitary gland. Complete hydrolysis of this tripeptide gives glutamic acid, histidine, and proline. The formula is

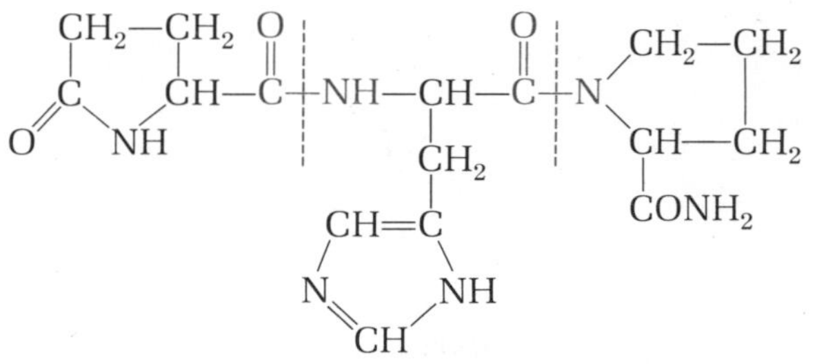

TRH (Pyroglutamylhistidylproline amide)
Pyro—Glu┼His┼Pro—NH₂

In this formula, one notices several minor variations from the usual peptide structure. For example, the carboxyl group at the C-terminal end is not present as such, but as a simple amide ($CONH_2$). This variation in terminal and side chain carboxyl groups is particularly common, and is designated by the abbreviation —NH_2. Furthermore, the free amino group of the N-terminal glutamic acid has cyclized to form an amide with the carboxyl group in the side chain of this acid (eq 17-12).

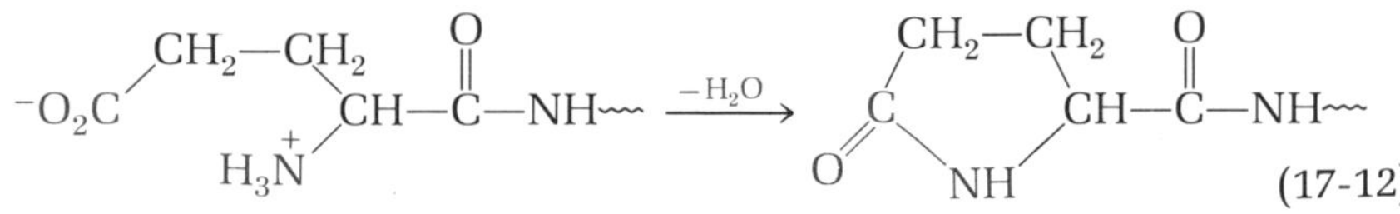

glutamic acid residue   pyroglutamic acid residue

(17-12)

Oxytocin and **vasopressin,** which are oligopeptides produced by the pituitary gland, were the first peptide hormones to be synthesized. Vincent du Vigneaud (Cornell University) was awarded the Nobel prize in 1955 for this accomplishment. Oxytocin, which regulates uterine contraction and lactation and is often administered at childbirth to induce delivery, has the structure shown. Note that the C-terminal glycine, as well as the carboxyl groups in the aspartic and glutamic acid side chains are present as simple amides. The structure is cyclic as a result of the disulfide link (which can easily be broken to two —SH groups by mild reduction).

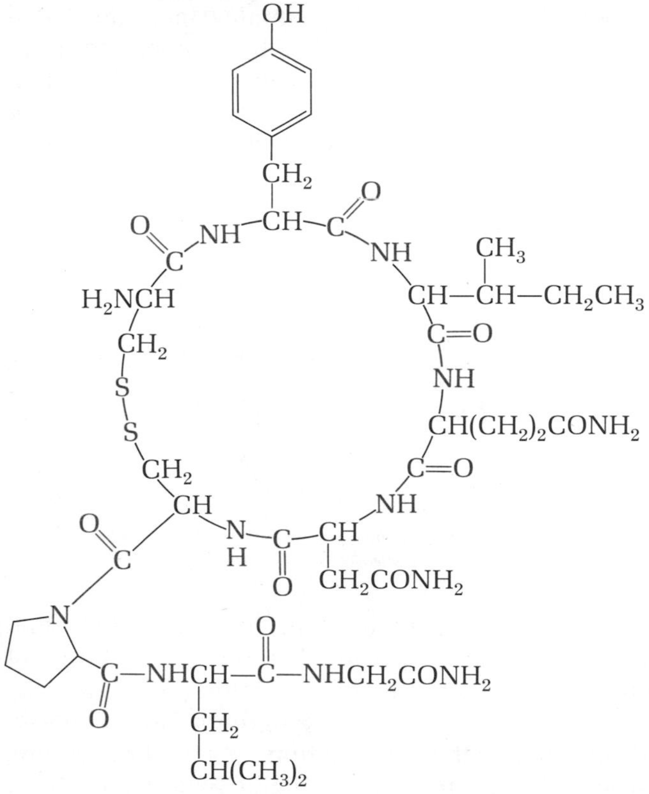

complete

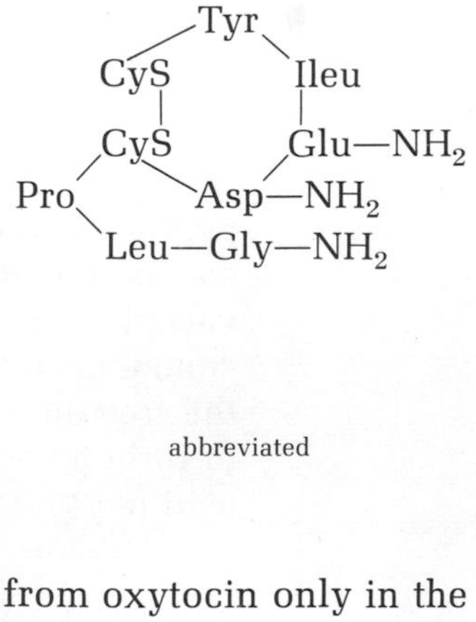

abbreviated

Structure of oxytocin

Vasopressin, which differs in structure from oxytocin only in the replacement of the isoleucine by phenylalanine, and the leucine by arginine, regulates the excretion of water by the kidneys and also affects the blood pressure. The disease *diabetes insipidus,* in which too much urine is excreted, is caused by its deficiency and can be treated by administering the hormone.

Many other physiologically active peptides are known, amongst them the antibiotics bacitracin and valinomycin and the polypeptide hormone insulin (sec 17.7).

17.7 PRIMARY PROTEIN STRUCTURE

There are several levels of sophistication at which the structure of a protein may be described. The first of these is the amino acid sequence. Once the number of each kind of amino acid present in a peptide or protein is determined, as already described (sec 17.3), it is necessary to determine the order in which they were originally linked.

The general strategy is to partially hydrolyze the polypeptide or protein to oligopeptides, perhaps containing from two to five amino acid residues each. These smaller peptides are then separated by column chromatography. If each of these oligopeptides can be identified, then by piecing together *overlapping sequences,* one can establish the original order. For example, suppose a hexapeptide gave, on complete hydrolysis, the six different amino acids Gly, Ala, Phe, Val, Leu, and Ser. If partial hydrolysis led to the isolation and identification of three tripeptides as

Gly—Val—Ala

Leu—Phe—Gly

Val—Ala—Ser

one could deduce that the original hexapeptide must have been

Leu—Phe—Gly—Val—Ala—Ser

Isolation of the tripeptide Phe—Gly—Val or dipeptides such as Ala—Ser, Leu—Phe, or Gly—Val, etc., would provide confirmatory evidence for the assigned structure.

This example illustrates how the complex problem of a hexapeptide sequence determination can be reduced to the simpler problem of *three* tripeptide sequence determinations. Eventually, however, one must solve the sequence problem, even for a simple di- or tripeptide. To illustrate one way in which this can be done, consider the tripeptide Gly—Val—Ala. Note that in the tripeptide, Gly has a free amino group (it is the N-terminal acid),

$$H_2N-CH_2-\overset{\overset{\displaystyle O}{\|}}{C}-NH-CH-\overset{\overset{\displaystyle O}{\|}}{C}-NH-CH-CO_2H$$

CH(CH₃)₂   CH₃

Gly————Val————Ala

Ala has a free carboxyl group (it is the C-terminal acid), and Val has no free functional groups. Advantage can be taken of these differences to determine the sequence.

The **N-terminal amino acid** is usually identified by allowing the peptide to react with **2,4-dinitrofluorobenzene** (DNFB; review sec 8.5). This reagent readily undergoes nucleophilic aromatic substitution with amines; fluoride ion is displaced, and a 2,4-dinitro-phenylamine is produced. Since the N-terminal amino acid has a free amino group, it reacts with the reagent. Amino acids in the middle of the chain or at the carboxyl end, not having a free amino group, do not react. (An exception would be lysine (entry 18, Table 17-1) which can react regardless of its position because of the amino group in its side chain). In the example, the reaction

O₂N— (benzene ring with NO₂) —F + Gly—Val—Ala ⟶

2,4-dinitrofluorobenzene

O₂N— (benzene ring with NO₂) —NHCH₂$\overset{O}{\overset{\|}{C}}$—NHCH—$\overset{O}{\overset{\|}{C}}$—NHCH—CO₂H $\xrightarrow[H_2O]{H^+}$
                                                        | |
                                                   CH(CH₃)₂   CH₃

DNP—Gly—Val—Ala

O₂N— (benzene ring with NO₂) —NHCH₂CO₂H + Val + Ala

DNP—Gly   (17-13)

leads to a tripeptide in which the glycine amino nitrogen is attached to a dinitrophenyl (DNP) group. When the modified tripeptide is hydrolyzed (eq 17-13) and the hydrolyzate is column chromato-graphed, DNP-Gly is obtained instead of Gly, although the other two amino acids are produced unchanged. The DNP derivatives of amino acids are yellow; they have all been independently synthesized and their behavior on chromatographic columns is established, so that it is easy to identify one which is isolated from a particular DNP-peptide hydrolyzate. In the example, the isolation of DNP—Gly establishes the structure as either Gly—Val—Ala or Gly—Ala—Val; the four structures in which Gly is *not* the N-terminal acid are eliminated. The complete sequence could be established by partial hydrolysis of the tripeptide and isolation of a glycine-free dipeptide. This would have to be either Val—Ala or Ala—Val, and once again DNFB could be used to establish which was the N-terminal amino acid.

Frederick Sanger (Cambridge, England) pioneered the use of DNFB for sequence determination. He received the Nobel prize in 1958 for elucidating the complete amino acid sequence of the hormone **insulin,** a natural peptide used to treat diabetes. The structure (Figure 17-2) contains 51 amino acid residues of 16 different kinds (of the acids in Table 17-1, only Met, HPro, and Try are missing), and has a molecular weight of 5734. The structure consists of two peptide chains, one with 30 amino acid residues and the other with 21, joined by two disulfide links between cysteine units; one of the chains is also looped by a disulfide link. Insulins isolated from beef, sheep, horse, or whale are nearly identical; they differ only in 1–3 amino acids in the "looped" portion of the shorter chain.

Figure 17-2

The primary structure of beef insulin

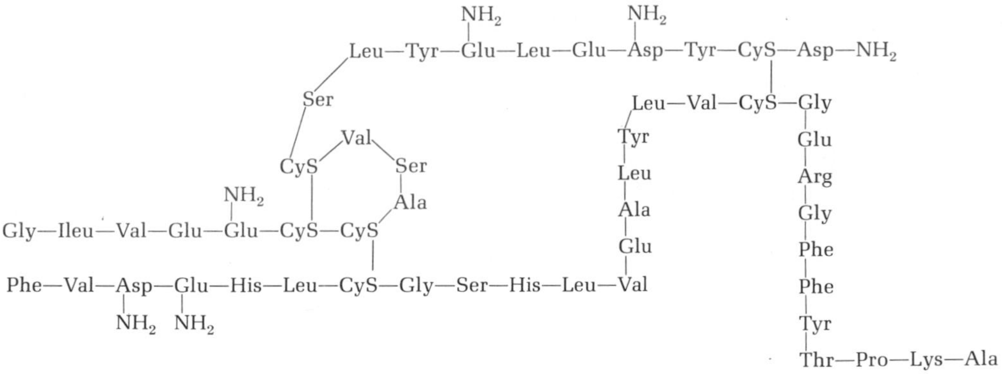

Chemical methods for "labeling" the **C-terminal amino acid** are not as effective as the method just described for N-terminal acids. However, certain enzymes are highly specific in peptide hydrolysis and can be used for sequence determination. **Carboxypeptidase** selectively hydrolyzes peptide bonds adjacent to free carboxyl groups and can degrade a peptide chain one amino acid residue at a time from the carboxyl end. One can follow the reaction as a function of time by withdrawing samples and determining the amino acid content of the hydrolyzate chromatographically. The amino acid which appears in highest concentration first is the C-terminal acid; the next to appear in large concentration is the second amino acid from the carboxyl end, etc. The method is not foolproof, but is extremely useful.

Other enzymes are also highly specific in their hydrolysis of peptide bonds. **Aminopeptidase** selectively attacks peptides from the N-terminal end. **Trypsin** hydrolyzes proteins at the carbonyl end of lysine and arginine residues, whereas **chymotrypsin** splits proteins at the carbonyl end of tyrosine and phenylalanine residues.

The determination of primary protein structure is a problem that is now solved in principle, though the task of determining the structure of any given protein may be arduous and difficult. But each year brings new successes. Among the proteins whose amino acid sequences have now been established are (the number of amino acid residues is given in parentheses) the enzymes ribonuclease (124) and chymotrypsinogen (246), the α- and β-chains of the protein portion of human hemoglobin (141 and 146, respectively), the muscle protein myoglobin (153) and the protein subunit (see sec 17.9) of tobacco mosaic virus (158).

17.8 SECONDARY PROTEIN STRUCTURE

One might expect a long polymer of amino acids to assume almost an infinite variety of shapes, yet nothing could be further from the truth. Many proteins have been isolated in pure crystalline form in which the polymer chain is held in a definite shape, and there is reason to believe that even in solution protein molecules assume certain characteristic geometries. These shapes are referred to as the **secondary structure** of the protein. The two factors which most affect secondary structure are the geometry of the peptide bond and the possibilities for hydrogen bonding.

The amide group is planar, and rotation about the abnormally short carbon-nitrogen bond is restricted (review sec 10.8). Thus, nmr studies of N,N-dimethylacetamide show that at room tempera-

$$CH_3-C\overset{\displaystyle O}{\underset{\displaystyle N}{\diagup\diagdown}}\,CH_3$$
$$\underset{\displaystyle CH_3}{|}$$

N,N-dimethylacetamide

ture the two methyl groups attached to the nitrogen have different chemical shifts. Only when the substance is heated to 150° does rotation about the C—N bond become sufficiently rapid to make the two N-methyl groups equivalent. Similar restriction of the rotation about the many —C—NH— bonds in a peptide or protein help

$$\overset{\displaystyle |}{\underset{\displaystyle O}{\|}}$$

impose a definite shape on the molecules.

X-ray studies, particularly by Linus Pauling (Figure 17-3) and his colleagues at Cal Tech, have established the precise geometry of the peptide bond. The characteristic bond angles and distances of the backbone, which are common to all peptides and proteins, are shown in Figure 17-4.

Figure 17-3
Linus Pauling (Cal Tech), out-standing theoretical chemist who has made many contri-butions to the knowledge of organic structures. He did fun-damental work on the theory of resonance, on the meas-urement of bond lengths and energies, and on the structure of proteins and the mecha-nism of antibody action. He received the Nobel prize in chemistry in 1954 and the Nobel peace prize in 1962.

Figure 17-4
The characteristic bond angles and distances in the —N—C—C— back-bone of a protein, shown in the fully extended form.

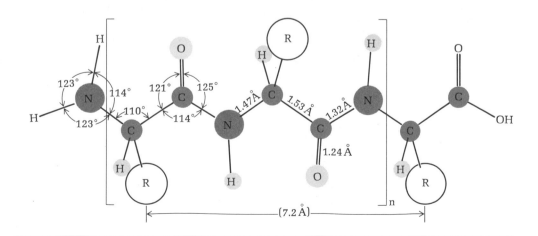

Although the bond angles and distances in Figure 17-4 are correct, the predicted distance of about 7 Å between alternate amino acid residues in the extended form of the chain is much larger than is observed experimentally. The chains are usually coiled into a spiral, or **helix.** The α-helix, first suggested by Pauling as a structure for **α-keratin,** the protein of unstretched wool, hair, horn, and nails, is now known to be present in segments of the structures of many other proteins as well. A portion of such a helix is shown in Figures 17-5 and 17-6. Both right- and left-handed helices are known. The structure derives its strength from the large number of hydrogen bonds of the type

$$\text{N—H----O=C}$$

between groups which are five amino acid residues apart along the chain, but are close enough in space to interact.

The helix is not the only type of secondary structure a protein may have. For example, when α-keratin is stretched it takes a new form called **β-keratin** in which the helices become uncoiled and the

---

Figure 17-5
One turn of an α-helix includes more than three amino acid units (colored N's) but less than four; to be exact, 3.6.

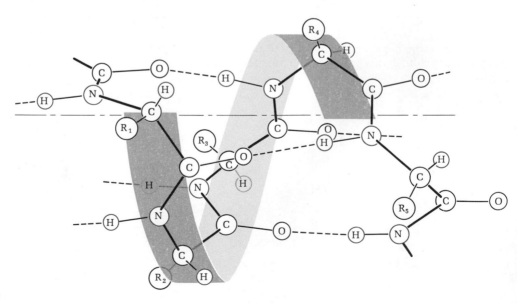

Figure 17-6

Representation of a polypeptide chain as an α-helix. The helix is held
together by hydrogen bonds between —NH and C=O, shown as colored
dotted lines in the right drawing.

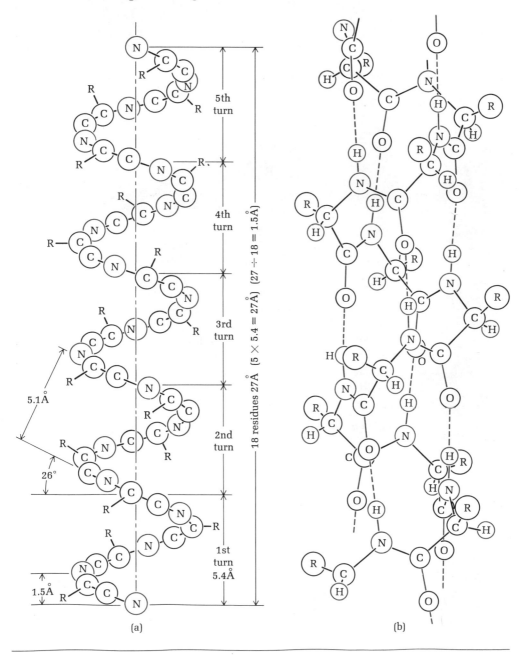

(a)

(b)

Figure 17-7
Segment of the pleated sheet structure of β-keratin.

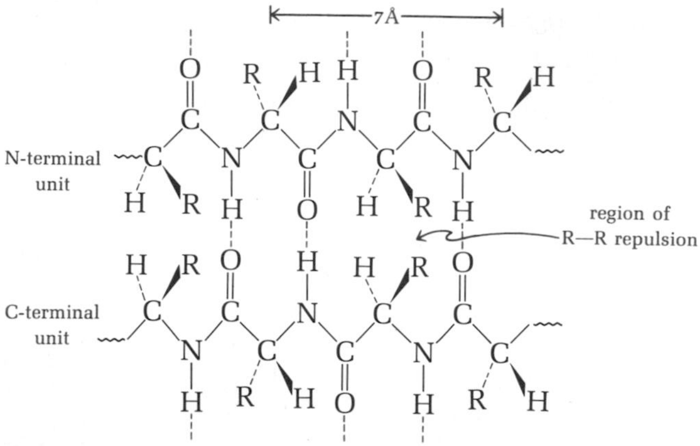

chains stretch side by side to give a sheet-like structure (Figure 17-7). Adjacent chains are aligned in opposite directions and are bound to one another by inter-chain hydrogen bonds (as contrasted with the intra-chain hydrogen bonds in the α-helix). This structure is less common than the helix because R-groups on adjacent chains come rather close to one another. Indeed, the structure is slightly pleated to reduce the R····R repulsion. The sheet structure is only common when the R-groups are small, as in silk fibroin where 46% of the amino acid residues are glycine (R=H).

17.9 TERTIARY AND QUATERNARY PROTEIN STRUCTURE

Most proteins do not have the long rod-like structure that might be expected if the entire molecule were in the form of a helix. Indeed, many proteins, called globular proteins, are roughly spherical in shape even though large segments of the molecule may be present as the helix. Figure 17-8 shows how this might occur. **Myoglobin,** an oxygen-carrying protein isolated from muscle tissue, provides an example. Though 70% of its 153 amino acid units are in right-handed α-helices, myoglobin has an overall hemispherical shape.

Some reasons for the deviation from a purely helical structure are obvious. Wherever a proline (Pro) unit occurs in the primary structure, there is no N—H group available for intra-chain hydrogen bonding. One frequently finds a proline residue in the "turns" of a

Figure 17-8
A hypothetical protein struc-
ture which, though mainly
helical, has an overall globu-
lar shape.

protein structure. Disulfide groups between cysteine units (see, for example, the structure of insulin, Figure 17-2) may link one portion of a protein chain covalently to another, thus interfering with the helical structure. Other R-group interactions associated with tertiary protein structure are "salt bonds" (for example, the terminal NH_2 group in a lysine residue and the terminal carboxyl group in an aspartic or glutamic acid residue may react electrostatically, as $-NH_3^+\cdots\,^-O_2C-$), and hydrogen bonds which involve the hydroxyl groups in the side-chains of serine, threonine, or tyrosine residues. Finally, there are the so-called hydrophobic interactions of nonpolar R groups. It is found, for example, that the amino acids which have non-polar (hydrocarbon) side chains lie mainly on the inside of the myoglobin structure, whereas those with polar or ionic side chains lie on the molecule's surface and enhance its water solubility. This situation is similar to that of an oil emulsion in water (sec 11.3c). When we refer to the **tertiary structure** of a protein, we mean all the modifications in its three-dimensional structure brought about by interactions between the side chains of the amino acid units.

Finally, some proteins exhibit a **quaternary structure,** in which several protein subunits may be joined together in a larger aggregate. Hemoglobin, the protein which carries oxygen in the red blood cells, affords a good example. It is composed of two pairs of two different (α and β) proteins, and has a molecular weight of about 68,000 in neutral solution. The units are held together by polar or ionic forces rather than covalent bonds, for if its solutions are diluted or acidified, the protein dissociates into two subunits, each with a molecular weight of about 34,000. Polio virus and tobacco mosaic virus also consist of agglomerates of protein subunits. Quaternary structures may influence the activity of certain enzymes; some are active only in their polymeric or agglomerate state and become inactive when split into smaller units, whereas for others the converse behavior is observed.*

17.10 PEPTIDE AND PROTEIN SYNTHESIS

It is important, for several reasons, that chemists be able to synthesize peptides of known primary structure. One may wish to verify a particular peptide structure or study the effect on the physiological properties of the peptide of exchanging one amino acid in a sequence for another. Possibly one may wish to synthesize a fragment of a natural enzyme to determine whether it, too, will be physiologically active. The need for linking amino acids in a controlled manner has long been recognized and many methods have been devised to accomplish the task.

Consider the problem of synthesizing glycylalanine as illustrative. It is not practical to heat an equimolar mixture of the two amino acids, for one would obtain a hopeless product mixture which would include not only Gly—Ala but Ala—Gly, tripeptides, etc., and various diketopiperazines (sec 17.5). A more subtle approach is essential.

One might consider "activating" the carboxyl group of the glycine first by converting it to the acid chloride, which could then be treated with alanine to form the peptide bond.

$$\underset{\text{H}_2\text{NCH}_2\overset{\displaystyle O}{\overset{\|}{\text{C}}}-\text{Cl}}{} + \underset{\underset{\displaystyle \text{CH}_3}{|}}{\text{H}_2\text{NCHCO}_2\text{H}} \xrightarrow{-\text{HCl}} \underset{\underset{\displaystyle \text{CH}_3}{|}}{\text{H}_2\text{NCH}_2\overset{\displaystyle O}{\overset{\|}{\text{C}}}-\text{NHCHCO}_2\text{H}}$$

glycylalanine   (17-14)

Unfortunately, this method also fails, since the acid chloride is bifunctional and can polymerize.

$$\text{H}_2\text{NCH}_2\overset{\displaystyle O}{\overset{\|}{\text{C}}}-\text{Cl} + \text{H}_2\text{NCH}_2\overset{\displaystyle O}{\overset{\|}{\text{C}}}-\text{Cl} \xrightarrow{-\text{HCl}}$$

$$\text{H}_2\text{NCH}_2\overset{\displaystyle O}{\overset{\|}{\text{C}}}-\text{NHCH}_2\overset{\displaystyle O}{\overset{\|}{\text{C}}}-\text{Cl} \longrightarrow \text{etc.} \quad (17\text{-}15)$$

It is clear that one way to avoid these difficulties is to (a) protect the amino group of glycine, (b) activate the carboxyl group of the protected glycine, (c) use the protected, activated glycine to form

---

*For students who wish to read further, the book by R. E. Dickerson and I. Geis, "The Structure and Action of Proteins," Harper and Row, New York, 1969, gives a beautifully illustrated and lucid account of protein structure and chemistry.

the peptide bond with alanine, and (*d*) remove the protecting group.

$$NH_2CH_2CO_2H \xrightarrow[\text{reagent}]{\text{protecting}} P\!-\!NHCH_2CO_2H \xrightarrow[\text{reagent}]{\text{activating}} P\!-\!NHCH_2\overset{\overset{\displaystyle O}{\|}}{C}\!-\!A$$

glycine   protected glycine   protected and activated glycine

$$H_2NCH_2\overset{\overset{\displaystyle O}{\|}}{C}\!-\!\underset{\underset{\displaystyle CH_3}{|}}{NHCHCO_2H} \xleftarrow[\text{protecting group}]{\text{remove the}} P\!-\!NHCH_2\overset{\overset{\displaystyle O}{\|}}{C}\!-\!\underset{\underset{\displaystyle CH_3}{|}}{NHCHCO_2H} \xleftarrow{\text{alanine}}$$

glycylalanine   protected glycylalanine   (17-16)

If one wished to synthesize a longer peptide, the last step could be deferred. The protected glycylalanine could be activated, then treated with a third amino acid. The process would be repeated until the desired peptide was synthesized, the protecting group being removed in the final step. The sequence outlined in equation 17-16 is quite general for peptide synthesis.

The **protecting group** must convert the amino group to a form which will not react with the activated carboxyl group, and it must be easily removed at the end of a peptide synthesis by a method which does not hydrolyze the peptide bonds or interact with any of the functional groups in the amino acid side chains (sometimes side chain functional groups must also be protected during peptide synthesis). Many protecting groups are known; the most widely used are the **carbobenzoxy** (or benzyloxycarbonyl) and the **t-butyloxycarbonyl** groups. The reactions which protect the amino group and which are used to remove the protecting group are given in equations 17-17 and 17-18. In each case the amino group is protected by converting it to the less nucleophilic amide group.

$$R\!-\!NH_2 + Cl\!-\!\overset{\overset{\displaystyle O}{\|}}{C}\!-\!OCH_2C_6H_5 \xrightarrow[\text{dilute base}]{\text{protection}} R\!-\!NH\!-\!\overset{\overset{\displaystyle O}{\|}}{C}\!-\!OCH_2C_6H_5$$

amino acid   benzyl chloroformate   a carbobenzoxyamino acid

(17-17)

$$R\!-\!N\!\!\begin{matrix} H \\ \diagdown \\ \diagup \\ H \end{matrix} + CO_2 + C_6H_5CH_3 \xleftarrow{H_2/Pd} \text{removal}$$

                        toluene

$$R\!-\!NH_2 + N_3\!-\!\overset{\overset{\displaystyle O}{\|}}{C}\!-\!O\!-\!C(CH_3)_3 \xrightarrow[-HN_3]{\text{protection}} R\!-\!NH\!-\!\overset{\overset{\displaystyle O}{\|}}{C}\!-\!O\!-\!C(CH_3)_3 \quad (17\text{-}18)$$

amino acid   t-butyl azidoformate   a t-butyoxycarbonyl amino acid

$$R\!-\!NH_2 + CO_2 + (CH_3)_2C\!=\!CH_2 \xleftarrow{CF_3CO_2H} \text{removal}$$

Figure 17-9

The basic plan for solid phase peptide synthesis.

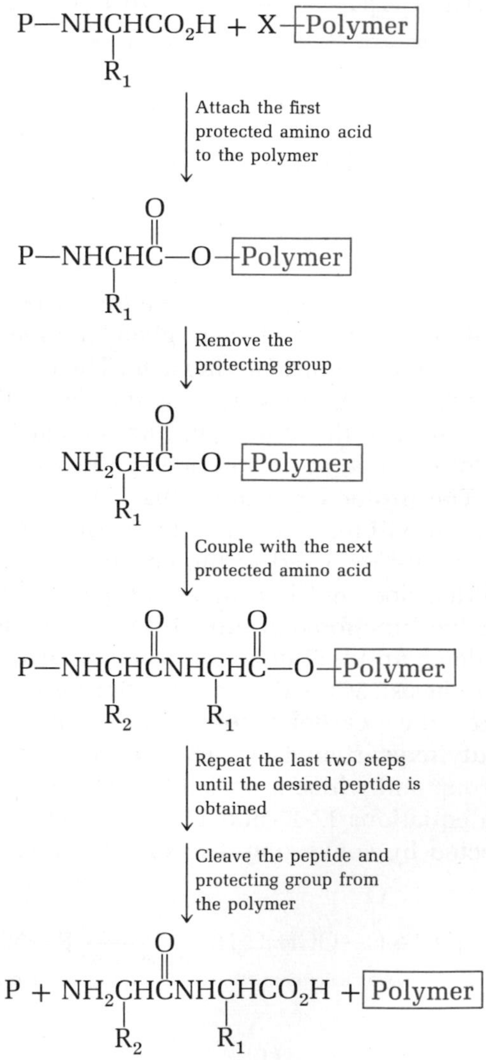

$$P-NHCHCO_2H + X-\boxed{Polymer}$$
$$\underset{R_1}{|}$$

Attach the first
protected amino acid
to the polymer

$$\overset{O}{\overset{\|}{P-NHCHC}}-O-\boxed{Polymer}$$
$$\underset{R_1}{|}$$

Remove the
protecting group

$$\overset{O}{\overset{\|}{NH_2CHC}}-O-\boxed{Polymer}$$
$$\underset{R_1}{|}$$

Couple with the next
protected amino acid

$$\overset{O}{\overset{\|}{P-NHCHC}}\overset{O}{\overset{\|}{NHCHC}}-O-\boxed{Polymer}$$
$$\underset{R_2}{|}\qquad\underset{R_1}{|}$$

Repeat the last two steps
until the desired peptide is
obtained

Cleave the peptide and
protecting group from
the polymer

$$P + \overset{O}{\overset{\|}{NH_2CHC}}NHCHCO_2H + \boxed{Polymer}$$
$$\underset{R_2}{|}\qquad\underset{R_1}{|}$$

Since the carboxyl group itself is not very susceptible to nucleophilic attack (say, by an amine to form an amide), it is necessary to convert it to a more reactive form. Usually an acid anhydride, but sometimes an acyl halide or azide are used (review sec 10.7). The main restriction on activating groups is that they not lead to racemization at the α-carbon which, in all amino acids except glycine, is chiral and optically active.

Usually large proteins are synthesized by preparing smaller peptides (say, containing 6–15 amino acid units each), then linking these together. In this way an error in the sequence at some stage does not ruin the entire synthesis.

In 1965 R. B. Merrifield and J. M. Stewart of the Rockefeller Institute developed an automatic method for synthesizing peptides which has truly simplified what was otherwise a difficult, tedious, multistepped process.* One difficulty with the methods just described is that at each stage one must purify intermediates and remove or wash away excess reagent or unwanted by-products (HCl, NaCl, NH_4Cl, etc.). The **solid phase peptide synthesis** surmounts these problems in a clever way. The basic plan is illustrated in Figure 17-9. The idea is to assemble, stepwise, a peptide chain while one end of it is attached chemically to an insoluble, easily filterable, solid particle. In this way the desired product remains attached to the solid throughout all the reactions and can readily be purified from excess reagents, by-products, etc. simply by washing and filtering the solid. At no intermediate stage need the growing peptide chain be purified. When the desired peptide is constructed it is cleaved chemically from the polymer.

Several methods have been used to attach the first amino acid residue to the polymer. Typically, the solid support is a co-polymer of styrene and divinylbenzene in which the benzene rings have been converted to benzyl chloride groups. These react (eq 17-19) with the

$$\boxed{\text{Polymer}}\!\!-\!\!\bigcirc\!\!-\!CH_2Cl + {}^-O-\overset{\overset{\displaystyle O}{\|}}{C}-\underset{\underset{\displaystyle R_1}{|}}{CH}-NH-P \longrightarrow$$

$$\boxed{\text{Polymer}}\!\!-\!\!\bigcirc\!\!-\!CH_2-O-\overset{\overset{\displaystyle O}{\|}}{C}-\underset{\underset{\displaystyle R_1}{|}}{CH}-NH-P + Cl^- \quad (17\text{-}19)$$

protected amino acid to form a benzyl ester. At the end of the synthesis, the finished peptide-polymer is suspended in anhydrous trifluoracetic acid, and hydrogen bromide is slowly bubbled through the suspension. This removes the peptide from the resin (eq 17-20)

---

*An excellent brief book on their method has been written by J. M. Stewart and J. D. Young *Solid Phase Peptide Synthesis,* W. H. Freeman, 1969.

without cleaving any of the peptide bonds.

$$\boxed{\text{Polymer}}\!-\!\!\left\langle\bigcirc\right\rangle\!\!-\!\text{CH}_2-\text{O}-\overset{\overset{\text{O}}{\|}}{\text{C}}\!\sim\quad\xrightarrow[\text{HBr}]{\text{CF}_3\text{CO}_2\text{H}}$$

polymer-bound peptide

$$\boxed{\text{Polymer}}\!-\!\!\left\langle\bigcirc\right\rangle\!\!-\!\text{CH}_2\text{Br}\ +\ \text{HO}-\overset{\overset{\text{O}}{\|}}{\text{C}}\!\sim\quad (17\text{-}20)$$

free peptide

A key component in the automated synthesis system is a programmer which automatically controls about 80 different operations that permit the synthesis of one peptide bond in about four hours. Working around the clock, the programmer can incorporate six amino acids into a polypeptide in a day. In 1969, Merrifield used the automated synthesizer to prepare the enzyme ribonuclease (124 amino acid residues). The synthesis required 369 chemical reactions and 11,931 steps of the automated peptide synthesis machine. The largest protein synthesized at the time of this writing was prepared in 1971 by C. H. Li (University of California) and his associates, using the solid phase technique. They synthesized a protein (188 amino acid residues) which has the biological properties of the human growth hormone (HGH), a pituitary hormone which is both growth-promoting and lactogenic.

17.11 THE CHEMICAL PROPERTIES OF PROTEINS

Proteins, like amino acids, are amphoteric. This behavior is due not only to the terminal amino and carboxyl groups, but to acidic and basic groups in certain side-chains (Table 17-1, entries 16–20). Proteins have isoelectric points which depend on the number of amino acids present with these acidic or basic side chains. The isoelectric points range from acidic (the milk protein, casein, 4.6) to neutral (hemoglobin, 6.8) to basic (ribonuclease, 9.5).

When placed in an electric field, proteins will migrate toward one of the electrodes. The direction and rate at which they move depends on the isoelectric point and the pH of the medium. This technique, called **electrophoresis,** is used to separate proteins from mixtures and purify them.

Denaturation can be defined as any change which disrupts the configuration (secondary or tertiary structure) of a protein but does not cleave any of the peptide bonds. For example, pH changes can alter hydrogen bonding or "salt bonding" which may disrupt helical or other three-dimensional aspects of a protein structure. Reduction of disulfide links to —SH groups can open loops in protein structures (see, for example, insulin, Figure 17-2). Advantage is taken of the latter reaction in "permanent" waving of hair. The protein in hair contains a high percentage of the amino acid cystine. First, a reducing lotion is applied to cleave the —S—S— bonds. The reduced hair is then placed on curlers and constrained to the desired arrangement, after which an oxidizing lotion is applied to reconstruct the —S—S— bonds in a different pattern. This holds the hair in the desired curl, although the wave cannot be truly permanent since the hair grows.

Some denaturation processes are reversible. Certain enzymes can, for example, be precipitated from aqueous solution (and purified) by ammonium sulfate. When re-dissolved, the enzyme may re-assume its original shape and catalytic activity. In other cases, such as the coagulation of egg-white on heating, denaturation is irreversible.

17.12 PROTEIN METABOLISM

Proteins are hydrolyzed by digestive enzymes to peptides and amino acids which are absorbed and transported by the blood to various body tissues. Tissue proteins are continually being broken down and resynthesized, ingested amino acids being incorporated and those already present being eliminated. Certain amino acids necessary for growth and maintenance must be included in the diet (see the footnote to Table 17-1), but others can be synthesized by the cells from non-protein precursors.

Due to the great diversity in the structures of the common amino acids, it is difficult to generalize about their metabolic pathways. The nitrogen present in amino acids is eliminated in the urine in the form of urea. Enzymes in the liver oxidatively deaminate amino acids to ketoacids (which may be oxidized directly, supplying the body with energy, or indirectly through storage as glycogen) and ammonia which is detoxified by conversion with carbon dioxide to urea.

$$\underset{\substack{|\\ NH_2 \\ \text{an amino acid}}}{R-CH-CO_2H} \underset{\text{enzyme}}{\rightleftharpoons} \underset{\substack{\parallel \\ NH}}{R-C-CO_2H} \xrightarrow{H_2O} \underset{\substack{\parallel \\ O \\ \text{a ketoacid}}}{R-C-CO_2H} + NH_3 \qquad (17\text{-}21)$$

In another important biochemical process, α-amino acids and α-keto acids may be interconverted. For example, alanine may be converted to pyruvic acid (eq 17-22) by the enzyme *transaminase*.

$$CH_3\underset{\underset{NH_2}{|}}{C}HCO_2H + HO_2C(CH_2)_2\underset{\underset{O}{||}}{C}CO_2H \xrightarrow{\text{transaminase}}$$

alanine   α-ketoglutaric acid

$$CH_3\underset{\underset{O}{||}}{C}CO_2H + HO_2C(CH_2)_2\underset{\underset{NH_2}{|}}{C}HCO_2H \quad (17\text{-}22)$$

pyruvic acid   glutamic acid

Since ketoacids are intermediates in carbohydrate metabolism, there is a close relationship in the body between carbohydrate and protein metabolism.

17.13 THE ORIGIN OF AMINO ACIDS AND PROTEINS

In recent years there has been considerable speculation, accompanied by some experimentation, regarding the events which may have occurred about 4.5–3.5 billion years ago that led to the appearance of the first living cell on earth. The subject is sometimes called "chemical evolution," the chemical precursor of Darwinian evolution.*

It is now generally agreed that the primitive earth's atmosphere was hydrogen-dominated, or reducing. The elements carbon, nitrogen, and oxygen were present in their reduced froms (methane, ammonia, water). In 1953 S. L. Miller, working in Professor H. C. Urey's laboratory, subjected a mixture of methane, ammonia, water vapor, and hydrogen to an electric discharge and demonstrated that the amino acids glycine, alanine, and aspartic acid were present in the complex mixture of reaction products. In subsequent experiments, other forms of high energy radiation such as ultraviolet light and γ-rays, and high temperatures have been studied. Other gases such as carbon dioxide, formaldehyde, and hydrogen cyanide have been tried and most of the common amino acids have been detected among the reaction products. Thus, the possibility (not necessarily the actuality) of amino acid synthesis from the earth's primordial atmosphere has been established experimentally.

Mixtures of amino acids, when heated to 150–200° (high temperatures possibly present in volcanic areas), give products with protein-like properties. Certain "primordial" C—H—N compounds will convert amino acids to di-, tri-, and even tetrapeptides in aqueous

*For two recent reviews, the interested reader is referred to an article by Richard M. Lemmon in *Chemical Reviews*, Volume 70, (1970), p. 95 and a book by M. Calvin, *Chemical Evolution*, Oxford University Press, 1969.

solution at room temperature. In still other experiments, polypep-
tides have been isolated from the hydrolysis of hydrogen cyanide
polymers (hydrogen cyanide is almost always formed in experiments
of this type). The mechanisms of these reactions for protein synthesis
are less well understood than those for amino acid synthesis.

The field of chemical evolution is a fascinating area of current
research in chemistry.

New Concepts, Facts, and Terms

1. α-Amino acids with the L configuration are the building blocks of
   proteins; they are linked through amide, or peptide, bonds.
2. Table 17-1 lists the 20 most common amino acids; certain of these,
   essential for life, must be ingested.
3. Amino acids are dipolar and amphoteric; isoelectric point.
4. Separation and analysis uses a combination of chromatographic and
   spectroscopic methods; ninhydrin reacts with amino acids to give a blue
   dye.
5. Synthesis (a) from α-halo acids and ammonia (b) from aldehydes, ammonia, and HCN (the Strecker synthesis)
6. Reactions are characteristic of the two functional groups present.
7. Peptides (di-, tri-, oligo-, poly-); N- and C-terminal amino acids; oxytocin,
   vasopressin
8. Primary protein structure; amino acid sequence determination through
   overlapping sequences; use of 2,4-dinitrofluorobenzene as a reagent for
   N-terminal amino acids; enzymatic methods
9. Secondary protein structure; restricted rotation and planarity of amide
   bonds; the α-helix and the pleated sheet; strength through multiple
   hydrogen bonds
10. Tertiary and quaternary protein structure; disulfide links, salt bonds,
    hydrophobic interactions, aggregation
11. Peptide and protein synthesis; protecting and activating groups; solid
    phase peptide synthesis
12. Chemical properties of proteins; isoelectric points, electropheresis;
    denaturation
13. Protein metabolism; deamination, transamination
14. The origin of amino acids and proteins on earth

Exercises and Problems

1. Define or illustrate with a structural formula

   a. peptide bond   f. dipolar ion
   b. basic amino acid   g. L-configuration
   c. acidic amino acid   h. dipeptide
   d. essential amino acid   i. oligopeptide
   e. isoelectric point   j. polypeptide

2. Draw a three-dimensional structure for L-alanine. What is the Cahn-Ingold-Prelog priority of groups around the chiral carbon atom? Does L-alanine have the **R** or **S** configuration? Repeat for L-aspartic acid.

3. Draw all the structures which contribute to the resonance hybrid when arginine is protonated on the side-chain (i.e., on the guanidine residue). Explain why the guanidine group is more basic than a simple amino group.

4. Write equations which describe the reaction of alanine with (a) hydrochloric acid, and (b) sodium hydroxide. If electrodes are present, in which solution will the alanine migrate toward the cathode (negative electrode)?

5. Write equations which show the likely changes when a strongly alkaline solution of lysine is made acidic. Explain why the isoelectric point for lysine is on the alkaline side of neutrality (pH 9.7).

6. A peptide hydrolyzate contains the three amino acids: glycine, arginine, and glutamic acid. The mixture is placed on the column of an automatic amino acid analyzer and is eluted with an aqueous buffer at pH 4.5. In what order would you expect the acids to be eluted from the column, and why?

7. Write the equations which describe what occurs when phenylalanine is treated with ninhydrin.

8. Write equations which describe the mechanisms of the reactions in equations 17-4 and 17-5.

9. Give equations for the preparation of
   a. alanine from propionic acid
   b. valine from isobutyraldehyde
   c. glycine from acetic acid
   d. leucine from isobutyl chloride

10. Write equations for the reaction of alanine with the following reagents:
    a. nitrous acid
    b. benzoyl chloride
    c. heat
    d. acetic anhydride
    e. ethanol, H^+

11. Write the structural formulas for the following peptides:
    a. glycylalanylglycine
    b. histidylvaline
    c. isoleucylleucylserine
    d. aspartyltryptophan

12. Draw the structures for all possible tripeptide isomers of glycylalanylserine (sec 17.6).

13. Draw the complete and abbreviated formula of vasopressin (sec 17.6).

14. Bradykinin is a nonapeptide obtained by the partial hydrolysis of blood serum protein. It causes a lowering of blood pressure and an increase in capillary permeability. Its structure is abbreviated as Arg—Pro—Pro—Gly—Phe—Ser—Pro—Phe—Arg. Draw its complete structure.

15. The following compounds are isolated as hydrolysis products of a peptide: Ala—Gly, Tyr—Cys—Phe, Phe—Leu—Try, Cys—Phe—Leu, Val—Tyr—Cys, Gly—Val, and Gly—Val—Tyr. Complete hydrolysis of the peptide shows that it contains one unit of each amino acid. What is the structure of the peptide, what are its N- and C-terminal amino acids, and what is its name?

16. Write equations for the reaction of 2,4-dinitrofluorobenzene with glycine, which clearly show the reaction mechanism.

17. Give the structure of the first products obtained when the peptide Ala—Lys—Glu—Leu—Phe—Lys—Tyr—Val—Arg—Gly is treated with
    a. trypsin   c. carboxypeptidase
    b. chymotrypsin   d. aminopeptidase

18. List the factors which are responsible for the secondary, tertiary, and quaternary structures of proteins.

19. Using equations 17-16 and 17-17 as a guide, write equations for the synthesis of glycylalanine. Use benzyl chloroformate as the protecting reagent, and activate the carboxyl group by making its p-nitrophenyl ester.

20. A pentapeptide was converted to its DNP-derivative, then completely hydrolyzed and analyzed quantitatively. It gave DNP-methionine, two moles of methionine, and one mole each of serine and glycine. The peptide was then partially hydrolyzed, the fragments were converted to their pure DNP-derivatives, and each of them was hydrolyzed and analyzed quantitatively. Two tripeptides and two dipeptides isolated in this way gave the following hydrolysis products:

    Tripeptide A: DNP-methionine and one mole each of methionine and glycine.
    Tripeptide B: DNP-methionine and one mole each of methionine and serine.
    Dipeptide C: DNP-methionine and one mole of methionine.
    Dipeptide D: DNP-serine and one mole of methionine.

    Deduce the structure of the original pentapeptide and explain your reasoning.

CHAPTER EIGHTEEN

NATURAL PRODUCTS

One major goal of organic chemists is to determine the structures of compounds which occur in nature and to understand their chemistry. Many plants have been used for centuries as medicines or herbs. Chemists have isolated, identified, and synthesized the active ingredients of many such preparations in order to study precisely their physiological activity and also to synthesize new drugs with improved properties over those found naturally. In this chapter we examine a few of the major classes of natural products.

Most compounds which occur in nature have rather complex structures, often with several functional groups. It is therefore difficult to classify them according to functional group as has been done with both naturally occurring compounds and those synthesized in the laboratory up to this point. They can, however, be considered under several categories which, though a bit arbitrary, do point up certain gross structural features. In this chapter, natural products will be considered in three major groups: (*a*) **terpenes and steroids,** whose main skeleton consists solely of carbon atoms in either open-chain or cyclic conformations; (*b*) **heterocyclic natural products,** compounds in which an atom other than carbon—usually oxygen or nitrogen—is an important structural feature; and (*c*) **nucleic acids** which, although they belong in group (*b*) will be considered separately and in greater detail because of the intense current interest in these genetic materials. Throughout the chapter, compounds may be discussed which, although they do not occur in nature, have an important effect on natural processes. Among these are certain **drugs** and **agricultural chemicals.** Finally, we shall have something to say, at the end of the chapter, about the general nature of organic reactions as they occur in cells.

18.1 TERPENES

For many years chemists have realized that a large group of compounds which occur in plants could be considered as having been formed by linking together C_5 units with the **isoprene** carbon skeleton. Isoprene is the C_5H_8 diene obtained from the pyrolysis of rubber (sec 4.10), and an **isoprene unit** is defined as a five-carbon array consisting of a four-carbon chain with a one-carbon branch. Two

isoprene   isoprene unit

or more isoprene units may be joined together in many ways (head-to-head, head-to-tail, with multiple links, etc.) to form a wide variety of hydrocarbon structures. Cyclic as well as acyclic carbon skeletons are common, and functional groups (double bonds and hydroxyl and carbonyl groups being the most common) may be added to form a large number of interrelated structures.

    Terpenes themselves (from *terpentin,* an old form of the word *turpentine*) are unsaturated hydrocarbons with the molecular formula $C_{10}H_{16}$. They are composed of two isoprene units usually joined at carbons one and four. Examples are **myrcene** (from bay or verbena oils) and **limonene** (in oils of citrus fruits, pine leaves, and peppermint). **α-Pinene,** a bicyclic terpene, is the chief constituent of turpentine and is probably the most abundant hydrocarbon in nature. The colored and black dots in the following formulas show how these molecules can be considered to have been derived from two isoprene units.

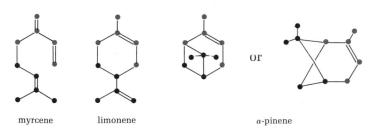

myrcene   limonene   α-pinene

Other natural hydrocarbons related to the terpenes but with three, four, or more isoprene units are common. C_{15} compounds are known as **sesquiterpenes,** C_{20} as **diterpenes,** and C_{30} as **triterpenes. α-Farnesene** is an acyclic sesquiterpene obtained from oil of citronella. The sesquiterpenes **caryophyllene** (from oil of cloves) and **cedrene** (from oil of cedar), with their unusual ring sizes and junctures, illustrate some of the more intricate ways in which isoprene units may be combined. The unsaturated acyclic triterpene **squalene**

α-farnesene   caryophyllene   cedrene

(sec 4.10) is an important intermediate in the biosynthesis of steroids (sec 18.3).

Perhaps more common than the terpene hydrocarbons themselves are their many oxygenated derivatives. The primary alcohol **geraniol** occurs in oil of geranium, whereas its partially reduced derivative **citronellol** occurs in rose oil. Geraniol is a colorless liquid with a very pleasant rose odor. Its esters are often more stable than the alcohol itself and are used in perfumes. The aldehydes corresponding to these alcohols, **geranial** and **citronellal,** occur in lemon oil and oil of citronella, respectively. Natural rose oil, extracted from rose petals, contains 40–60% geraniol and 20–40% citronellol, together with a mixture of odorless hydrocarbons. Synthetic rose oil, manufactured by suitable reduction of oil of citronella, is considerably less expensive.

| geraniol | citronellol | geranial | citronellal |

The close structural relationship between these alcohols, aldehydes, and hydrocarbons such as myrcene suggests that they are all formed in the plants by rather minor variations on a single biosynthetic path.

Farnesol, a sesquiterpene primary alcohol with a lily-of-the-valley aroma, is an important intermediate in the biosynthesis of squalene and steroids (sec 18.3). **Vitamin A** is a diterpene alcohol necessary for synthesis of certain pigments essential to sight.

farnesol   vitamin A alcohol

Menthol, a saturated alcohol related structurally to the hydrocarbon limonene, is the chief constituent of peppermint oil. The corresponding ketone, menthone, is also present in the oil. **Camphor** is a particularly well-known bicyclic terpene ketone. It is synthesized commercially from pinene at a price which competes favorably with natural camphor isolated from the camphor tree grown almost exclusively in the Orient. The synthetic material is racemic, whereas the natural product is optically active.

menthol   menthone   camphor

18.2 STEROIDS

The steroids constitute a biologically important class of compounds which, though derived biogenetically from terpene precursors (sec 18.3), do not usually have a complete isoprenoid carbon skeleton. Included among the steroids are cholesterol, the hormones of the adrenal cortex, the sex hormones, the bile acids, cardiac stimulants, toad poisons, the D vitamins, and oral contraceptives.

The common structural feature of steroids is a system of four fused rings, usually designated by the letters A–D. The ring positions are numbered as shown:

the steroid ring system

The rings are usually not aromatic, and most steroids contain methyl groups (called "angular" methyls) at C10 and C13 and a side chain at C17. Other functionality, especially double bonds and hydroxyl and carbonyl groups, is often present. Most commonly the rings are *trans* fused. By this we mean that if, say, rings A and B were *trans* fused, then substituents attached at the ring junctures (C5 and C10) would be *trans* to one another. *Trans* ring junctures and the usual chair conformation of the cyclohexane rings gives steroids a relatively flat shape:

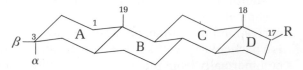

general steroid shape

Angular methyl groups (numbered C18 and C19) usually occupy axial positions; substituents at other ring positions are designated as α- or β-, depending upon whether they are "down" or "up" respectively with regard to the molecular "plane" as shown.

18.2a CHOLESTEROL AND OTHER STEROLS

Sterols are steroids with an alcohol functionality. The best known is cholesterol. It is present in rather large amounts in the brain and spinal cord and in smaller amounts in all cells of animal organisms. Cholesterol was first isolated from and is the chief constituent of gallstones. In recent years, it has received attention in connection with circulatory ailments, especially hardening of the arteries. The total cholesterol extractable from all body tissues, including blood, amounts to about half a pound for the average person. If excess cholesterol is present in the body, it tends to precipitate from solution in the gall bladder as gallstones, and in the blood vessels where the constriction thus formed reduces blood flow, causing high blood pressure.

Cholesterol has 27 carbon atoms, three fewer than a triterpene.

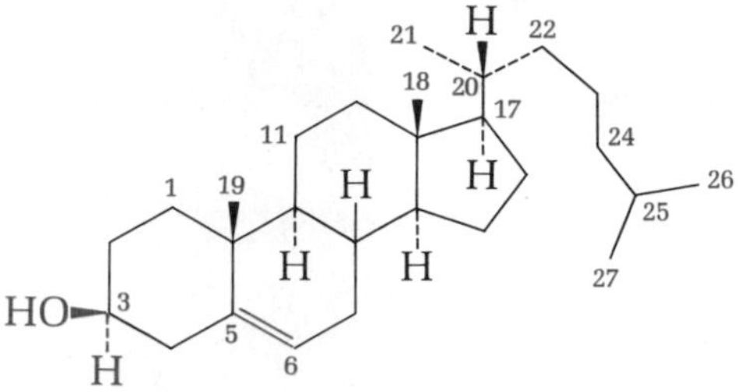

cholesterol

All ring fusions are *trans,* and the steroid ring system is substituted with a β-hydroxyl group at C3, a C5–C6 double bond, and a saturated side chain at C17. The molecule has eight different chiral centers; the configuration at each of these has been established and a total synthesis of the natural product was achieved in 1951 by Professor R. B. Woodward (Harvard University) and co-workers (Figure 18-1).

Although cholesterol occurs only in animals, some sterols may be obtained from plants. Among the more common is **ergosterol,** originally obtained from ergot but also present in yeast. Ergosterol differs from cholesterol by one methyl group (in the side chain) and two double bonds. Irradiation of ergosterol with ultraviolet light opens the B ring and leads to calciferol (vitamin D₂), a substance which can control the amount and ratio of calcium and phosphorus in the blood.

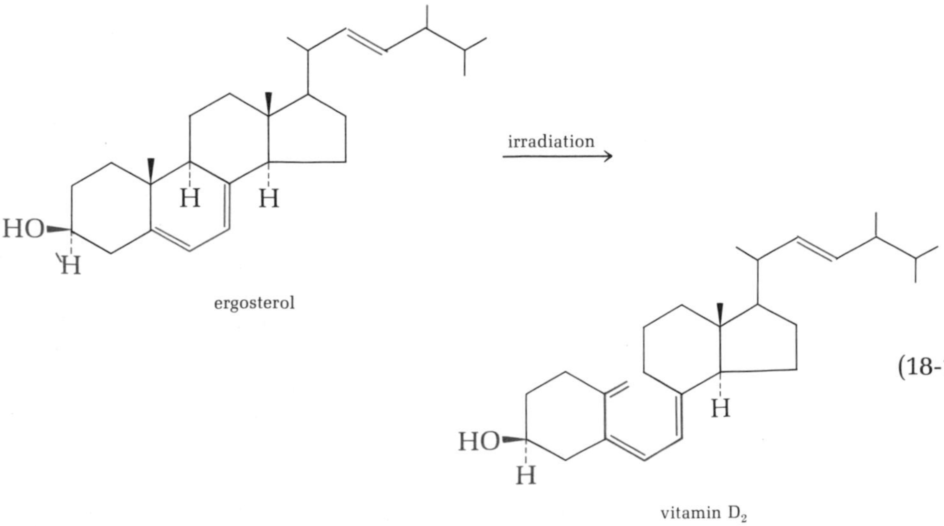

ergosterol

irradiation

(18-1)

vitamin D₂

Figure 18-1
Robert B. Woodward, Harvard chemistry professor and one of the world's outstanding organic chemists. At age 27 he, with W. Doering (also at Harvard), synthesized quinine. His synthesis of steroids and work on the structure of strychnine and other natural products have been particularly noteworthy. Lysergic acid (ergot), strychnine, reserpine (used to treat mental disorders), and chlorophyll are on his list of synthetic triumphs. Professor Woodward received the Nobel prize in chemistry in 1965 for his contributions to the "art" of organic synthesis.

18.2b   SEX HORMONES AND ADRENAL CORTICAL HORMONES

Substances produced in the gonads (ovaries and testes) which control the physiology of the reproductive process and the secondary sex characteristics are known as *sex hormones*. Although primarily involved with reproductive mechanisms, sex hormones are also significant as growth substances of general importance to body health and well-being.

The female sex hormones are of two types: the **estrogens** which are essential for the changes of estrus (menstrual cycle) and **progesterone,** which prepares the uterus for fertilization and is required for normal pregnancy. One of the estrogens is **estradiol,** which controls the development of female characteristics and initiates the first phase of the menstrual cycle. It has an aromatic (phenolic) A-ring, only one angular methyl group, and a hydroxyl group at C17.

Certain synthetic compounds, some of which may be only remotely related structurally to the natural estrogens, such as **stilbesterol,** possess estrogenic activity and are used medically to alleviate difficulties which may arise from a deficiency of the natural hormone.

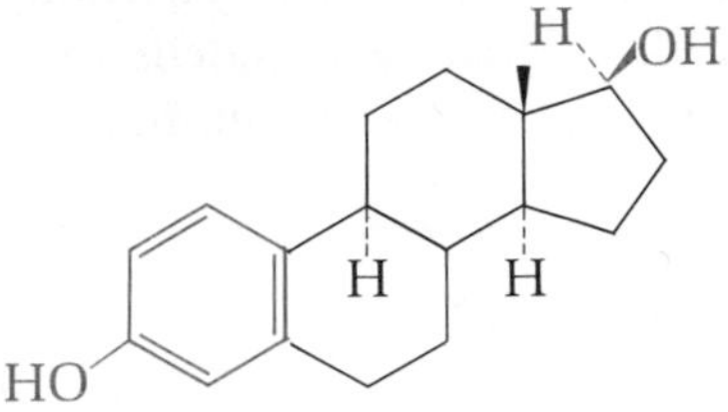

estradiol

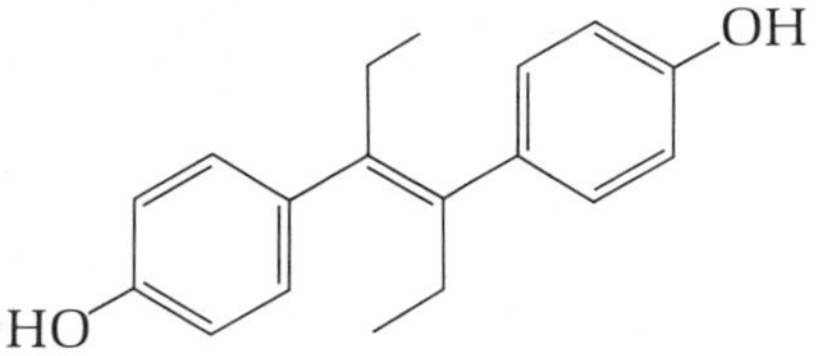

stilbesterol

Progesterone, now produced synthetically, is used clinically to prevent abortion in certain difficult pregnancies. It differs from the estrogens in that the A-ring is not aromatic. On the other hand, a number of closely related steroids, also produced synthetically, are used to prevent conception. An example is the acetylenic ketoalcohol **norethynodrel** (marketed as *Enovid*).

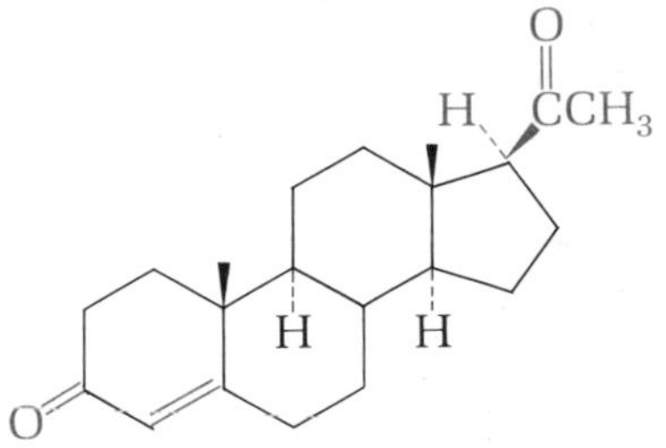

progesterone

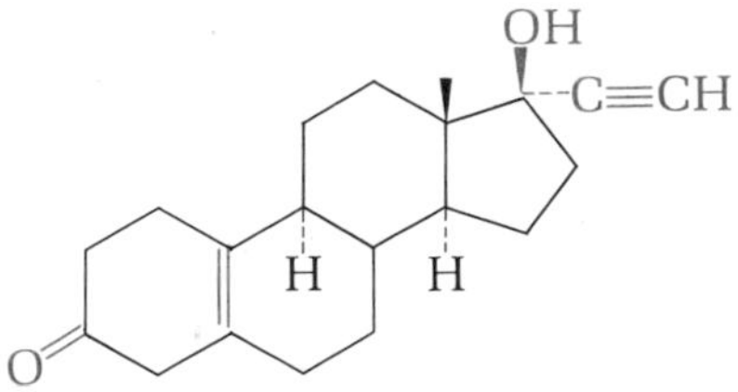

norethynodrel

The male sex hormones, or **androgens,** regulate the development of reproductive organs and secondary sex characteristics (beards, deep voices, etc.). The amount of these hormones present in the genital organs is extremely small. However, small amounts are also present in urine, and the first pure materials for study were isolated from this source (15 mg from some 15,000 liters!). **Testosterone** and **androsterone,** two of these hormones, bear an interesting structural relationship. The carbonyl and hydroxyl functions have exchanged positions (also, testosterone is unsaturated).

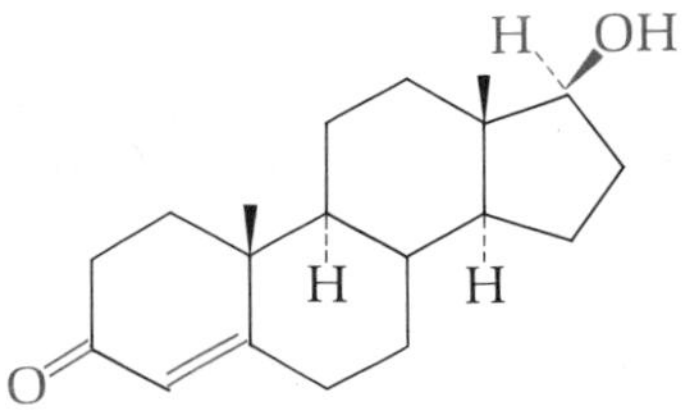

testosterone

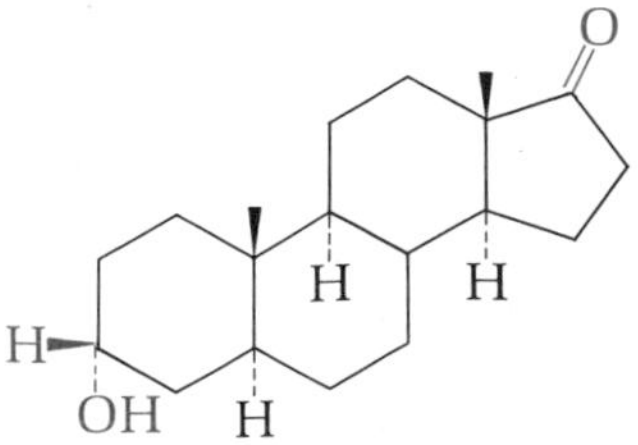

androsterone

About 30 hormones have been isolated from extracts of portions of the adrenal cortex, small glands located one above each kidney. They affect various body functions. One of these compounds, **cortisone,** is used clinically to treat skin diseases and rheumatoid arthritis. A unique structural feature of cortisone is the carbonyl function at C11. The reduced form (hydrocortisone, or cortisol) is more active and is used to treat allergies and tissue inflammations.

cortisone   cortisol

18.3   THE BIOSYNTHESIS OF TERPENES AND STEROIDS

Though terpenes and steroids are composed totally or in part of isoprene units, they are not derived from isoprene itself; indeed, isoprene is not a naturally occurring compound. Whence, then, does the C_5 unit arise? The ways of nature are often devious, and a six-carbon atom compound, **mevalonic acid,** appears to be the progenitor of terpenoid compounds, giving rise to an "isoprene" unit by simultaneous loss of water and carbon dioxide.

Mevalonic acid (3-methyl-3,5-dihydroxypentanoic acid) is derived by the condensation of three moles of acetic acid. The first stage

$$CH_3\overset{O}{\overset{\|}{C}}CH_2\overset{O}{\overset{\|}{C}}SCoA + CH_3\overset{O}{\overset{\|}{C}}SCoA \xrightarrow[-CoA-SH]{H_2O} CH_3\overset{OH}{\underset{CH_2\overset{\|}{\underset{O}{C}}SCoA}{\overset{|}{C}}}CH_2CO_2H$$

acetoacetyl
coenzyme A

β-hydroxy-β-methylglutaryl coenzyme A

2 reduction steps

$$CH_3\overset{OH}{\underset{CH_2CH_2OPP^*}{\overset{|}{C}}}CH_2CO_2H \xleftarrow[\text{triphosphate}]{\text{adenosine}} CH_3\overset{OH}{\underset{CH_2CH_2OH}{\overset{|}{C}}}CH_2CO_2H \quad (18\text{-}2)$$

mevalonic acid-
5-pyrophosphate

(ATP)

mevalonic acid

of the process is identical with the steps in fatty acid synthesis. Acetoacetylcoenzyme A is formed from two moles of acetyl-coenzyme A (eq 11-13). Terpenoid metabolism then departs from fatty acid metabolism, with an aldol-type condensation to give β-hydroxy-β-methylglutaryl coenzyme A which, in two successive reductions, is converted to mevalonic acid (eq 18-2). The pyrophosphate of this acid loses CO_2 and H_2O, probably via a phosphate of the tertiary alcohol, to give **isopentenyl pyrophosphate** which is the biogenetic isoprene unit.

$$CH_3-\underset{\underset{CH_2CH_2OPP^*}{|}}{\overset{\overset{OH}{|}}{C}}-CH_2-\underset{O-H}{\overset{O}{C}} \quad \xrightarrow[-HOH]{-CO_2} \quad CH_3-\underset{\underset{CH_2CH_2OPP}{|}}{C}=CH_2 \qquad (18\text{-}3)$$

isopentenyl pyrophosphate

An isomerizing enzyme equilibrates isopentenyl pyrophosphate with its double bond isomer, dimethylallyl pyrophosphate.

$$CH_3\underset{\underset{CH_2CH_2OPP}{|}}{C}=CH_2 \quad \underset{\xrightarrow{\text{isomerase}}}{\rightleftharpoons} \quad CH_3\underset{\underset{CHCH_2OPP}{||}}{C}CH_3 \qquad (18\text{-}4)$$

isopentenyl pyrophosphate   dimethylallyl pyrophosphate

These two C_5 compounds provide the necessary intermediates for the key reaction in terpenoid biosynthesis. Nucleophilic attack by the terminal double bond of isopentenyl pyrophosphate on the primary alkyl pyrophosphate group of dimethylallyl pyrophosphate gives the terpene geranyl pyrophosphate.

$$\xrightarrow{-HOPP} \qquad (18\text{-}5)$$

geranyl pyrophosphate

Hydrolysis of the pyrophosphate gives the monoterpene geraniol. The essential correctness of this scheme has been established by

$$*P = -\underset{\underset{OH}{|}}{\overset{\overset{O}{||}}{P}}-OH; \quad PP = -\underset{\underset{OH}{|}}{\overset{\overset{O}{||}}{P}}-O-\underset{\underset{OH}{|}}{\overset{\overset{O}{||}}{P}}-OH.$$

isotopic labeling experiments. For example, if acetate labeled with radioactive carbon (C^{14}) in the methyl group (i.e., $C^{14}H_3CO_2H$ or CH_3CO_2H) is fed to a plant which synthesizes geraniol, the product is labeled as predicted.

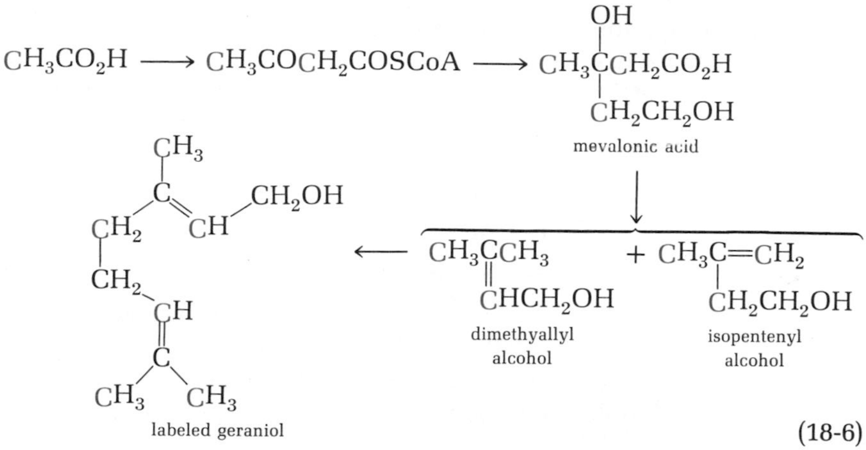

mevalonic acid

dimethylallyl alcohol

isopentenyl alcohol

labeled geraniol

(18-6)

Other monoterpenes can be derived from geraniol via ordinary carbonium ion type reactions. Limonene, for example, can be obtained from the dehydration-cyclization of geraniol.

Larger terpenoid chains are built up in an analogous manner. Extension of the geranyl unit by reaction with another mole of isopentenyl pyrophosphate gives farnesyl pyrophosphate; repetition yields geranylgeranyl pyrophosphate. In this way, the basic carbon skeletons of the sesquiterpenes (C_{15}) and diterpenes (C_{20}) are pieced together.

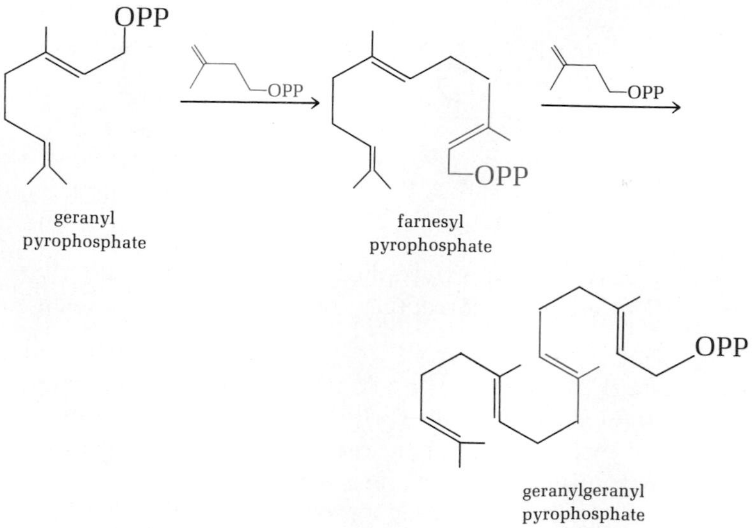

geranyl pyrophosphate

farnesyl pyrophosphate

geranylgeranyl pyrophosphate

(18-7)

The central role played by cholesterol in many biological processes has long been recognized, and almost as soon as radioactive isotopes became readily available (after World War II) tracer experiments were initiated to determine its biogenetic source. By 1950 it had been established that cholesterol was synthesized from two-carbon fragments (acetic acid). This fact, together with the observation that the end of the C17 side chain appears to be isoprenoid, suggested that cholesterol might be derived from terpene precursors (but that in the later stages it was modified to lose three carbon atoms, since it contains only 27, rather than 30). As a result of exhaustive studies, it is now known that the biosynthetic sequence is

$$\text{farnesol} \longrightarrow \text{squalene} \longrightarrow \text{lanosterol} \longrightarrow \text{cholesterol}$$

Two moles of farnesyl pyrophosphate unite in a tail-to-tail fashion to produce squalene. This C_{30} hydrocarbon, first isolated from shark

PPO

PPO

enzyme

2 farnesyl pyrophosphates

squalene

(18-8)

liver and thought to be an exotic substance, is now known to be a universal metabolite. The formation of squalene from two moles of farnesyl pyrophosphate is a multi-step process and requires the presence of a reducing agent. If methyl-labeled acetate is metabolized, the resulting squalene is labeled in the predicted fashion as shown by the colored dots in the formula (eq 18-8).

Squalene has a chain of 24 consecutive carbon atoms, with six methyl branches. In the next step of this biosynthetic scheme, this acyclic structure cyclizes in a highly stereospecific manner to produce the four fused rings of lanosterol. The reaction is initiated by epoxidation of the terminal double bond in squalene, followed by acid-catalyzed ring opening of the epoxide. The cyclization results in a carbonium ion with the positive charge on C20 of the steroid ring system (ion A, eq 18-9).

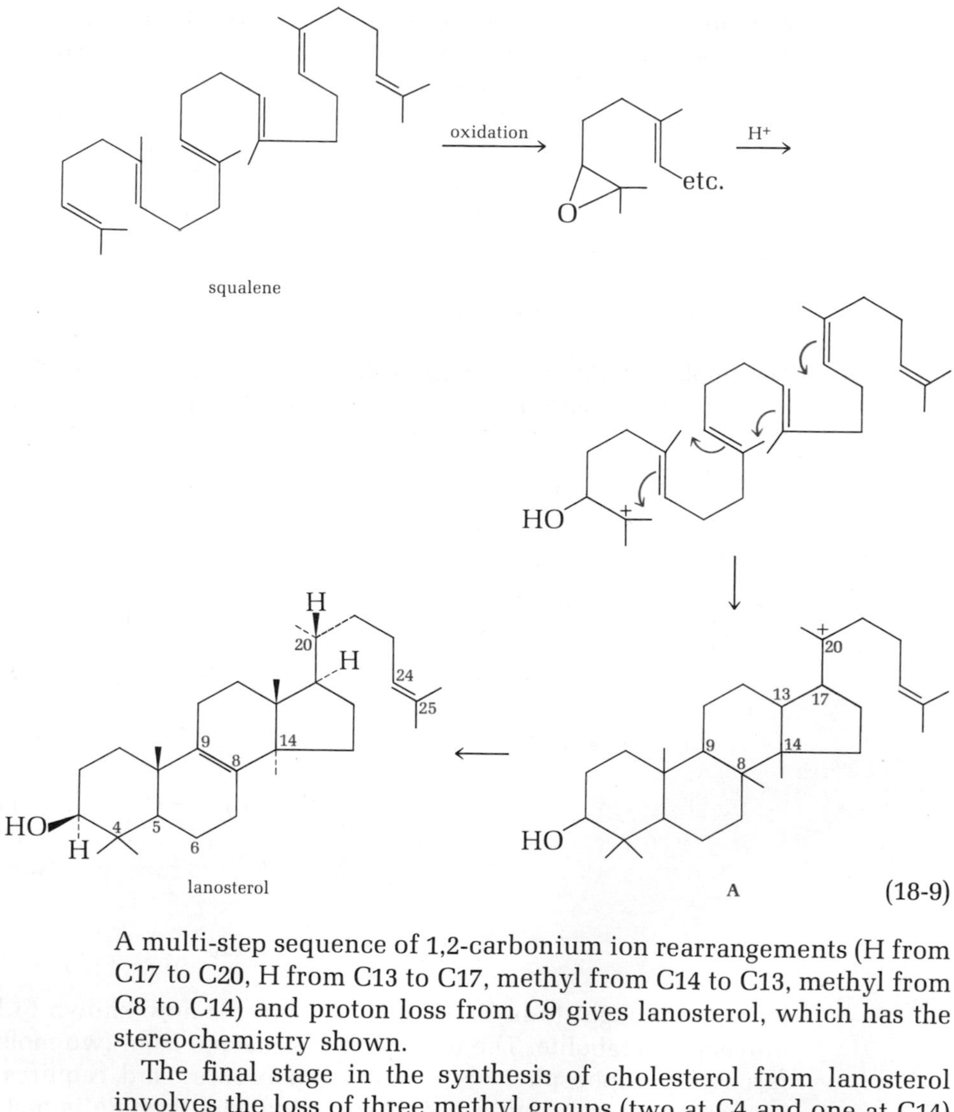

squalene

lanosterol

A (18-9)

A multi-step sequence of 1,2-carbonium ion rearrangements (H from C17 to C20, H from C13 to C17, methyl from C14 to C13, methyl from C8 to C14) and proton loss from C9 gives lanosterol, which has the stereochemistry shown.

The final stage in the synthesis of cholesterol from lanosterol involves the loss of three methyl groups (two at C4 and one at C14) through oxidation to carbon dioxide, reduction of the double bonds at C8–C9 and C24–C25 and the introduction of a double bond between C5 and C6.

Cholesterol is then the precursor of many other steroids. The sex hormones arise from total or partial loss of the C17 side chain, whereas ergosterol requires the introduction of an "extra" methyl group in the side chain at C24. Enzymes are essential biochemical catalysts for virtually all of these transformations.

18.4 OXYGEN HETEROCYCLES

Carbohydrates probably constitute the largest class of natural products which can be classed as oxygen heterocycles. **Vitamin C or ascorbic acid,** is a carbohydrate derivative found in citrus fruits, tomatoes, and fresh vegetables. Its deficiency in the diet causes scurvy, and as long ago as the eighteenth century British sailors were required to eat fresh limes (hence, "limey") to prevent outbreak of the disease. The biosynthetic precursor of ascorbic acid is D-glucose, but during the process, C1 of glucose (colored) becomes C6 of ascorbic acid. Consequently the stereochemistry at C5 of ascorbic

$$\text{(18-10)}$$

acid is related to that of L-glyceraldehyde. Ascorbic acid is easily and reversibly oxidized, and it functions in biochemical redox processes.

Coumarin is a naturally occurring oxygen heterocycle which can be found in clover and grasses, and is responsible for the pleasant fragrance given off when they are freshly cut. It is widely used as a perfume and flavoring agent. Compounds with a similar structure, but with the carbonyl group in the four-position, are called **chromones.** Derivatives of 2-phenylchromone (flavones) are responsible for many of the beautiful colors of flowers, plant leaves, fruits, and berries.

coumarin　　　　　chromone

The active ingredients in hemp (*Cannabis sativa*), from which one obtains hashish and marihuana, are relatively simple oxygen heterocycles. They are **tetrahydrocannabinols** (THC), derivatives of the aromatic compound cannabinol. The $\Delta^{1,2}$- and $\Delta^{1,6}$-*trans* isomers are believed to be the most active components.

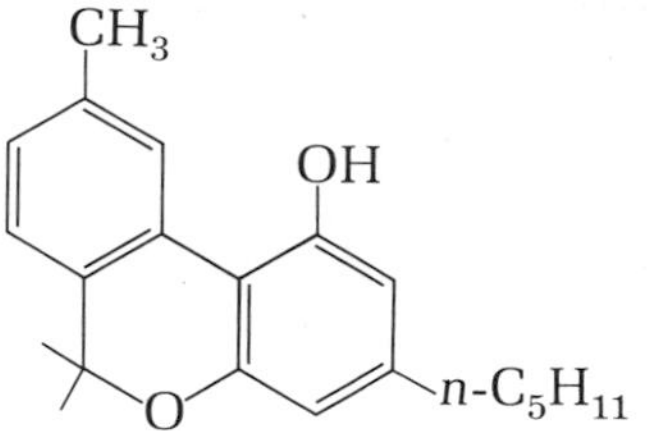

cannabinol

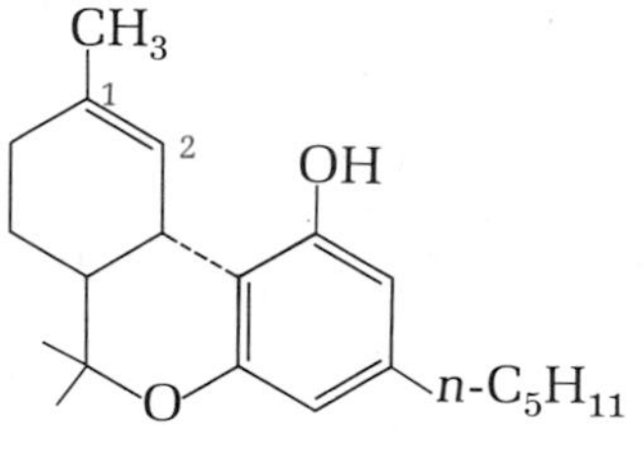

$\Delta^{1,2}$-*trans*-
tetrahydrocannabinol*

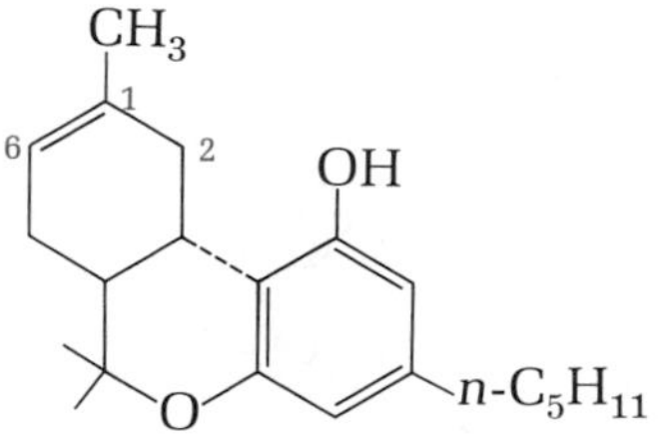

$\Delta^{1,6}$-*trans*-
tetrahydrocannabinol*

18.5 NITROGEN HETEROCYCLES

Some of the major nitrogen-containing rings were discussed in section 12.7. Natural products which contain those (and other) rings include the alkaloids, the porphyrins, certain vitamins and antibiotics, and the nucleic acids.

18.5a THE ALKALOIDS

The largest group of natural nitrogen-containing heterocyclic compounds are the **alkaloids,** basic substances which occur chiefly in plants and which often have associated with them a marked physiological activity. The alkaloids are derived biosynthetically mainly from certain of the amino acids, especially lysine, ornithine (similar to lysine, but with one less methylene group in the chain), phenylalanine, tyrosine, and tryptophan. Two aromatic amino acids, nicotinic acid (eq 12-51) and anthranilic acid (eq 12-66) are also alkaloid precursors. Some alkaloids arise biosynthetically from the terpenoid (sec 18.3) or other routes.

Several alkaloids are present in tobacco (*Nicotinia tabacum*), the predominant one being **nicotine.** It constitutes about 5% by weight

---

* $\Delta^{x,y}$ is a symbol used to locate, in this case, a double bond between atoms x and y in a complex structure.

of the dry plant leaves. Nicotine is a violent poison, but in small amounts it causes initial stimulation, followed by depression. In high dilution, nicotine salts are useful as insecticides. Oxidation of nicotine gives **nicotinic acid,** an important vitamin required for the

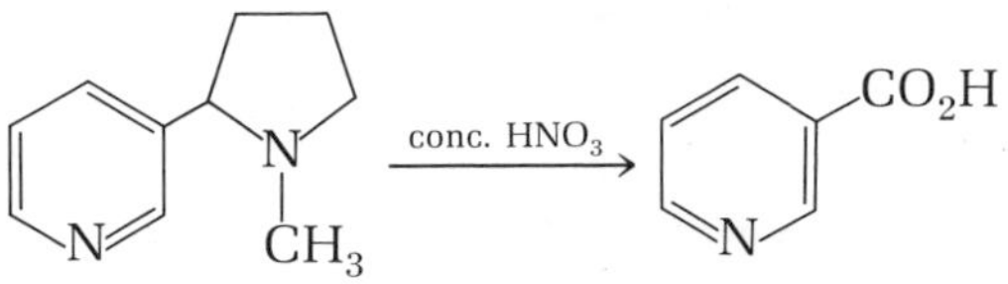

conc. HNO_3

nicotine   nicotinic acid

(18-11)

production of nicotinamide adenine dinucleotide (NAD), an essential component of many biochemical redox reactions. Nicotinic acid (also called *niacin*) is manufactured commercially from 3-methylpyridine (eq 12-15) for use as a food supplement.

The **tropane** alkaloids contain a reduced pyridine (piperidine) ring with a two-carbon bridge from C2 to C6. **Cocaine** is the chief alkaloid of the leaves of the coca bush. In small doses, it decreases

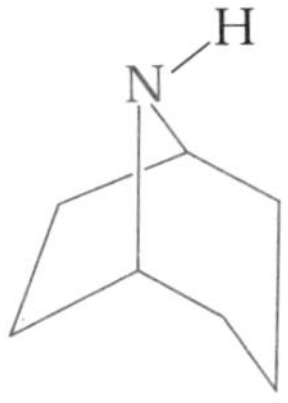

tropane 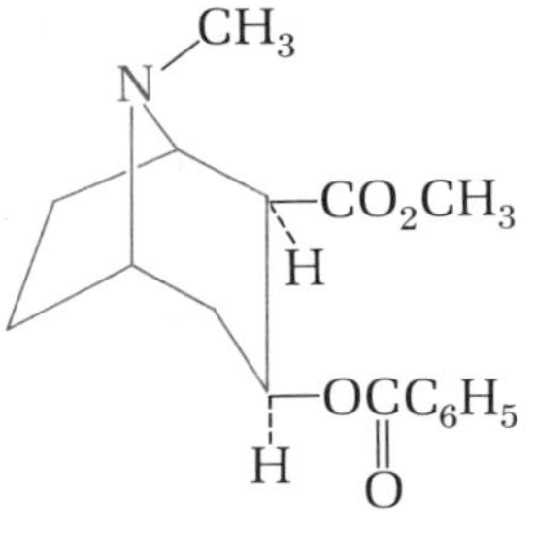   cocaine 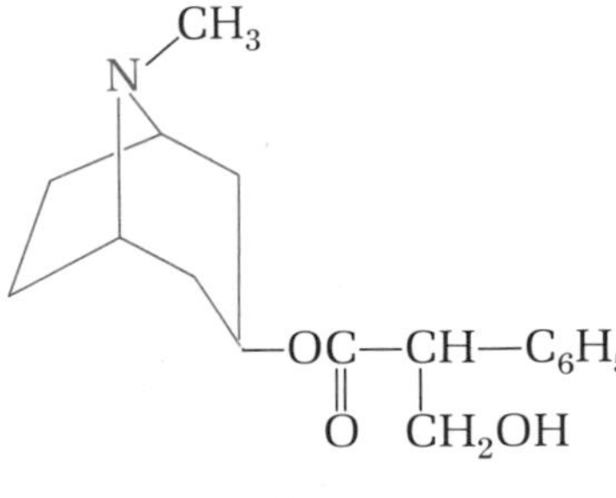   atropine

fatigue, increases mental activity, and affords a feeling of calm and happiness. But these benefits are short-lived, being followed by periods of strong depression in which the person demands more of the drug. The observation that the methoxycarbonyl group was responsible for the toxic, habit-forming properties of the molecule whereas the anesthetic activity was due to the benzoate ester of an amino alcohol led to the development of **novocaine,** an important local anesthetic.

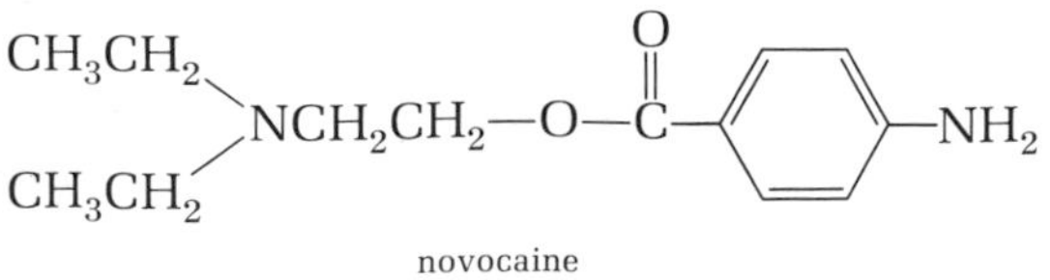

novocaine

Atropine, an alkaloid from the belladonna plant, is used in dilute solution to dilate the eye pupil prior to ophthalmic examinations.

Mescaline, the well-known hallucinogen, is one of several cactus alkaloids. It is derived biogenetically from the amino acid tyrosine. The synthetic drugs **amphetamine** (or benzedrine) and epinephrine (or adrenalin) are closely related structurally to mescaline.

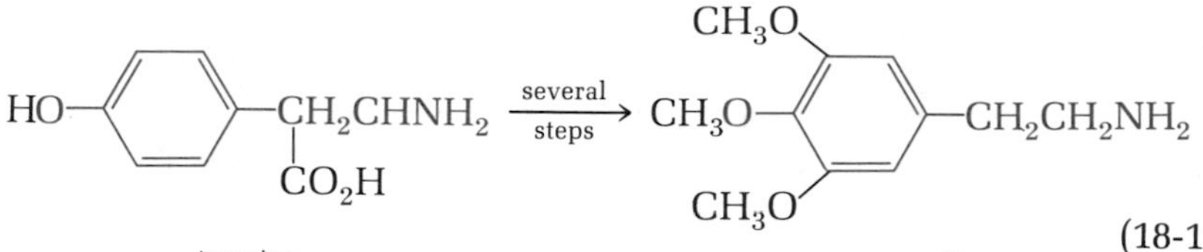

tyrosine

mescaline

(18-12)

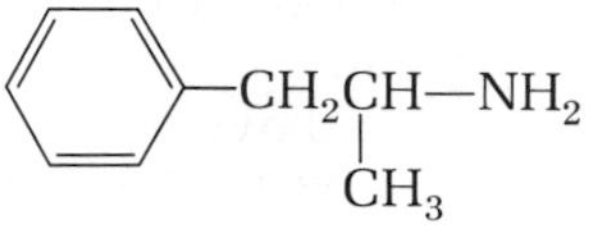

amphetamine

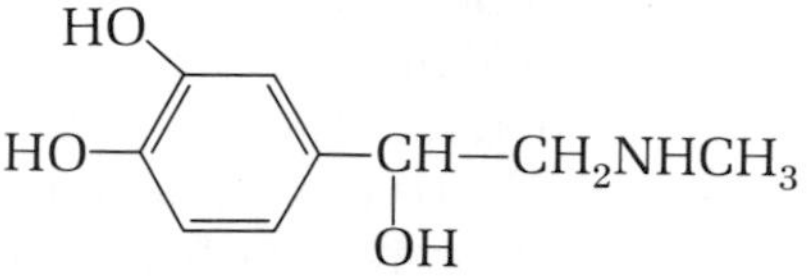

epinephrine

Neosynephrine, used to relieve nasal congestion, is similar structurally to epinephrine, but lacks the *para* hydroxyl group.

Though mescaline is not a nitrogen heterocycle, it bears some structural relationship to the isoquinoline alkaloids which are also derived in part, biogenetically, from tyrosine. These include the opium alkaloids **papaverine** (where the isoquinoline ring is aromatic) and **morphine** and **codeine** (where the rings are reduced). **Heroin**

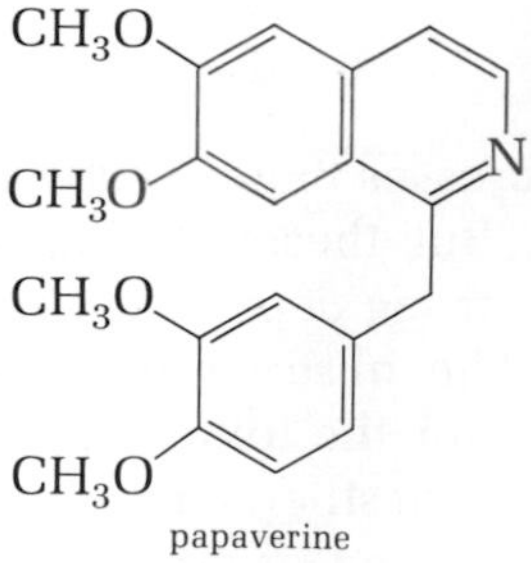

papaverine

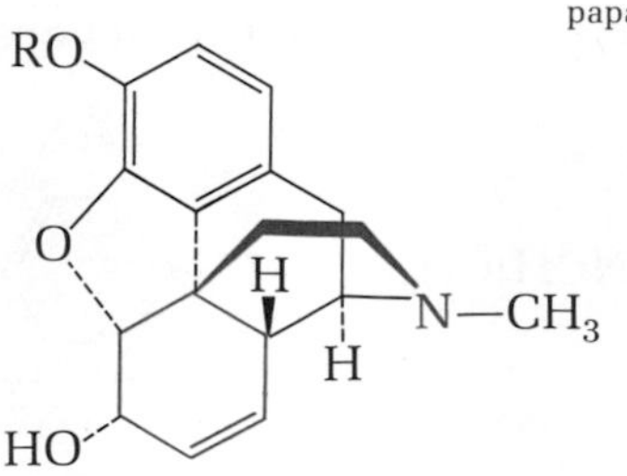

morphine (R=H)
codeine (R=CH₃)

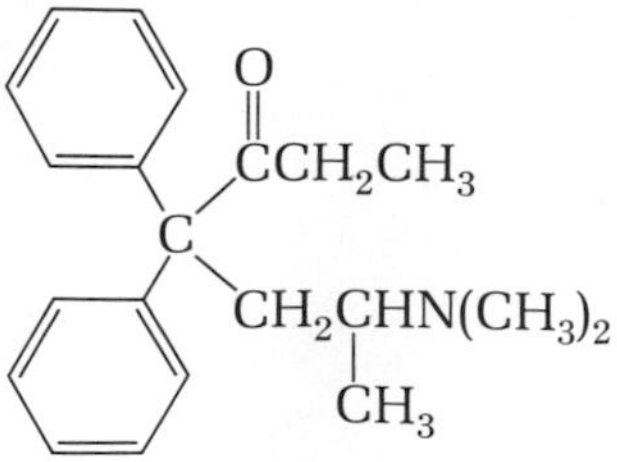

methadone

is the diacetyl derivative of morphine. **Methadone,** though not closely related to morphine structurally, is even more effective as an analgesic. Recently it has been used to treat heroin addiction. Though methadone is also addictive, one can function while using it.

Many alkaloids have an indole ring system as part of their structure, either as such or in a partially reduced form. They are all derived from the amino acid tryptophan (whose skeletal structure is shown in color in the formulas) and include not only serotonin and bufotenin (sec 12.7) but more complex derivatives such as reserpine, lysergic acid, and strychnine.

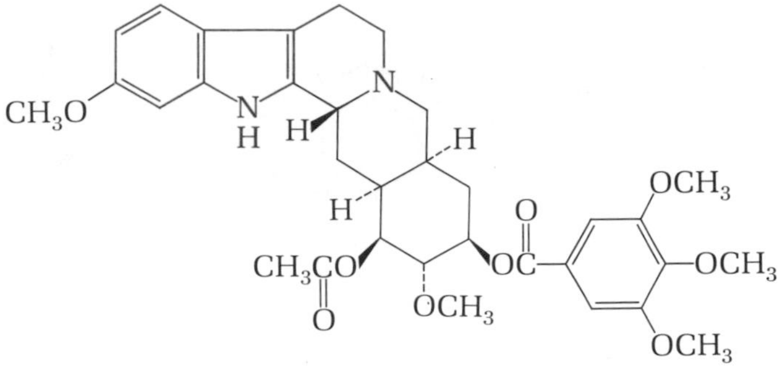

reserpine

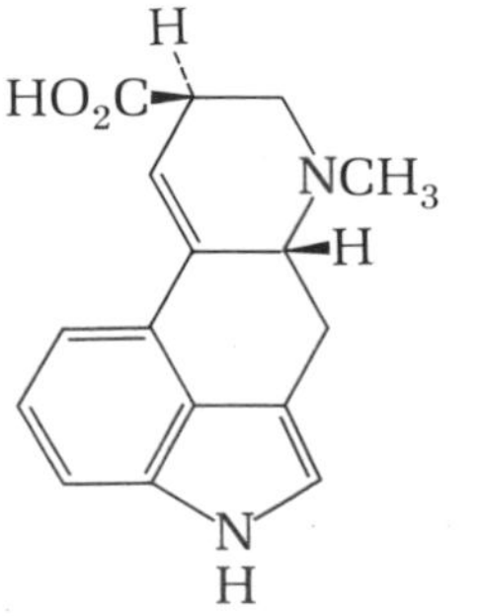

lysergic acid

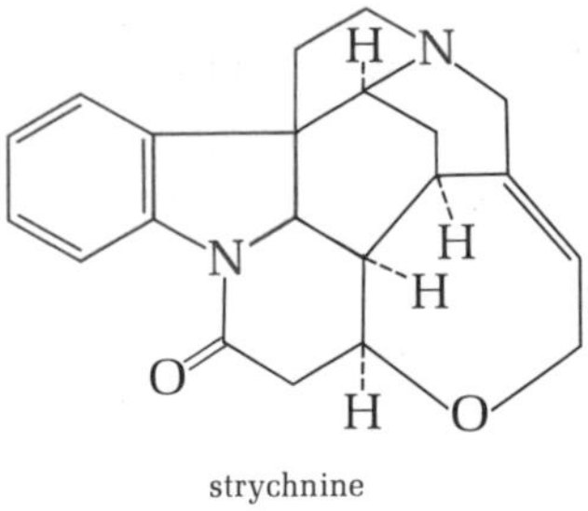

strychnine

Reserpine is the active alkaloid in *Rauwolfia serpintina* (Indian snake root), which grows wild on the foothills of the Himalayas. It has been used medically for centuries, and is now used to calm schizophrenics and make them amenable to psychiatric treatment. Despite its stereochemical complexity, the structural determination and total synthesis of reserpine were accomplished in the short interval 1952–56 (R. B. Woodward, Figure 18-1). **Lysergic acid,** in the form of amides, is a constituent of the ergot alkaloids, present in the fungus ergot which grows on rye and other cereals. The diethylamide (LSD-25) produces hallucinations when ingested in extremely minute amounts. **Strychnine,** the exceptionally strong poison, was first synthesized by Woodward and his co-workers in 1954. The nitrogen-containing ring in this indole alkaloid is reduced. **Quinine** occurs in the cinchona bark and is used to treat malaria. Though it contains a quinoline ring, quinine is derived biogenetically from tryptophan.

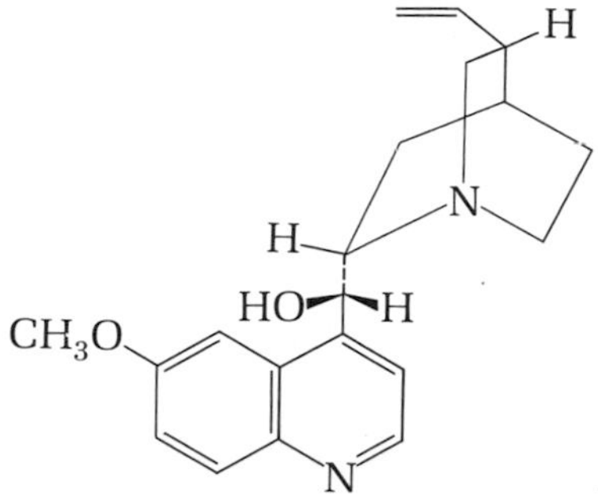

quinine

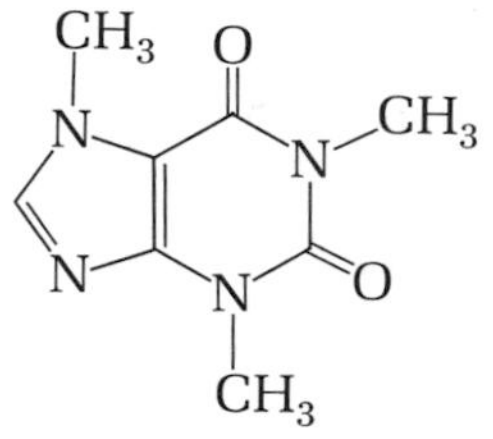

caffeine

Caffeine is the best known purine alkaloid. It occurs in tea, coffee, and cola nuts, and is the substance responsible for the stimulation derived from drinking the corresponding beverages.

18.5b THE PORPHYRINS

Several biochemically important pigments contain four pyrrole rings joined in the two- and five-positions by one-carbon bridges. The parent compound **porphyrin** does not occur naturally but several

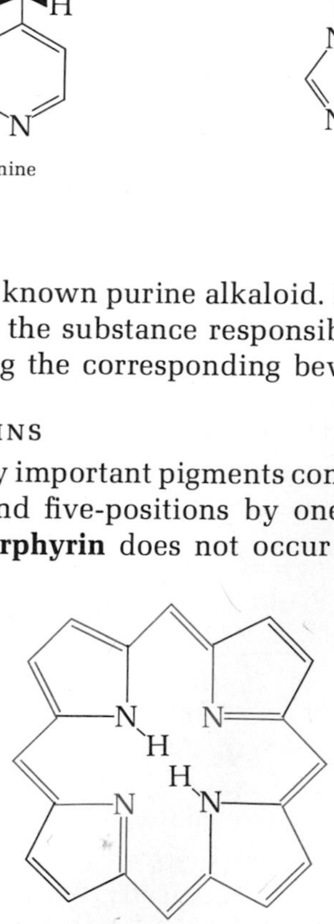

porphyrin

important natural products contain this ring system with various side-chains, different central metal ions, and other slight modifications. The porphyrin ring system is flat and aromatic (note the 18π electrons in the portion of the ring drawn in color fulfills the $4n + 2$ rule, where $n = 4$; sec 5.7). Chlorophyll, the green plant pigment essential to photosynthesis, has the four pyrrole nitrogens coordinated to a magnesium ion, whereas in **heme,** the red blood pigment, the central metal is iron. The structural determination of a complex

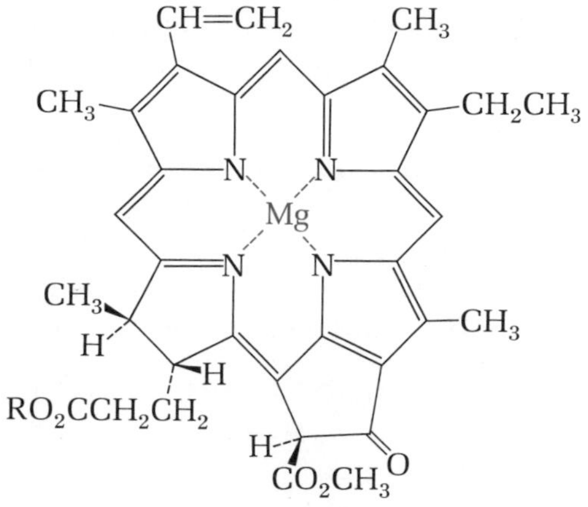

chlorophyll a

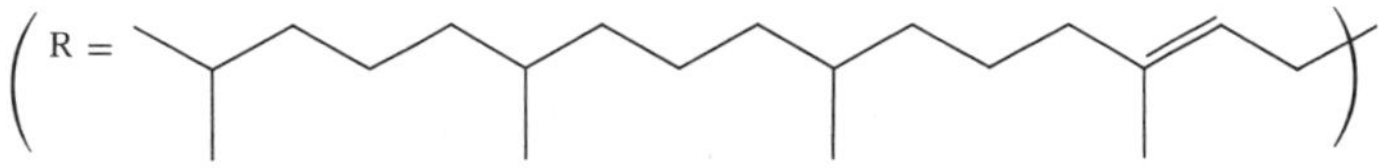

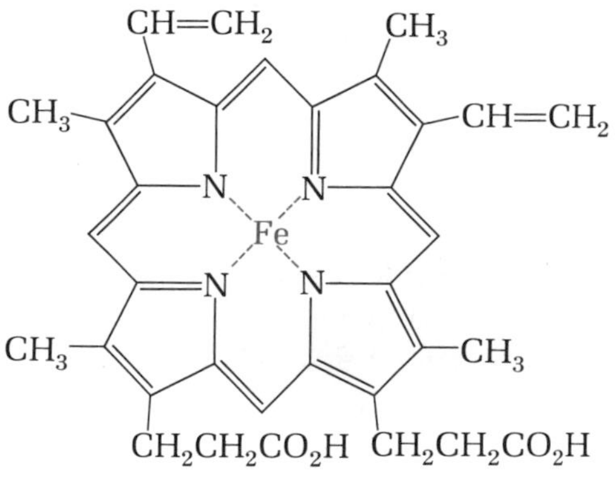

heme

molecule such as chlorophyll required many years of brilliant work by chemists in Germany (R. Wilstätter, H. Fischer) and the USA (J. B. Conant); total synthesis was elegantly achieved by R. B. Woodward and co-workers in 1960. **Vitamin B$_{12}$,** effective in the treatment of pernicious anemia, contains a similar ring system but with a central cobalt ion.

Porphyrins also occur naturally in petroleum where they may complex trace metals; this sometimes interferes with catalytic processes used in the refining operation.

18.5c   VITAMINS

Vitamins are substances which cannot be synthesized by an organism but which nevertheless are essential for its normal metabolism. They were discovered largely through the study of diseases caused by dietary deficiencies. The function of many of the vitamins is to serve as structural units for certain coenzymes required in metabolic reactions. Several of them are nitrogen heterocycles.

Thiamin (vitamin B_1) is essential to human nutrition, its deficiency causing the disease beri-beri. Thiamin pyrophosphate (the ester at the primary alcohol group) is the coenzyme of carboxylase, which catalyzes the decarboxylation of α-keto acids—for example, the conversion of pyruvate to acetaldehyde (Figure 14-2). **Riboflavin** (vitamin B_2) forms an essential part of the coenzyme flavin adenine

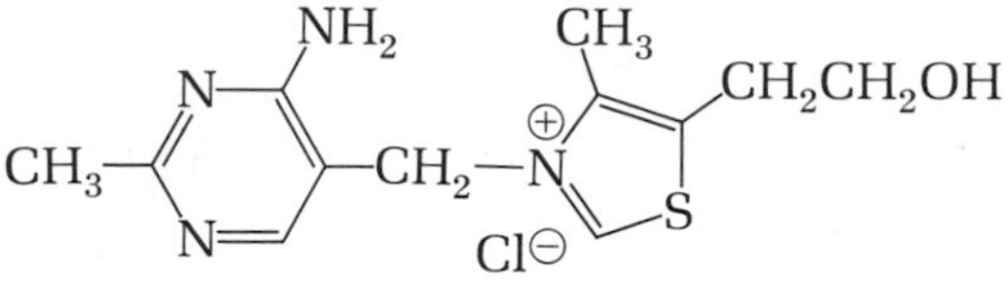

thiamin

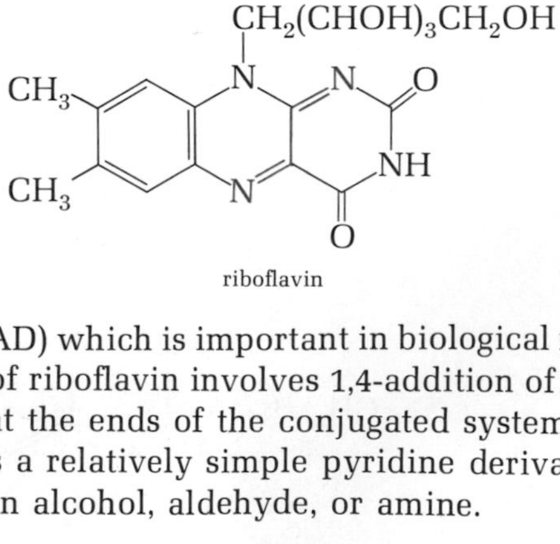

riboflavin

dinucleotide (FAD) which is important in biological redox reactions. The reduction of riboflavin involves 1,4-addition of hydrogen to the two nitrogens at the ends of the conjugated system.

Vitamin B_6 is a relatively simple pyridine derivative. The group at C4 may be an alcohol, aldehyde, or amine.

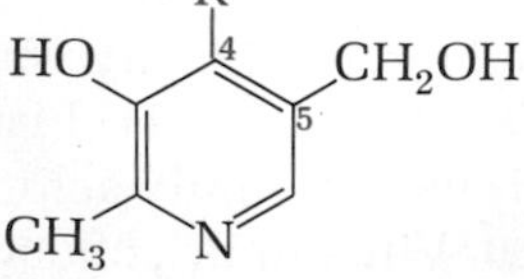

R = CH_2OH (pyridoxine)
$CH{=}O$ (pyridoxal)
CH_2NH_2 (pyridoxamine)

vitamin B_6

As the phosphate of the primary alcohol at C5, the aldehyde and amine constitute the coenzymes which function reversibly in the interconversion of α-keto and α-amino acids (transamination, sec 17.12).

Certain of the other important vitamins (A, B_{12}, C, D) have been mentioned elsewhere in the text.

18.5d ANTIBIOTICS

Antibiotics are chemicals which are antagonists to the metabolic processes of certain microorganisms and either inhibit their growth or kill them. The best known, and the first antibiotics used in medicine, are the **penicillins.** The most commonly administered form is penicillin-G, where R = benzyl, but many other R-groups can be used and still retain the physiological activity. The main structural features are a five-membered heterocyclic ring containing nitrogen and sulfur, and a four-membered cyclic amide group. Penicillins are obtained commercially from a mold (Figure 18-2).

penicillins

Figure 18-2

Trays at the bottom of the picture (left) contain the sodium salt of penicillin ready for drying operations after extraction from Penicillium notatum mold. Flasks containing experimental quantities of the crude mold are shown on the right. (Courtesy of Merck Company, and Corning Glass Works.)

Perhaps the most striking feature of antibiotics is the tremendous variety and complexity of their structures. Examples now in common use are the **tetracyclines** and the **macrolides;** the latter have unusually large lactone rings.

tetracycline (R_1=R_2=H)
terramycin (R_1=H, R_2=OH)
aureomycin (R_1=Cl, R_2=H)

erythromycin, a macrolide

18.6 NUCLEIC ACIDS

The development in basic science which has attracted the greatest public interest during the last decade is without doubt the chemistry associated with genes. The "double helix" has become a household phrase. In this section we will describe the structures of the nucleic acids, the key compounds responsible for transfer of genetic information from one generation to another and for protein synthesis.

Though nucleic acids were first isolated from cell nuclei (hence the name), it is now known that there are two types. These are **DNA** or *deoxyribonucleic acids,* which are found in the chromosomes of the cell nucleus and are responsible for the transfer of genetic information, and **RNA** or *ribonucleic acids,* which are found both in the ribosomes of the cell and in the cytoplasm, and participate in protein biosynthesis. Both are polymeric molecules. DNA molecules may have molecular weights as large as 100 million, enabling them to be "seen" with an electron microscope. Ribosomal RNA molecules are also large, with molecular weights in the 1 million range, but cytoplasmic RNA molecules are much smaller (ranging from 20,000 to 30,000) and therefore soluble in the cellular fluid.

Nucleic acids can be degraded hydrolytically into smaller fragments whose structures are more amenable to chemical investigation. The steps in these hydrolyses are

$$\text{nucleic acids} \xrightarrow[\text{H}_2\text{O}]{\text{enzymes}} \text{nucleotides} \xrightarrow[\text{base}]{\text{dilute}}$$

$$\text{H}_3\text{PO}_4 + \text{nucleosides} \xrightarrow{\text{H}_3\text{O}^+} \text{organic bases} + \text{sugars} \quad (18\text{-}13)$$

18.6a   THE COMPONENT PARTS

The **sugars** are pentoses. RNA, regardless of its source, gives D-ribose, whereas DNA yields 2-deoxy-D-ribose. The R and the D in the names of these nucleic acids call attention to this distinction.

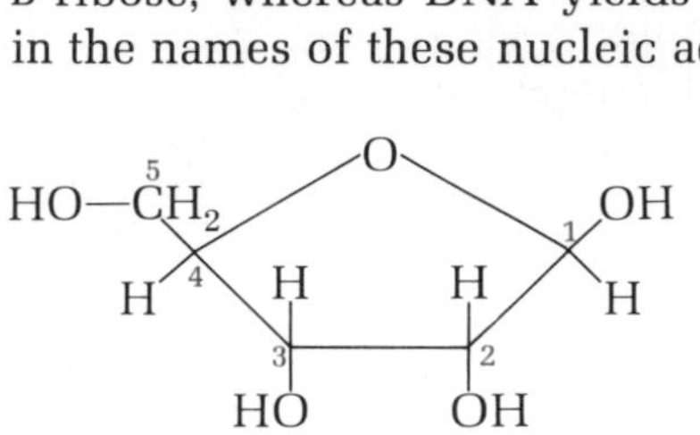

D-ribose 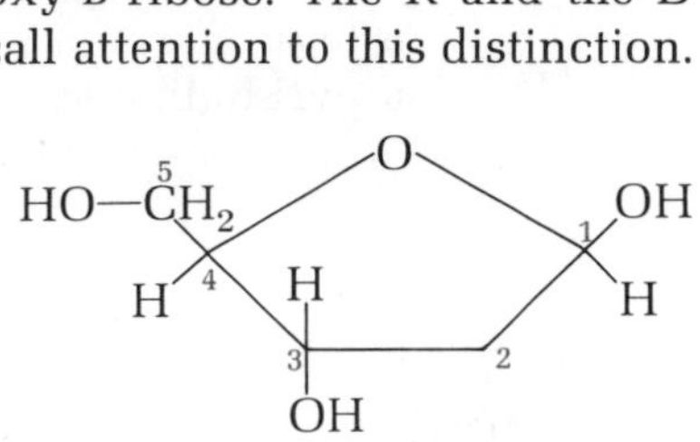   2-deoxy-D-ribose

Both sugars occur in the furanose ring form; they differ structurally only at C2.

The **organic bases** are of two structural types, pyrimidines and purines (sec 12.7).

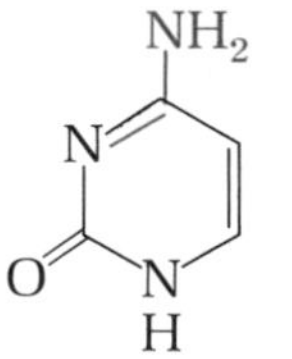

Cytosine 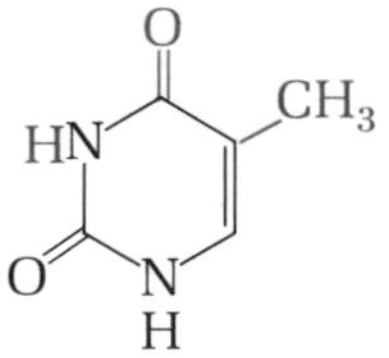   Thymine 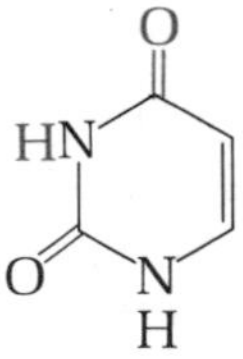   Uracil

the pyrimidine bases

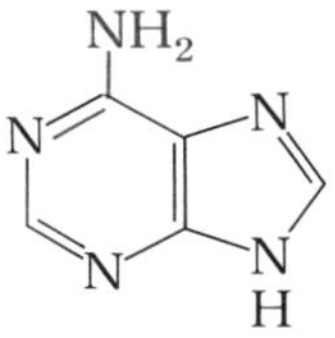

Adenine 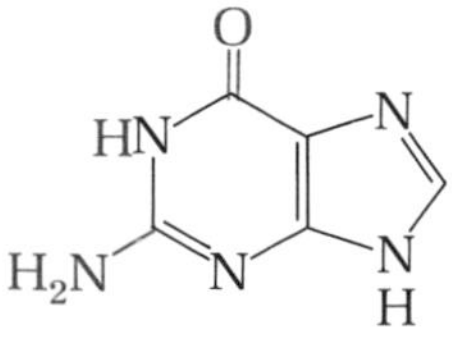   Guanine

the purine bases

Each type of nucleic acid furnishes, on complete hydrolysis, four bases—two pyrimidines and two purines. RNA differs from DNA in that one of the four bases is uracil, in place of thymine.

$$\text{DNA} \longrightarrow \text{A, G, C, T*}$$

$$\text{RNA} \longrightarrow \text{A, G, C, U}$$

Several other bases occur, in minor amounts, in RNA. Complete hydrolysis of nucleic acids from different biological species gives different relative percentages of the four bases. There is, however, a special relationship in the mole ratios of certain base pairs from DNA (sec 18.6c).

The **nucleosides** are N-glycosides, and furnish one mole each of a base and a sugar on hydrolysis (review sec 16.5c). Examples of a pyrimidine and a purine nucleoside are

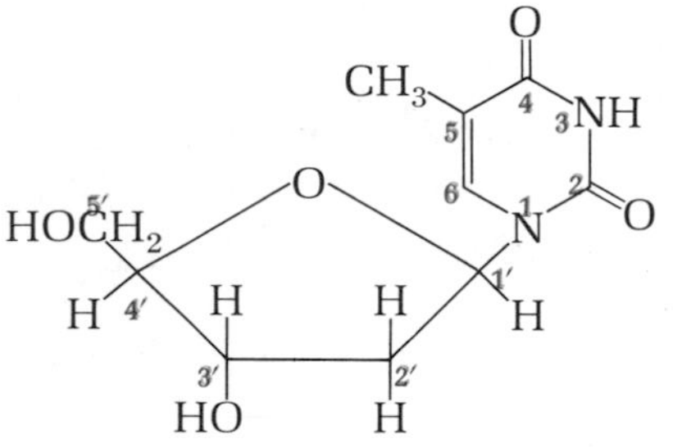

thymidine
(thymine deoxyriboside)

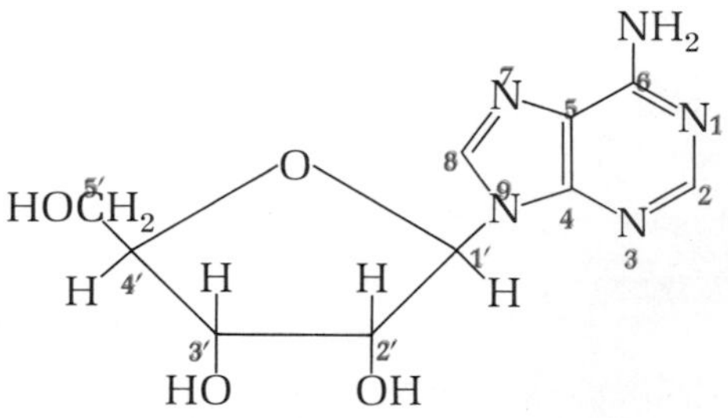

adenosine
(adenine riboside)

The four common *ribo*nucleosides are adenosine, guanosine, cytidine, and uridine; the prefix *deoxy* (e.g. deoxyadenosine) is used to designate the corresponding deoxyribonucleosides, except for thymidine, which occurs only in DNA. The pyrimidine and purine bases are attached to the sugars at N1** and N9 respectively. The sugar carbon atoms are numbered with primes, as shown. Carbon 1' of the sugar has two electronegative atoms (O, N) attached to it, and behaves chemically very much like an acetal. For this reason, nucleosides are stable toward dilute base, but are readily hydrolyzed to the component parts (sugar plus nitrogen base) when treated with aqueous acid.

---

*The first letters of the names are used as abbreviations.
**Some authors begin numbering with the other nitrogen, in which case the nitrogen attached to the sugar is N3. *Chemical Abstracts* uses the numbering shown. Regardless of numbering, the structures as shown here are correct.

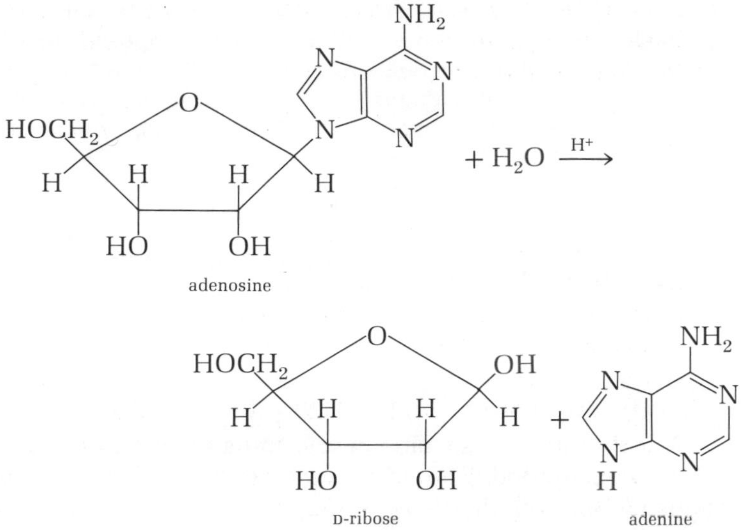

adenosine

$+ H_2O \xrightarrow{H^+}$

D-ribose

adenine

(18-14)

Nucleotides are phosphate esters of nucleosides. Esters at the primary alcohol function (C5′) are most common, and if no number is specified in the name, the 5′ phosphate structure is implied. Being esters, the nucleotides are readily hydrolyzed by base.

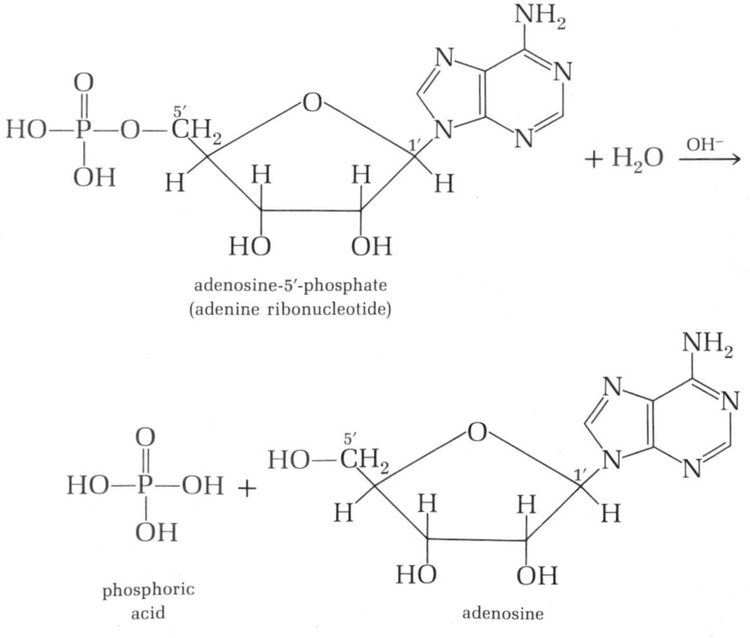

adenosine-5′-phosphate
(adenine ribonucleotide)

$+ H_2O \xrightarrow{OH^-}$

phosphoric
acid

adenosine

(18-15)

During this process the N-glycosidic link at C1′ remains intact.

In addition to being structural components of nucleic acids, certain nucleotides or their derivatives play other important roles in biological processes. The nucleotide shown in equation 18-15 is sometimes called adenosine monophosphate (abbreviated AMP) to distinguish it from the di- and triphosphates (abbreviated ADP or ATP), which have the groups

$$\underset{\underset{\displaystyle OH}{|}}{HO-\overset{\overset{\displaystyle O}{\|}}{P}}-O-\underset{\underset{\displaystyle OH}{|}}{\overset{\overset{\displaystyle O}{\|}}{P}}-O- \quad\text{or}\quad \underset{\underset{\displaystyle OH}{|}}{HO-\overset{\overset{\displaystyle O}{\|}}{P}}-O-\underset{\underset{\displaystyle OH}{|}}{\overset{\overset{\displaystyle O}{\|}}{P}}-O-\underset{\underset{\displaystyle OH}{|}}{\overset{\overset{\displaystyle O}{\|}}{P}}-O-\quad\text{attached to C5'. These}$$

substances, either free or combined with other moieties (see, for example, the structure of coenzyme A, sec 11.7), are essential to many of the steps in carbohydrate and fat metabolism, terpene and steroid biosynthesis, etc.

18.6b   THE PRIMARY STRUCTURE OF NUCLEIC ACIDS

Nucleotides are the monomeric units from which the nucleic acids are constructed. The units are linked together as phosphate esters, using the C3' hydroxyl group of one nucleoside and the C5'-hydroxyl group of another.

DNA
    R = H, B = A,G,C,T

RNA
    R = OH, B = A,G,C,U

One acidic proton remains on each phosphate link, hence the term nucleic *acid*. These protons may, of course, dissociate when the nucleic acid is in solution or is associated with a protein in the cell.

A complete description of the primary structure of a particular DNA or RNA molecule would require at the very least a knowledge of the exact sequence of the bases B_1, B_2, etc. along the chain. The problem is similar to that of protein structure, where a knowledge of amino acid sequence is required. Since the chain length of a nucleic acid may involve as many as 1,000,000 units per molecule, the problem is much more difficult than that of protein structure.

A major breakthrough in this problem occurred in 1965, when

R. W. Holley and co-workers (Cornell University) determined the complete nucleotide sequence for **alanine transfer RNA,** the molecule which binds and later delivers the amino acid alanine, in protein synthesis. The technique resembled that used in determining amino acid sequences in proteins. The RNA was *partially* hydrolyzed and the fragments thus obtained were examined for overlapping sequences. The structure, which has 77 nucleotide units and a molecular weight of 26,600, is shown in Figure 18-3. Holley shared the 1968 Nobel prize for this contribution. The structures of other transfer RNA's have since been determined. The molecular weights are not uniform, but are similar (about 80 ± 10 nucleotides per molecule). The sequence determination of a DNA molecule, which is $10,000 \times$ longer than these RNA molecules, is a frontier area of current research.

18.6c THE SECONDARY STRUCTURE OF NUCLEIC ACIDS

As with proteins, nucleic acids have a secondary structure. That the long polymeric DNA molecules must have some discrete shape has been known since 1938 when X-ray studies on DNA threads showed a regular stacking pattern with some periodicity. A key observation was made in 1950 by E. Chargaff (Columbia University), who noted that the mol ratio of adenine to thymine, and of guanine to cytosine was approximately one, regardless of the absolute amounts of each present in the particular DNA sample. That is, $A/T = 1$ and $G/C = 1$ even though $(A + T)/(G + C)$ differed from one organism to another. The explanation for this observation, and the double helix structure for DNA was put forth by Watson and Crick in 1953, and received simultaneous supporting X-ray data from Wilkins; all three shared the 1962 Nobel prize for their contribution.

Examination of hydrogen-bonding between base pairs showed that the structural relationship in A-T and G-C was uniquely favorable:

T–A base pair   C–G base pair

Figure 18-3

The base sequence of alanine tRNA as determined by Holley and co-workers in 1965. The Symbols A, G, C, U stand for the four most common bases present in RNA. The molecule also contains one T unit and eight units of six less common bases: MeG = 1-methylguanosine, DiMeG = guanosine with a dimethylamino group at C-2, DiHU = 2,6-dihydrouridine, I = inosine (guanosine, but lacking the amino group at C-2). MeI = 1-methylinosine, and ψ = pseudouridine (uridine, but with the base attached to the sugar at C-5 instead of at N1.

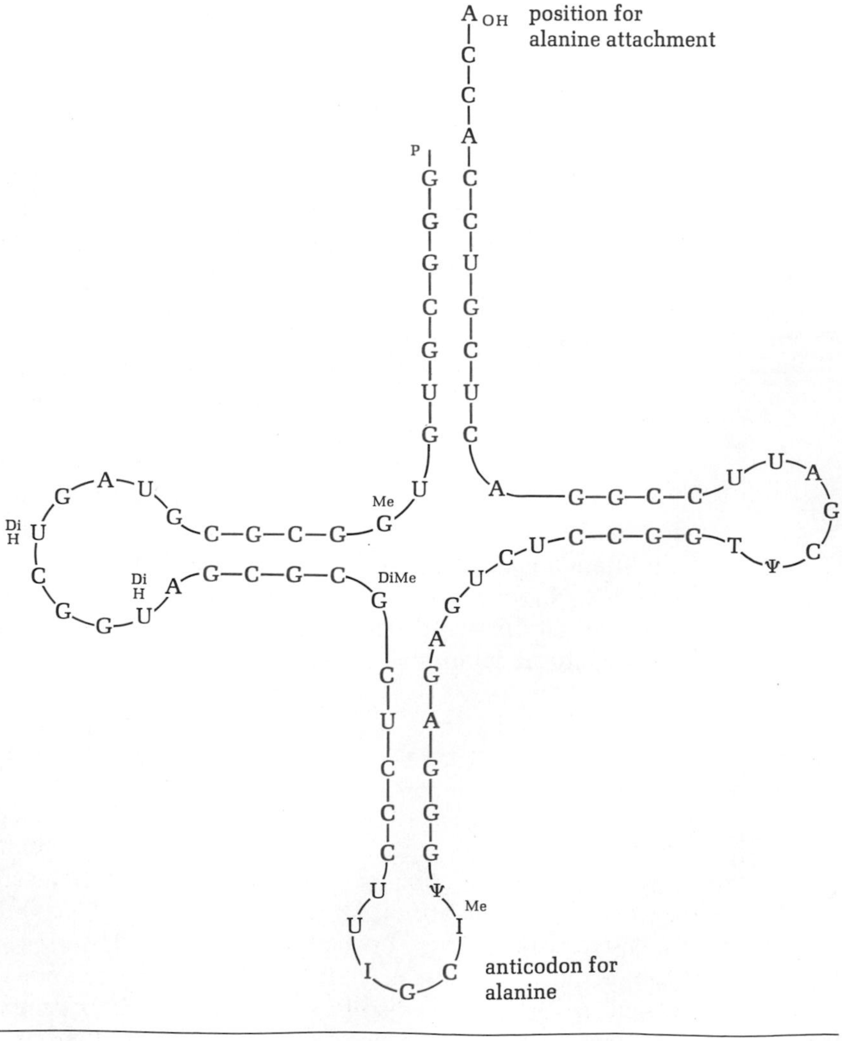

In both cases the bases are about 2.9 Å apart and the sugar moieties are also separated by a nearly equal distance, 11 Å. The pairs are

Figure 18-4
The double helix structure of
DNA. (D = deoxyribose,
P = phosphate, and A, T, G, C
are the bases).

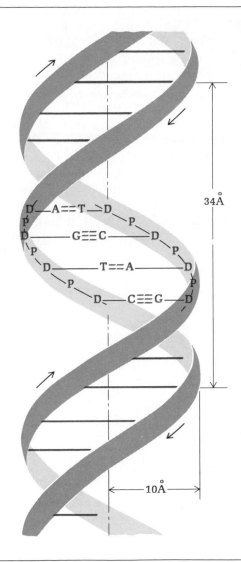

stabilized by two and three hydrogen bonds, respectively. Other hydrogen-bonded pairs (such as T-G, etc.), though possible, are much less favored geometrically and energetically.

Watson and Crick proposed that DNA consists of two intertwined deoxyribose-phosphate chains; the base on each sugar unit of one strand is directed toward the center of the helix where it is hydrogen-bonded to the corresponding base on the other strand, A always bonded to T and G always bonded to C (Figure 18-4). There are ten base pairs for one complete 360° twist of the chain. The protein moiety of nucleoproteins is probably wrapped around the helix.

In contrast with DNA, base-pairing appears to be less important in RNA, though some pairing occurs in the structures of soluble, "low" molecular weight RNA. Figure 18-3 illustrates this pairing (the units are shown in color). This pairing imparts an overall cloverleaf shape to the molecule, and seems to be general for all *t*RNA's.

18.6d THE BIOLOGICAL FUNCTIONS OF NUCLEIC ACIDS

A detailed discussion of the mechanisms of genetic information transfer and protein biosynthesis is beyond the scope of this text,[*] but a brief description of the principles involved can be presented.

A method for DNA replication was proposed by Watson and Crick in 1954. As the two strands of the double helix separate, a new complementary strand is synthesized on each single strand, using nucleotides available in the cell. The process is shown schematically in equation 18-16. Each single strand is a template for the formation

the double helix uncoils

nucleotides in the cell bond to the separate strands, following the base pairing rules; a polymerizing enzyme links the nucleotides in the new strands to one another

the process continues to completion, giving two new double helices.

$$(18\text{-}16)$$

*For a more thorough discussion, see E. Harbers, G. F. Domagk, and W. Müller, *Introduction to Nucleic Acids—Chemistry, Biochemistry and Functions*, Reinhold, 1968, or chapters in any modern biochemistry text.

of its counterpart. A beautiful experiment by Meselson and Stahl (USA, 1958) confirmed the essential correctness of the Watson-Crick proposal. They grew bacteria in an N^{15}-rich medium (N^{15} is a heavy isotope of ordinary nitrogen, N^{14}), and isolated the DNA in which the bases of both strands were rich in the heavy isotope. They then allowed this "heavy-heavy" DNA to replicate in a medium containing ordinary nitrogen (N^{14}) nutrients. After one generation, the DNA isolated was of only one type, with one "heavy" and one "light" strand, but after a second generation they could isolate equal amounts of two types of DNA. One was identical with the ordinary (N^{14}) DNA of the bacterium grown on N^{14} nutrients, and the other was of the mixed heavy-light type. Further generations gave greater and greater dilution of the "heavy" strands, as predicted by the Watson-Crick proposal. The Meselson-Stahl experiment is illustrated schematically in equation 18-17.

$$\|\| \xrightarrow[\text{pool}]{N^{14}\ \text{nucleotide}} \|\| + \|\| \xrightarrow[\text{pool}]{N^{14}\ \text{nucleotide}} \|\| + \|\| + \|\| + \|\| \qquad (18\text{-}17)$$

N^{15}-rich   1st generation   2nd generation
double helix

The biosynthesis of specific proteins requires that the language of the genetic material (i.e., the base sequences in DNA) be translated to specific amino acid sequences in the protein. This is accomplished indirectly, after transcription from DNA to RNA.

$$\text{DNA} \xrightarrow{\text{transcription}} \text{mRNA} \xrightarrow{\text{translation}} \text{proteins} \qquad (18\text{-}18)$$

The translation occurs at the ribosomes, using messenger RNA (*mRNA*) as a template. The genetic code can be defined as the relationship between the base sequence in a particular *mRNA* and the amino acid sequence of the protein which is synthesized from that *mRNA* template.

There are only 4 common bases in RNA, but there are 20 common amino acids. A 1:1 or even 2:1 correspondence is not possible (2:1 would give only $4^2 = 16$ possible combinations of base pairs). It is now known that the genetic code involves triplets of bases: that is, three bases in a sequence correspond to one amino acid in the protein. Since there are 64 possible triplets (4^3), each amino acid can have more than one *codon* (triplet of bases). The initial experiment which "cracked" the genetic code was done by M. Nirenberg (USA, 1961), who shared the 1968 Nobel prize for his work. He noted that addition of a synthetic RNA consisting only of linked uridine nucleotides (polyuridine) to a cell-free system of ribosomes, transfer

RNA, and nutrients, resulted in an enormous increase in the utilization of the amino acid phenylalanine, for protein synthesis, and in the formation of polyphenylalanine. Clearly the codon UUU, the only codon in polyuridine, corresponded to phenylalanine. Khorana (Wisconsin and MIT; shared the 1968 Nobel prize) succeeded in synthesizing many RNA chains with different but known repeating base sequences, and these were useful in unequivocally establishing the entire genetic code.

Transfer RNA's (tRNA) are essential to protein biosynthesis. They contain, in one of the "loops" in their structure, an amino acid anticodon (triplet of bases which can hydrogen-bond with the codon). Another portion of the tRNA molecule carries the amino acid which corresponds to the codon. Figure 18-3 shows both these sites for alanine tRNA; the alanine codon is GCC and this is "read" by the anticodon CGI (I can sometimes replace one of the other bases, in decoding). When two tRNA molecules, carrying their appropriate amino acids, are aligned at the mRNA, an enzyme joins the two amino acids in a peptide bond and releases them from the tRNA. The process is shown schematically in Figure 18-5.

The code in the mRNA, shown vertically at the left, is being "read" from bottom to top. Two tRNA units are hydrogen-bonded to the

---

Figure 18-5

Schematic representation of protein synthesis showing tRNA molecules "translating" the message carried by mRNA.

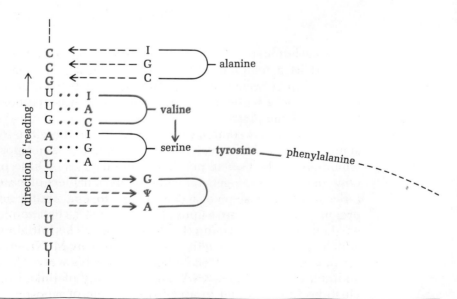

mRNA; one of these carries the growing peptide chain, and the other carries a valine unit which is to be added as the next amino acid in the sequence. Below, a tyrosine *t*RNA molecule is leaving after having delivered its amino acid, and at the top an alanine *t*RNA is coming into the proper position for bonding with the mRNA prior to delivering its amino acid to the growing protein chain. In this way the enzymes and other proteins essential to the living cell are synthesized.

18.7 ORGANIC CHEMISTRY IN LIVING ORGANISMS

We conclude this text with a few general comments about organic reactions as they occur in the living cell. The number of different reactions used by the cell is considerably smaller than that available to the synthetic organic chemist; in a sense, biochemical reactions constitute a subset of the larger set of organic reactions. Reactions in living cells are usually required to proceed approximately at room temperature ($\pm$ perhaps 25°), at or near one atmosphere of pressure, often at nearly neutral pH, and frequently in the presence of water, the major component of most living organisms. These constraints eliminate many of the variables which are accessible to the synthetic chemist—one cannot use strong acid, strong base, heat, or pressure to catalyze reactions in living organisms, since these conditions would destroy the cell. One reason why biochemistry is so fascinating is that *despite these constraints* cells carry on a truly remarkable array of organic syntheses and degradations.

At least three techniques are used by cells to overcome the limitations in reaction conditions. (*a*) Molecules are often **activated** prior to reaction. Typical is equation 11-10 in which a relatively inert fatty acid is converted to a more reactive form, in this case a thiol ester, before the metabolic reactions which degrade it to acetate can occur. (*b*) Molecules are brought into position through binding with enzymes, so that **polyfunctional catalysis** can bring about reactions which might otherwise require more severe reaction conditions. Consider, for example, the hydrolysis of a peptide by the enzyme chymotrypsin. A serine residue and a histidine residue on the enzyme act in concert to affect hydrolysis, as shown in equation 18-19. The serine and histidine residues, though they are not consecutive units in the amino acid sequence in chymotrypsin, come close to one another in space, as a consequence of the secondary and tertiary structure of the enzyme. In the first step, the histidine nitrogen and the serine hydroxyl group act in concert to attack the carbonyl group of the peptide; other steps in the sequence also require a cooperative interaction of the two groups. (*c*) Processes which might involve a single step for the synthetic chemist are usually **multi-step processes** in the cell. Note in equation 18-19 that the amine and the acid are generated in separate steps, though the overall reaction is simply the hydrolysis of an amide. Often the multistep processes are **cyclical.** This is also illustrated by equation 18-19, where it is seen that the enzyme is regenerated in the final step (marked by the colored arrow); the overall reaction in equation 18-19 is $RCONHR' + HOH \longrightarrow RCOOH + R'NHH$, the enzyme being recovered unchanged. Many biochemical reactions have this cyclical character.

(18-19)

Finally one may observe that there is nothing magical nor mystical about reactions in cells. They involve the same kinds of chemical processes that characterize organic chemistry.

New Concepts, Facts, and Terms

1. Terpenes; isoprene unit. Myrcene, limonene, α-pinene, geraniol, farnesol, vitamin A, menthol, camphor

2. Steroids and their numbering; cholesterol and ergosterol; sex hormones (estradiol, progesterone, testosterone, androsterone), cortisone
3. Biosynthesis of terpenes and steroids; mevalonic acid and isopentenyl pyrophosphate furnish the isoprene unit. Geranyl and farnesyl pyrophosphates are intermediates; farnesol $\longrightarrow$ squalene $\longrightarrow$ lanosterol $\longrightarrow$ cholesterol $\longrightarrow$ other steroids; acetate labeling was important in establishing this mechanism
4. Oxygen heterocycles include ascorbic acid (vitamin C), and tetrahydrocannabinol
5. Alkaloids—nitrogen bases, mainly from plants, with physiological activity. Nicotine, cocaine, mescaline, morphine, reserpine, strychnine, lysergic acid, quinine, caffeine. Biosynthesis is mainly from certain amino acids.
6. Porphyrins; chlorophyll, heme
7. Vitamins B_1 (thiamin), B_2 (riboflavin), B_6 (pyridoxal)
8. Antibiotics; penicillins, tetracyclines, macrolides
9. Nucleic acids: DNA, RNA. Components are a sugar (2-deoxyribose or ribose), phosphoric acid, and nitrogen bases (cytosine, thymine, uracil, adenine, and guanine). Nucleosides (sugar-base glycosides); nucleotides (phosphate esters of nucleosides). Primary structure (base sequence); secondary structure (base pairing through hydrogen bonds), the double helix, the cloverleaf. The biological functions are genetic information transfer and protein biosynthesis. The genetic code, codons, anticodons, messenger and transfer RNA.
10. Organic reactions in cells; activation, polyfunctional catalysis and multistep, cyclical processes

Exercises and Problems

1. Explain briefly the meaning of each of the following terms, and write a structural formula which illustrates each:
   a. terpene
   b. steroid
   c. sterol
   d. alkaloid
   e. porphyrin
   f. vitamin
   g. hormone
   h. nucleic acid

2. In the formulas of caryophyllene, cedrene, vitamin A, menthol, and camphor (sec 18.1), indicate with colored dots or in some other way that the molecules are constructed from isoprene units.

3. In a steroid in which rings A and B are fused *trans*, tell whether a substituent in the following positions is equatorial or axial:
   a. β at C1
   b. α at C2
   c. α at C6
   d. β at C7
   What pattern emerges for positions 1–4, 6, 7?

4. Which carbon atoms of cholesterol are chiral? How many optically active isomers are possible for this structure? How many stereoisomers of ergosterol are possible?

5. Limonene can be synthesized in the laboratory by heating isoprene to about 280–300°. It can also be obtained by treating geraniol with strong acid. Write equations for each of these reactions.

6. If one were to start with methyl-labeled acetate ($C^{14}H_3CO_2H$), what labeling pattern would be predicted for each of the following compounds: myrcene, limonene, farnesol, geranylgeraniol?

7. Suggest a possible carbonium ion mechanism for the formation of humulene and caryophyllene (both found in oil of cloves) from farnesyl pyrophosphate.

humulene =

8. What labeling pattern is expected in lanosterol and cholesterol synthesized from $C^{14}H_3CO_2H$?

9. Will lanosterol synthesized from $C^{14}H_3CO_2H$, when converted to cholesterol, give labeled (i.e. $C^{14}O_2$) or ordinary carbon dioxide? What would be the result if the lanosterol were synthesized from $CH_3C^{14}O_2H$?

10. What is the Cahn-Ingold-Prelog designation (**R** or **S**) for the absolute configuration at C3 in cholesterol? C20 in cholesterol? C3 in androsterone? C17 in cortisone?

11. Draw the intermediate **A** in equation 18-9 showing its stereochemistry, and describe the rearrangements which lead from **A** to lanosterol.

12. Write equations describing the action of warm, aqueous sodium hydroxide on
    a. cocaine   c. novocaine
    b. atropine   d. reserpine

13. The (—)R isomer of epinephrine occurs naturally; draw its structure.

14. Write a possible mechanism by which pyridoxamine might convert pyruvic acid to alanine. To what is the pyridoxamine converted?

15. How many stereoisomers of the penicillins are possible?

16. Draw the structures of each of the following nucleosides:
    a. cytidine   d. guanosine
    b. uridine   e. deoxycytidine
    c. deoxyadenosine

17. Show the steps in the mechanism for
    a. acid-catalyzed hydrolysis of a nucleoside (eq 18-14).
    b. base-catalyzed hydrolysis of a nucleotide (eq 18-15).

18. Draw a segment of an RNA chain, with four nucleotide units, each different. Name each part of the structure.

19. Draw the best possible T-G and C-A base pairs, and compare them with the structures of the observed T-A and C-G pairs. Explain why the latter are preferred.

20. Draw the expected composition of 3rd generation DNA in the Meselson-Stahl experiment (eq 18-17). Is it ever possible for a "heavy-heavy" strand to appear among the products?

INDEX

I Love You!

THat's way I'm doing all this shit work for you.